Lecture Notes in Civil Engineering 382

Lecture Notes in Civil Engineering (LNCE) publishes the latest developments in Civil Engineering—quickly, informally and in top quality. Though original research reported in proceedings and post-proceedings represents the core of LNCE, edited volumes of exceptionally high quality and interest may also be considered for publication. Volumes published in LNCE embrace all aspects and subfields of, as well as new challenges in, Civil Engineering. Topics in the series include:

- Construction and Structural Mechanics
- Building Materials
- Concrete, Steel and Timber Structures
- Geotechnical Engineering
- Earthquake Engineering
- Coastal Engineering
- Ocean and Offshore Engineering; Ships and Floating Structures
- Hydraulics, Hydrology and Water Resources Engineering
- Environmental Engineering and Sustainability
- Structural Health and Monitoring
- Surveying and Geographical Information Systems
- Indoor Environments
- Transportation and Traffic
- Risk Analysis
- Safety and Security

To submit a proposal or request further information, please contact the appropriate Springer Editor:

- Pierpaolo Riva at pierpaolo.riva@springer.com (Europe and Americas);
- Swati Meherishi at swati.meherishi@springer.com (Asia—except China, Australia, and New Zealand);
- Wayne Hu at wayne.hu@springer.com (China).

All books in the series now indexed by Scopus and EI Compendex database!

Ping Xiang · Liangdong Zuo

Editors

Novel Technology and Whole-Process Management in Prefabricated Building

Conference Proceedings of The 5th International Prefabricated Building Seminar on Frontier Technology and Talent Training

 Springer

Editors
Ping Xiang
Central South University
Changsha, China

Liangdong Zuo
Shanghai Jiao Tong University
Chongqing, China

ISSN 2366-2557   ISSN 2366-2565 (electronic)
Lecture Notes in Civil Engineering
ISBN 978-981-97-5107-5   ISBN 978-981-97-5108-2 (eBook)
https://doi.org/10.1007/978-981-97-5108-2

This Springer imprint is published by the registered company Springer Nature Singapore Pte Ltd.
The registered company address is: 152 Beach Road, #21-01/04 Gateway East, Singapore 189721, Singapore

If disposing of this product, please recycle the paper.

Preface

The 5th International Prefabricated Building Seminar on Frontier Technology and Talent Training (PBSFTT 2023) was held from September 21 to 22, 2023, in Chongqing, China. With PBSFTT 2023, the conference series International Prefabricated Building Seminar on Frontier Technology and Talent Training (PBSFTT) completed its fifth edition.

The general topics of PBSFTT 2023 mainly deal with researches on the application, difficulties, and frontier technology in the development of prefabricated building. Topics at the conference included but were not limited to: building quality inspection and reinforcement technology, new building technology and green building materials, intelligent building and intelligent construction technology, advanced engineering technology, etc.

All papers were exhaustively reviewed by program committee members and peer-reviewers, who took into account the breadth and depth of the research topics that fall under the scope of PBSFTT. The most promising and PBSFTT mainstream-relevant contributions were selected from all the submissions for presentation and inclusion in this volume, which presents innovative original ideas or results of general significance supported by clear and rigorous reasoning and compelling new evidence, as well as methods.

The conference included two days of technical parts consisting of leader speeches, keynote speeches, oral presentations, poster presentations, and academic investigation. We were delighted and honored to invite Shilin Dong (Chinese Academy of Engineering (CAE), China) and Andre Dienest (Curetec Shanghai New Energy Technology Co., Ltd., Germany) to share their insightful ideas and mind-blowing researches as keynote speakers. Shilin Dong is an expert in space structure and a member of the China Democratic League, who has long carried out teaching and scientific research on space structure and has published more than 200 academic papers and several books. Besides, his research results have found successful applications in multiple engineering projects, such as the Capital Gymnasium, Islamabad Gymnasium in Pakistan, the National Centre for the Performing Arts, and the National Swimming Center "Water Cube" for the 2008 Olympic Games. Andre Dienest is an expert in the fully automated, semi-automatic, and manual production of the prefabricated building and the robot modeling system.

We would like to thank all the keynote and invited speakers, authors, program committee members, and anonymous reviewers for their efforts in making PBSFTT 2023 a conference of the highest standard. We are also grateful to Springer, for its help in publishing this paper volume.

The Committee of PBSFTT 2023

Committee Members

Conference General Chair

Dongyan Liu   Chongqing College of Architecture and Technology, China

Vice Chair of the Conference

Peibin He   Chongqing University, China
Junhua Zhou   Chongqing College of Architecture and Technology, China

Publishing Chairs

Wei Xu   Chongqing Jiaotong University, China
Jike Ge   Chongqing University of Science and Technology, China
Wei Xie   Shanghai Architectural Design Institute, China
Nana Liu   Chongqing College of Architecture and Technology, China

Local Organizing Committee Chair

Jing Wang   Chongqing College of Architecture and Technology, China

Organizing Committee Chairs

Jian Xia   Fujian Province Construction Science Study Institute Co., Ltd.
Jing Huang   China Academy of Building Research, China

Technical Program Committee Chairs

Huancheng Jiang   East China Architectural Design and Research
                              Institute, China
Tsuji Naoki   Shenzhen Wanqian Construction Technology Co.,
                              Ltd., China
Sophia Slingerland   Urban and Architectural Office, Germany
Ping Xiang   Central South University, China

Committee Members

Guofu Wu   Chongqing University, China
Yajuan Chen   Chongqing College of Architecture and
                              Technology, China
Wenwen Luo   Chongqing University of Science and
                              Technology, China
Xiaoxi Zhou   Chongqing College of Architecture and
                              Technology, China
Chengming Liu   Chongqing Jundao, China
Junxia Sun   Chongqing College of Architecture and
                              Technology, China
Alessandro Bianchi   Politecnico di Milano, Italy
Alfrendo Satyanaga   Nazarbayev University, Singapore
Edyta Plebankiewicz   Cracow University of Technology, Polish
Yang Wu   Guangzhou University, China
Bingxiang Yuan   Guangdong University of Technology, China
Noor Akmal Adillah Ismail   Universiti Teknologi MARA, Malaysia
Xiangmin Liu   Fujian Academy of Architectural Sciences Co.,
                              Ltd., China
Hua Yan   The Chinese People's Liberation Army General
                              Logistics Department, China
Xingyi Fan   Chongqing University, China
Yongyi Zhang   Chongqing College of Architecture and
                              Technology, China
Kang Liu   Daiwa House (China) Investment Co., Ltd., China
Andre Dienest   Curetec Shanghai New Energy Technology Co.,
                              Ltd., Germany
Liangdong Zuo   Chongqing Nearspace Innovation R&D Center,
                              China
Lili Peng   Chongqing College of Architecture and
                              Technology, China

Contents

Load Calculation and Strength Analysis of Prefabricated Floating Dock

Yu Zeng[1], Jianting Guo[1(✉)], and Kai Zheng[2]

[1] School of Naval Architecture and Ocean Engineering, Jiangsu University of Science and Technology, Zhenjiang 212003, Jiangsu, China
guojianting11@163.com
[2] Marine Design and Research Institute of China, Shanghai 200011, China

Abstract. This paper proposes a multi-module assembly floating dock, which is assembled by a pin-jointed hinge structure between modules. Based on the design wave method, the three-dimensional potential flow theory is used to calculate the loads and analyze the strength of the assembly floating dock, and the configuration is optimized and analyzed. The results show that the longitudinal torque load in the wave load of the floating dock is the largest, and the maximum stress area is the connection between the platform and the column. Based on numerical results, structural strengthening design and analysis were carried out for multiple high-stress areas, which significantly reduced structural stress, with a maximum reduction of 54%, making the structural strength meet the design requirements. The research results of this paper can provide a reference for the design and strength optimization of floating dock configurations in the future.

Keyword: Floating dock · Design wave · Wave load prediction · Strength analysis

1 Introduction

With the development of maritime production and transportation industries in the South China Sea, the construction of offshore and deep-water ports and wharves on islands and reefs has become increasingly important. Designing and constructing a simpler, more convenient, and cost-effective floating dock is crucial. Compared to traditional fixed docks, floating docks require less sand, stone, and steel materials, which reduces construction costs and makes them more economically viable. Moreover, floating docks are more flexible and mobile, better able to adapt to changes in the marine environment and the varying needs of ships. However, the marine environment where prefabricated floating docks are located is very complex and subjected to long-term wave loads, which may lead to a combination of various external forces such as shear and torsion, resulting in significant stress on local structural components and making them prone to failure. Once the entire structure of a floating dock fails, it will cause significant economic losses and pose a threat to the life safety of workers on the platform. Therefore, it is essential to study the overall strength of large offshore floating structures and ensure their safety.

© The Author(s) 2024
P. Xiang and L. Zuo (Eds.): PBSFTT 2023, LNCE 382, pp. 1–17, 2024.
https://doi.org/10.1007/978-981-97-5108-2_1

Currently, research on the structural performance of floating docks and large floating structures has been conducted both domestically and internationally. Shahrabi and Bargi [1] provided optimal dimensions for different float widths that meet design constraints based on various modes and loads acting on the floating dock. Tajali [2] found that the response of a floating bridge is largely dependent on the size of the floating bridge, connector stiffness, wave conditions, and characteristics of the interaction between the floating bridge. Ji Chunyan et al. [3] analyzed the ultimate strength of a typical semisubmersible platform's critical structures using the progressive failure method and finite element calculation. They predicted the design wave that the platform faced based on the theory of wave loads and computed the overall strength of a typical semisubmersible platform using the direct calculation method. Zheng Zhiguo [4] found that the wave resistance of the container combination floating body was poor due to its shape not being a double-curved surface by calculating the motion response of a single floating dock and different container combination floats. Kim [5] improved the vertical motion performance of the floating dock structure and optimized the structure's shape to adapt to the coastal environment of Korea. Lin Sinan [6] designed a large floating dock with an asphalt deck surface and conducted a mechanical response study. Zhang Jingyi [7] calculated the strength of two configurations of ultra-large floating structures using the direct calculation method and obtained different high-stress regions. Based on the structural characteristics of these high-stress regions, they proposed improvement schemes. Lee et al. [8] calculated the overall strength of a traditional semisubmersible platform based on the theory of wave loads and found three main high-stress regions: the connection between the column and the float, the connection between the column and the cross-brace, and the connection between the column and the lower float. Lin Sinan [6] designed a large floating dock with an asphalt deck and conducted a mechanical response study. Zhang Jingyi et al. [7] proposed various improvement schemes based on the structural characteristics of the high stress areas of longitudinal and transverse large floating boxes and compared the results of three schemes to obtain the optimal configuration. Su Changnan[9] conducted a detailed study on the horizontal fixed tilt center height and freeboard height of the buoyancy tank through numerical simulation to ensure the reliability of the lateral stability of the buoyancy tank.

From the research of the scholars mentioned above, it can be seen that there is currently limited research on the strength of floating dock structures, and most studies focus on analyzing a single module of a floating dock. In this paper, we design a multi-module floating dock connected by hinge structures. This design is cost-effective and can be quickly assembled according to different needs. Using three-dimensional potential flow theory and DNV's SESAM software, we establish a numerical model of the floating dock and analyze its wave load and structural strength. The results of this study can provide reference for future floating dock configuration design and strength optimization analysis.

2 Theoretical Basis

2.1 Three-Dimensional Potential Flow Theory

Wave loads on marine structures can be divided into three types: diffraction forces, inertia forces, and drag forces. Diffraction forces are generated by the diffraction effect of water flow on the structure, while inertia forces consist of added mass forces and added damping forces. Drag forces are generated by the disturbance of the structure on the water flow.

For different structural configurations and sizes, the proportion of wave load components can vary greatly. For small structures, drag forces and inertia forces are equally important. However, for large floating structures, inertia forces and diffraction forces are more important. Therefore, when calculating wave loads, it is necessary to first determine which calculation method is more appropriate based on the scale of the structure. Generally, three-dimensional potential flow theory is used to calculate the first-order wave load of large structures such as some large floating platforms. The size of the floating dock designed in this paper can reach 60 m, and the structural scale is relatively large, so three-dimensional potential flow theory is used for calculation.

In the case of large structures, serious reflection and diffraction phenomena will occur when incident waves are present, so they cannot be ignored when calculating wave loads. Using potential flow theory to describe the motion state of the floating body can simplify the problem. When the ratio of wave height to wavelength is small, linear theory can further simplify the problem. Three-dimensional potential flow theory assumes that the fluid is an ideal fluid with no viscosity, no vorticity, and incompressibility. According to potential flow theory, the velocity potential satisfies the Laplace equation in the flow domain:

$$\nabla^2 \Phi = 0 \tag{1}$$

The velocity potential can be further decomposed linearly into:

$$\left.\begin{array}{c} \phi = \phi_O + \phi_R + \phi_D \\ \phi_O = \frac{iAg}{\omega} \times \frac{\cosh k(z+H)}{\cosh kH} e^{-ik(x\cos\beta + y\sin\beta)} \\ \phi_R = i\omega \sum_{j=1}^{6} \xi_j \phi_j \end{array}\right\} \tag{2}$$

In the formula, ϕ_O represents incident potential; ϕ_R represents radiated potential; ϕ_D represents diffracted potential; A represents wave amplitude; ω represents circular frequency of the wave; β represents wave direction angle; k represents wave number; H represents water depth; ξ_j represents the amplitude of motion in the six degrees of freedom of the object; ϕ_j represents unit radiated potential.

The diffraction potential and the radiation potential are satisfied:

$$\frac{\partial \phi_j}{\partial n} = n_j, \frac{\partial \phi_D}{\partial n} = 0 \tag{3}$$

By determining the velocity potential of the flow field through the distribution of sources and sinks on a wet surface, and solving the boundary conditions using Green's formula, the total velocity potential can be obtained. Then, using numerical discretization, the pressure distribution acting on the floating dock can be calculated, and the wave forces and moments acting on the floating dock can be determined.

2.2 Finite Element Analysis Process

The finite element analysis process for a floating dock is shown in Fig. 1. The assembly-type floating dock is modeled, analyzed for hydrodynamics, and long-term forecasted using the GeniE, Wadam, and Postresp modules of the ship and ocean engineering large-scale finite element analysis software SESAM. The quasi-static analysis method is used to transmit the calculated wave loads and design wave parameters to the floating dock in the form of pressure loads through the Sestra module, and the overall structural stress information can be obtained in the Xtract module.

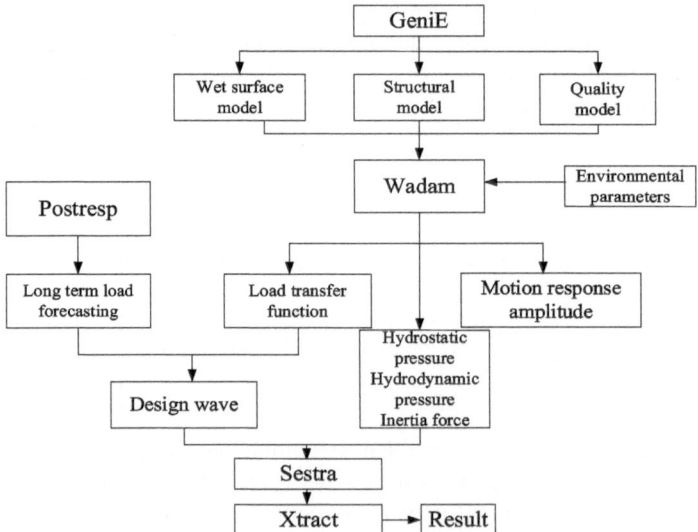

Fig. 1. Finite element analysis process of prefabricated floating dock

3   Design Scheme for Floating Dock

3.1   Design of the Main Structure of a Floating Dock

As shown in Fig. 2, a single assembly-type floating dock is a floating structure composed of three parts: platform, column, and float. The platform adopts a box-type structure design, made of steel plate, and has strong bearing capacity and stability. The lower float is composed of four floating cylinders with a closed structure, and eight columns are used as supporting structures to connect the platform with the lower float, making the entire floating dock have strong durability and stability. For easy connection and disassembly, adjacent two single-section floating docks are assembled using a two-pin hinge connection structure. As shown in Fig. 3, it includes a bearing ring and a pin shaft, with a simple form, and the connection and release of the float can be achieved by inserting and pulling the pin shaft into and out of the bearing ring. Table 1 shows the relevant parameters of a single assembly-type floating dock.

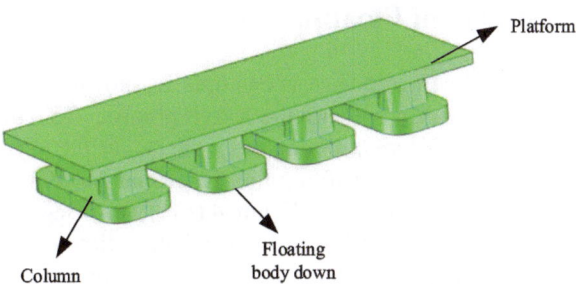

Fig. 2.   3D Rendering of an assembled floating dock

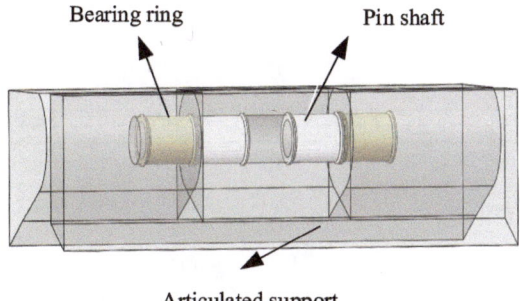

Fig. 3.   Shaft pin connection structure

Table 1. Main parameters of a single floating dock

parameters	numerical value	Main parameters	numerical value
Length/(m)	60	Lower float box width/(m)	10
Width/(m)	20	Draft/(m)	4
Model depth/(m)	7	Displacement/(T)	1949.82
Column height/(m)	4.5	Rolling moment of inertia/(kg·m^2)	9.172×10^7
Column length × width/(m)	4 × 6	Pitching moment of inertia/(kg·m^2)	4.463×10^8
Lower float box height/(m)	2	Yawing moment of inertia/(kg·m^2)	4.827×10^8

4   Coordinate System of Floating Dock

Taking a three-connected floating dock as an example, as shown in Fig. 4 (a), the coordinate system of the floating dock is set with the X-axis pointing from the stern to the bow as the positive direction, the Y-axis pointing from the centerline to the port side as the positive direction, and the Z-axis satisfying the right-hand coordinate system with the upward direction as the positive direction. Figure 4 (b) is a top view of the floating dock. When the waves propagate along the X-axis in the positive direction, the wave direction is 0°, and when the waves propagate along the Y-axis in the positive direction, the wave direction is 90°. When the waves change from the X-axis in the positive direction to the Y-axis in the positive direction in a counterclockwise direction, the wave angle gradually increases from 0° to 90°.

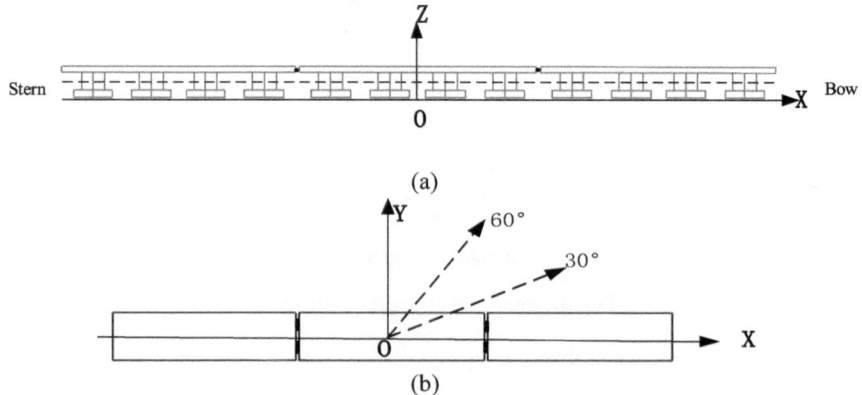

Fig. 4. Definition of coordinate system

5   Calculation of Load and Stress Analysis for Floating Dock

5.1   Hydrodynamic and Structural Model

The strength analysis of the structure was carried out using SESAM software, and the finite element model included hydrodynamic finite element model and mass finite element model. The length of a single floating dock model is 60 m, width 20 m, designed water depth 20 m, seawater density $1.025 \times 10^3 \text{kg/m}^3$, displacement 1949 t, and draft 4 m. The wet surface model includes the lower floating body and column, and the wave load is calculated using diffraction theory. The wet surface as a whole adopts a 0.8 m × 0.8 m grid, with a total of 3452 nodes and 3378 elements. Figure 5 shows the structural model consisting of three docks connected together, and the corresponding hydrodynamic model is shown in Fig. 6.

Fig. 5. Structural model

Fig. 6. Wet surface model

5.2   Motion Response and Analysis of Floating Dock

To investigate the effects of wave direction and period on the motion response and profile load transfer function of the modular floating dock, while referencing DNV regulations and considering the structure's symmetry about the X and Y axes, a range of wave directions from 0 to 90° with a step size of 15° was selected. A total of 7 wave directions were chosen. The wave period range was from 1 to 40 s with an interval of 1 s, resulting in a total of 40 periods.

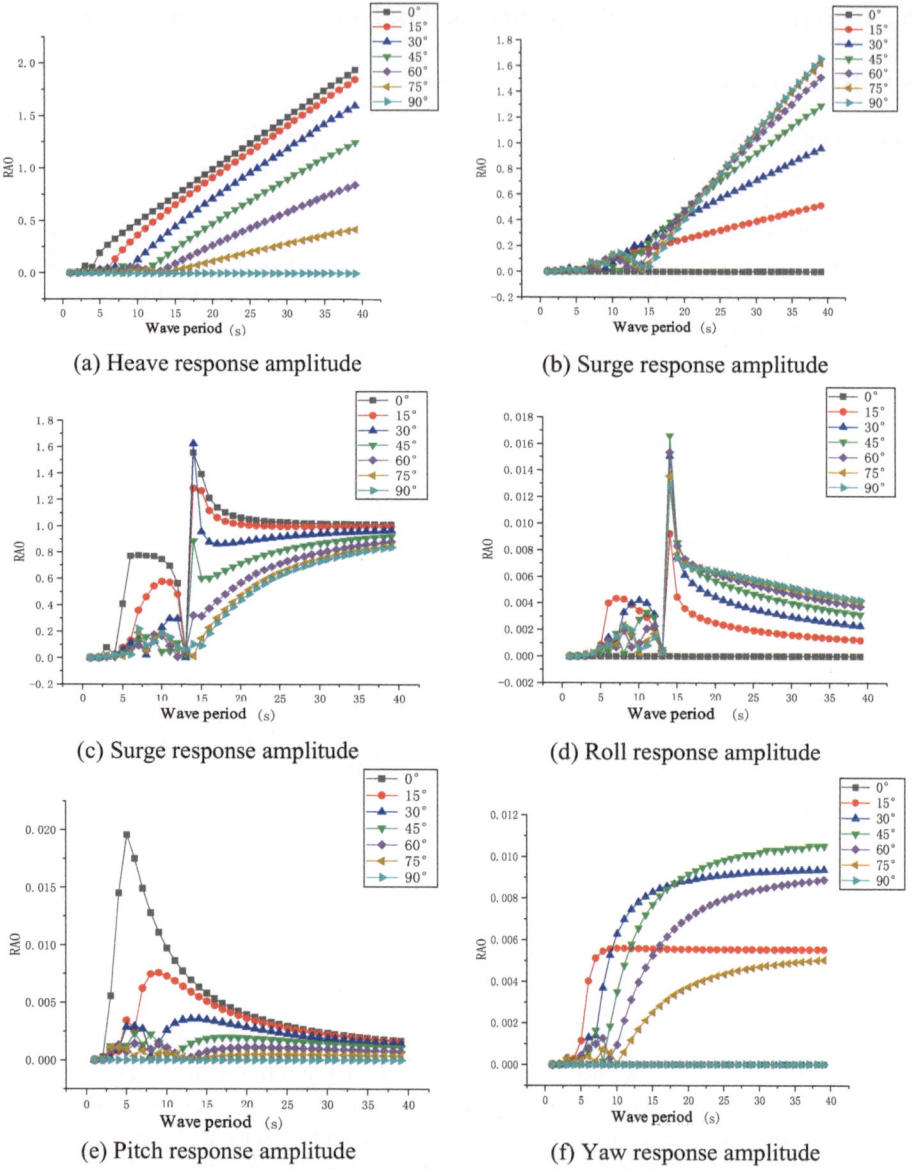

(a) Heave response amplitude

(b) Surge response amplitude

(c) Surge response amplitude

(d) Roll response amplitude

(e) Pitch response amplitude

(f) Yaw response amplitude

Fig. 7. Motion response amplitude of modular floating docks

The motion response amplitude of the floating dock is shown in Fig. 7. It can be seen from the graph that the heave motion response reaches the maximum amplitude when the wave heading angle is 0°, and the sway motion response reaches the maximum amplitude when the wave heading angle is 90°. The pitch and roll motion responses reach the maximum amplitude at a period of 14 s, while the yaw motion response reaches the maximum amplitude at a period of 5 s. Compared to other motion responses, the motion response amplitude of the bow yaw is the smallest.

5.3   Sectional Load Transfer Function

The assembly-type floating dock is a new type of marine floating structure, and the industry currently does not have unified standards or regulatory references for its strength calculation. Therefore, it is necessary to refer to relevant structural strength analysis specifications for offshore platforms and determine the wave parameters of dangerous working conditions to conduct a comprehensive strength analysis. Firstly, the Wadam module of SESAM is used to conduct an overall structural analysis of the floating dock, selecting typical sections and predicting wave loads on these sections. Typical working conditions for offshore platforms are referenced to determine the design wave. Nine transverse sections between the two floats are selected, as shown in Fig. 8. SESAM is used to calculate the longitudinal force, transverse force, vertical force, longitudinal torque, vertical bending moment, and horizontal bending moment for the nine sections, and the design wave is finally determined based on the predicted wave loads.

Fig. 8. Select the cross-section location

As shown in Fig. 9, the longitudinal force shows a decreasing trend, reaching its maximum at cross-Sect. 1. The transverse force shows a symmetrical trend, reaching its maximum at cross-Sect. 1 and its minimum at cross-Sect. 4. The vertical force is distributed more evenly and reaches its maximum at cross-Sect. 3. The longitudinal torque increases and then decreases, with a clear trend of growth and decline, reaching its maximum at cross-Sect. 5. The vertical bending moment shows a symmetrical trend and is distributed relatively evenly, reaching its maximum at cross-Sects. 3 and 7. The horizontal bending moment shows an overall decreasing trend and reaches its maximum at cross-Sect. 1. Therefore, the maximum wave load is the longitudinal torque, which reaches a maximum of 1.49×10^8 N·m, and should be a focus in the strength analysis of the assembled floating dock. The maximum loads for different profiles are shown in Table 2.

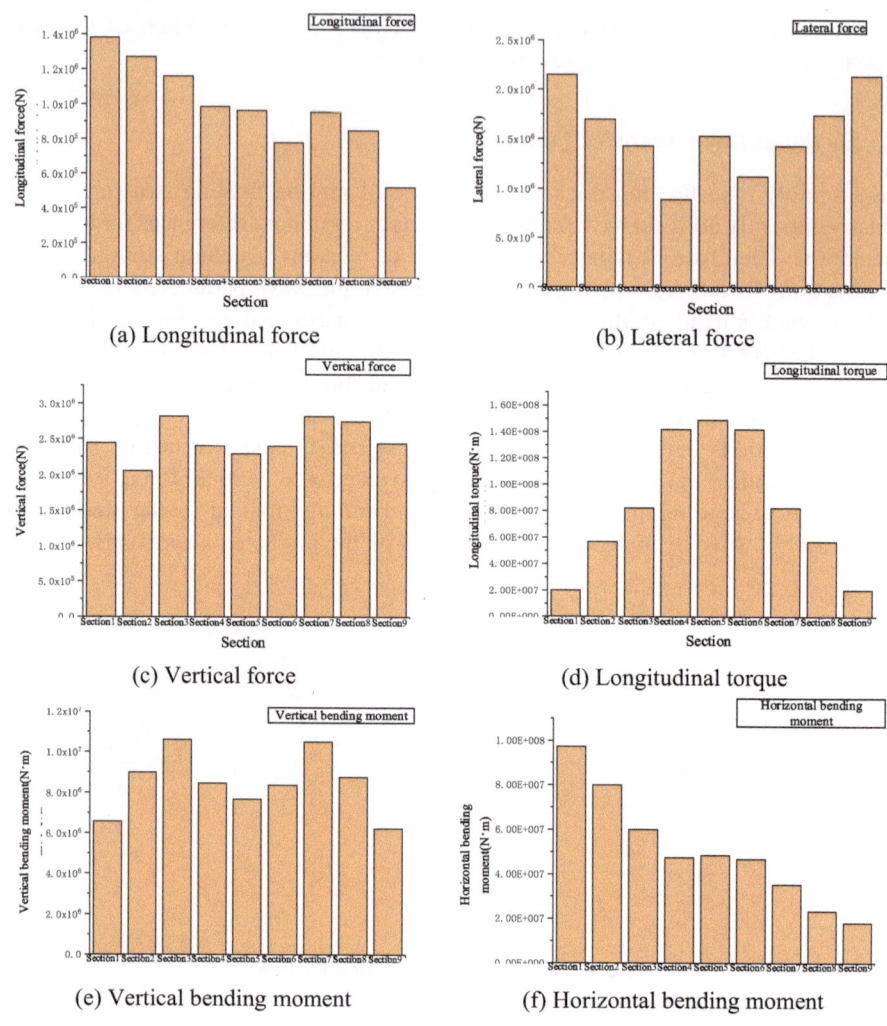

(a) Longitudinal force

(b) Lateral force

(c) Vertical force

(d) Longitudinal torque

(e) Vertical bending moment

(f) Horizontal bending moment

Fig. 9. Different sectional forces and sectional bending moments

Table 2. Maximum loads for different profiles

Transverse load	Cycle (s)	Wave direction (°)	Maximum
Longitudinal force	5	0	1.38×10^6 N
Lateral force	5	90	2.2×10^6 N
Vertical force	7	45	2.82×10^6 N
Longitudinal torque	7	30	1.49×10^8 N·m
Vertical bending moment	6	15	1.06×10^7 N·m
Horizontal bending moment	5	0	9.92×10^7 N·m

5.4 Long-Term Forecast for Floating Dock

The significant wave height in the working water area of the floating dock is 4 m, with a peak wave period of 7.64 s. For long-term forecasting, the Jonswap spectrum is used to fit the long-term sea conditions, with a peak enhancement factor of 2. According to DNV regulations, the long-term forecast values for each profile load are selected for a return period of 50 years to calculate the design wave parameters. The long-term forecast values for each profile load are shown in Table 3.

Table 3. Long-term forecast value of profile load.

Returning cycle/(Year)	50
Longitudinal force	3.14×10^6 N
Lateral force	6.73×10^6 N
Vertical force	1.22×10^7 N
Longitudinaltorque	7.77×10^8 N·m
Vertical bending moment	3.13×10^7 N·m
Horizontal bending moment	6.75×10^7 N·m

5.5 Determine Design Wave Parameters

The design wave amplitude is the long-term forecast value of profile load divided by the corresponding profile load response function's extreme value. The wave height is twice the wave amplitude, and the wave direction, period, and phase are the same as the maximum value of the corresponding profile load transfer function. The design wave parameters for various typical working conditions are calculated and shown in Table 4.

Table 4. Typical design wave parameters for operating conditions

Condition number	Wave direction (°)	Wave period (s)	Phase (°)	Wave amplitude (m)	Wave height (m)
101	0	5	−47.54	2.28	5.56
102	90	5	125.53	3.06	6.12
103	45	7	88.6	4.35	8.7
104	30	8	−32.2	5.22	10.44
105	30	5	12.13	2.96	5.92
106 ·	0	5	4.93	0.65	1.3

6 Overall Strength Calculation and Result Analysis of Floating Dock

6.1 Overall Strength Calculation of Floating Dock

In order to carry out the strength analysis of floating structures, it is necessary to select the most dangerous working condition as the calculation condition. Calculation conditions can be divided into two categories: static water and wave conditions, according to the wave situation. Static water condition refers to the force situation when the floating body is in calm water surface, while wave condition refers to the force situation when the floating body is under the action of waves. At this time, the force on the floating body is not only gravity and buoyancy, but also the impact force and inertial force of the waves. In actual marine environment, floating structures will produce complex dynamic response under the action of waves, which will cause deformation and stress concentration of the structure. Therefore, in order to accurately calculate the strength of floating structures, it is necessary to consider the combined effect of static water condition and wave condition. The wave combination conditions are shown in Table 5.

Table 5. Wave combination working conditions

Combined working conditions	Description of the operating condition
LC101	Static water condition + longitudinal force load condition
LC102	Static water condition + lateral force load condition
LC103	Static water condition + vertical force load condition
LC104	Static water condition + longitudinal torque load condition
LC105	Static water condition + vertical bending moment load condition
LC106	Static water condition + horizontal bending moment load condition

Figure 10 shows the stress cloud diagram of the floating dock under the most dangerous working condition, and Fig. 11 shows the intensity stress cloud diagram of the high stress area. Table 6 shows the maximum Von Mises stress values at key locations of the dock under various working conditions, and Table 7 shows the location and maximum stress values of the high stress area.

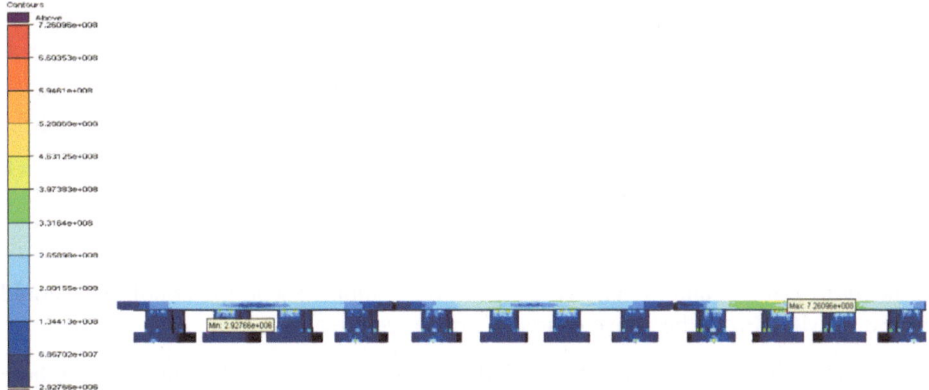

Fig. 10. Overall maximum Von Mises stress contour plot of the structure

Fig. 11. Von Mises stress contour plot of high stress areas

Table 6. Maximum Von Mises stress of critical parts of the structure

Combined operational condition	Float/(Mpa)	Platform/(Mpa)	Connection between platform and column/(Mpa)
LC101	628	423	652
LC102	604	347	556
LC103	356	322	483
LC104	415	362	726
LC105	375	428	569
LC106	511	435	603

The study adopts the equivalent stress Von Mises for overall strength evaluation. According to ABS MODU2008 specifications, the allowable stress for Von Mises stress

Table 7. Location and Maximum Value of High Stress Area

Number	Location	Maximum Stress Value/(Mpa)
High Stress Area1	Float	628
High Stress Area2	Platform	435
High Stress Area3	Connection between platform and column	726

Table 8. Ocean Structure Allowable Stress Balance Standard

Working Condition	Safety Factor	Allowable Stress(355 Mpa)
Combination of Working Conditions	1.05	338.09

is shown in Table 8, and the allowable stress for combined working conditions should be 338.09 MPa. From the above charts, it can be seen that:

1) The stress distribution of the assembled floating dock is generally uniform, but there are still some high stress areas, which are the connection between the float and the column, the connection between the platform and the column, and the float.
2) High stress area 1 is located at the float on both sides, with a maximum value of 628 MPa for Von Mises stress; high stress area 2 is located at the connection between the column and the float, with a maximum value of 435 MPa for Von Mises stress; high stress area 3 is located at the connection between the platform and the column, with a maximum value of 726 MPa for Von Mises stress, all of which are greater than the maximum allowable stress;
3) These high stress areas are the hotspots most likely to experience fatigue fracture in the future, and the structure of these parts must be strengthened and improved to improve the reliability and safety of the structure.

6.2 Configuration Improvement and Analysis

According to the calculation results of the overall strength of the floating dock, reinforcement structures such as stiffeners and thicker wall panels should be added in the high stress areas to enhance the strength and stiffness of the floating dock. The specific improvement measures are shown in Fig. 12, and the improvement plan for structural strengthening in the high stress areas mentioned above is shown in Table 9. The maximum overall Von Mises stress cloud diagram of the improved structure is shown in Fig. 13.

Fig. 12. Specific Improvement Measures

Table 9. Improvement Measures for High Stress Areas

Location	Structural Form
Float	Add a circle of reinforcing ribs to the bottom and top of the original structure of the buoyancy tank, and add two longitudinal bulkheads inside the buoyancy tank
Column	Add a reinforcement ring and two horizontal steel bars to the top of the column
Platform	Increase the thickness of the platform I-beam

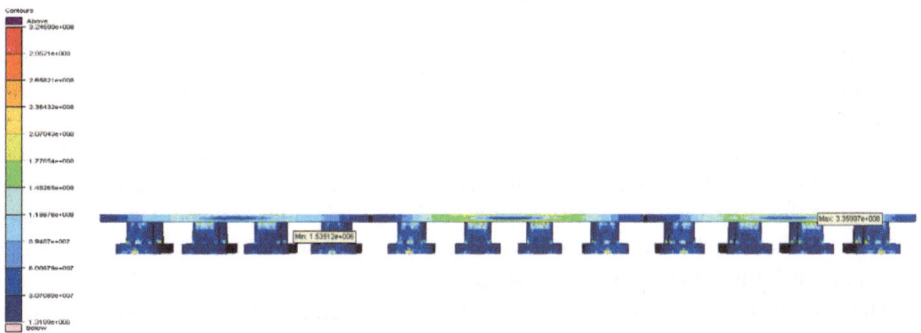

Fig. 13. Improved Structure Maximum Overall Von Mises Stress Cloud Map

The calculation results in Table 10 show that the overall strength of the improved floating dock structure is significantly better than the original design, with a notable decrease in stress values in high-stress areas. However, areas with relatively high stress values still exist, such as the connection between the float and column, the connection structure intersecting with the y-axis, and the float on both sides of the connection. In terms of the overall strength of the float, the maximum Von Mises stress value of the float is 324 MPa, the maximum stress value of the platform's outer edge is 332 MPa, and the maximum stress value of the connection between the platform and the column is 335 MPa. The overall structural strength meets the design requirements.

Table 10. Maximum Von Mises Stress in Key Areas After Structure Improvement

Working Condition	Float/(Mpa)	Platform/(Mpa)	Connection between Platform and Column/(Mpa)
Loc101	322	317	315
Loc102	302	280	329
Loc103	281	297	268
Loc104	306	326	335
Loc105	294	332	318
Loc106	324	325	328

7   Conclusion

This paper presents a new type of modular floating dock structure assembled using a pin connection system. Based on the design wave method, three-dimensional potential flow theory is used to calculate the wave loads and perform strength analysis on the floating dock. Based on numerical results, the configuration is optimized and analyzed. The following conclusions are drawn from numerical calculations and theoretical analysis:

1) The external load calculation results show that the longitudinal torque under the condition of a wave period of 7 s and a wave direction of 30° is the most critical condition for the floating dock, with a maximum stress of 726 MPa. This condition should be given attention in future research.
2) The overall stress distribution of the floating dock is reasonable, and some high-stress areas exist mainly at the connection between the platform and the column, the outer edge of the platform, and the floatation tank on both sides. The maximum stress is at the connection between the platform and the column, and these areas are prone to structural failure due to excessive stress.
3) The overall strength of the floatation body is improved by adding reinforcing ribs and thickening the wall after structural improvement of high-stress areas. The maximum stress at the connection between the platform and the column is reduced by 54% compared with the maximum stress before improvement, from 726 MPa to 335 MPa, due to structural reinforcement of the column. The maximum stress values at the outer edge of the platform and the floatation tank decrease by 23% and 48%, respectively, with a maximum decrease of 54%. The overall strength of the structure meets the design requirements. The research results of this paper have certain reference and guiding significance for the configuration design and strength optimization design of floating dock structures.

References

1. Shahrabi, M., Bargi, Kh.: Numerical simulation of multi-body floating piers to investigate pontoon stability. Front. Struct. Civ. Eng. 7(3), 325–331 (2013)

2. Tajali, Z., Shafieefar, M.: Hydrodynamic analysis of multi-body floating piers under wave action. Ocean Eng. **38**(17–18), 1925–1933 (2011). https://doi.org/10.1016/j.oceaneng.2011.09.025
3. Ji, C., Yu, W., Shen, Q.: Research on calculation method of total strength of semi-submersible platform and ultimate strength of key structure. Shipbuilding Technolo. (1), 8–12 (2012)
4. Zheng, Z., Dong, Y., Tang, Y.: Discussion on the technical scheme of Modular Floating Dock. Ship Sci. Technol. (05), 7–8+11 (2003)
5. Kim, D.-M., Heo, S., Koo, W.: A study on the improvement of the motion performance of floating marina structures considering Korea coastal environment. J. Ocean Eng. Technol. **33**(1), 10–16 (2019)
6. Sinan, L.: Study on Mechanical Response of large Floating Wharf with Asphalt Deck Surface. Dalian Maritime University (2014)
7. Zhang, J.: Study on the overall Strength and Structural Improvement of large Floating Structures with Different Configurations. Jiangsu University of Science and Technology (2019)
8. Lee, C.H., Jang, C.H., Jun, S.H., et al.: Global structural analysis for semi-submersible drill rig. Int. Ocean Polar Eng. **91**(3), 15–20 (2014)
9. Su, C., Liu, F.: Preliminary discussion on transverse stability calculation of floating tank in yacht wharf. Pearl River Transp. **08**, 85–86 (2020)

Mechanical Characteristics of Assembled Overhanging Composite Roadbed Structure

Rui Tang[1], Hao Li[1], Minglong You[2(✉)], Gongchang Li[1], and Fei Tan[2]

[1] Anhui Construction Engineering Investment Group Co., Ltd., Hefei 230031, People's Republic of China
[2] Faculty of Engineering, China University of Geosciences, Wuhan 430074, People's Republic of China
{minglongyou,tanfei}@cug.edu.cn

Abstract. In order to solve the problems of narrow roads, difficult construction, long construction period and high impact on the surrounding environment encountered during the construction of mountainous highways, railroads and other infrastructures, this paper proposes an assembled overhang composite roadbed structure. The mechanical analysis of the assembled overhang is carried out, and the three-dimensional finite element numerical model of the assembled roadbed structure is established, and the force characteristics of the overhang structure are analyzed. The results show that there is stress concentration between the two spans of the road panel, and the deformation of the road panel is mainly concentrated in the overhang side, and the deformation of the side span is larger than that of the middle span. The stress concentration phenomenon appeared on the outer side of the connection end of the middle overhanging and column, and the horizontal displacement of the middle column was larger, and the displacement was about twice that of the edge column. The research results can provide useful guidance for the design and optimization of assembled overhanging roadbeds.

Keyword: Overhang structure · Assembled construction · Finite Element Method analysis · Mechanical characteristics

1 Introduction

With the implementation of major national strategies such as Sichuan-Tibet Railway and Transportation Power, the construction technology of mountainous highways, railroads and other infrastructure has become a hot concern in the field of engineering construction [1, 2]. In the changing environment of mountainous terrain, the cast-in-place concrete construction scheme is not convenient. And the levelness and stability of the ground in mountainous areas are not easily guaranteed, which has an impact on the construction parts that require support formwork or other enclosure measures.

The emergence of overhanging structures has led to a richer and more efficient construction solution for mountainous roadbeds. Wang [3] first proposed the design concept of widening mountain roads with monolithic overhanging structures in 2007. This concept forms the embankment with monolithic poured wall-post retaining wall without

P. Xiang and L. Zuo (Eds.): PBSFTT 2023, LNCE 382, pp. 18–32, 2024.
https://doi.org/10.1007/978-981-97-5108-2_2

destroying the original ecological environment, and completes the road under-width with overhanging structure. For the problem of narrow roads in mountainous areas, Liu et al. [4, 5] considers the joint action of the overhanging structure and the geotechnical body, and proposes a widening scheme for mountainous roads. Zhuo [6] takes the re-expansion project of the tertiary road from Ju Shui Town to Gao Chuan Township, An County, Sichuan Province as an example, and introduces in detail the application of overhanging slabs in the widening of the roadbed of a mountainous cliff section with high and steep slopes. Other scholars proposed a road widening scheme of buried overhang structure for a mountainous road modification and expansion project [7, 8]. In terms of theoretical research, Tian [9] proposed the mechanical model and calculation principle of the overhanging roadbed based on the simplified single-bay overhang structure. Guo et al. [10] carried out a systematic study on the structural form, external load type and load combination method of overhanging U-shaped roadbed, and proposed the design and analysis methods of structural bearing capacity, fatigue resistance, crack control and deflection deformation of overhanging U-shaped roadbed. In terms of numerical simulation, You [11] used ABAQUS to analyze the working performance of the overhanging structure of slope lattice foundation with various foundation conditions, various side slope angles and various overhanging beam lengths. Wei et al. [12] established a two-dimensional finite element model of overhanging structure and proposed the appropriate structural dimensions of overhanging slab. Yao et al. [13] established a three-dimensional finite element model of overhanging-support roadbed structure and proposed the optimal size selection of counterweight blocks, support columns and other components. In terms of engineering application, Yuan [14, 15] proposed the general design idea of assembled overhang structure and supporting construction plan for the problems faced by overhang composite roadbed. Wang et al. [3] conducted a study on the application of monolithic overhang structure in Tibetan class Chang highway, and summarized the processing techniques of structural support columns and anchorage points of overhanging beams under different geological conditions and different widening requirements. Li [16] presents detailed construction schemes and technical points for the engineering application of composite roads with integral overhanging structures. Previous studies on overhanging structures stay on the structural calculation principles and the influence of site conditions on overhanging structures, lacking a more systematic study on the mechanical characteristics of assembled overhanging structures, and not revealing the mechanical behavior characteristics of assembled overhanging roadbeds under different overhanging lengths and anchor placement methods.

The assembled structure has many advantages such as improving production efficiency, shortening construction period, reducing resource consumption, and having little impact on the environment [17, 18], so it is widely used in the process of engineering construction. The application of assembled overhang structure in road foundation engineering can achieve less ecological damage, reduce excavation and filling, low cost, simple construction, and apply to different terrain and geological conditions, which can effectively reduce the safety hazards of road foundation construction in mountainous areas. Therefore, new road construction solutions suitable for mountainous areas are gradually mentioned and increasingly in demand. This paper adopts finite element analysis method to study the main influencing factors of overhanging roadbed, combine with engineering Conditions, and propose an optimized construction technology scheme of assembled overhanging roadbed.

2 Working Principle and Mechanical Analysis of Assembled Overhang Structure

The assembled overhang structure is a structural form consisting of columns, outer longitudinal beams, overhanging beam, precast slabs, inner longitudinal beams, retaining slabs and anchors in one piece. The column and the soil are often in contact with each other by filling foam concrete, so the column is subjected to the pressure from the foam concrete, which is recorded as No. 1 soil pressure. The lower end of the column is embedded in the soil body, and the soil pressure generated by the interaction between the soil body and the column is recorded as No. 2 soil pressure. Therefore, the column is mainly subjected to the two types of earth pressure as shown in Fig. 1. The part above the embedded section of the column is calculated according to the geotechnical lateral pressure, and the internal force below the embedded section can be calculated according to the bending moment and shear force at the ground level of the embedded end combined with the foundation coefficient method.

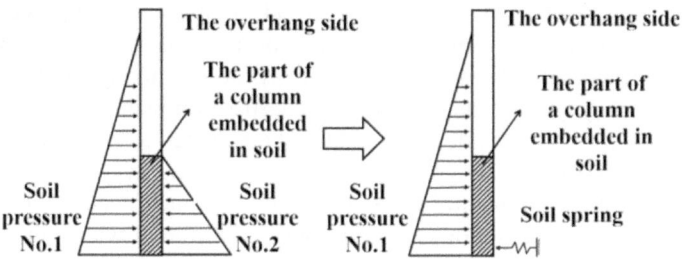

Fig. 1. Simplified schematic diagram of assembled overhang structure subjected to soil pressure

As shown in Fig. 2, the connection between the column and the soil can be regarded as rigid connection, and the overhanging beam is bolted to the column, which can be regarded as rigid connection. The left side of the overhanging beam is reinforced by anchor rods, which provide anchoring tension, and the force is influenced by the structural load, and the connection between the anchor rods and the overhanging beam is regarded as hinged. The loads on the overhanging structure are mainly the self-weight of the structure, the lane load consisting of inner and outer longitudinal beams and precast slabs, and the vehicle load on the road.

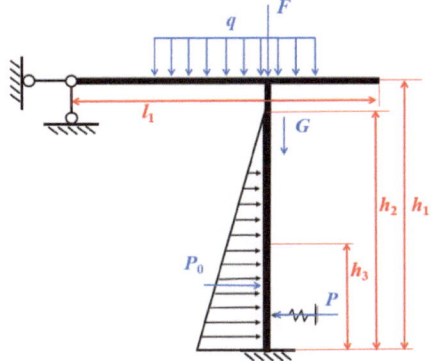

Fig. 2. Simplified calculation diagram of overhanging structure

3 3D Numerical Model of Assembled Road Base Structure

3.1 Introduction of the Overhanging Structure Scheme

The relying project is located in the section of G4012 Liyang-Ningde Expressway from Huangshan to Qiandao Lake, starting and ending at pile number k9 + 258–k15 + 482, with a total length of 6.244 km, using assembled overhanging structure roadbed and high and low mountain bridge to widen the road surface. Figure 3 is the schematic diagram of the assembled overhanging roadbed.

Fig. 3. Schematic diagram of assembled overhanging roadbed

3.2 Model Building

ABAQUS finite element software is used to simulate the overhanging structure, and the most steeply inclined slope in the geographic location of the overhanging structure is selected as the model cross section, and the soil is restrained by the normal displacement around and vertical displacement at the bottom, as shown in Fig. 4 below. The anchor rod adopts truss unit, and the built-in area function is used to simulate the contact between the anchor rod and soil. The Mohr-Coulomb principal structure is used for the geotechnical body, and the geotechnical material parameters of the slope are shown in Table 1 below.

Table 1. Rock parameters of the slope

Soil layer	Density (10^3 kg/m^3)	Young's modulus (MPa)	Poisson's ratio	Friction angle (°)
Strongly weathered siltstone slate	2.65	1200	0.26	30
Medium weathered siltstone slate	2.72	1530	0.23	30

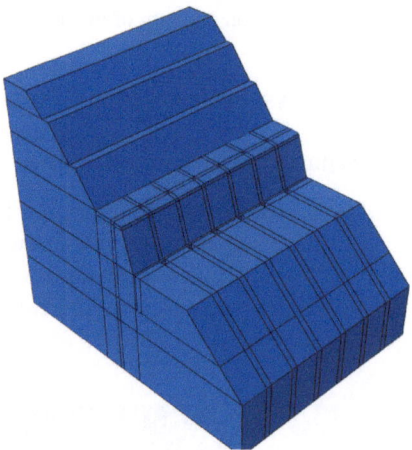

Fig. 4. Schematic of the slope model

The dimensions of the model are: the spacing between the overhanging beam and columns is 8 m; the overhanging length is 12.75 m, the beam height is 0.75 m–1.5 m, and the overhanging length is 4.25 m; the column cross-section size is 1.5 m × 1.5 m, and the columns are connected by reinforced concrete internal and external longitudinal beams to form a whole frame structure; the width of the external longitudinal beam and retaining plate is 50 cm, and the width of the internal longitudinal beam is 100 cm. The overhanging structure and the effect are shown in Fig. 5 The overhang structure and the effect are shown in Fig. 5. The precast overhanging beam are made of C50 concrete, the precast hollow slab, outer longitudinal beam and retaining slab are made of C30 concrete, and the inner longitudinal beam, inner longitudinal beam foundation, column and retaining slab foundation are made of C30 cast-in-place concrete. Foam concrete is filled between the permanent protection surface of the slope and the retaining slab, and each concrete parameter is shown in Table 2.

Table 2. Concrete parameters

Materials	Density (10^3 kg/m^3)	Young's modulus (MPa)	Poisson's ratio
C30	2.5	30000	0.2
C50	2.5	35000	0.2
Foam concrete	0.9	250	0.2

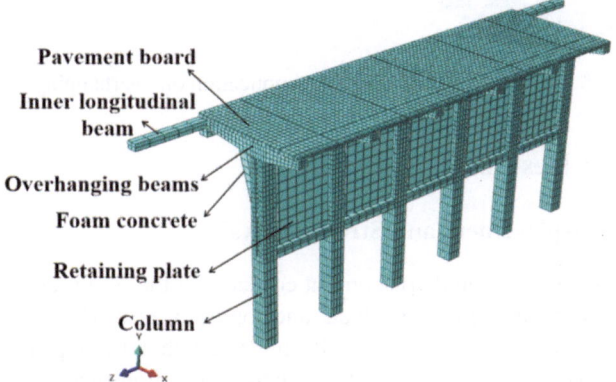

Fig. 5. Finite element model of the overhang structure

3.3 Load Application

According to the General Design Specification for Highway Bridges and Culverts (JTG D60-2018), the relevant loads are applied in combination with the most unfavorable position of the overhanging structure. The narrow strip area of the overhanging pavement is selected to convert the line load and concentrated load imposed in the code into a uniform load to achieve the effect of simulating the lane load and vehicle load.

Analyzing the characteristics of the overhang structure and the assembly type, there are two most unfavorable load arrangements for the overhang structure, which are set as Condition 1 and Condition 2. Condition 1 is to distribute the lane load on the whole overhang road according to the specification spacing, under this arrangement, the over-hang structure has adverse effects on the pressure of the embedded section at the bottom of the column, the pressure of the assembly node and the side slope soil. In this arrange-ment, the overturning moment inside the overhanging structure is the largest, and the connection part with the soil such as the anchorage end and the inner side of the column is the most unfavorable. The load application method of the overhang structure is shown in Fig. 6.

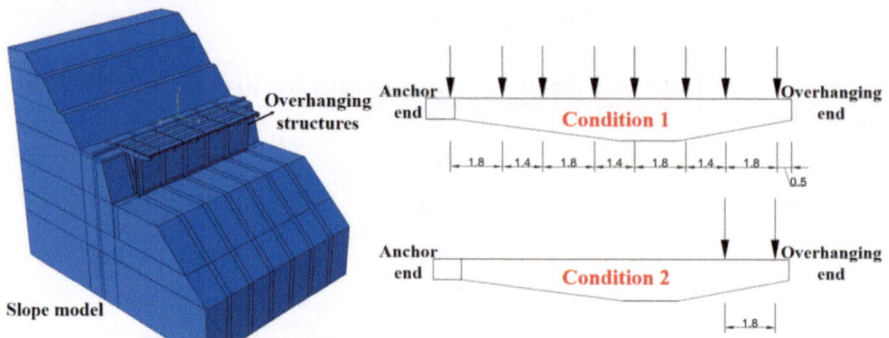

Fig. 6. Schematic illustration of load application on overhanging structure

4   Analysis of Results

4.1   Slope Soil Displacement and Stress Analysis

Figure 7(a) and (b) shows the displacement contours in the X-direction for condition 1 (overhanging structure under full lane load) and condition 2 (overhanging structure under overhanging side lane load), respectively. It can be seen that the slope under the action of Condition 1 has increased displacement and there is a sliding surface. As the embedded foundation is driven by the overhanging beam, the soil in the embedded section shows an obvious displacement in X direction. Comparing condition 1 and condition 2, it can be found that because the soil under the action of full lane load in condition 1 is subjected to the largest force, the largest deformation and a larger range of sliding surface, is used to analyze the deformation and stability of the soil under the most unfavorable load in. In condition 2, the overhang is prone to overturn under the lane load on the overhang side, and the column deformation is the largest. Therefore, condition 2 can be used to analyze the stresses and displacements of the overhanging structure under the most unfavorable conditions.

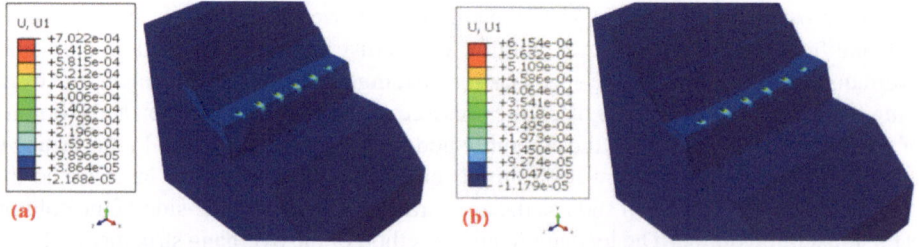

(a) Condition 1(full lane load action)   (b) Condition 2(overhanging side lane load action)

Fig. 7. Displacement contour on the X-direction of the slope

Figure 8 shows the stress contour in the X-direction under the action of condition 1. The soil at the embedded end portion of the side slope of the overhanging structure is subjected to a full lane load and the self-weight of the structure. As shown in the black dashed box in Fig. 8, the soil in this area presents a large tensile stress along the side slope, which leads to a slip surface on the side slope. Therefore, foam concrete is often filled between the column and the slope in practical engineering, thus containing the slip damage on the overhanging slope.

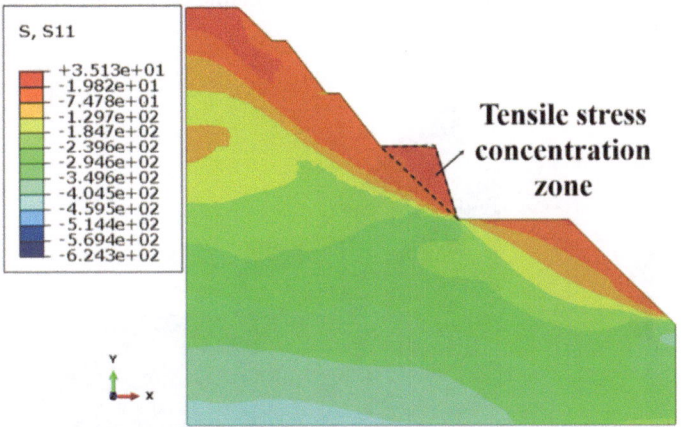

Fig. 8. Stress contour on the X-direction of the slope

4.2 Soil Stability Analysis

The reduction factor F_s in the strength reduction method is defined as the ratio of the maximum shear strength of the soil in the slope to the actual shear stress produced by the external load in the slope, while the external load is kept constant. In the finite element calculation, as the reduction factor F_s increases, the shear strength of the soil decreases until it is reduced to a certain degree and the calculation cannot be converged, which means that the shear strength of the soil at this time is not enough to support the stability of the soil and the soil is in an unstable state. The reduced shear strength index is calculated according to Eq. (1) and Eq. (2)

$$C_F = C/F_S \tag{1}$$

$$\varphi_F = \tan^{(-1)}(\tan \varphi/F_S) \tag{2}$$

where: C is the cohesion of soil before reduction, ϕ is the angle of friction within the soil before reduction, C_F is the cohesion of soil after reduction, and ϕ_F is the angle of friction within the soil after reduction.

Based on the strength reduction method, the internal friction angle and cohesion of the slope soil are reduced according to a certain proportion, and the following Fig. 9 is calculated by the reduction analysis contour map of the plastic zone of the slope soil. It can be seen that after the sliding surface is reduced by strength, the sliding surface of the slope extends from the location of the excavation and slope release of the overhanging structure to the top of the slope. To extract the calculation results, take the horizontal displacement at the top of the slope as the vertical axis and the reduction coefficient as the horizontal axis, and draw the reduction process curve. Figure 10 below shows the curve of reduction coefficient and displacement of slope top. From the Figure, we can see that the inflection point of the curve is located at about 1.71, and the stability coefficient of the slope is 1.71 according to the principle of strength reduction method, which means that the slope is stable.

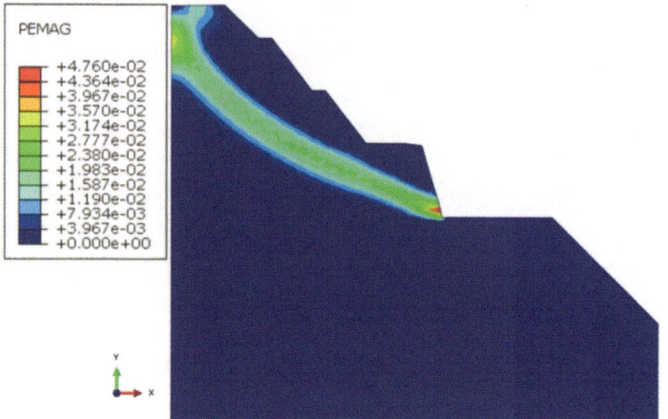

Fig. 9. Contour of plastic zone of slope soil

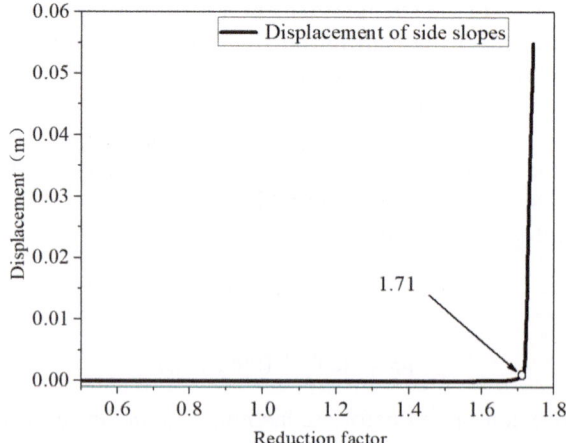

Fig. 10. Plot of reduction factor and slope top displacement

4.3   Stress and Displacement Analysis of the Overhanging Structure

The main stress-bearing elements in the whole overhanging structure are the road panel, overhanging beam and columns, so these parts are also the parts with large deformation. Figure 11 below is the road panel stress contour. As the concentrated load of the lane is applied to the middle part of each span of the overhanging structure, a stress concentration phenomenon occurs between the two spans. This is due to the middle three span road panel each span is supported by both sides of the overhanging beam, bearing a large negative bending moment to resist the road panel downward depression. The stress concentration also occurs in the road panel directly above the overhanging beam, which is caused by the bending moment generated by the interaction between the overhanging beam section and the road panel after the lane load is applied to the road panel. Figure 12 shows the deformation cloud diagram of the roadway panel. From this figure, we can see that the deformation is mainly concentrated in the overhang side, and the side span deformation is larger than the middle span. The reason is that the middle span overhanging plate is subjected to large negative bending moment at both pivot points, which plays a role in resisting plate deformation. While the side span only near the middle span side by the same size of the negative bending moment. In the most edge side of the restraint is smaller, so the deformation of the plate is relatively large. And the road panel in the overhanging beam between the obvious sag, overhanging beam upper road panel due to the overhanging beam support deformation is small, and the outermost side of the road panel lack of restraint, resulting in the road panel upward bending. The actual project should strengthen the outermost sides of the road panel restraint.

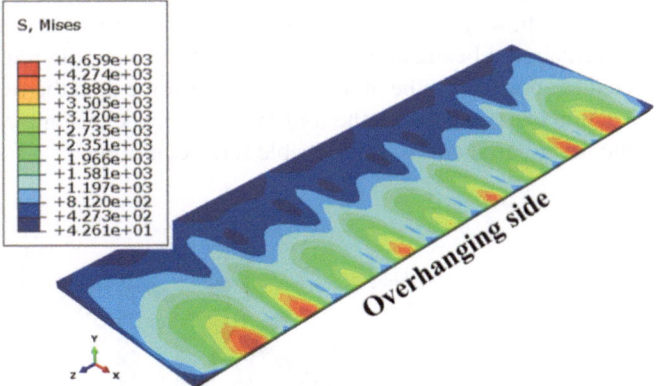

Fig. 11.   Stress contour of the road panel

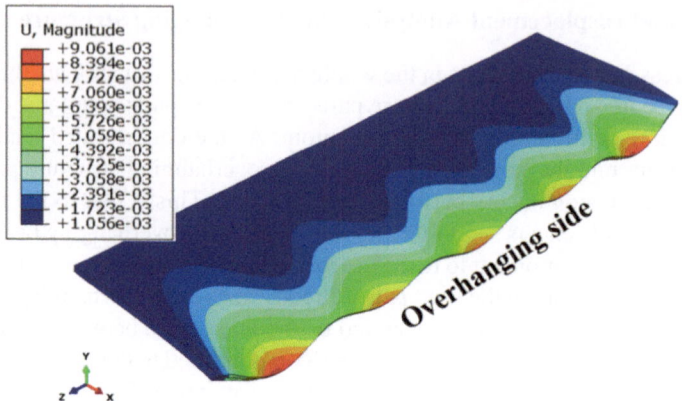

Fig. 12. Deformation contour of the road panel

Figures 13 and 14 below show the stress and displacement contours of the overhanging skeleton, which mainly consists of three parts: overhanging beam, columns and external longitudinal beams. According to the stress cloud diagram, it can be seen that the stress concentration appears on the outer side of the connection between the four middle overhanging beams and the column, indicating that the stress concentration occurs due to the extrusion of the four middle overhanging beams at the connection with the outer edge of the column due to the large bending moment. In Fig. 14, it can be seen that the displacement of the overhanging section of the middle four overhanging beam is larger, about 5 mm, while the displacement of the two spans at the edge is smaller, about 3 mm. Figure 15 shows the moment-shear diagram of the connection between the column and the overhanging beam, and it can be seen that the middle four overhanging beam are subjected to about twice the moment and shear force than the two spans at the edge. Under the action of external load, the four overhanging beams in the middle of the overhanging structure are in the most unfavorable force condition.

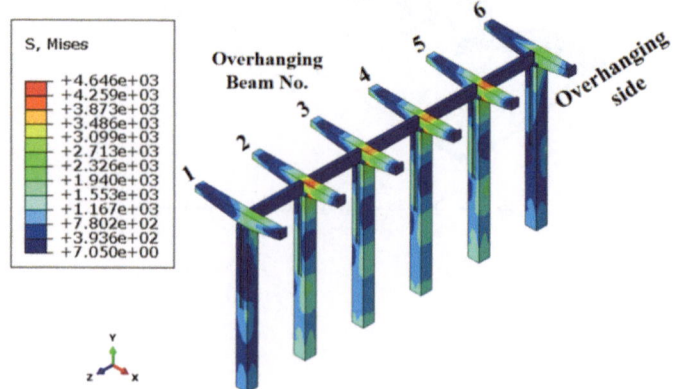

Fig. 13. Overhanging structures skeleton stress contour

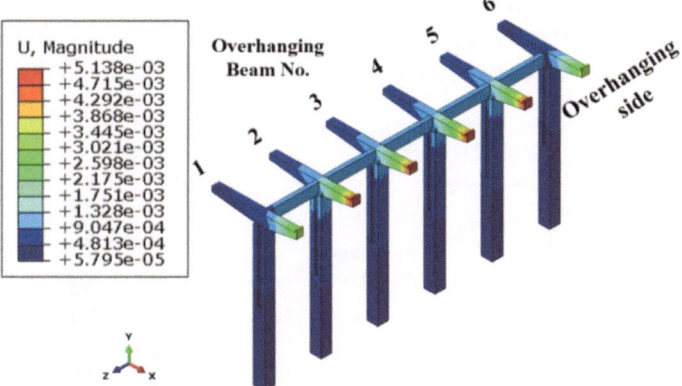

Fig. 14. Deformation of overhanging structures skeleton.

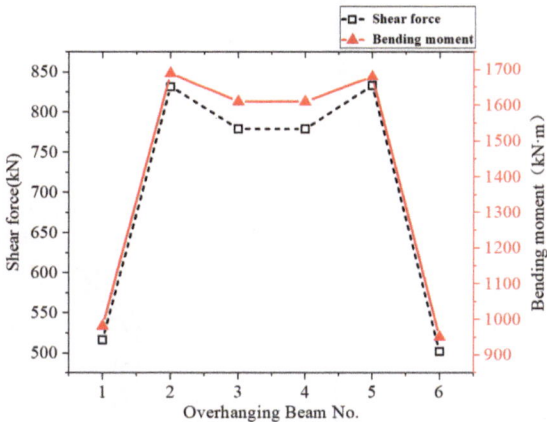

Fig. 15. Shear moment diagram of the connection point between the overhanging beam and the column

The upper end of the column is connected to the overhanging beam, and the lower part is embedded into the interior of the soil body of the slope, bearing the vertical load and bending moment transmitted from the upper part, the deflection and deformation of the column will directly affect the stability of the upper structure, especially the X-direction displacement of the column directly affects the stability of the overhanging structure. Figure 16 shows the X-direction displacement contour diagram of the column, and Fig. 17 shows the horizontal displacement curve from top to bottom of the unembedded soil section of the column. It can be seen that the X-direction displacement of the middle four columns is larger, especially at the top. According to the displacement curve, the displacement of the middle column is twice that of the edge column, which is mainly due to the outward bending of the column caused by the transfer of the bending moment of the upper beam and the deflection deformation caused by the action of the upper load, and the superposition of the two produces a deformation curve similar to the "S" shape of the column.

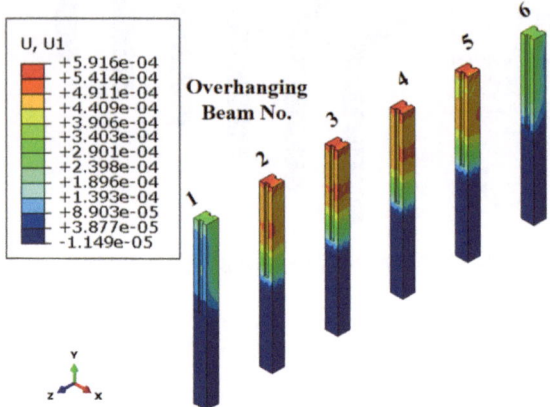

Fig. 16. X-direction displacement contour of column

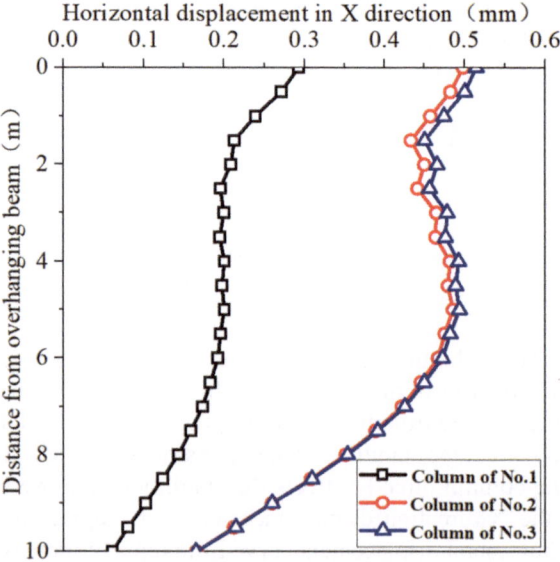

Fig. 17. Horizontal displacement curve of the unembedded soil section of the column

5 Conclusion

In this paper, a three-dimensional finite element numerical model of the assembled overhang roadbed structure is established, and the mechanical characteristics of the overhang structure under the slope stress and displacement, slope soil stability and load are analyzed. The research findings are summarized as follows.

(1) After strength reduction, the sliding surface of the slope extends from the location of the excavated and released slope of the overhanging structure to the top of the slope. The stability coefficient of the slope is 1.71, which indicates that the slope is stable.

(2) Stress concentration phenomenon between the two spans of road panels, road panel deformation is mainly concentrated in the overhang side, and side span deformation is larger than the middle span, road panels in the part between the overhanging beam appeared obvious sagging deformation, the actual project should strengthen the outermost sides of the road panel restraint.

(3) The stress concentration phenomenon appeared on the outer side of the middle overhanging and column connection end because the middle overhanging was subjected to about twice the bending moment and shear force than the two spans at the edge. Under the action of external load, the middle overhanging beam of the overhanging structure is in the most unfavorable state of force.

(4) The horizontal displacement of the middle column in X direction is larger, especially the displacement at the top of the column is larger, and the displacement of the middle column is about twice of the edge column. This is mainly due to the bending of the column caused by the transfer of the bending moment of the upper beam and the deflection deformation caused by the action of the upper load, under the superposition of the two, the column produces a deformation curve similar to "S" type.

References

1. Liu, Y.: Research on Optimization and Comprehensive Evaluation of Slope Protection Technology for Mountain Highway Roadbeds. Shijiazhuang University of Railways (2017)
2. Bao, X., Li, H.: Construction risk analysis of difficult and mountainous road foundation project. J. Railway Eng. **39**(07), 109–115+121 (2022)
3. Wang, X.: Research on the Application of Integral Suspension Structure Road Construction Technology in Tibetan Class Chang Highway. Chongqing Jiaotong University (2007)
4. Liu, C., Zhou, Z., Li, S.: New technology for widening mountain roads with integral overhang structure. Prestressing Technol. **04**, 10–13 (2008)
5. Liu, C.: Research on the Joint Action of Overhanging Structure and Geotechnical Soil in Mountain Roads. Chongqing Jiaotong University (2009)
6. Zhuo, B.: Application of overhanging slabs in roadbed widening of mountain cliff high steep slope sections. Northern Transport. **08**, 37–41 (2009)
7. Wang, K.: Design and application of overhanging slabs in the improvement and expansion of roads in mountainous and heavy hilly areas. Nonferrous Metals Digest **30**(03), 122–124 (2015). https://doi.org/10.19534/j.cnki.zyxxygc.2015.03.064

8. Su, R.: Research on the design of mountain highway reconstruction and expansion project under complex terrain. Transport. Stand. **42**(22), 44–47 (2014). https://doi.org/10.16503/j.cnki.2095-9931.2014.22.013

9. Tian, H.: Analysis of Mechanical Behavior of Assembled Overhanging Roadbed Structures in Mountainous Areas. Hefei University of Technology (2019)

10. Guo, S., Song, X., Chen, H.: Research on structural design and analysis methods of overhanging U-shaped roadbed for high-speed railroad. Railway Standard Des. **65**(04), 14–19 (2021). https://doi.org/10.13238/j.issn.1004-2954.202004070005

11. You, T.: Research on the Calculation Method of Road Overhanging Structure Based on the Joint Structure-Geotechnical Interaction. Chongqing Jiaotong University (2011)

12. Wei, Y., Li, H., Nie, W.: Computational analysis of overhanging slab roadbeds. Roadbed Eng. **211**(04), 184–188 (2020). https://doi.org/10.13379/j.issn.1003-8825.201908040

13. Yao, Y., Sun, G., Chen, Z., et al.: Parameter optimization design of new "overhang-supported" roadbed structure. Highw. Eng. **45**(05), 228–233 (2020). https://doi.org/10.19782/j.cnki.1674-0610.2020.05.038

14. Yuan, R.: Discussion on the design of separated stacked roadbed scheme for mountain highways. Eng. Constr. **32**(02), 190–191+194 (2018)

15. Yuan, R., Cheng, Y.: Discussion on the application of assembled suspension structure in new roads in mountainous areas. Eng. Constr. **31**(04), 547–549 (2017)

16. Lai, A. Exploration on the Application of Composite Road with Overhanging Structure in Kerai Highway. Chongqing Jiaotong University (2012)

17. Jiang, Q.: A review of the development of assembled concrete buildings at home and abroad. Construct. Technol. **41**(12), 1074–1077 (2010)

18. Fan, L.: Research on the Seismic Performance of Assembled Precast Concrete Frame Structures. Tongji University (2007)

Research on Existing Bridges Rapid Load-Lifting Technology with UHPC in Highways Reconstruction and Expansion Projects

Jintao Shi[1]([⊠]) and Zhijiang Chen[2]

[1] CCCC Highway Bridges National Engineering Research Centre Co., Ltd., No. 23, Huangsi Street, Xicheng District, Beijing, China
jintao_2_@126.com
[2] College of Aerospace, Chongqing University, Chongqing, China

Abstract. There are a large number of small and medium-span bridges in the highway reconstruction and expansion projects. These bridges have been used for at least 20 years. Due to the design according to the old code the ultimate bearing capacity is generally insufficient after the reconstruction and expansion. This paper selects hollow-core slab bridges that account for more than 80% of the reconstruction and expansion projects as the research object, carefully evaluates the specific reasons for the insufficient bearing capacity of small and medium-span bridges, designs a plan for rapid loading-lifting of existing bridges, and uses theoretical analysis. Then, the comprehensive method of finite element calculation analyzes and studies on the reinforced bridge's bearing capacity. The research shows that the existing small and medium-span bridges in the reconstruction and expansion project are mainly manifested by insufficient shear-bearing capacity after the load is improved. Using the bridge deck to add the UHPC cast-in-place layer (participating in the structural internal force) can effectively improve the existing old bridge Shear resistance capacity, and UHPC cast-in-situ layer shear reinforcement efficiency is much greater than that of ordinary concrete.

Keywords: Reconstruction and expansion of Highway · Shear capacity · UHPC · Bridge reinforcement

1 Introduction

At present, a large number of small and medium-sized span concrete bridges are involved in domestic highway renovation and expansion projects, with a service life of about 20 years. After testing, most of the bridges have good overall performance. However, due to the old code used in the bridge's initial design, the load standards could be higher. If the technical standards used in the reconstruction and expansion project are followed (Several Opinions on Handling Technical Issues in Highway Reconstruction and Expansion Project ([2013] No. 634), "Technical Standards for Highway Engineering" (JTG B01-2014), and Design Rules for Highway Reconstruction and Expansion

© The Author(s) 2024
P. Xiang and L. Zuo (Eds.): PBSFTT 2023, LNCE 382, pp. 33–47, 2024.
https://doi.org/10.1007/978-981-97-5108-2_3

(JTG L11-2014) [1], the ultimate bearing capacity cannot meet the requirements. Therefore, it is necessary to carefully select whether the existing bridge should be reinforced
or demolished for reconstruction.

If the existing old bridge is demolished in situ and then newly built, it will cause
significant financial waste; When using conventional reinforcement utilization schemes,
there are drawbacks such as low reinforcement efficiency, complex construction techniques, and significant impact on existing high-speed traffic. In response to the above
issues, mainstream China design units currently recommend using the method of adding
ordinary concrete. However, adding ordinary concrete is inefficient in improving the
cross-sectional bearing capacity and requires complex processes such as planting steel
bars on existing bridge decks.

In order to effectively utilize existing old bridges and save engineering investment,
it is urgent to study how to quickly strengthen existing concrete bridges, a practical
problem faced in renovation and expansion projects. This article relies on a China highway renovation and expansion project to address the problem of insufficient ultimate
bearing capacity of existing small and medium-sized span concrete bridges. It proposes
a rapid load-bearing plan for pouring ultra-high performance concrete (UHPC) bridge
deck layer (participating in structural internal stress).

2 Project Overviews

A certain Highway runs from east to west and is an essential economic artery transportation channel within the province. Since its opening, with the rapid development
of the regional economy and society and the further improvement of the highway network, the traffic volume has overgrown. The existing road capacity of highways can no
longer meet the needs of economic, social, and transportation developments. The project
urgently needs to change the existing 4 lanes to 8 lanes, and the total route mileage of
the renovation and expansion project is about 310 km. The bridges along the line include
a large number of small and medium-sized span bridges designed based on the 1985
bridge code According to the original design and construction drawings and on-site field
investigations of this project, the main types of bridge superstructure are: prefabricated
reinforced concrete solid slabs, prefabricated reinforced concrete hollow slabs, prefabricated prestressed concrete hollow slabs, prefabricated prestressed concrete I-shaped
composite beams and prefabricated prestressed concrete T-beams. The spans of hollow
slab (solid slab) bridges with a length of 5–16 m account for 80% of the total spans of
the bridge.

In order to investigate the performance of existing small and medium-span bridges,
the testing unit conducted load tests on several typical bridges in this project. The vehicle
loads in the load tests were taken as Automobile – Class 20, Trailer -120, and the spans
of typical bridges were selected as 5 m, 8 m, 10 m, 13 m, 16 m, and 25 m. The test results
showed that the calibration coefficients of all bridges were less than 1, the relative residual
displacement was less than 20%, the measured fundamental frequency was greater than
the theoretical fundamental frequency, and the measured impact coefficient was less than
the theoretical impact coefficient, No new cracks were found in the inspection results;
The inspected bridge is in an elastic working state under the load of "car over 20 ton",

and the overall performance is good. In this renovation and expansion project, based on multiple plan reviews and expert discussions, it was unanimously agreed to continue to fully utilize or reinforce small and medium-sized span bridges with good performance. Because the proportion of hollow slab bridge beams in this project reaches 80%, research should focus on the rapid reinforcement scheme of existing hollow slab bridge beams.

In order to concise this article, which only evaluates the bearing capacity of reinforced concrete hollow slabs with typical spans of 10 m, 13 m, and 16 m, and studies the rapid reinforcement scheme. Typical hollow slab bridge section structure is shown as in Fig. 1.

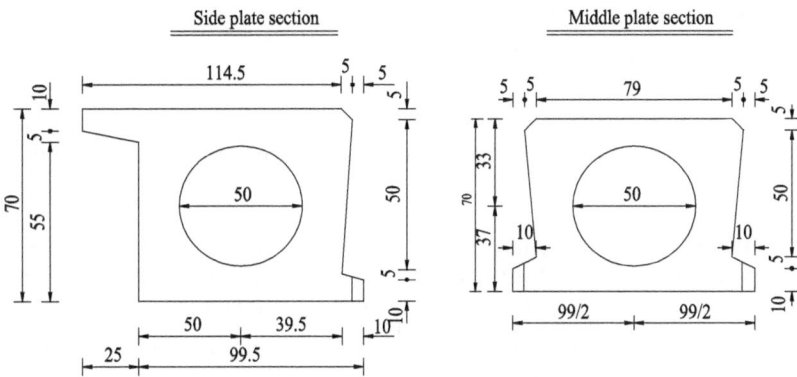

Fig. 1. Typical hollow slab bridge section structure (cm)

3 Assessment of Bearing Capacity of Existing Bridges Before Reinforcement

3.1 Technical Status of Bridges

According to Specification for Inspection and Evaluation of Load-bearing Capacity of Highway Bridges (JTGIT J21-2011) and the inspection report, the statistical analysis of the technical status of prestressed concrete solid (hollow) slab bridges along the entire line is shown in Table 1. From the statistical analysis results in Table 1, it can be seen that the prestressed concrete hollow slab bridge is mainly composed of 10 m and 16 m bridges of Class II and Class III, with the most representative bridges having a technical level of Class III.

3.2 Calculation of Bearing Capacity Before Reinforcement (According to JTG B01–2014 Code)

In the case of directly utilizing the old bridge, there is basically no lifting space for the old bridge. According to Highway Class I (JTG B01-2014 load), the bearing capacity of the old bridge after widening was checked, and the calculation results are shown in Tables 2 and 3.

Table 1. Statistics of technical status of prestressed concrete solid (hollow) slab bridge

| Class | Span distribution | | | | | | Total | |
| | 10 m | | 13 m | | 16 m | | | |
	Quantity	ratio	Quantity	ratio	Quantity	ratio	Quantity	ratio
I	4/163	2%	2/52	4%	2/111	2%	8/326	2%
II	71/163	44%	26/52	50%	41/111	37%	138/326	42%
III	88/163	54%	24/52	46%	66/111	59%	178/326	55%
IV	1/163	0%	0/52	0%	2/111	2%	2/326	1%
V	0/163	0%	0/52	0%				

Table 2. Calculations of flexural bearing capacity of 1/2 section in the middle of 10, 13, 16 m slabs

Span(m)	Bridge wide(m)	Position of slab	Impact of cast-in-place layer	Allowable value (km.m)	Design value (km.m)	Safety factor	Meet the requirements Yes or No
10	26	Side slab	do not consider	483.7	560.4	0.8631	No
			consider	608.5	568.2	1.0709	Yes
		Middle slab	do not consider	473.2	662.9	0.7138	No
			consider	597.9	679.2	0.8803	No
13	26	Side slab	do not consider	889.1	954.7	0.9313	No
			consider	1047.9	986	1.0628	Yes
		Middle slab	do not consider	873.4	1032.7	0.8457	No
			consider	1028.9	943	1.0911	Yes
16	26	Side slab	do not consider	1306.1	1484.7	0.8797	No
			consider	1522.7	1386.9	1.0979	Yes
		Middle slab	do not consider	1273.2	1561.7	0.8153	No
			consider	1512.1	1549.2	0.9761	No

The calculation results of whether the required section bending and shear resistance are met indicate that when the thickness of the old pavement layer does not participate in the section stress, the bearing capacity of all hollow slabs does not meet the requirements; When the thickness of the old pavement layer is involved in the force acting on the section,

Table 3. Calculation of shear bearing capacity at support of 10, 13, 16 m slabs

Span(m)	Bridge wide(m)	Position of slab	Impact of cast-in-place layer	Allowable value (km.m)	Design value (km.m)	Safety factor	Meet the requirements Yes or No
10	26	Side slab	do not consider	282.5	302.9	0.9327	No
			consider	342.3	426.4	0.8028	No
		Middle slab	do not consider	263	454.1	0.5792	No
			consider	403.5	459	0.879	No
13	26	Side slab	do not consider	345.3	357.9	0.9648	No
			consider	397.3	489	0.8125	No
		Middle slab	do not consider	309.8	480.4	0.6449	No
			consider	434.6	489.7	0.8875	No
16	26	Side slab	do not consider	474.1	605.8	0.7826	No
			consider	526.6	601.2	0.8759	No
		Middle slab	do not consider	433.7	670.7	0.6466	No
			consider	603.4	667.6	0.9039	No

the bending bearing capacity of the section can basically meet the requirements of the new code, but the safety factor is low. At the same time, the increase in shear bearing capacity is limited and still does not meet the requirements of the new code.

Based on engineering experience, the bridge deck pavement layer in actual engineering is involved in bearing capacity of structure. Therefore, the bending bearing capacity of existing small and medium-sized span bridges can basically meet the requirements, but the safety factor is relatively low. On the other hand, the reserve of shear bearing capacity for existing small and medium-sized span bridges still needs to be increased even after considering the participation of the pavement layer in the bearing capacity. Therefore, when formulating the design plan for lifting and strengthening existing bridges, the primary consideration should be to enhance the shear-bearing capacity of existing bridges.

4 Research on Load Lifting and Reinforcement Scheme for Existing Bridges

4.1 Overall Principles and Objectives of Reinforcement

(1) General principles of reinforcement

1) Scientific and reasonable design with economic and environmental protection;
2) Increasing the self-weight of the structure is not significant and does not cause damage or slight loss to the original components;
3) Convenient and fast construction with minimal impact on the surrounding environment;
4) The bridge reinforcement can meet the bearing capacity requirements of the new code.

(2) Reinforcement target
1) The ultimate limit state calculation of the bearing capacity of the original bridge after reinforcement should meet the requirements of Highway – Class I (JTG B01–2014 load), and the ultimate limit state calculation of normal use should meet the load level requirements of "Automobile – Class 20, Trailer -120";
2) Damage that occur in bridge structures or components should be repaired or improved through reinforcement measures to meet the needs of expansion.

4.2 Comparison and Selection of Reinforcement Methods

Based on current engineering practice [2], the methods for strengthening the ultimate bearing capacity of bridges include: ① increasing the cross-section (increasing the primary reinforcement, increasing the beam ribs, thickening the bridge deck, etc.); ② pasting steel plates (carbon fiber, fiberglass, etc.); ③ external prestressing (steel bars, carbon plates, etc.) reinforcement; ④ adding auxiliary components or changing the structural system method. Among them, ②–④ reinforcement methods have high requirements for construction technology, require professional teams for construction, have high requirements for the construction environment, and have high construction costs.

In order to ensure the unity of the later management and maintenance of newly built and existing bridges in the renovation and expansion project, this project proposes to use secondary reinforcement technology for existing bridges, so that when constructing and expanding bridges, simple and rapid reinforcement method is used for existing old bridges; When the performance of the bridge structure deteriorates in the later stage, the secondary reinforcement methods described in ②–④ can be adopted to maximize the service life of the existing old bridge.

Most of the existing old bridges in this project can meet the requirements for flexural bearing capacity, but their shear bearing capacity needs to be significantly improved. Using simple and rapid reinforcement methods for shear reinforcement is a technical challenge faced by engineering.

In terms of shear-bearing capacity, the main factors affecting shear bearing capacity are shown in Fig. 2 mainly including Vc (shear bearing capacity of concrete in the compression zone at the top of the inclined section), Vs (shear bearing capacity of stirrups), and Vsb (shear bearing capacity of bent steel bars).

Since the shear reinforcement of existing old bridges cannot generally be supplemented or strengthened, it is essential to focus on strengthening the compression zone Vc of the inclined section. Based on the principle of rapid construction, the method of "increasing the beam section" is considered to increase the shear area of the section. Specifically, the technique of "thickening the bridge deck" can be selected to use the method of adding a concrete layer to the bridge deck.

Regarding selecting the cast-in-place concrete layer for the bridge deck, two schemes, C50 ordinary concrete and UHPC, have been preliminarily formulated.

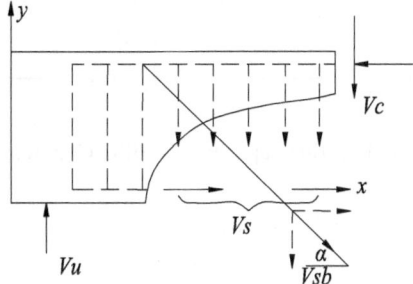

Fig. 2. The bearing capacity influencing factors of the oblique section

4.3 Basic Principles of Bridge Deck Reinforcement for Existing Bridges Based on UHPC

There are few cases and research results of using UHPC bridge deck to reinforce existing bridges in engineering practice [3]. This study draws on the relevant research ideas of the Swiss standard "Recommendation: Ultra High Performance Fiber Reinforced Cement based composites (UHPFRC) Construction Material, Dimensioning and Application" (SIA-2052) [4].

$$M_d \leq \sigma_{cb}\frac{(x-h_U)}{2}\left(h_C - d_{sc} - \frac{1}{3}(x-h_U)\right) + \sigma_{sc}A_{sc2}(d_{scc} - d_{sc}) + (\sigma_{Uc} +$$
$$\frac{E_{UC}}{E_C}\sigma_{cb})b\frac{h_U}{2}\left(h_U + h_C - d_{sc} - \frac{1}{2}h_U\right) + \sigma_{sU}A_{sU}(d_{sU} - d_{sc}) \tag{1}$$

The height x of the concrete compression zone is determined by the following formula:

$$f_{sd}A_{sc1} = \sigma_{cb}\frac{(x-h_U)}{2} + \sigma_{sc}A_{sc2} + (\sigma_{Uc} + \frac{E_{UC}}{E_C}\sigma_{cb})b\frac{h_U}{2} + \sigma_{sU}A_{sU} \tag{2}$$

(1) Calculation of bending bearing capacity

Basic assumptions for calculation: a) The cross-sectional strain remains plane b) Do not consider the tensile strength of concrete in the tensile zone; c) Take the corresponding stress-strain for the UHPC section; d) The compressed area is distributed in a triangular shape.

The bending bearing capacity of the composite section is jointly provided by ordinary concrete in the compression zone, existing steel bars in the compression zone, UHPC and supplementary reinforcement in the compression zone. Refer to Fig. 3 and Eqs. 1 and 2.

The formula for calculating the shear bearing capacity of UHPC ordinary concrete composite section is as Fig. 4 and Eqs. 3 and 4.

$$V_{Rd} = V_{Rd,c} + V_{Rd,s} + V_{Rd,U} \tag{3}$$

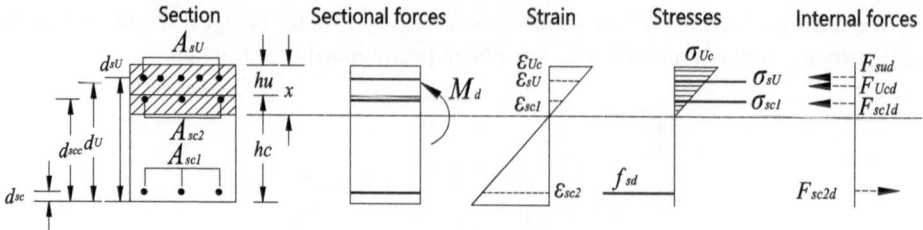

Fig. 3. Calculation of flexural bearing capacity of UHPC-Common concrete composite layer

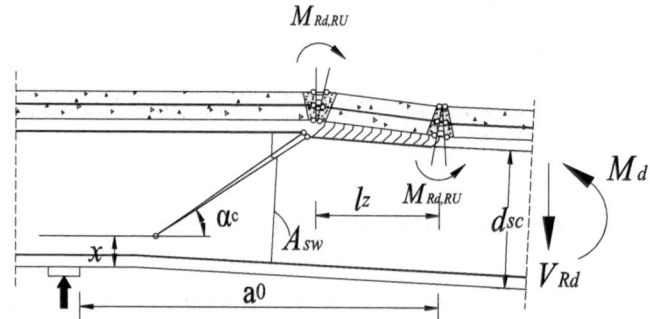

Fig. 4. Calculation of Shear Resistance of Cross Section of UHPC Concrete

a) Ordinary concrete part:

$$V_{Rd,c} = \frac{f_{cd} \cdot b_w}{2} \left[\frac{x}{\sin \alpha_c} \cdot (1 - \cos \alpha_c) \right] \qquad (4)$$

The inclination angle of diagonal cracks caused by bending moment and shear in ordinary reinforced concrete α_c. There are the following boundaries: $20° \ll \alpha_c \ll 60°$. The first approximation can be assumed to be $\alpha_c = 35°$. Calculate the height of the concrete compression zone according to the following expression: $x = 0.9 \cdot \omega_M \cdot d_{Eq}$. Among them: ω_M is the content of composite section steel bars; d_{Eq} is the equivalent height of the composite cross-section.

b) Vertical shear reinforcement:

$$V_{Rd,s} = A_{sw} \cdot f_{sd} \cdot \cot \alpha \qquad (5)$$

c) Reinforced UHPC part:

$$V_{Rd,U} = \frac{2 \cdot M_{Rd,RU}}{l_z} \text{ where } l_z = a_0 - \frac{d_{sc}}{\tan \alpha_c}$$

Among them, $M_{Rd,RU}$ is the ultimate flexural bearing capacity of UHPC.

The above analysis shows that after the combination of UHPC and ordinary concrete sections, the shear bearing capacity of the combined section exceeds that of VRD and U. The core of the UHPC reinforcement section is that the ultra-high strength UHPC layer limits the propagation of oblique cracks in the cracking zone.

4.4 Design of Reinforcement Plan for Existing Bridges

In this project, the elevation needs to be adjusted through asphalt concrete pavement. Hence, the reinforcement design plan also considers the thickness matching of the cast-in-place concrete layer and the asphalt concrete layer. The specific design scheme is shown in Table 4.

Table 4. Reinforcement plan for cast-in-situ layer of 10 m, 13 m, 16 m span hollow slab bridge deck

number	design scheme	Overview of the plan	Reinforcement of cast-in-place layer	Interface connection
1	Original design	10 cm C40 concrete + 6 cm asphalt concrete	15 cm Φ 16 double-layer bidirectional steel mesh	Cross section chiseling, Φ 16 steel bar planting
2	Reinforcement Plan 1	16 cm C50 concrete + 10 cm asphalt concrete		
3	Reinforcement Plan II	16 cm UHPC + 10 cm asphalt concrete		Cross section chiseling to form concave and convex surfaces
4	Reinforcement Plan III	8 cm UHPC + 8 cm C50 concrete + 10 cm asphalt concrete		

The main construction steps of using UHPC to reinforce the bridge deck include: (1) chiseling off the old asphalt concrete and concrete pavement layer of the bridge deck; (2) Using high-pressure water jet and other methods to chisel and remove debris; (3) The UHPC joint is connected with steel bars as auxiliary connections, and the anchoring length of the joint steel bars should be > 15d (d is the diameter of the overlapping steel bars), as shown in Fig. 5.

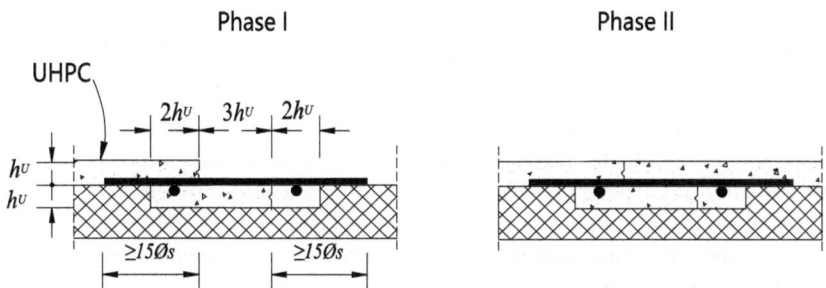

Fig. 5. Joint treatment of cast in situ layer of UHPC bridge deck

Many studies have shown that [5–11] UHPC has good bonding performance with conventional concrete, so there is no need to use complex processes such as planting steel bars for interface connection. Due to the omission of the original bridge deck reinforcement planting process, construction efficiency can be significantly improved.

4.5 Effectiveness Evaluation of Reinforcement Schemes Based on Swiss Code (SIA-2052)

In the above reinforcement plan, due to the reliable interface connection between the cast-in-place layer and the existing old bridge, the UHPC (or C50 concrete) cast-in-place layer can participate in the stress as a part of the structure.

The bending and shear bearing capacity of hollow slabs with spans of 10 m, 13 m, and 16 m were calculated and analyzed using the UHPC reinforcement design theory in the Swiss code (SIA-2052) and combined with three-dimensional finite element analysis as an auxiliary analysis.

Only the calculation results of the middle board are displayed. The analysis results are shown in the following Figure.

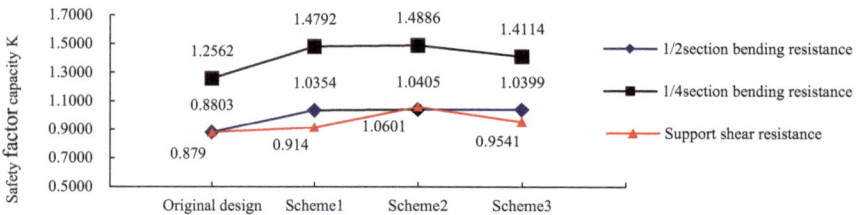

Fig. 6. Calculation result after reinforcement of 10 m span low slab (medium slab)

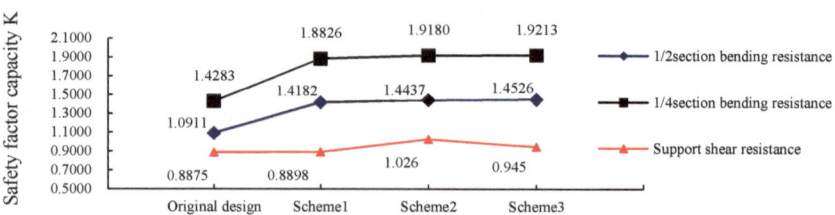

Fig. 7. Calculation result after reinforcement of 13 m span low slab (medium slab)

From the calculation results in Figs. 6, 7 and 8, it can be seen that compared to ordinary concrete C50, the UHPC layer does not significantly improve the bending bearing capacity of existing bridges, and the coefficient of improvement is about 1.0% to 2% under the same thickness of reinforcement layer; The UHPC layer significantly improves the shear bearing capacity of existing bridges, with an increased coefficient of about 15%–16% under the same thickness of reinforcement layer.

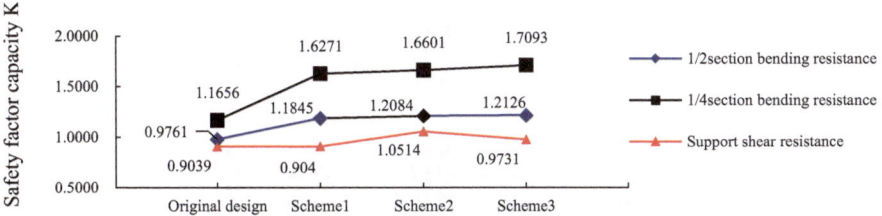

Fig. 8. Calculation result after reinforcement of 16 m span low slab (medium slab)

4.6 Evaluation of Reinforcement Effect Based on 3D Solid Finite Element Method

In order to fully reveal the influence of pavement thickness and materials (ordinary concrete, UHPC) on the shear bearing capacity of the section, Abaqus three-dimensional universal finite element method was used for precise elastic-plastic numerical simulation. A 10 m span hollow slab was taken as a specific research object.

(1) Model establishment

Establish two finite element models: 'original hollow slab', 'hollow slab + 10 cmC50 reinforcement layer', 'hollow slab + 15 cmC50 reinforcement layer', and 'hollow slab + 10 cmUHPC reinforcement layer'.

The concrete slab is simulated using solid unit C3D8R; Ordinary steel bars and prestressed steel bars are simulated using rod element T3D2.

Due to the connection between the reinforcement layer and the original hollow slab through planting bars, in the model analysis, the reinforcement layer and the top surface of the original hollow slab are constrained by binding constrain (tie), without considering the shear slip between the new and old concrete surfaces.

The load is displaced, with a loading position of 0.8 m from the fulcrum and a maximum loading displacement of 80 mm. Apply simple support at the four corners of the bottom surface of the box beam, as shown in the Fig. 9.

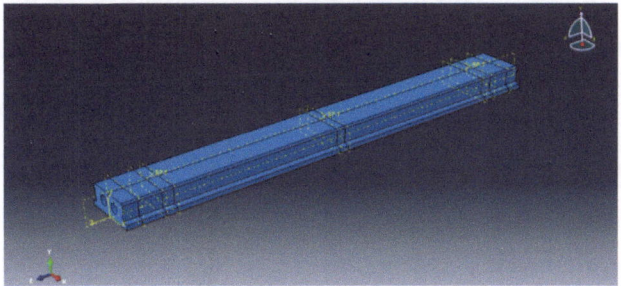

Fig. 9. Three dimensional solid finite element model of low slab

In this simulation, numerical analysis was conducted on the shear bearing capacity of different overlay heights and materials as shown in Fig. 10.

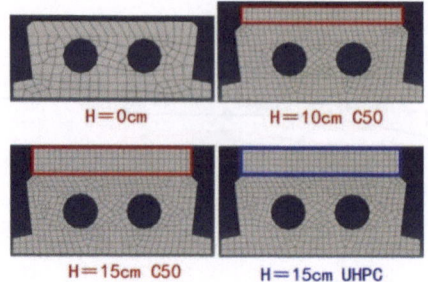

Fig. 10. Shear reinforcement plan for low slab

(2) Main calculation parameters

The original hollow slab C40 concrete and the reinforcement layer C50 both use the ordinary concrete damage plastic mode, and its constitutive model adopts the relevant provisions of the "Code for Design of Concrete Structures" (GB 50010-2010); Ordinary steel bars adopt a bilinear elastic-plastic model; The prestressed reinforcement adopts an ideal elastic-plastic model and is prestressed by the cooling method.

The material indicators of UHPC adopt the relevant regulations of the Technical Code for Ultra High Performance Lightweight Composite Bridge Deck Structures (GDJTG/TA01–2015) and its detailed parameters are shown in Table 5.

Table 5. UHPC damage plasticity model parameters

Elastic modulus $E_0(N/mm^2)$	Poisson's ratio ν	Uniaxial compressive strength $f_c(N/mm^2)$
37600	0.2	77.4
Uniaxial tensile strength $f_t(N/mm^2)$	Expansion angle ψ	Flow potential offset η
22	38	0.1
σ_{b0}/σ_{c0}	Invariant stress ratio K_c	Viscosity coefficient μ
1.16	0.66667	0.0005

(3) Analysis results

The calculation results are shown in the following Figure:

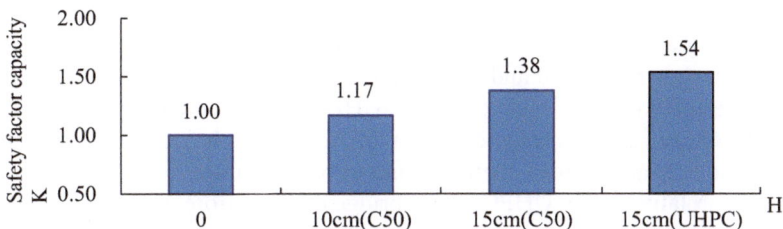

Fig. 11. Influence of payment thickness and material on shear capacity

From the above three-dimensional elastic-plastic finite element analysis results (Fig. 11), it can be seen that when using the same thickness of reinforcement layer, the shear bearing capacity of UHPC layer is increased by about 12% compared to C50 layer, which is close to the calculation results using Swiss standard (SIA-2052) in Sect. 4.3.

5   Conclusion

This article takes the hollow slab with small and medium-sized spans, which accounts for many existing highways, as the research object. In response to the problem of insufficient bearing capacity brought about by the renovation and expansion of highways, a rapid reinforcement scheme using a UHPC layer on the bridge deck is proposed. The feasibility of this scheme is preliminarily verified through theoretical analysis and three-dimensional finite element simulation results. The main research conclusions are as follows:

(1) After the renovation and expansion of existing small and medium-sized span bridges, when using new standards to evaluate their ultimate bearing capacity performance, their insufficient bearing capacity is mainly manifested as insufficient shear bearing capacity.
(2) According to the on-site construction characteristics of the renovation and expansion project, a scheme of cast-in-place UHPC layer on the bridge deck and planting reinforcement to form a structural whole with the existing bridge deck is adopted, which has significant advantages such as simplicity, speed, and minimal traffic impact.
(3) The effect of the UHPC layer on improving the bending bearing capacity of small and medium-sized span bridges is not significant, and it only increases by 1% to 2% compared to ordinary C50 concrete of the same thickness; The main reason is that the bending bearing capacity of bridges is not only related to the concrete in the compression zone but also directly determined by the reinforcement configuration in the tension zone.

(4) The UHPC layer has a significant effect on improving the shear bearing capacity of small and medium-sized span bridges, with an increase of 12%–16% compared to ordinary C50 concrete of the same thickness; The main reason is that UHPC, which is rich in steel fibers and has high compressive and tensile strength, can significantly limit the development of cracks in inclined sections under shear limit states.

(5) When using a UHPC bridge deck to reinforce existing bridges, as only the compression area is reinforced and the tension area is not correspondingly reinforced, it should be noted that the thickness of the UHPC reinforcement layer should be manageable. Otherwise, it is easy to form a few reinforced beams so that the structural failure mode changes from ductile failure to brittle failure.

(6) There are relatively few UHPC-based bridge deck reinforcement cases for existing bridges worldwide. When researching design schemes, relevant calculations and analysis can be made by referring to the Swiss standard (SIA-2052); In the later stage, systematic research should continue to be conducted through theoretical analysis and model experiments, ultimately improving the appropriate design methods.

Acknowledgments. Authors wishing to acknowledge assistance or encouragement from colleagues, special work by technical staff or financial support from organizations should do so in an unnumbered Acknowledgments section immediately following the last numbered section of the paper.

References

1. Zhejiang Provincial Department of Transportation: Guidelines for Design of HighwayReconstruction and Extension: JTG/TL11-2014. China Communications Press Co. Ltd, Beijing (2014)
2. CCCC First Highway Consultants Co., Ltd.: Code for Strengthening Design of Highway Bridges: JTG/TJ22-2008. China Communications Press Co. Ltd., Beijing (2008)
3. 2020 China Ultra High Performance Concrete (UHPC) Technology and Application Development Report CHINA CONCRETE 2021 4 pp. 20–29
4. EPFL Wiss Federal Institute of Technology: Recommendation: Ultra High Performance Fiber Reinforced Cement based composites (UHPFRC): SIA-2052. MCS-EPFL Lausanne, Switzerland (2016)
5. Shangshun, L., Kedan, C., Chen, W., Xiaoming, X.: Application of UHPC reinforced concrete beams. J, Fujian Univ. Technol. **19**, 255–260 (2021)
6. Shangmeng, Z., Wei, W., Conglong, H.: Study on UHPC $1/9$ NC Composite Interface Connection and Its Shear Performance Railway Engineering 61, pp. 10–13 (2021)
7. Xudong, S., Wei, F., Zhengyu, H.: Application of ultra high performance concrete in engineering structures China civil. Eng. J. **54**, 1–13 (2021)
8. Peijun, Z., Dashan, L.: Analysis and Application on Reinforcement of Steel Structural Footbridge with UHPC Urban roads and bridges and flood control 12, pp. 84–86+115+15 (2020)
9. Xiaofang, K.: Study on bonding performance of reinforced concrete beams reinforced by UHPC Earthq. Resident Eng. Retrofitting 42, 46–54+17 (2020)
10. Zhongke, S.: Experimental Study on the Improvement of Flexible Behavior of Rectangular Beams by UHPC with High Flow State. Inner Mongolia University of Science & Technology, Inner Mongolia (2020)

11. Daqing, Z.: Application Research of UHPC Pavement in Improving Bearing Capacity of Existing Concrete Beam Bridge. Kunming University of Science and Technology, Kunming (2020)

Research on Key Technologies for Translation and Dismantling in Tunnel of Large-Diameter TBM

Haiyang Li[(⊠)], Ti Zhang, Lei Huo, YiJie Wang, Shun Zhang, and Lianjie Jin

China Railway Engineering Equipment Group Co., Ltd., Zhengzhou City, China
haiyangli@crectbm.com

Abstract. Based on actual construction cases, this paper focuses on the design of main machine translation tooling and the main machine translation process in the tunnel of large-diameter TBM. Aiming at the design and dismantling tooling of the large-diameter screw machine in the tunnel, the key work of the tooling and the dismantling process was expounded. According to the requirements of the Back-up system retreat to the launching shaft, all Gantries retreat method is designed, the special tooling design is carried out, and the method is explained. SolidWorks 3D modeling software was used to build the disassembly model and conduct finite element analysis to check the strength of the tooling. This case ensures the safety of construction and improves the efficiency of TBM dismantling, which has reference significance for the future dismantling of large-diameter TBM in tunnel.

Keywords: Main machine move · Special tools design · Finite element analysis

1 Introduction

The TBM must be dismantled after the completion of the construction of the TBM, taking Lot of Beijing New Airport Line as an example, the shield construction uses two Φ9040 mm EPB TBM from China Railway Engineering Equipment Group Co., Ltd. For construction. The screw machine weighs about 55 T and must be dismantled in the tunnel, dismantled tooling will be build in tunnel, and it is difficult to control risks in the tunnel and demolition process. The shield main machine and the shield receiving frame weigh about 870 T and must be translated to the disassembly area for dismantling, and it is difficult to make translation tooling and control the translation process. Gantries retreated to the launching shaft, and the retreat distance is about 3.6 km, and it is difficult to dismantle and retreat. In view of the difficulties in the construction process of the project, such as the demolition of the screw machine, the translation of the main engine, and the retreat of the trailer, the key technologies are studied.

2 Dismantling Sequence of TBM

After the TBM breaking throughout, it is prepared for shutdown, and then the belt of the belt conveyor is cut. Disconnect the high-voltage electricity, retract all thrust cylinders, discharge the hydraulic oil in the main hydraulic tank and the gear oil in the main drive

P. Xiang and L. Zuo (Eds.): PBSFTT 2023, LNCE 382, pp. 48–58, 2024.
https://doi.org/10.1007/978-981-97-5108-2_4

and the water in the system, etc. [1]. Disconnect Gantry1 and erector, and then move Back-up system backward 30, reserving space for dismantling screw conveyor, and the disassembly flow chart [2] (as shown in Fig. 1).

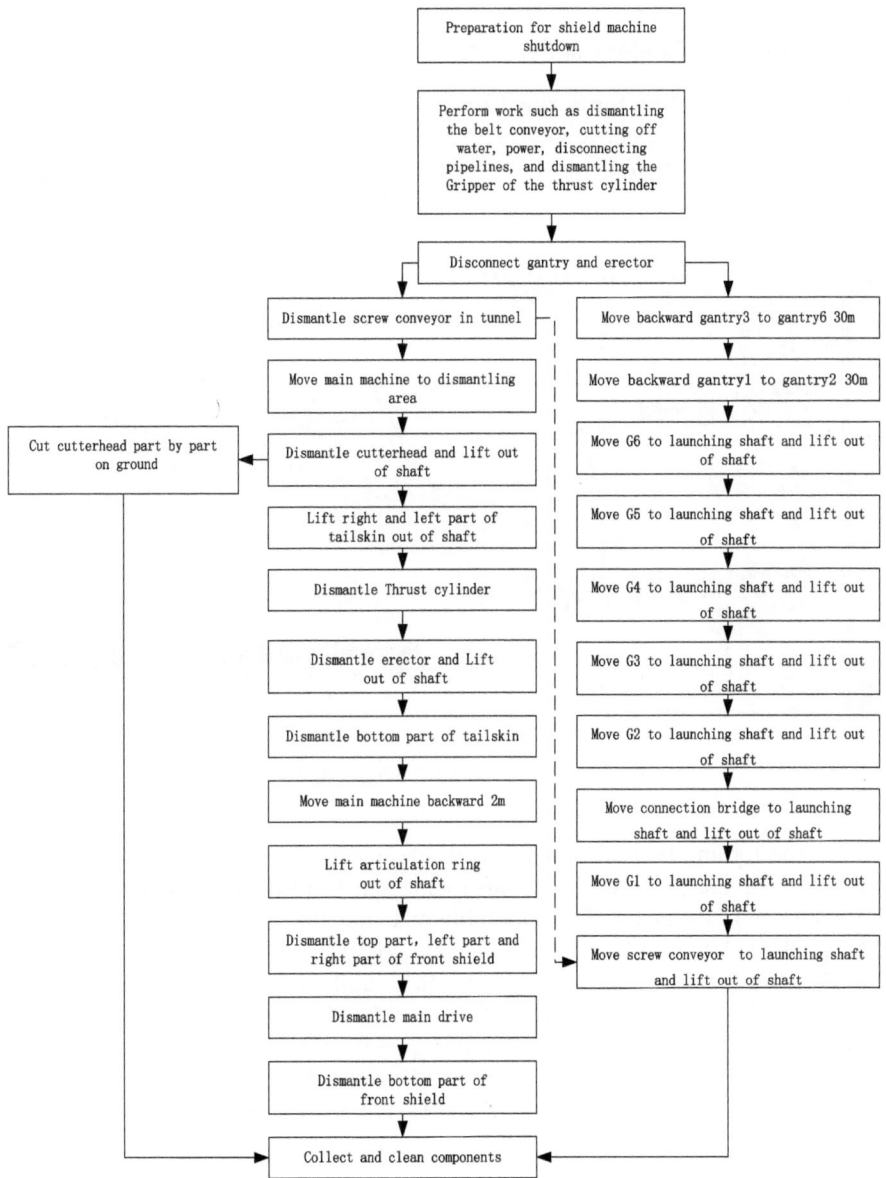

Fig. 1. Sequence of TBM disassembly

3   Research on Key Technologies for the Main Machine

There are two main technical difficulties when removing the main machine, ①Demolition technology of screw conveyor in tunnel translation technology of main machine.

3.1   Demolition Technology of Screw Conveyor in Tunnel

Design Removal Tooling for Screw Conveyor

The large-diameter screw conveyor weighs about 55 tons, and due to the space constraints for demolition in the tunnel, the demolition tooling had to be made in advance [3] (as shown in Fig. 2).

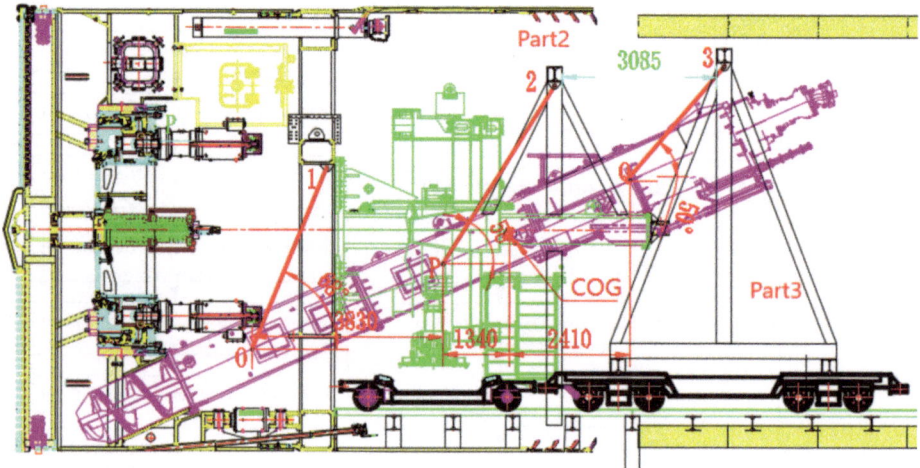

Fig. 2. Screw conveyor removing tooling

Since the tooling is used in the tunnel, the transportation space is limited, so the individual parts are prepared in advance and then welded in the tunnel as a whole. Part2 is welded to the beam of the assembly machine, the overall width of the fixture is 2810 mm, the height is 3000 mm, and the angle of tilting fixed support is 65°(as shown in Fig. 3). Part3 is welded to the flat plate, the overall width of the fixture is 2200 mm, the height is 6630 mm, and the angle of inclined fixed support is 75°. The width of the vertical beam is 1600 mm, and the maximum φ of the screw machine parts is 1400 mm, which can pass through the screw machine. Through the calculation of the force, the main beam of the tooling is made of HW300 section steel, and the material is S355JR (as shown in Fig. 4).

Description of the Screw Conveyor Dismantling Process

During the dismantling of the screw conveyor, 3 lifting points are set up, lifting points 2 and 3 use 4 × 20 T chain hoists respectively, and lifting point 1 uses 2 × 10 T chain hoists. When the screw conveyor is slowly pulled out, move the trolley of lifting point

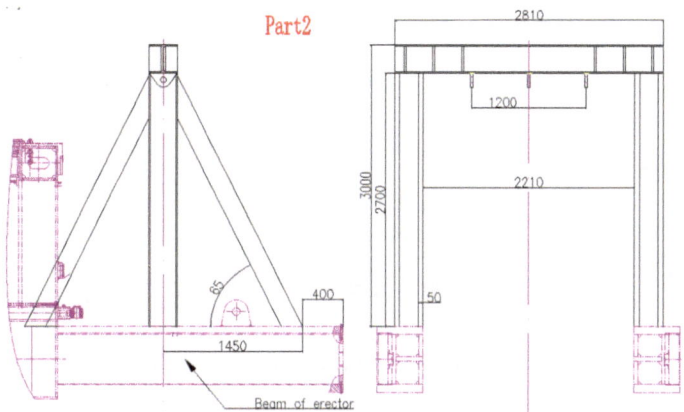

Fig. 3. Demolition tooling of lifting pint 2

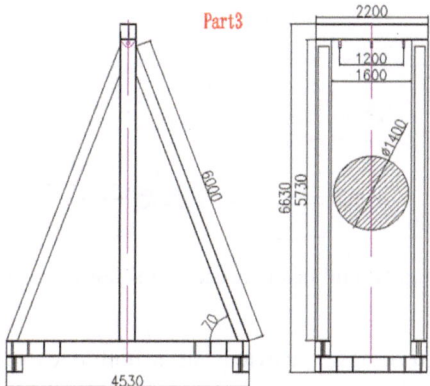

Fig. 4. Demolition tooling of lifting piont3

3 backwards, so that the center of gravity of the screw conveyor cannot exceed the top force point of lifting point 3 [4]. After the screw shaft completely leaves the front shield body, release the chain hoist at lifting points 2 and 3 to make the front part of the screw conveyor fall, and at the same time continue to tighten the chain hoist at lifting point 1 to make the tail of the screw conveyor continue to lift (as shown in Fig. 5).

When adjusting the attitude of the screw conveyor is close to the horizontal state, and the height exceeds the rotary structure of the erector by about 100 mm, use the support to fix the screw machine support on the transport trolley (as shown in Fig. 6). The chain hoist is removed and the screw conveyor is transported to launching shaft for lifting.

3.2   Main Machine Move and Dismantle

Design Moving Tooling for Main Machine

First, four rows of single 3050 × 625 × 40 mm steel plates are laid on the concrete floor,

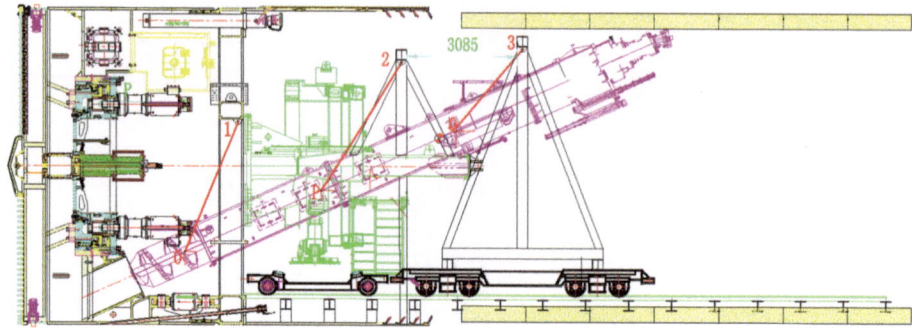

Fig. 5. Lift screw conveyor out of front shield

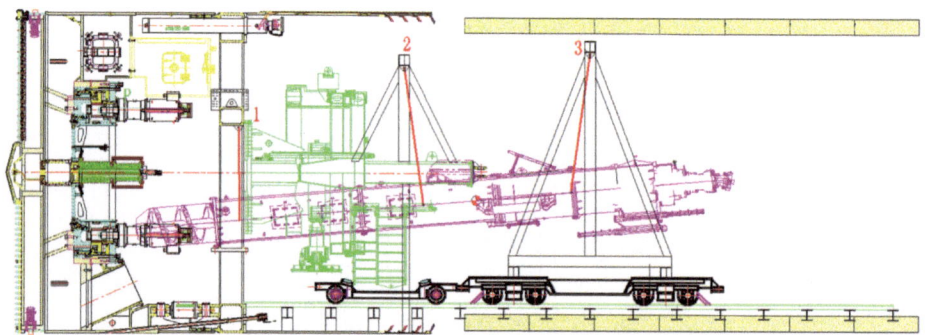

Fig. 6. Pulling out process of screw conveyor

and the dovetail grooved design is carried out at both ends of each steel plate [5] (as shown in Fig. 7), so that the connection between the steel plates is firmer, and the total laying length of the single row is 51.85 m (as shown in Fig. 8).

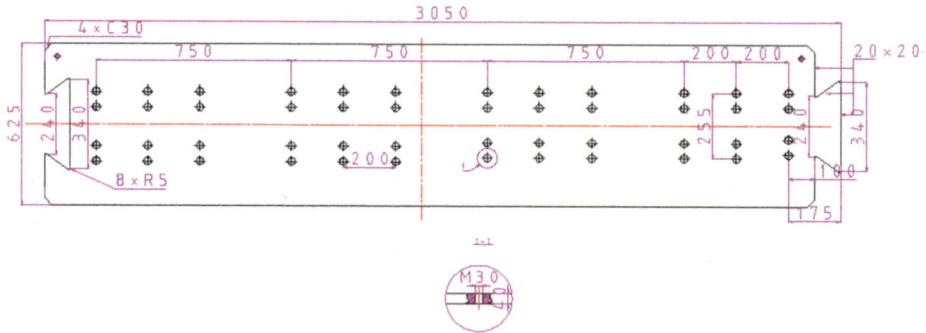

Fig. 7. Steel plate

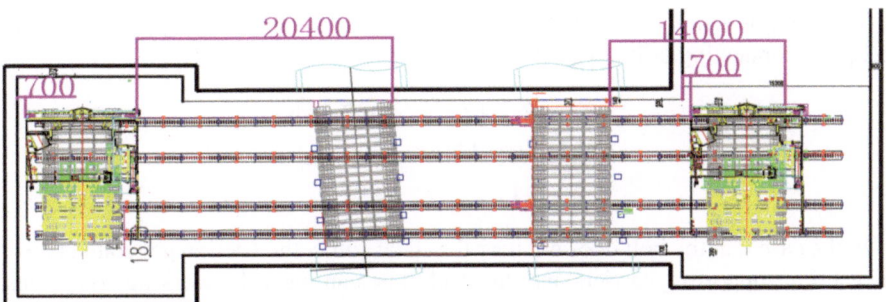

Fig. 8. Steel plate laying

Calculate Force for Moving Main Machine

The large-diameter TBM main machine weighs about 820 tons, and the shield receiving frame weighs about 50 tons. The shield translation force supporting carriers are connected to the steel plate using M30 × 60–12.9 bolts, the preload of the bolt is 432.5 kN, each reaction support seat has a total of 12 bolts (as shown in Fig. 9), and 2 reaction supporting carriers are used when the main engine translation[6].

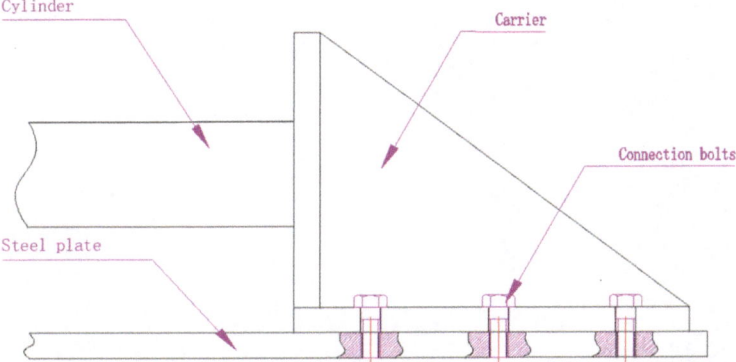

Fig. 9. Connection between reaction support seat and steel plate

According to the working conditions, take the friction coefficient of the supporting carriers and the steel plate u = 0.1. The friction force $f1$ when the shield is moved, and two cylinders with thrusting force 200 T and stroke of 1000 mm are used when the shield body is moved;

$$f1 = G \times u$$
$$f1 = 870 \, kN$$

The friction generated when the main machine slides with the shield receiving frame is 870 kN, and the thrust of the cylinder meets the requirements of the main machine moving.

The friction generated by the pretension between the reaction support seat and the steel plate connecting bolt is $f2$;

$$f2 = 2 \times 12 \times u \times 432.5\,kN$$
$$f2 = 1038\,kN$$
$$f2 > f1$$

In the case that the bolt does not bear the shear force, the friction generated between the reaction supporting carriers and the steel plate can meet the working conditions of shield moving.

Main Machine Moves Process

Install the translation cylinder and reaction support seat, and push forward the receiving frame of the TBM pushed by the cylinder [7] (as shown in Fig. 10). Two cylinders push the receiving frame, and the cylinder rod is retracted when each cylinder stroke is about 800 mm, and then the cylinder is retracted to dry forward and the jacking supporting carriers are retracted, and the main machine is translated to the disassembly area after repeated pushing.

Fig. 10. Main machine moving

Main Machine Removal and Precautions

After the main machine is moved to the disassembly area, use the crane to lift out the shield blocks one by one, and disassemble each part of the main machine according to the disassembly sequence of the main machine. Note other items during the removal of the main machine;

1) In order to reduce the disassembly time, some connecting bolts between the cutter head and the main drive M48-10.9 can be removed in advance. The weight of the

cutter head is 95 T. By calculation, there are 16 bolts reserved, such as top, bottom, left and right respectively. The pretight force of M48-10.9 is F1 = 1050 kN, and the friction generated by the connection bolts between the main drive and the cutterhead is like cosmos.

$$f1 = F1 \times u \times 16 = 1680\,kN > 950\,kN \tag{1}$$

By calculating that the friction force generated by the connecting bolt between the main drive and the cutterhead is about 1.8 times of the cutter head weight, it can meet the requirements of full operation.

2) Since the overall weight of the articulation ring and the thrust cylinders exceeds the lifting capacity of the crane, the grippers of the thrust cylinder is removed first when the main machine is disassembled, and then the thrust cylinder is pulled out one by one. The thrust cylinders is removed in two stages. The upper thrust cylinders are removed first, and the lowerthrust cylinders are removed after the erector is removed.

4 Research on Gantries Removal and Backing Technology

4.1 Gantries Removal Special Tooling Design

Gantries Tooling Design and Manufacture
In the dismantling of the Gantries, the connecting pins should be removed first. At the same time, 4 supporting seats should be welded on the main vertical beam of the Gantry (as shown in Fig. 11). Use a jack to lift the Gantry 100 mm [8].

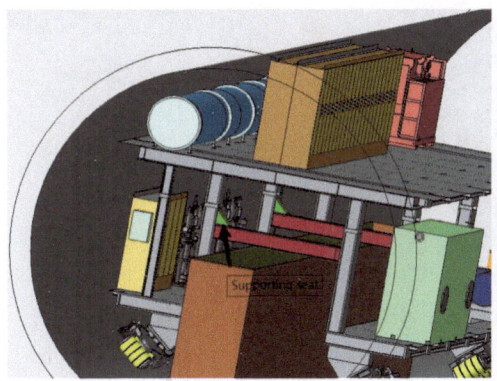

Fig. 11. Gantry supporting structure

Force Calculation of Gantry Supporting Beam
According to the supporting force of Gantry on the muck truck, the model was simplified into a simple beam stress model (Fig. 12). F1 = F2 = 17.5 T was calculated, and L1 = L3 = 320 mm and L2 = 2270 mm were obtained according to the structure of the muck truck.

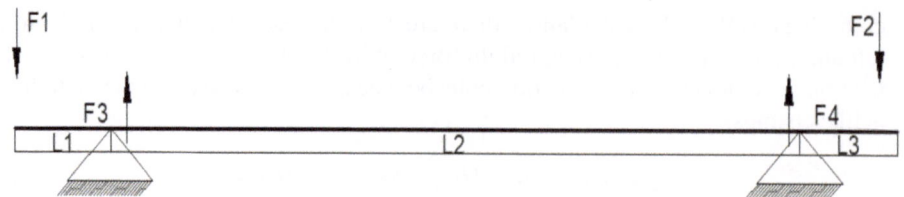

Fig. 12. Simple beam model

According to the law of balance of couples.

$$F1 \times (L1 + L2) - F3 \times L2 = F2 \times L + \tag{2}$$

$$F3 \times (L3 + L2) - F4 \times L2 = F1 \times L1 \tag{3}$$

So F3 = F4 = 17.5 T, that is, the pressure of the support seat against the beam is 17.5 T.

SolidWorks was used to build a three-dimensional model, and the model was divided into grids and loaded with 175 kN by external force. The calculation results were obtained through finite element analysis (Fig. 13).

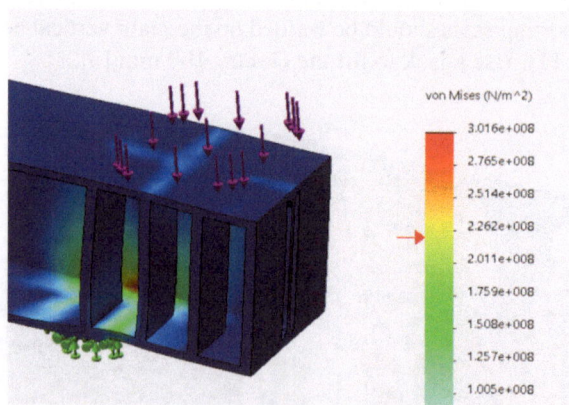

Fig. 13. Results of finite element analysis

Through finite element analysis, the maximum force of the beam is about 300 MPa, and the material of the beam is Q355B, so it can meet the requirements of the Gantry support strength.

4.2 Gantries Move Backward

After the Gantry is jacked up, the locomotive drives the muck car to move to the bottom of the Gantry, and the muck car beam is supported under the support seat, and then the jack is unloaded and the Gantry support seat is supported on the beam. Then the locomotive drives the Gantry slowly. When the Gantry is driving, an observer is set before and after the locomotive to pay attention to the posture of the Gantry at all times.

5 Conclusion

By analyzing and summarizing the dismantling process of two soil pressure balancing TBM with Φ9040 mm in Sect. 6 of Beijing New Airport Line, the following conclusions are drawn.

1. The combination of muck car formation and special tooling can safely, efficiently and quickly solve the problem of long-distance retreat of Gantries;
2. The dismantling of screw screw tunnel is the risk point. The screw conveyor weighing about 55 T can be safely dismantled in the tunnel through the dismantling tools welded on flat car and chain block;
3. When the main machine is translated, the bottom of the shield receiving frame is laid with steel plate. The reaction support seat and the jacking cylinder can be used to solve the translation problem of the large diameter shield main machine efficiently and quickly.

The successful implementation of the project has accumulated rich construction experience for dismantling TBM in the tunnel.

References

1. Qu, X., Li, Z.: Disassembly process and key points of shield screw conveyor in shield tunnel. Eng. Technol. **31** (2017)
2. Li, S.: Research on New Technology Assembly of Urban Large-diameter Shield Machine. South China University of Technology (2014)
3. Zhao, H., Ge, B,. Ding. H.: Assembly of large slurry balance shield machine. Theoretical Res. 42–43 (2012)
4. Wang, F., Liu, M., Chen, L., Liu, W., Tang, L.: Integral hoisting technology of tunnel boring machine in limited space. J. Shenzhen Univ. Sci. Eng. **33**(3), 317 (2016). https://doi.org/10. 3724/SP.J.1249.2016.03317
5. Ni, Y.: On the construction technology of shield machine passing subway station. Smart City **11**, P172-173 (2019)
6. Pu, X., Chen, L., Zhao, Q., Li, H.: Site assembly process and key technology of super large diameter shield machine. Construct. Mach. 172–173 (2019)
7. Bista, S.: The disassembly of Annabell and Joyce. Aust. J. Mech. Eng. **14**(1), 10–25 (2016). https://doi.org/10.1080/14484846.2015.1093208
8. Ng, P.L., Polycarpe, S., Barrett, T.N.D.R., Roux, G., Vallon, F.: Development of a novel tunnel dismantling machine for the MTR West Island Line Construction. All J. HKIE Trans. **12**, 151–168 (2017)

Research on the Removal Sequence of Arch Rib Assembly Supports for Concrete-Filled Steel Tubular Tied Arch Bridges

Zhihui Peng[1,2(✉)] and Jiaqi Li[1,3]

[1] CCCC Second Harbor Engineering Co., Ltd., Wuhan 430040, China
496143661@qq.com
[2] CCCC Highway Bridge National Engineering Research Centre Co., Ltd., Beijing 100120, China
[3] Research and Development Center of Intelligent Manufacturing Technology for Transportation Infrastructure in Transportation Industry, Wuhan 430040, China

Abstract. The main bridge of the Bayi Bridge project in Yangxin County is a concrete-filled steel tubular tied arch bridge with a calculated span of 127.04 m. The steel pipe arch ribs adopt a dumbbell-shaped cross-section, are manufactured in seven longitudinal sections, and are constructed using the support method. The tie beams are fully prestressed concrete structures. The design guidance plan is to remove the arch rib assembly bracket after completing the arch rib steel pipe concrete construction. Considering that during the construction process of steel pipe concrete top pressure injection, the arch rib assembly support bears a large load and has high safety risks, an optimization plan is proposed to remove the arch rib assembly support first after the steel pipe arch rib assembly is completed, and then top pressure injection of steel pipe concrete. To analyze the optimization effect, a finite element model was established using Midas Civil analysis software to analyze the stress of the two schemes. The results showed that the maximum stress of the steel pipe arch rib in the optimized scheme increased by 11.7 MPa, reaching 128.4 MPa, which is less than the material design strength; The compressive stress of the upper and lower chord concrete decreased by 1.8 MPa and 2.4 MPa respectively; The difference in arch rib deformation has little impact on the selection of the two schemes, and can be solved by setting pre camber; The tensioning of N2 and N7 prestressed steel tendons has little effect on the structural stress, and the optimized plan meets the stress requirements.

Keywords: Concrete-filled steel tubular tied arch bridge; Arch rib assembly; Falseworkt for Assembly; Removal sequence of falsework

1 Introduction

The main bridge of the Bayi Bridge project in Yangxin County adopts a 1–130 m through concrete-filled steel tubular tied arch bridge [1–4], with a calculated span of 127.04 m, a rise span ratio of 1/5 of the arch ribs, and a rise height of 25.40 m, as shown in Fig. 1. The

P. Xiang and L. Zuo (Eds.): PBSFTT 2023, LNCE 382, pp. 59–67, 2024.
https://doi.org/10.1007/978-981-97-5108-2_5

main bridge is located on a 2.3% symmetrical longitudinal slope and a convex vertical curve with a radius of 3000 m. The transversal arrangement of the bridge deck is 2.5 m (sidewalk) + 2.0 m (arch rib area) + 0.5 m (collision barrier) + 11.5 m (three lanes) + 1.0 m (facility strip) + 11.5 m (three lanes) + 0.5 m (collision barrier) + 2.0 m (arch rib area) + 2.5 m (sidewalk) = 34 m.

The main bridge adopts a full-width design with two arch ribs, each of which adopts a dumbbell-shaped steel tube concrete structure. The diameter of the upper and lower chords of the bridge is 1264 mm, with a wall thickness of 28 mm, a web height of 0.684 m, a wall thickness of 28 mm, and a total height of 3 m for the arch ribs. It is made of Q390D steel. The chord of the arch rib and the interior of the abdominal cavity are filled with C60 micro expansive concrete, and the concrete is poured using the pumping-up method. Each steel pipe arch rib is manufactured in 7 sections, with the middle section being the closure section. The steel pipe arch rib is constructed using the support method [5–7].

A total of 5 wind braces are set up for the entire bridge, including an "I" type wind brace and four K-type wind braces. The "I" type wind bracing uses two steel pipes with a diameter of 900 mm to form a space frame, and the web pipe uses a steel pipe with a diameter of 600 mm. K-type wind bracing is composed of one horizontal steel pipe with a diameter of 1200 mm and two slant support steel pipes with a diameter of 900 mm. All wind braces have a wall thickness of 20 mm. The wind support is manufactured in the factory and constructed by on-site lifting and welding.

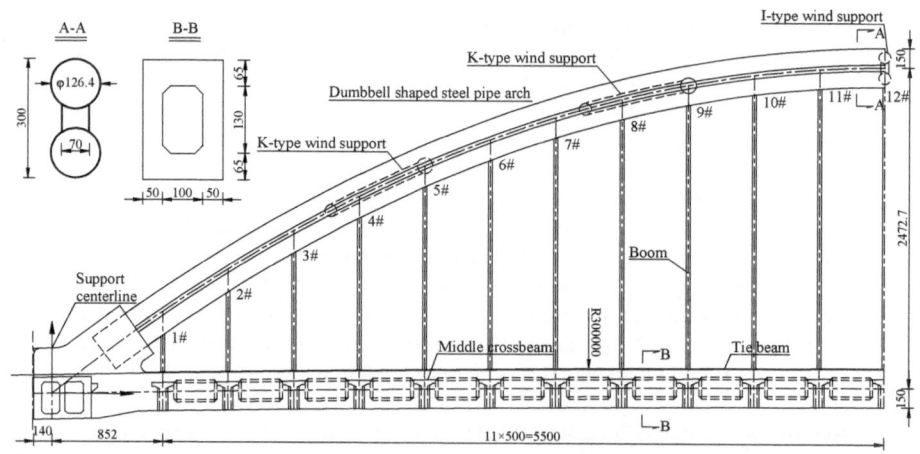

Fig. 1. Layout of Main Bridge Type 1/2 of Bayi Bridge Project (cm)

The tie beam adopts a prestressed concrete structure with a height of 3.0 m and a width of 2 m. The thickness of the top and bottom plates is 0.65 m, and the thickness of the web plate is 0.5 m. A chamfer of size 0.2 × 0.2 m is set inside the box. A total of 28 prestressed steel strands of 19–15.2 and 1860 MPa are set up for a single tie beam, and the tie beam is constructed using the full support method. The middle and end crossbeams are set between the tie beams, both of which are prestressed concrete structures, and a concrete bridge deck is set above the crossbeam.

In the design guidance plan of the bridge, the arch rib assembly bracket is removed after the completion of the steel pipe concrete construction of the arch rib. Considering that the arch rib assembly support bears a high load and poses a high safety risk during the construction process of steel tube concrete pressure injection jacking, the dismantling sequence of the arch rib assembly support is optimized, and a comparative analysis is conducted on the pre and post optimized schemes from the perspective of structural stress.

2   Dismantling Plan for Steel Pipe Arch Rib Assembly Support

2.1   Original Plan

This article mainly studies the removal sequence of steel pipe arch rib assembly brackets, so only attention needs to be paid to the removal of steel pipe arch rib brackets and previous construction steps. According to the design drawings, the construction process of the original plan is shown in Fig. 2. The specific steps are ① Sequentially construct the pile foundation, bearing platform, pier column, and cover beam, set up brackets at the corresponding positions of the tie beam, pour concrete for the tie beam and arch foot on the brackets, and embed the steel pipe arch rib embedded section. ② Symmetrically tensioned tie beams with longitudinal prestressed steel tendons (N1, N8). ③ Symmetrically lift the arch ribs of the first, second, and the arch top sections in sequence, place the connections on the arch rib assembly support platform, complete the closure of the arch ribs, and symmetrically install the wind braces. ④ Pour concrete into the upper chord. ⑤ Pour concrete into the lower chord (after the concrete in the upper chord forms strength). ⑥ Pour concrete into the abdominal cavity of the chord (after the concrete in the lower chord forms strength). ⑦ The second batch of prestressed steel tendons (N2 and N7) for symmetric tensioning tie beams. ⑧ Cast in place the end crossbeam, remove arch rib assembly support.

2.2   Optimization Plan

In order to reduce the stress on the arch rib assembly bracket, reduce construction safety risks, optimize the construction process, and facilitate on-site construction organization, an optimization plan for dismantling the steel pipe arch rib assembly bracket is proposed. Compare with the guidance plan and advance the two steps of "symmetric tensioning of the second batch of prestressed steel beams (N2 and N7) and dismantling of the arch rib assembly support" until the arch rib closure. The specific steps are ① sequentially construct the pile foundation, bearing platform, pier column, and cover beam, set up brackets at the corresponding positions of the tie beam, pour concrete for the tie beam and arch foot on the brackets, and embed the steel pipe arch rib embedded section. ② Symmetrically tensioned longitudinal prestressed steel tendons on tie beams (N1, N8). ③ Symmetrically lift the first, second, and arch ribs in sequence, connect and dispose of them on the arch rib assembly support platform, complete the closure of the arch ribs,

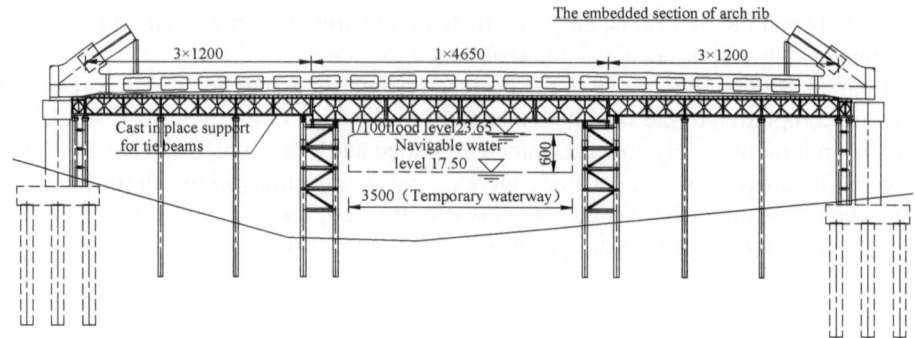

a) Cast-in-place and embedded arch foot section on the tie beam support (cm)

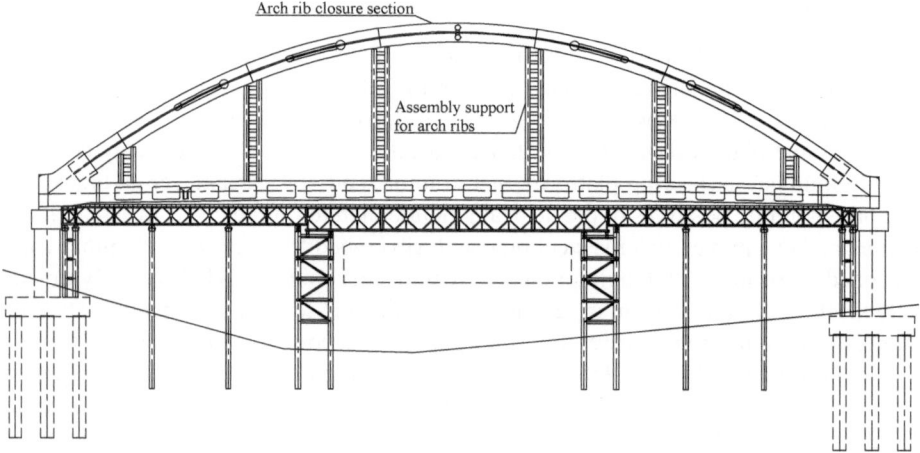

b) Assembly of arch rib support on the support and pressure injection of concrete inside the lifting pipe

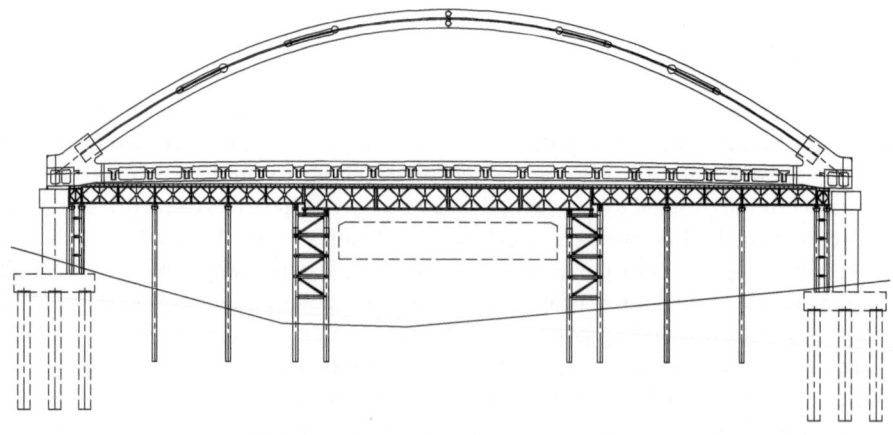

c) Remove the arch rib assembly bracket

Fig. 2. Schematic diagram of the construction process of the original plan

and symmetrically install the wind braces. ④ Symmetrically tensioned the second batch of longitudinal prestressed steel tendons on tie beams (N2 and N7). ⑤ Cast-in-place end crossbeam; remove arch rib assembly support. ⑥ Pour concrete into the upper chord. ⑦ Pour concrete into the lower chord (after the concrete in the upper chord forms strength). ⑧ Pour concrete into the arch rib abdominal cavity (after the concrete in the lower chord has strength).

3   Structural Calculations

To verify the rationality of the optimization plan and compare the stress conditions of the original plan and the optimized plan, the finite element analysis software Midas Civil was used to establishing an analysis model and calculate the two plans.

3.1   Model and Construction Conditions

The prestressed tensioning of the tie beam, the assembly of arch ribs, and the pouring of concrete in the pipe are all constructed symmetrically upstream and downstream; The construction of intermediate tie beams and bridge decks does not need to be considered in the proposed construction steps; The wind braces and end tie beams are only subjected to self-weight and have relatively small loads. From the above factors, it can be seen that the construction simulation of wind braces and end tie beams can be ignored. Therefore, only one side of the arch rib calculation can reflect its regularity.

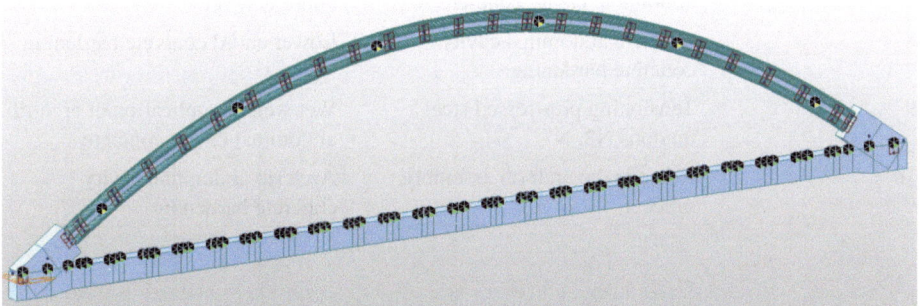

Fig. 3. Finite element model diagram

The construction process of steel pipe arch ribs was simulated using Midas Civil to establish a finite element model, and the stress differences between the original plan and the optimized plan were compared and analyzed. The steel pipe arch rib and concrete inside the pipe are simulated using beam elements, and the construction process of concrete inside the pipe is simulated using the dual element method [8–10]; The arch rib assembly support is simulated using only compression node elastic support, and the elastic stiffness of each support from the side span to the middle span is 1097568 KN/m, 487808 KN/m, and 365856 KN/m; The tie beam adopts beam element simulation, and the full support of the tie beam adopts general support simulation to simulate the prestressed tensioning construction process of the tie beam; Elastic connection is used between the arch rib and the tie beam. The finite element model is shown in Fig. 3.

According to the construction steps, ten construction stages are divided. The main difference between the original plan and the optimized plan is that the two construction stages of "Tensioning prestressed N2 and N7; Dismantling the arch rib assembly support" in the optimized plan have been advanced to the lifting and closure of the arch rib. The specific construction steps are shown in Table 1.

Table 1. Recommended and Optimized Plan Construction Stage Working Conditions Table

Construction stage	Original plan	Optimization plan
1	Pouring tie beams, tensioning prestressed steel tendons N1, N8	Pouring tie beams, tensioning prestressed steel tendons N1, N8
2	Hoisting and closing of arch ribs	Hoisting and closing of arch ribs
3	Wet weight application of upper chord concrete	Tensioning prestressed steel tendons N2, N7
4	Upper chord concrete hardening	Remove the arch rib assembly bracket
5	Wet weight application of lower chord concrete	Wet weight application of upper chord concrete
6	Lower chord concrete hardening	Upper chord concrete hardening
7	Wet weight application of arch rib abdominal cavity concrete	Wet weight application of lower chord concrete
8	Arch rib abdominal cavity concrete hardening	Lower chord concrete hardening
9	Tensioning prestressed steel tendons N2, N7	Wet weight application of arch rib abdominal cavity concrete
10	Remove the arch rib assembly bracket	Arch rib abdominal cavity concrete hardening

3.2 Results

The calculation results are shown in Table 2. Compared with the recommended design scheme, the maximum steel pipe stress in the optimized scheme increased by 11.7 MPa, while the compressive stress of the upper and lower chord concrete decreased by 1.8 MPa and 2.4 MPa, respectively. This is mainly due to the transfer of some concrete weight to the steel arch ribs, resulting in an increase in the stress on the arch ribs and a decrease in the stress on the concrete inside the pipes. According to the design calculation sheet, the maximum compressive stress of the arch rib steel pipe under frequent combined actions is 116.7 MPa. Considering an increase of 11.7 MPa, the maximum steel pipe stress in the optimized plan is 128.4 MPa. Since the steel pipe uses Q390 material, the maximum stress is less than the design strength of Q390 material, which is 335 MPa [11].

The arch crown deform 3.6 mm upwards in the design scheme and 4.6 mm downwards in the optimization scheme, with significant differences between the two. The main reason is that after the prestressed tension of the tie beam, the tie beam is shortened, which can cause the arch ribs to deform upwards. In the design plan, the bracket was removed late, the arch rib stiffness was high, and the downward deformation was small, so the final deformation is still upward; In the optimization plan, the support was removed early, and the concrete of inside the pipe was constructed in stages. The arch rib stiffness was relatively small, and the downward deformation was large, which was still more significant than the deformation caused by the prestressed tension of the tie beam. Therefore, the final deformation was downward. Although there are significant differences in the deformation of the arch ribs, the difference in deformation has little impact on the selection of the two schemes. It can be solved by setting the pre-camber.

In the optimization plan, the compressive stress of the beam concrete during the tensioning stage of N2 and N7 prestressed steel tendons is −11.5 MPa, the maximum stress of the steel pipe is 20.9 MPa, and the vertical deformation of the arch crown is 12.4 mm, all of which are not significant. This indicates that the pre-tensioning of N2 and N7 prestressed steel tendons in the proposed optimization plan has little impact on the structural stress.

Table 2. Recommended and Optimized Construction Stage Working Conditions Table

Working condition	Original plan	Optimization plan	Difference
Maximum steel pipe stress in the final stage (MPa)	23.4	35.1	11.7
Final stage upper chord concrete stress (MPa)	−3.1	−1.3	−1.8
The stress of the lower chord concrete in the final stage (MPa)	−2.8	−0.4	−2.4
Deformation of the apex of arch in the final stage (mm)	3.6	−4.3	−7.9
The stress of concrete during the stage of tensioning prestressed steel tendons N2、N7 in the optimization plan (MPa)	/	−11.5	/
The stress of steel pipe during the stage of tensioning prestressed steel tendons N2、N7 in the optimization plan (MPa)	/	20.9	/
The deformation of the apex of arch during the stage of tensioning prestressed steel tendons N2、N7 in the optimization plan (mm)	/	12.4	/

4 Conclusion

In the design guidance plan for the main bridge of the Bayi Bridge project in Yangxin County, the arch rib assembly support is removed after the completion of the arch rib steel pipe concrete construction. Considering that during the construction process of steel pipe concrete top pressure injection, the arch rib assembly support bears a large load and has high safety risks, an optimization plan is proposed to remove the arch rib assembly support first after the steel pipe arch rib assembly is completed, and then top pressure injection of steel pipe concrete. By using Midas Civil to establish an analysis model for the stress analysis of the two schemes, the results showed that the maximum stress of the steel pipe arch rib in the optimized scheme increased by 11.7 MPa to 128.4 MPa, which is less than the material design strength; The compressive stress of the upper and lower chord concrete decreases by 1.8 MPa and 2.4 MPa respectively; The difference in arch rib deformation has little impact on the selection of the two schemes, and can be solved by setting pre camber; The tensioning of N2 and N7 prestressed steel tendons has little effect on the structural stress. In summary, the optimization scheme meets the stress requirements.

References

1. Chen, C., Wei, J., Zhou, J., Liu, J.: Application of concrete-filled steel tube arch bridges in china: current status and prospects. China Civil Eng. J. **50**(06), 50–61 (2017)
2. Li, D., Chen, K., Luo, F., Yang, S.: Structural stress analysis of long span mid through steel tube concrete arch bridge. Highway **68**(01), 194–199 (2023)
3. Lederman, G., You, Z., Glišić, B.: A novel deployable tied arch bridge. Eng. Struct. **70**, 1–10 (2014)
4. Sun, J., Cao, H., Xi, D.: Construction control technology of concrete filled steel tube tied arch bridge. Construction Technology **51(15),** 67–71+77 (2022)
5. Vlad, M., Kollo, G., Marusceac, V.: A modern approach to tied-arch bridge analysis and design. Acta Technica Corviniensis-Bulletin of Engineering **8**(4), 33 (2015)
6. Mo, J.: Long-span Concrete Filled Steel Tube Tie Arch Bridge Research on Key Technology of Arch Before Beam Construction. Kunming University of Science and Technology, KunMing (2022)
7. Zhao, P.: Research on Construction Technology of Concrete-filled Steel Tube Tied Arch Bridge New Technology & New Products of China **2016**(16), 113–114 (2016)
8. Zhong, Y., Cui, Z.: Study on calculation and experimental modal analysis of long-span concrete-filled steel tube arch bridge Highway **67**(10), 245–250 (2022)
9. Zhou, J.: Research on Cable Force Calculation Method of Cable-stayed Fastening Hanging Methods of Long Span Concrete Filled Steel Tube Arch Bridge. NanNing Guangxi University (2021)
10. He, F.: Reasonable Construction Cable Force Calculation and Key Working Procedure Analysis of Long-span Concrete-filled Steel Tubular Arch Bridge. Changsha University of Science and Technology, ChangSha (2021)
11. Industry Recommended Standards of the China: Recommended Industry Standards of the People' s Republic of China(JTGT-D65-06-2015). China Communications Press, Beijing (2015)

Design Strategy and Application Research of Small Steel Structure Prefabricated Building

Yihui Du[✉], Zijing Zhang, and Yuchao Hu

Architectural and Urban Planning Department, Chongqing College of Architecture and Technology, Chongqing 401331, China
ivygogo@qq.com

Abstract. Under the background of building industrialization, small steel structure prefabricated buildings can achieve quality improvement, efficiency improvement, technological upgrading, resource conservation, and environment friendly demand, so that low-rise houses can get rid of the public impression of cheap and low-tech. Combined with the common problems of small prefabricated buildings in southwest China, this paper sorts out four design strategies for small prefabricated buildings. Based on the case of "Four Seasons of Heshan", this paper expounds the practical application of design strategies for small steel structure Prefabricated building, including quick construction and environmental protection; Free space and diverse decoration; High safety and complete functionality; Resource cycle, strong sustainability, etc.

Keywords: Prefabricated Architecture · Small Steel Structure buildings · Architectural Design Strategy

1 Introduction

1.1 Small Steel Structure Prefabricated Building

Low-rise housing has long been criticized for its traditional cheap and low-skilled construction. With the technological progress of the industry, the values of consumers and practitioners have changed, the small steel structure prefabricated buildings have gradually entered the public life. Under the background of building industrialization, small steel structure prefabricated buildings can achieve quality improvement, efficiency improvement, technology upgrading, resource conservation, and environmental friendly demand is one of the best construction methods [1, 2].

Compared with concrete structure, it is found that the steel structure prefabricated building only has higher material consumption and construction tool usage costs, but has obvious advantages in labor costs and construction period input. In addition, the prefabricated structure building site construction is simple, the construction speed is fast, the construction efficiency is high, and it can be recycled, which can play a lot of economic benefits for on-site cost control and has a broader economic benefit prospect [3, 4] (Table 1).

© The Author(s) 2024
P. Xiang and L. Zuo (Eds.): PBSFTT 2023, LNCE 382, pp. 68–74, 2024.
https://doi.org/10.1007/978-981-97-5108-2_6

Table 1. Proportion of prefabricated residential construction cost of the project [5].

Section	Compose	Proportion of expenses
Steel structure	Main structure system	31.28%
Roof part	Polyurethane roof board, PVC external hanging	12.12
Wall part	Board, PVC thatch, asphalt tile, resin tile, integrated board, roof waterproof, etc	8.13%
Ground part	Wood grain wallboard, sprayed wallboard, integrated board, etc	7.63%
Door and window	Doors, windows, top glass, etc	12.09%
Water and electricity part	Water and electricity materials, switches, sockets, lamps, etc	2.81%
Auxiliary materials	Decorative parts, stairs, waterproof coiled materials, angle steel, structural adhesive, etc	16.72%
Other parts	Installation fee, long-distance transportation fee and material handling fee	9.22%

1.2 Status of Small Prefabricated Residential Buildings

Southwest China is mostly mountainous and hilly, with a large rural area. With the improvement of living standards, in the process of urban-rural integration development, consumers have put forward different levels of demand for small residential buildings, diversified product designs and prosperous markets. Through the team's investigation, it is found that the common small prefabricated houses in southwest China have the following problems:

The construction is arbitrary, causing a certain degree of damage to the ecological environment. Some small residential buildings directly choose local building materials, the construction is arbitrary, there is no planning, increase construction waste and pollution emissions, and have a greater impact on the ecological environment.

The interior space structure is single, and additional interior decoration is required. The indoor space combination and scale of small houses are relatively single, and they are blindly seeking large. At the same time, after the completion of construction, more interior space decoration is required, and the construction period is extended.

Building safety and weather resistance are uneven, and the living function is not perfect. The construction technology content is low, the building weather resistance is poor, the use comfort is poor, and the occupants have a bad experience.

Poor economy, insufficient sustainability of building resource utilization. Today's products are expensive, require high maintenance, and have low reusability. When changes or adjustments are required for objective reasons, it is difficult to reorganize and use, and the green ecological sustainability is not strong.

2 Small Prefabricated Housing Design Strategies

2.1 Adapt to the Site and Respect Nature

Southwest China is mountainous with many high slopes, rugged roads, great climate change, cold winters and hot summers. Most of the local houses originated from the dry column type, and the building complex varied from place to place, forming a rich courtyard layout, including: stilt building, concave shape, one seal and one zigzag. The building complex is built along the mountain, does not pay attention to strict orientation and geometric symmetry, and has a flexible layout; Focusing on ecology and nature, it embodies the concept of "the unity of nature and man" [6, 7].

2.2 Recycling and Ecological Protection

Small steel structure prefabricated buildings, in the process of construction and use, pay attention to the combination with the ecological environment, to avoid continuous damage to the environment; The construction period is short, and the assembly and decoration are convenient; Building safety guarantee, can provide a comfortable living experience; The recycling of building resources is highly sustainable. In addition, considering the differences in different construction conditions, it also takes into account the characteristics of standardized design and mass production [8].

2.3 Standard Space and Perfect Function

Carefully analyze the function of residential buildings, design standard modules such as living room, bedroom, kitchen, bathroom and balcony, and the modules can be independently spliced in combination with the design theme, which can be extended horizontally and widened horizontally, and can also be heightened vertically. At the same time, the building itself light control system, water supply system and sewage treatment system and other functions are combined with design.

In the process of combining space modules, it can also be adapted to the different use objects of the room by replacing furniture and soft furnishings, such as redefining its use function; Lightweight partition walls can also be used to change the position of the original wall door openings in the room, so that the function of the room can be changed to adapt to different stages of the family [9].

2.4 High Security and Diverse Styles

The main body of the steel structure prefabricated building structure should have good force performance, easy to disassemble and replace, and fatigue resistance; Pay attention to the small thermal conductivity of the roof and wall, and the thermal insulation performance should be good. While taking into account the comfort and aesthetic effect, reduce construction waste and avoid environmental pollution. In addition, by choosing a combination of various styles, the transformation of exterior design, material and interior decoration can be realized, forming a personalized and independent style effect, avoiding the dilemma of single similarity.

3 Application Case of Small Steel Structure Prefabricated House – "Four Seasons of Heshan"

Located in the Heshan Four Seasons project located in Heshanping, Jiangjin, Chongqing, the main body of the building adopts a small steel structure assembly building, which has the advantages of fast and environmental protection, free and efficient, safe and comfortable, and strong sustainability (Fig. 1).

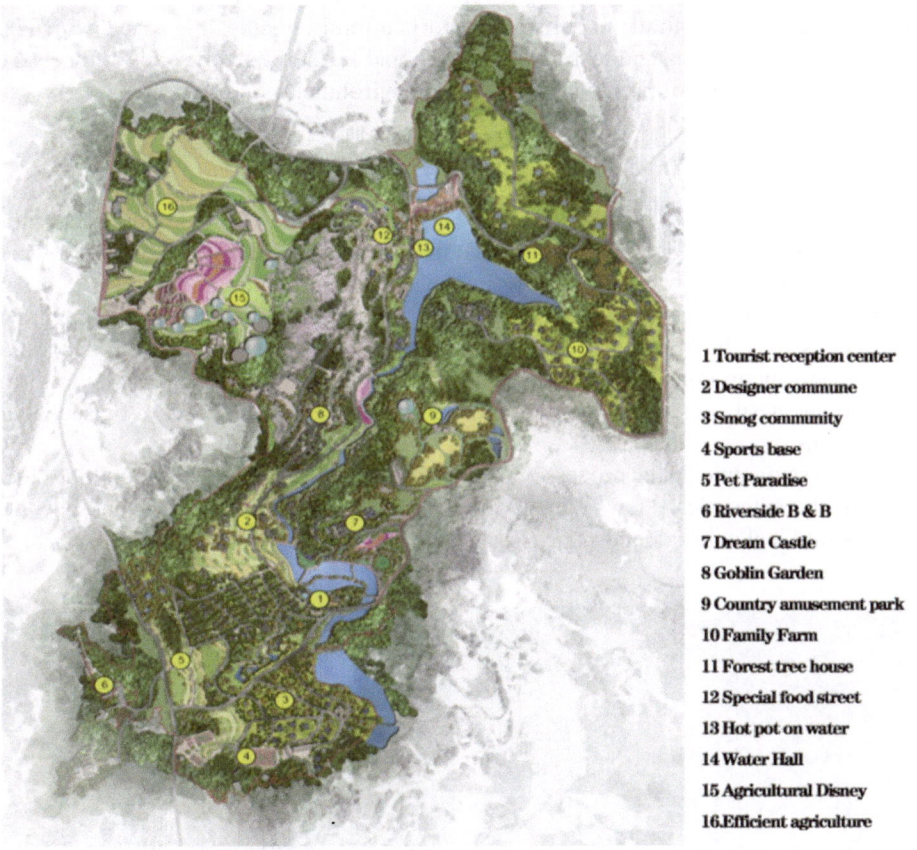

1 Tourist reception center
2 Designer commune
3 Smog community
4 Sports base
5 Pet Paradise
6 Riverside B & B
7 Dream Castle
8 Goblin Garden
9 Country amusement park
10 Family Farm
11 Forest tree house
12 Special food street
13 Hot pot on water
14 Water Hall
15 Agricultural Disney
16.Efficient agriculture

Fig. 1. General layout of the project.

This paper takes this project as an example to explain the practical application of small steel structure prefabricated buildings in southwest China.

3.1 Build Quickly and Protect the Environment

The architecture of Heshan Four Seasons implements the design concept of modularization and lightweight into the whole process of design and production, maximizing industrialization, convenient transportation, and rapid assembly, so as to adapt to the topographic characteristics of the mountains in southwest China.

Through full investigation and practice summary, the structural design of the project imitates the traditional dry-column building in the southwest region, applies the traditional building frame structure system to the prefabricated building, is based on independent small piles, and uses steel square pipe as the structural frame, forming an overall overhead building form, without ground hardening, without occupying surface green space, so as to adapt to various complex terrain and landforms in southwest China. In addition, under the condition that the overhead height of the building is sufficient, the overhead layer can continue to cultivate, or form a semi-open flexible space, which can be used for drinking tea and resting after work, and further meet the daily life needs of residents on the basis of protecting the natural environment.

3.2   Free Space and Diversified Decoration

Through the subdivision of functions, five basic modules are formed: living room module, bedroom module, kitchen module, bathroom module and balcony module, which are spliced horizontally and vertically independently in combination with the design theme. In addition, the interior and exterior decoration insulation composite integrated board is used as the maintenance system, and the interior and exterior wall decoration is realized synchronously through one installation and basic molding. While ensuring comfort and aesthetics, avoid the cost and environmental pressure caused by secondary renovations.

At the same time, in the process of design and production, the project combines local cultural characteristics, shows the charm of local cultural characteristics, and becomes a beautiful local cultural landscape (Figs. 2 and 3).

Fig. 2. Courtyard renderings 1.   **Fig. 3.** Courtyard renderings 2.

3.3   High Security and Perfect Functions

The project building is based on steel structure, which has good force performance, removable and replaceable, and fatigue resistance; All steel structure surface thermosetting powder coating anti-corrosion treatment, oxidation corrosion resistance, can guarantee the design service life of more than 50 years.

Fig. 4. Real scene of project courtyard.   **Fig. 5.** Interior decoration.

The building itself integrates the whole house intelligent light control system, water supply system and sewage treatment system, and at the same time can be integrated into energy (solar energy, wind energy), heating (floor heating, radiator), weak current (remote control system), artificial intelligence and other humanized scientific and technological details according to demand, intelligent cameras, intelligent door locks, etc. are unified into the building's IOT system, which is convenient and safe. After installation, you can move in, making it easy to achieve a comfortable and smart living experience (Figs. 4 and 5).

3.4 Resource Circulation and Strong Sustainability

The whole system of the small steel structure assembly building of Heshan Four Seasons is composed of frame structure system, wall system, roof system and ground system, each of which is an independent system, and the structure is a large frame structure, bolts are actively connected, few connection points, and easy and quick disassembly. The main steel adopts a large size, thickness and strength, which ensures the complete rate of disassembly.

According to statistics, in the process of dismantling such buildings, the damage rate of floor and roof decoration materials is about 10%, and the damage rate of floor and roof base slabs is less than 5%. Therefore, the overall disassembly damage rate is about 10%, which reflects the advantages of relocation and reuse.

4 Conclusion

Through research and comparison, the paper expounds the advantages of steel structure in the prefabricated construction industry. Combined with the common problems of small prefabricated buildings in southwest China, four design strategies for small steel structure prefabricated buildings are proposed. Finally, taking the project "Heshan Four Seasons" in Chongqing, China as an example, the practical application of methods such as fast construction, environmental protection, free space, diversified decoration, high safety, perfect function, and strong sustainability of resource circulation is expounded, and strives to provide a reference for the application of small steel structure prefabricated buildings in southwest China.

References

1. Qi, C.L., Fang, B.: Development status of new light steel residential structures. Building Materials World **28**(3), 15–17 (2007)
2. Gad, E.F.: Lateral performance of cold-formed steel-framed domestic structures. Engineering Structures **21**(1), 83–95 (1999)
3. Wang, L.: Cost analysis of prefabricated steel prefabricated building. Value Engineering **42**(14), 38–40 (2023)
4. Shen, L., Zhang, H.M.: Evaluation of residential structure system based on life cycle cost analysis. J. Shenyang Univ. Technol. **28**(05), 568–572 (2006)
5. Du, Y.H., Zhang, Z.J., Yin, Z.X., Li, Y.: Research on design strategy and application of prefabricated tourism residential building – Taking Chongqing agricultural tourism complex project "Heshan four seasons" as an example (2022). https://www.taylorfrancis.com/cha pters/edit/10.1201/9781003305026-31/research-design-strategy-application-prefabricated-tourism-residential-building-taking-chongqing-agricultural-tourism-complex-project-heshan-four-seasons-example-yihui-du-zijin-zhang-zixiang-yin-yi-li?context=ubx&refId=b0175b93-3d47-4e01-b626-53ec3e5afd66
6. Fu, Y.: Research on the design of residence for the aged in Chengdu and Chongqing. Chongqing University (2016)
7. Cheng, Y.H.: Study on integrated planning and design of living and providing for the aged in Southwest Mountain Resort -a case study of shangtianchi resort in jiuba Town. Chongqing University, Guizhou Province (2019)
8. Luo, N.: Analysis of architectural design strategies under the concept of low carbon. Manage. Technol. Small and Medium Sized Enterpr. **8**, 130–131 (2021)
9. Hu, R., Mu, E.H.: Exploration and analysis of residential design in new rural areas-taking hangjia lake as an example. Anhui Architect. **23**(01), 83–85 (2016)

Cost Comparison and Analysis of Prefabricated Frame Structures Under Different Assembly Rates

Ping Xu[1](✉) and Chengming Liu[2]

[1] Chongqing College of Architecture and Technology, Chongqing, China
kassr8828@163.com
[2] Chongqing Jundao Yucheng Green Building Technology Co., Ltd., Chongqing, China

Abstract. Cost control is one of the important factors in the promotion and application of Prefabricated building. Prefabricated frame structures are widely used in public buildings such as hospitals and schools because of their strong adaptability. Therefore, it is particularly important to conduct cost research and scheme comparison for prefabricated frame structures with different assembly rates. This article takes a single building project in a hospital in Chongqing as a research example to demonstrate and analyze the relationship between different assembly rates and construction costs, in order to provide reference for related projects.

Keywords: Prefabricated building · Frame structure · Construction cost · Assembly rate

1 Introduction

Prefabricated building is a kind of building that is fabricated by prefabricated components on the construction site [1]. Because of its fast construction speed, environmental protection and energy saving, Prefabricated building have obviously become an important driver of the development of building industrialization.

With the continuous improvement of urban construction, in addition to building a large number of residential buildings, it is also necessary to construct supporting public service buildings. The supporting public service buildings, such as shopping malls, hospitals, schools, are mostly frame structures.

Compared with traditional cast-in-place mode, prefabricated frame structure has the problem of higher cost. The prefabricated frame structure system is a standardized design that separates components such as columns, beams, slabs, stairs, balconies, and exterior walls according to the characteristics of the building and structure. It is standardized prefabricated and produced in the factory, and mechanically installed and reliably connected on site to form a frame structure building [2].

Against the backdrop of the country vigorously promoting the transformation and upgrading of the construction industry, it is a challenge for most projects to meet the requirements of national or local construction administrative authorities for building assembly rates while effectively controlling project construction and installation costs.

© The Author(s) 2024
P. Xiang and L. Zuo (Eds.): PBSFTT 2023, LNCE 382, pp. 75–82, 2024.
https://doi.org/10.1007/978-981-97-5108-2_7

This article takes the comprehensive building project of a certain hospital in Chongqing as a research example, quantitatively compares and analyzes the changes in the construction cost of each component of the prefabricated frame structure under different assembly rate schemes, and proposes corresponding cost control measures.

2 Project Background

2.1 Project Overview Description

The construction scale of a hospital project in Chongqing is approximately 42000 square meters. Among them, the comprehensive building has 9 floors above ground (including 3 podiums and corridors), totaling approximately 18400 square meters; The second floor of the underground garage is approximately 13000 square meters, with a planned height of 40.9 m.

2.2 Basis for Determining Assembly Rate

Assembly rate refers to the ratio of prefabricated component concrete volume to all concrete volume used within the scope of a building unit [3]. So the size of the assembly rate will inevitably affect the construction cost.

Therefore, this paper takes a hospital complex building project in Chongqing as the object, and according to the Detailed Rules for Calculation of Assembly Rate of Prefabricated building in Chongqing (Version 2021) [4], three sets of assembly schemes under different prefabricated assembly rates are formulated.

3 Assembly Plans with Different Assembly Rates

3.1 Cast-In-Place Frame Structure Scheme

The comprehensive building project adopts a traditional cast-in-place construction scheme. The specific methods are as follows: Beams, columns, and slabs are fully cast-in-place; 50% of the exterior walls are made of self insulating blocks, and 50% of the exterior walls are curtain walls; 50% exterior wall decoration adopts plastering + real stone paint; The interior wall adopts traditional masonry and plastering; The kitchen and bathroom are constructed using traditional techniques.

3.2 Plan of Assembly Rate 50%

According to the Evaluation Standard for Prefabricated building [5], those with an assembly rate of less than 50% cannot be called Prefabricated building. In order to meet the requirements of building assembly, an assembly scheme with a minimum standard assembly rate of 50% has been formulated.

The specific assembly scheme adopted is as follows: Using vertical components such as columns, supports, load-bearing walls, and ductile wall panels with a 15% ratio, score 10 points; Using 75% proportion of horizontal components such as floor slabs, stairs, balcony panels, and air conditioning panels, scoring 7 points; The proportion of standardized components used for prefabricated components reaches 70%, with a score of 2 points; 80% of the formed steel bar processing and distribution are integrated, with a score of 1 point; The non load-bearing enclosure wall adopts a thin masonry process wall with self insulation function, and scores 3 points; The internal partition adopts an 80% proportion of prefabricated internal partition integrated with pipelines, with a score of 7 points; 6 points for fully decorated public buildings; Using a 77% proportion of integrated toilets, scored 4 points; Using a 90% pipeline separation ratio, score 6 points; Effective transmission of BIM [6] data in design, production, and construction, with a score of 1 point; Digitize the identity of on-site management personnel by using electronic signatures and seals, with a score of 1 point; Digitize construction operation behavior and management behavior, score 2 points.

3.3 Plan of Assembly Rate 65%

On the basis of the minimum required assembly rate of 50%, increase the assembly type indicators of some projects.

The specific plan is as follows: Increase the proportion of prefabricated components used for horizontal components such as floor slabs, stairs, balcony panels, and air conditioning panels from 75% of the original plan by 76%, and increase the score by 1 point; The plan for non load-bearing retaining walls has been changed from 100% using thin masonry walls with self insulation function in the original plan to 50% using prefabricated retaining walls integrated with insulation, insulation, and decoration, with an increase of 7 points; Increase the integrated kitchen project by 90% and increase the score by 6 points; Add real-time generation of digital archive projects, increase score by 1 point.

Through the adjustment of the above four parts of the project, the assembly rate of the project can reach 65%.

3.4 Plan of Assembly Rate 90%

In order to further investigate the impact of increased assembly rate on project cost, the assembly index of the project was further improved on the basis of the 65% assembly rate plan.

The specific plan is as follows: Increase the proportion of vertical component prefabrication from the original 15% prefabricated columns to 75% prefabricated columns, with a score increase of 15 points; Increase the use of 50% prefabricated beams and increase the score by 5 points; Increase the standardized proportion of prefabricated components from 70% to 80%, and increase the score by 1 point; Increase the use of 70% overhead, dry laying, or thin pasting processes, with a score increase of 2 points; Increase the proportion of integrated bathrooms to 80% and increase the score by 2 points;

Through the adjustment of the above five parts of the project, the assembly rate of the project can reach 90%.

4 Comparative Analysis of Frame Structure Costs Under Different Assembly Rates

4.1 Scope of Cost Analysis

To reflect the cost differences under different assembly rate schemes, floors above ± 0.00 were selected as the object, including the main body, rough decoration, external wall insulation and facade decoration engineering fees, measure fees for individual buildings, temporary construction construction fees, and project management fees; But it does not include the cost of roof, fine decoration, electromechanical, door and window, and railing engineering.

4.2 Principles of Cost Analysis

This project adopts the bill of quantities pricing model [7] to calculate the cost of each sub project of the project. Firstly, use BIM modeling software to calculate the quantities of each sub project. Secondly, the comprehensive unit price [8] of each sub item shall be determined in combination with the documents such as Chongqing Housing Construction and Decoration Engineering Pricing Quota [9] and Chongqing Prefabricated building Engineering Pricing Quota [10]. Finally, determine the cost of each sub project according to formula (1), and obtain the unit cost index of each sub project according to formula (2).

$$\text{Cost of sub projects} = \Sigma(\text{Bill of quantities for sub items} \times \text{comprehensive unit price}) \tag{1}$$

$$\text{Divisional and sub item unit cost} = \text{Divisional and sub item project cost/building area} \tag{2}$$

4.3 Cost Comparison Analysis

Comparative Analysis of Main Structure Costs
Based on the pricing principle mentioned above, under different assembly schemes, the unit cost data of the main structure of the comprehensive building project is shown in Table 1:

Cost Comparison and Analysis of Outer Envelope Structures
The unit cost data of the external maintenance structure of the comprehensive building project is shown in Table 2:

Comparative Analysis of Single Unit Cost of Comprehensive Building Project
The trend of project cost per square meter cost changes for different assembly rate schemes based on traditional schemes is shown in Fig. 1:

Table 1. Comparison Table of Cost Differences in Main Structure

Entry name	Traditional construction		Assembly rate 50%		Assembly rate 65%		Assembly rate 90%	
	Construction technology	Cost per square meter (yuan/square meter)	Construction technology	Cost per square meter (yuan/square meter)	Construction technology	Cost per square meter (yuan/square meter)	Construction technology	Cost per square meter (yuan/square meter)
Columns, shear walls	Fully cast-in-place	252.09	15% vertical components	273.46	15% vertical components	273.46	75% vertical components	358.95
beam	Fully cast-in-place	258.73	Fully cast-in-place	258.73	Fully cast-in-place	258.73	50% prefabricated beams	318.11
board	Fully cast-in-place	158.15	75% prefabricated horizontal components	196.38	76% prefabricated horizontal components	198.25	76% prefabricated horizontal components	198.25

Table 2. Comparison Table of Cost Differences in Peripheral Structures

	Entry name	Traditional construction		Assembly rate 50%		Assembly rate 65%		Assembly rate 90%	
		Construction technology	Cost per square meter (yuan/square meter)	Construction technology	Cost per square meter (yuan/square meter)	Construction technology	Cost per square meter (yuan/square meter)	Construction technology	Cost per square meter (yuan/square meter)
External maintenance structure	Exterior wall	50% external wall self insulation block	28.91	50% external wall self insulation block	28.91	50% external wall self insulation block	28.91	Uninsulated block	0
	Exterior wall decoration	50% exterior wall plastering + real stone paint	24.09	50% exterior wall plastering + real stone paint	24.09	50% exterior wall plastering + real stone paint	24.09	Without decoration	0
	Exterior wall	50% exterior wall curtain wall	289.11	50% exterior wall curtain wall	289.11	50% prefabricated enclosure wall integrated with insulation, insulation, and decoration	216.83	100% prefabricated enclosure wall integrated with insulation, insulation, and decoration	433.66

4.4 Analysis of the Relationship Between Assembly Rate and Cost

After integrating project data, it was found that there is a certain relationship between the increase in construction cost and assembly rate. Therefore, Origin software was used for analysis, hoping to find a regression formula to describe the changes in construction cost under different assembly rates.

According to the Origin fitting results, the relationship between the construction cost increase (y) and the assembly rate (x) of the Prefabricated building system project can be expressed as formula (3):

$$y = -Ax^4 + Bx^3 - Cx^2 + Dx - E \tag{3}$$

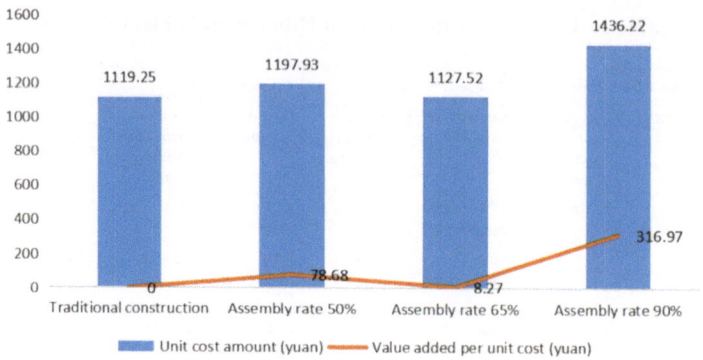

Fig. 1. Run chart of change of single square cost of complex building under different assembly rates

According to relevant research [11], the assembly rate of Prefabricated building is between 10% and 20%, and the curve shows an upward trend with a large slope, that is, the construction cost increases significantly with the increase of the assembly rate; The assembly rate is in the range of 20% to 40%, and the curve is almost horizontal, indicating a stable increase in construction costs, maintaining around 35%; The assembly rate is in the range of 40% to 65%, and the curve shows an upward trend again, reaching the peak of construction cost increase when the assembly rate reaches 65%. The slope is relatively large but smaller than the slope value of the assembly rate in the range of 10% to 20%; When the assembly rate exceeds 65%, the curve shows a downward trend, meaning that the increase in construction costs slows down with the increase in assembly rate.

Therefore, based on the trend of the curve and the use of regression formulas, the possible increase in construction costs can be estimated simply based on the application scale or assembly rate of prefabricated components. There are differences in the cost increases corresponding to different projects at the same assembly rate. There are differences in the construction techniques used in different projects. Previous studies have shown that achieving more efficient utilization of tower cranes on construction sites can achieve the effect of saving construction costs. Under the same assembly rate, the more efficient utilization of mechanical equipment and the increase in project schedule will have a certain impact on construction costs.

5   Conclusion and Suggestions

5.1   Conclusion

Essentially, there are significant differences in construction processes between prefabricated and cast-in-place construction methods. The different construction processes and organizational management methods result in different economic, social, and environmental benefits.

The construction costs corresponding to different prefabricated construction schemes vary greatly, and the incremental and reduced costs must be based on the technical system, which cannot be simply measured by assembly rate.

There is no strong linear positive correlation between the level of assembly rate and cost increment, and selecting an appropriate assembly plan is crucial for cost control of assembly construction; After selecting the appropriate assembly plan, the higher the assembly rate, the lower the construction cost. Similarly, when implementing fully assembled dry construction in low-rise and multi-story buildings, the cost is more advantageous.

On the basis of reducing costs through large-scale production, the maturity of the technical system can also significantly reduce construction costs. At the same time, the quality improvement, construction period reduction and policy dividend brought by the cost increment of Prefabricated building should be comprehensively considered.

5.2 Suggestions

The scheme comparison and selection of Prefabricated building should be considered at the scheme stage. In the scheme design stage, only by adjusting the prefabricated scheme and solving the problems of standardization, modularization and cost optimization from the source can the problem of high cost of Prefabricated building be fundamentally solved.

Summarize the cost reduction path of the design and coordination system by comprehensively balancing the safety, quality, cost, and progress of the project. For example, standardization of household products, PC components, molds, production layout optimization, and final assembly plan optimization can further reduce costs throughout the entire process.

References

1. Fa, G.B.: No. 71. Guiding Opinions of the General Office of the State Council on Vigorously Developing Prefabricated building (2016). https://www.gov.cn/
2. Zhihu homepage: Technical System of Prefabricated Concrete Frame Structure (2018). https://zhuanlan.zhihu.com/p/38267808
3. Consumption quota of Prefabricated building: Ministry of Housing and Urban-Rural Development. China (2016)
4. Detailed Rules for Calculation of Assembly Rate of Prefabricated building in Chongqing (Version 2021). Chongqing Housing and Urban Rural Construction Technology Development Center. Chongqing (2021)
5. Evaluation Standard for Prefabricated building: Ministry of Housing and Urban Rural Development of the People's Republic of China. China (2017)
6. Unified standard for application of Building information modeling: Ministry of Housing and Urban Rural Development of the People's Republic of China. China (2017)
7. Shen, Y.: The cost of construction and installation engineering under the pricing model of bill of quantities. J. Quality & Market. **04**, 184–186 (2022)
8. Code of bills of quantities ande valuation for construcion works (GB 50500-2013). China Planning Press. Beijing (2013)
9. Pricing Quota for Housing Construction and Decoration Engineering in Chongqing. Chongqing Urban Rural Development Commission. Chongqing (2018)
10. Pricing quota of Prefabricated building in Chongqing: Chongqing Urban Rural Development Commission. Chongqing (2018)

11. Huang, C.F.: Quantitative analysis of the impact of prefabricated assembly on construction costs. Construction Materials & Decoration **11**, 192–193 (2018)

Research on Promotion Strategy
of Prefabricated Buildings in Rural Areas
# of Chongqing

WanQing Chen[1], Yu Wang[2(✉)], JunXia Sun[1], and QiLi Gan[1]

[1] Chongqing College of Architecture and Technology, Chongqing, China
[2] Chongqing Vocational College of Science and Technology, Chongqing, China
850425529@qq.com

Abstract. In order to promote the development of prefabricated buildings in rural areas of Chongqing, this paper sorted out the influencing factors of the promotion of prefabricated buildings in rural areas of Chongqing through literature analysis, policy analysis and social research statistics. A hierarchical structure model based on SWOT analysis theory was constructed, and the weight of each factor in the index layer was determined by AHP combined with SPSSAU online analysis tool, and the strength of each factor in the index layer was determined by questionnaire survey. Factor weight and factor strength together determine the comprehensive strength of each criterion layer, and then construct the promotion strategy model. The research results show that the rural prefabricated buildings in Chongqing belong to the pioneering strategic regional opportunity strategy type, which provides a basis for the formulation of prefabricated buildings policy and the promotion of prefabricated buildings.

Keywords: prefabricated building · rural construction · swot · promotion strategy

1 Introduction

1.1 Research Background

The Opinions of the Central Committee of the Communist Party of China and The State Council on Comprehensively Promoting the Key work of Rural Revitalization in 2023 emphasize that: solidly promote the construction of livable and viable industries and beautiful villages [1], and rural housing construction has always been one of the key areas of rural construction. Traditional housing construction methods have such problems as long construction period, low level of construction technology, and difficult to guarantee construction quality, which affect the realization of the goal of beautiful countryside construction [2]. Prefabricated buildings have the characteristics of factory production and standardized assembly, and can quickly build a number of new buildings with unified style, so prefabricated buildings can be widely used in rural construction. However, due to a series of problems such as limited production conditions and transportation conditions [3], lack of professional institutions and talents, prefabricated buildings are not widely used in rural buildings.

© The Author(s) 2024
P. Xiang and L. Zuo (Eds.): PBSFTT 2023, LNCE 382, pp. 83–92, 2024.
https://doi.org/10.1007/978-981-97-5108-2_8

1.2 Research Status

Many scholars have conducted research on rural prefabricated buildings from various angles. L Jankovi [4] proposed that the use of modular prefabricated wood structures for public urban and rural areas would ensure the sustainability of the natural environment and continue the Croatian architectural tradition. Dianbo W [5] introduced the modular design process of ALC building external wall light steel structure, and formed a standardized prefabricated farm house, providing a reference for the large-scale promotion of prefabricated buildings in rural areas Matsumoto A [6] analyzed the current situation of reinforced concrete system apartments and PC system apartments and concluded that "it is necessary to gradually use higher prefabricated buildings". Wang Zhaohui [7] summarized the existing studies related to rural prefabricated buildings, and Lin Guohai et al. [8] believed that prefabricated passive house construction technology could promote the construction of beautiful villages and the implementation of rural revitalization strategy, but they did not give clear views on how to promote prefabricated buildings. Wu Lingling [9] and Zhang Heng [10] affirmed the necessity of the application of prefabricated buildings in rural areas and analyzed the influencing factors of the development of prefabricated housing in rural areas, but did not conduct quantitative analysis on the influencing factors. Dong Xuanxuan [11] analyzed the development of prefabricated farm houses based on the SWOT AHP, but only clarified the weights of each influencing factor and did not determine the final development strategy. Lv Lingxue [12] constructed a strategic positioning for the development of prefabricated buildings based on PEST-SWOT analysis, but the analysis was mainly carried out on a nationwide scale rather than in rural areas. He Pengwang [2] established a strategic selection model for rural extension prefabricated buildings by combining SWOT quantitative analysis method. However, only the literature method was used to extract the influencing factors, which was from a single source and not perfect enough.

On the whole, the existing literature affirms the necessity of promoting prefabricated buildings in rural areas, but there are some problems in the analysis of promotion factors, such as insufficient quantitative analysis, insufficient research, weak pertinancy and incomplete influencing factors. This paper aims to explore a suitable strategy for the promotion of prefabricated buildings in rural areas of Chongqing, so as to help rural revitalization.

2 Analysis on the Influencing Factors of the Popularization of Prefabricated Buildings in Rural Areas

In order to realize the deep popularization of prefabricated buildings in rural areas, it is necessary to analyze the factors that affect the popularization of prefabricated buildings in rural areas and establish the strategic model of prefabricated buildings in rural areas. This paper analyzes the influencing factors through the following three ways.

1. Literature analysis. Read the literature related to "rural revitalization" and "prefabricated buildings" to extract the influencing factors.
2. Policy analysis. Read policy documents related to prefabricated buildings, rural revitalization, rural housing construction, and rural construction to understand favorable policy factors.

3. Social research. Survey for industry practitioners to understand the development of prefabricated buildings in rural areas and existing problems.

3 Establishment of Strategic Model for Promoting Prefabricated Buildings in Rural Areas of Chongqing

According to the above, 10 favorable factors and 10 unfavorable factors are sorted out. These factors not only have the influence of external environment but also the characteristics of prefabricated buildings themselves, which is highly matched with the internal and external advantages and disadvantages of SWOT analysis theory. As SWOT analysis is mainly qualitative research, there are problems such as insufficient quantitative analysis and strong subjectivity [13], this paper will use the polar coordinate method combined with SWOT to conduct quantitative evaluation of SWOT factors.

Step 1: Construct the hierarchy diagram of rural extension prefabricated buildings, and make a quantitative analysis of each factor of the index layer.

1. Determine the target layer, criterion layer and index layer of the hierarchy chart according to the influence factors sorted out above and the theoretical elements of SWOT analysis. (Table 1 Target layer, criterion layer and index layer).

2. Develop a survey questionnaire. A Likert scale was developed based on the influence of each index layer on the target layer. The scale was scored from 1 to 5. The higher the score, the greater the influence. Relevant personnel of the whole industry chain of prefabricated buildings were invited to score according to the actual situation, 123 valid questionnaires were collected, and the reliability and validity of the questionnaires were tested. After passing the test, the average score of each factor was calculated. (Table 1 Average score)

Step 2: Use analytic hierarchy process to determine the weight of each index layer to describe the impact of each factor on the result.

1.Construct a pair comparison matrix. Taking each criterion layer as a dimension, the judgment matrix is obtained by dividing the average score of the index in this dimension. The data in the judgment matrix in row i and column j are denoted as a_{ij}, $a_{ij} = a_i/a_j$, a_i --The average score of A_i items, a_j--The average score of A_j items. As shown in Table 2:

Take the criterion layer "internal advantage (S)" as an example, As shown in Table 3:

2. The feature vector and weight value were calculated with SPSSAU online analysis tool, and do consistency test (take the criterion layer "internal advantage (S)" as an example, As shown in Table 4):

Consistency test index. Formula: $CI = \frac{\lambda_{max}-n}{n-1}$, among λ_{max}-- Determine the largest eigenroot of the matrix, n -- the order of the judgment matrix (number of terms), RI--ratio coefficient of random consistency, RI for details see Table 5:

In this case $CR = \frac{CI}{RI} = \frac{\lambda_{max}-n}{n-1}/RI = \frac{5-5}{5-1}1.12 = 0$, CR < 0.1, consistency check passed, the weight calculation results are valid.

3. The weights of factors under other criteria layers are calculated in the same way. (Table 6 Weights)

Step 3: Determine the degree of conformity between each factor of the index layer and the reality. (Table 6 Intensity factor)

Table 1. Rural extension prefabricated building hierarchy chart and index scores

Target layer	Criterion layer	Index level	Average score
Rural extension prefabricated building hierarchy diagram	Internal Advantage (S)	Good construction quality	4.78
		Short build cycle	4.51
		Living comfort	4.15
		Energy saving and environmental protection	4.06
		Construction process specification	4.01
	Internal Weakness (W)	High construction cost	4.67
		Unsound system	3.89
	External Opportunities (O)	Favorable policy	4.77
		Drive industrial upgrading	3.25
		Narrow the urban-rural gap	4.11
		Large market demand	4.67
		High enterprise enthusiasm	3.99
	External Challenges (T)	Great traffic obstruction	4.22
		The acceptance level is not high	4.84
		The supervision mechanism is not perfect	3.21
		Lagging policy implementation	4.69
		The whole industrial chain is imperfect	3.87
		The comprehensive benefit is not obvious	3.02
		Lack of regional characteristics	2.11
		Insufficient publicity	3.42

Prefabricated construction industry experts are invited to score the degree of conformity between the factors of the indicator layer and the current actual situation, using a 0–4 scale, advantages and opportunities are positive, disadvantages and challenges are negative, the larger the absolute value is, the more consistent the indicator layer is with the actual situation, 0 is not understood. The average score of each factor was calculated.

Step 4: Determine the comprehensive strength of each criterion layer.Multiply each intensity factor by the corresponding weight to get the factor strength, the sum of factors'

Table 2. AHP judgment matrix

Mean value	Item	A_1	A_2	A_3	...	A_j
a_1	A_1	1	a_1/a_2	a_1/a_3	...	a_1/a_j
a_2	A_2	a_2/a_1	1	a_2/a_3	...	a_2/a_j
a_3	A_3	a_3/a_1	a_3/a_2	1	...	a_3/a_j
...	...	...	...	...	1	...
a_i	A_i	a_i/a_1	a_i/a_2	a_i/a_3	...	a_i/a_j

Table 3. Internal dominance (S) layer pair comparison matrix

Average score	Item	Good construction quality	Short build cycle	Living comfort	Energy saving and environmental protection	Construction process specification
4.78	Good construction quality	1	1.060	1.152	1.177	1.192
4.51	Short build cycle	0.944	1	1.087	1.111	1.125
4.15	Living comfort	0.868	0.920	1	1.022	1.035
4.06	Energy saving and environmental protection	0.849	0.900	0.978	1	1.012
4.01	Construction process specification	0.839	0.889	0.966	0.988	1

Table 4. Results of internal advantage (S) AHP hierarchy analysis

Item	Eigenvector	Weighted value	Maximum eigenvalue	CI value
Good construction quality	1.111	22.222%	5.000	0.000
Short build cycle	1.048	20.967%		
Living comfort	0.965	19.293%		
Energy saving and environmental protection	0.944	18.875%		
Construction process specification	0.932	18.642%		

Table 5. Random consistent RI table

Rank n	1	2	3	4	5	6	7	8	9	10	11	12	13	14	15
RI value	0	0	0.52	0.89	1.12	1.26	1.36	1.41	1.46	1.49	1.52	1.54	1.56	1.58	1.59

strength under each criterion layer is the comprehensive strength of the criterion layer. (Table 6 Comprehensive strength).

Table 6. Weight, factor strength and comprehensive strength table

Target layer	Criterion layer	Index level	Weights	Intensity factor	Comprehensive strength
Rural extension prefabricated building hierarchy diagram	Internal Advantage (S)	Good construction quality	22.2%	3.27	2.93
		Short build cycle	21.0%	3.55	
		Living comfort	19.3%	2.73	
		Energy saving and environmental protection	18.9%	2.81	
		Construction process specification	18.6%	2.14	
	Internal Weakness (W)	High construction cost	54.6%	−2.96	−2.82
		Unsound system	45.4%	−2.65	
	External Opportunities (O)	Favorable policy	22.9%	3.58	2.95
		Drive industrial upgrading	15.6%	2.67	
		Narrow the urban-rural gap	19.8%	2.47	
		Large market demand	22.5%	3.11	
		High enterprise enthusiasm	19.2%	2.74	

(*continued*)

Table 6. (*continued*)

Target layer	Criterion layer	Index level	Weights	Intensity factor	Comprehensive strength
	External Challenges (T)	Great traffic obstruction	14.9%	−3.51	−2.93
		The acceptance level is not high	17.1%	−3.45	
		The supervision mechanism is not perfect	11.3%	−3.06	
		Lagging policy implementation	13.0%	−3.44	
		The whole industrial chain is imperfect	13.6%	−2.18	
		The comprehensive benefit is not obvious	10.6%	−2.44	
		Lack of regional characteristics	7.4%	−1.38	
		Insufficient publicity	12.1%	−3.01	

Step 5: Identify the type of strategy. Draw the development strategy quadrangle with the comprehensive strength of each criterion layer as the coordinate (Fig. 1). The strategic quadrilateral center of gravity coordinates, strategic azimuth, and strategic strength coefficient were calculated to determine the strategic direction and type pedigree (Fig. 2).

Center of gravity coordinate $P(x, y) = (\Sigma_i^x/4, \Sigma_i^y/4) = (0.027, 0.0257)$

Strategic type azimuth $\theta = \arctan(y/x) = \arctan(0.027/0.0257) = 46.4°$ ($\frac{\pi}{4} < \theta < \frac{\pi}{2}$).

Strategic strength coefficient ρ:Positive strategic strength $U = O*S$, Negative strategic intensity $V = W*T$, $\rho = \frac{U}{U+V} = \frac{2.95*2.93}{2.95*2.93+(-2.82)*(-2.93)} = 0.512$ ($0.5 < \rho < 1$)

The results show that the azimuth of the strategic type is 46.4°, belonging to the pioneering strategic area, between $\frac{\pi}{4}$—$\frac{\pi}{2}$, belonging to the opportunistic strategy type, and the strategic intensity coefficient $\rho > 0.5$. Active measures should be taken to speed up the development.

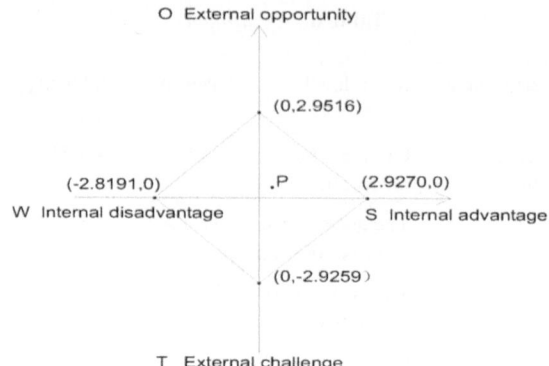

Fig. 1. Development strategy Quadrangl

Fig. 2. Strategic direction and type pedigree

4 Research Conclusions

The research shows that the development of rural prefabricated buildings in Chongqing belongs to the opportunistic strategy type of the pioneering strategic area, and the opportunity factor is the key factor affecting the promotion of rural prefabricated buildings in Chongqing. This conclusion provides a theoretical basis for the formulation of prefabricated building policies and the promotion of rural prefabricated buildings, provides practical suggestions for relevant departments and enterprises, and puts forward countermeasures and suggestions for the promotion of prefabricated buildings in rural areas of Chongqing.

To be specific, we should seize external opportunities, give full play to internal advantages, and avoid internal disadvantages and external threats. Seize the favorable national policies and huge market demand, increase the support and cultivation of local prefabricated construction leading enterprises, and fully mobilize the enthusiasm of enterprises; Strengthen the publicity, improve the social recognition of prefabricated buildings; Improve infrastructure construction, clear transport obstacles; Training township prefabricated building technical personnel; Formulate practical, reliable and easy to implement support policies, and establish a sound quality control system; Reduce construction and transportation costs. Through the combined application of various strategies, the development of prefabricated buildings in rural areas of Chongqing is effectively promoted.

References

1. Opinions of the Central Committee of the Communist Party of China and The State Council on Comprehensively Promoting the Key Work of Rural Revitalization in 2023, https://nrra. gov.cn/2023/02/14/ARTIs1YdeIMvWi3sUA3bPd89230214.shtml
2. He, P.W.: Research on the promotion strategy of prefabricated buildings in rural areas under the background of rural revitalization. Shandong construction university (2020). https://doi. org/10.27273/dcnki.Gsajc.2020.000245
3. Mei, W.P., Guan, Z.X., Xu, S.Z.: Based on beautiful rural construction of prefabricated farmhouse structure. China Construction Metal Structure 03, 107–109 (2023). https://doi.org/10. 20080/j.carolcarrollnki. ISSN: 1671-3362.2023.03.035
4. Jankovi, L.: Design of Sustainable Modular Wooden Booths Inspired by Revitalization of Croatian Traditional Construction and New User Needs Due to COVID-19 Pandemic. Sustainability 14 (2022). https://doi.org/10.3390/su14020720
5. Dianbo, W., Ruizeng, Y.: Research on the standardized household design of prefabricated light-steel structure based on ALC building external wall. Henan Sci. Technol. (2018)
6. Matsumoto, A.: Cost study of structual frame for school buildings (Part III): about the relation between the ordinary system and prefabricated system. Trans. Architect. Instit. Japan **194**, 71–80 and 98 (1972). https://doi.org/10.3130/aijsaxx.194.0_71
7. Wang, Z.H., Wang, Y.X.: Prefabricated construction in rural areas in China. Applied research process to a low-carbon world **13**(6), 115–117 (2023). https://doi.org/10.16844/j.carolcarroll nkicn10-1007/tk.2023.06.057
8. Lin, G.H., Zhai, H.Y., Zhang, S.B.: Fabricated passive home construction technology application in the country revitalization. J. Constr. Sci. Technol. **9**, 52–55 (2020). https://doi.org/ 10.16116/j.carolcarrollnkiJSKJ.2020.09.010
9. Wu, L.L.: Introduction to the application of the prefabricated building in the beautiful countryside construction. Modern Property (the ten-day) (01), 91 (2020). https://doi.org/10.16141/ j.carolcarrollnki.1671-8089.2020.01.078
10. Zhang, H., Wang, D.T., Zhang, Y.P.: The use of prefabricated construction feasibility analysis in the construction of rural. J. Anhui constr. **27**(6), 122–124 (2020). https://doi.org/10.16330/ j.carolcarrollnki.1007-7359.2020.06.062
11. Dong, X.X., Zeng, Q.: Based on AHP - a SWOT analysis of rural prefabricated construction research. J. Intell. City **9**(6), 107–110 (2023). https://doi.org/10.19301/j.carolcarrollnki ZNCS.2023.06.034
12. Lv, L.X., Huang, G.: Prefabricated construction in our country, the development of PEST, SWOT analysis. J. Eng. Econo. **28**(11), 45–48 (2018). https://doi.org/10.19298/j.carolcarroll nki.1672-2442.201811045
13. Liu, H.J., Sun, Q.Z.: Investigation and analysis of the mass sports policy execution at the grass-roots level of. J. Shanghai Sports Inst. **4**(4), 49–53 (2012). https://doi.org/10.16099/j. carolcarrollnkijsus.2012.04.011

Whole Process Management and Collaboration of Prefabricated Building Based on BIM

Jing Yang and Caixia Zuo[✉]

Department of Architectural Engineering, Chongqing College of Architecture and Technology, Chongqing 401331, China
jane451301513@163.com

Abstract. As a modern construction method, prefabricated building has been widely used in the construction field for its high efficiency, fast and sustainable characteristics. However, the traditional construction method has some problems such as low efficiency, high cost and difficult quality control. In order to solve these problems, this paper adopts Building Information Modeling (BIM) technology as the main application method, and discusses the application of BIM in prefabricated buildings. BIM technology can improve the efficiency of design and construction, reduce errors and conflicts, and improve building quality and sustainability. Through the application of BIM technology, prefabricated buildings can better meet people's needs for efficient, fast and sustainable construction. Through the research in this paper, the application of BIM technology enables the design and construction of building projects to be carried out more efficiently, and the application of BIM in prefabricated buildings can provide better user experience and satisfaction. This can promote the development of the construction industry in a digital and intelligent direction.

Keywords: Building Information Model · Prefabricated Building · Whole Process Management · Collaborative Research

1 Introduction

As a new construction method, prefabricated building has been widely concerned and applied in the construction industry in recent years [1, 2]. However, with the continuous promotion and development of prefabricated buildings, the whole process management and coordination are faced with many challenges [3].

In the past research, many scholars have carried out in-depth research on the development of prefabricated buildings and obtained some important conclusions. For example, Wen Y's research shows that based on BIM5D intelligent construction of prefabricated buildings and personnel training research, prefabricated buildings have significant advantages in reducing construction cycle, improving construction quality and reducing costs [4]. Through research on BIM technology integration in prefabricated buildings, Wang Y found that prefabricated buildings can effectively reduce the carbon emissions of construction sites, which has positive significance for environmental protection [5].

© The Author(s) 2024
P. Xiang and L. Zuo (Eds.): PBSFTT 2023, LNCE 382, pp. 93–99, 2024.
https://doi.org/10.1007/978-981-97-5108-2_9

Li Z studied the obstacles to the development of prefabricated buildings in China, and found that prefabricated buildings have obvious advantages [6]. However, although the research of these scholars has achieved some results, the current research still has some shortcomings.

This paper would analyze the relationship between prefabricated buildings and BIM.The paper put forward corresponding solutions and suggestions to promote the sustainable development of prefabricated buildings.

2   Prefabricated Building Overview

2.1   Definition of Prefabricated Buildings

Prefabricated building, also known as prefabricated building or modular building, is an emerging construction method [7, 8]. The prefabricated building adopts advanced design and manufacturing technology, industrializes the building process, and realizes the whole process management and collaboration of the building life cycle [9].

The core idea of prefabricated construction is to break down the construction process into modular parts, each of which is precisely manufactured in a factory and then quickly assembled on site [10, 11]. By studying the whole process management and collaboration of prefabricated buildings, they explored how to optimize the design, manufacturing, transportation, assembly and other links in order to improve the efficiency and feasibility of prefabricated buildings [12, 13].

2.2   Characteristics of Prefabricated Buildings

Prefabricated buildings adopt factory production mode, after the components or modules of prefabricated buildings are manufactured in the factory, they can be quickly transported to the site for assembly [14]. Prefabricated buildings can be flexibly designed and customized according to different needs. By adjusting and combining different components or modules, the architectural with various functions and styles can be realized [15, 16]. Prefabricated buildings adopt prefabricated components or modules, which can realize the efficient utilization of building materials and resource saving of [17]. Since prefabricated buildings are manufactured in factories, construction quality and construction safety can be better controlled [18]. The production environment in the factory is more stable, which can avoid being affected by the weather and other factors. [19, 20].

3   Application of BIM in the Construction Industry

3.1   Definition and Principle of BIM

The principle of BIM is to model all aspects of the construction project, including geometry, materials, construction, time schedule, cost budget, etc., in a digital form, and to store and manage this information in the model. With BIM, virtual design and construction simulation can be performed to reduce design errors and conflicts and optimize the performance of buildings.

3.2 Application Fields of BIM in the Construction Industry

BIM has a wide range of application fields in the construction industry, and several aspects of the application fields are shown in Table 1.

Table 1. Application fields of BIM in the construction industry

Sequence number	BIM application area	Represent
1	Design and modeling	Architectural design and modeling using BIM software for 3 D visualization and data management
2	Coordination and coordination	Multiple team members can simultaneously use the BIM platform for collaboration and coordination, reducing conflicts and errors
3	Data analysis and optimization	The data in the BIM model is used for building performance analysis and optimization to improve energy efficiency and comfort
4	Education and training	Application of BIM technology in architectural education and training to develop BIM skills for students and professionals

3.3 Combination and Application of Prefabricated Building and BIM

The combination of prefabricated construction and BIM technology can bring many advantages and improvements. Through BIM technology, designers can design and optimize building models in a virtual environment, thereby reducing design errors and conflicts. The combination of prefabricated construction and BIM can achieve synergies in design, construction and management, improving the efficiency and quality of projects. This combination can reduce errors and conflicts, provide accurate and comprehensive information, and reduce costs and risks.

3.4 Engineering Research and Analysis

In this paper, we use BIM technology and genetic algorithm to optimize the design of prefabricated buildings to improve their performance and efficiency. Using the BIM model and construction simulation software, the construction process of prefabricated buildings is simulated and analyzed to optimize the construction sequence, resource allocation and schedule planning. The formula of the genetic algorithm is as follows:

$$Fitness = f(x) \tag{1}$$

where x represents the chromosome of the individual, and f (x) represents the fitness assessment function as defined by the question.

$$P(i) = \frac{Fitness(i)}{\sum_{i=1}^{n} Fitness(j)} \tag{2}$$

where P (i) represents the probability that individual i is selected, Σ (Fitness (j)) represents the sum of the fitness values of all individuals, and n represents the population size.

4   Data Collection and Prefabricated Buildings

This paper will collect from the BIM model and project management software and BIM prefabricated building construction time, cost, error quantity data. Through descriptive statistical analysis, the evaluation value includes the average and standard deviation, so as to determine the advantages and disadvantages between the two. The traditional data were collected in Fig. 1, and the BIM-based data are shown in Fig. 2.

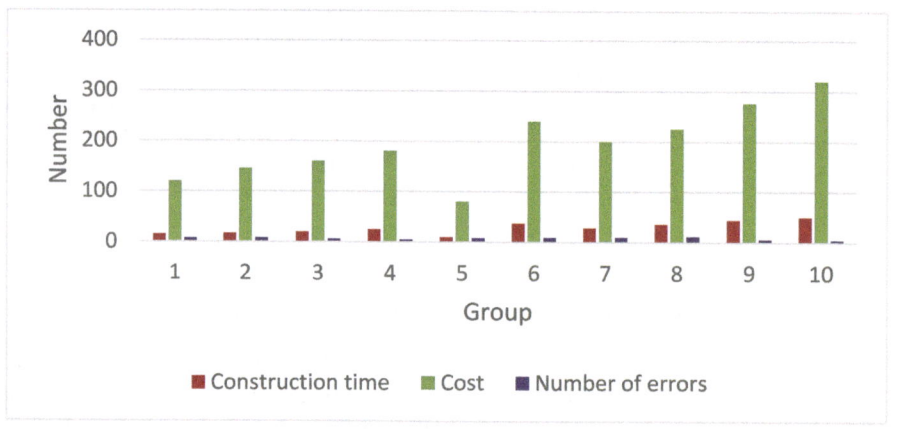

Fig. 1. Traditional prefabricated building data

As from the data from Figs. 1 and 2, it is clear that BIM based prefabricated buildings have significantly less construction time, cost and errors in terms of construction time and cost than traditional prefabricated buildings. BIM technology provides more accurate and comprehensive building information, including component dimensions, location and installation sequence. BIM models can help design and construction teams work better together to reduce errors and conflicts. During the design phase, BIM can detect and resolve design errors and conflicts, avoiding construction problems caused by design inconsistencies in traditional methods.

In order to comprehensively evaluate the advantages and disadvantages of traditional prefabricated buildings and BIM based prefabricated buildings in the construction process, this paper not only makes statistics on time and cost, but also collects data on user satisfaction.

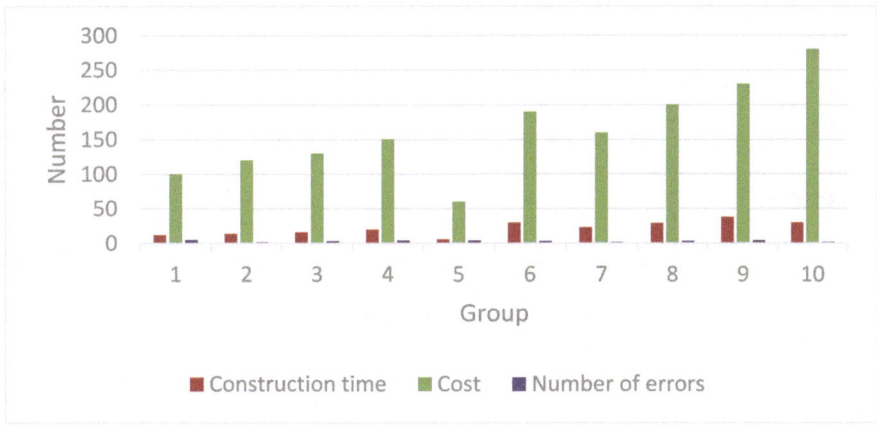

Fig. 2. BIM-based prefabricated building data

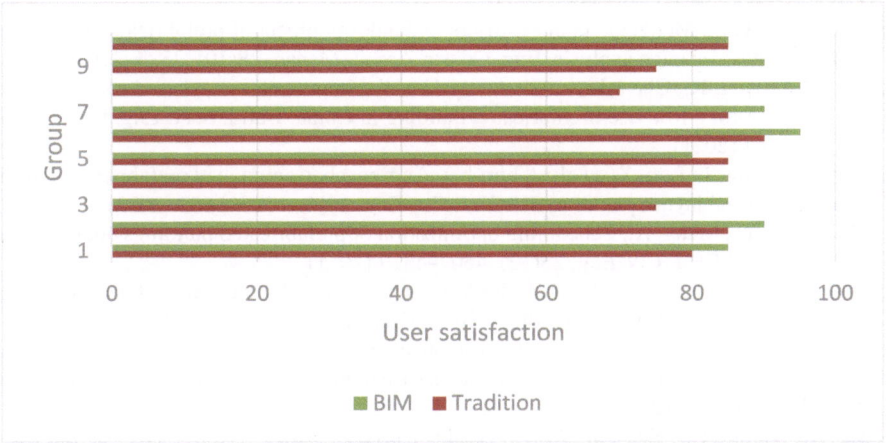

Fig. 3. User satisfaction survey data

The application of BIM in prefabricated buildings can improve user satisfaction. According to the data in Fig. 3, except for the tenth group of data, the BIM application methods of other groups have improved user satisfaction compared with traditional methods. This shows that the application of BIM in prefabricated buildings can provide users with a better experience and satisfaction.

5   Conclusions

When BIM technology is combined with prefabricated building management and collaboration, more efficient and precise design, construction and maintenance processes can be achieved. BIM based prefabricated buildings not only have obvious advantages in terms of construction time and cost, but also can improve user satisfaction. The application of BIM in the management and collaboration of prefabricated buildings has brought great advantages to the industry, driving the development and innovation of prefabricated buildings.

References

1. Xiao, Y., Bhola, J.: Design and optimization of prefabricated building system based on BIM technology. Int. J. Syst. Assurance Eng. Manag. **13**(Suppl 1), 111–120 (2022)
2. Gunawardena, T., Mendis, P.: Prefabricated building systems—design and construction. Encyclopedia **2**(1), 70–95 (2022)
3. Wang, Y., Xue, X., Yu, T., et al.: Mapping the dynamics of China's prefabricated building policies from 1956 to 2019: a bibliometric analysis. Build. Res. Inf. **49**(2), 216–233 (2021)
4. Wen, Y.: Research on the intelligent construction of prefabricated building and personnel training based on BIM5D. J. Intell. Fuzzy Syst. **40**(4), 8033–8041 (2021)
5. Wang, Y.: Research on the integration of BIM technology in prefabricated buildings. World J. Eng. Technol. **9**(3), 579–588 (2021)
6. Li, Z., Zhang, S., Meng, Q., et al.: Barriers to the development of prefabricated buildings in China: a news coverage analysis. Eng. Constr. Archit. Manag. **28**(10), 2884–2903 (2021)
7. Ma, M., Zhang, K., Chen, L., et al.: Analysis of the impact of a novel cool roof on cooling performance for a low-rise prefabricated building in China. Build. Serv. Eng. Res. Technol. **42**(1), 26–44 (2021)
8. Yang, B., Fang, T., Luo, X., et al.: A bim-based approach to automated prefabricated building construction site layout planning. KSCE J. Civ. Eng. **26**(4), 1535–1552 (2022)
9. He, W., Li, W., Meng, X.: Scheduling optimization of prefabricated buildings under resource constraints. KSCE J. Civ. Eng. **25**(12), 4507–4519 (2021)
10. Xue, H., Sun, T., Ling, F.Y.Y., et al.: Redesigning the virtual organisational structure for the management of prefabricated buildings. Int. J. Constr. Manag. **23**(6), 1069–1085 (2023)
11. Lee, P.C., Lo, T.P., Wen, I.J., et al.: The establishment of BIM-embedded knowledge-sharing platform and its learning community model: a case of prefabricated building design. Comput. Appl. Eng. Educ. **30**(3), 863–875 (2022)
12. Li, X., Wang, C., Kassem, M.A., et al.: Fairness theory-driven incentive model for prefabricated building development. Arab. J. Sci. Eng. **47**(10), 13487–13498 (2022)
13. Gong, C., Xu, H., Xiong, F., et al.: Factors impacting BIM application in prefabricated buildings in China with DEMATEL-ISM[J]. Constr. Innov. **23**(1), 19–37 (2023)
14. Issabayev, G, Slyambayeva, A., Kelemeshev, A., et al.: Development of the project of modular prefabricated buildings. EUREKA: Phys. Eng. (4), 36–45 (2022)
15. Zhou, Z., Luo, X.: Analysis on game evolution of latecomer prefabricated building firm from the perspective of population ecology. Alex. Eng. J. **61**(7), 5529–5537 (2022)
16. Wu, S., Jiang, X., Yao, X., et al.: A review on quality management of prefabricated buildings. Curr. Urban Stud. **10**(2), 212–223 (2022)
17. Ma, W., Sun, D., Deng, Y., et al.: Analysis of carbon emissions of prefabricated buildings from the views of energy conservation and emission reduction. Nat. Environ. Pollut. Technol. **20**(1), 39–44 (2021)
18. Zhao, N.: Research on the management mode of EPC project of prefabricated building based on BIM technology. Open Access Libr. J. **8**(7), 1–13 (2021)
19. Zhong, J., Zhang, P., Wang, L.: An energy consumption calculation model of prefabricated building envelope system based on BIM technology. Int. J. Global Energy Iss. **44**(2–3), 121–138 (2022)

20. Wang, C., Zou, F., Yap, J.B.H., et al.: System dynamics tool for entropy-based risk control on sleeve grouting in prefabricated buildings. Eng. Constr. Archit. Manag. **30**(2), 538–567 (2023)

Research on Prefabricated Building Wall Technology Based on SAR Theory

Jinwei Liu and Qiuna Li$^{(\boxtimes)}$

Chongqing Vocational College of Architecture and Technology, Chongqing, China
122056713@qq.com

Abstract. The article introduces the origin of SAR theory and its important guiding role in the research of prefabricated building technology. Based on SAR theory, this paper found that skeleton safety is fundamental in various technologies of prefabricated buildings, and the diversity, functionality, and sustainability of separable components, including wall systems, are the main research directions for future prefabricated building technologies in China. In this paper, the research on prefabricated building wall technology is conducted from the following aspects: Prefabricated building wall support system technology; prefabricated building exterior wall panel technology; Diversity of prefabricated building exterior walls.

Keywords: SAR theory · prefabricated buildings · Skeleton · Detachable components · Wall technology

1 Introduction

To illustrate the design and construction of architecture, we usually make a visual analogy, comparing architecture to living organisms. The building structure is the skeleton of an organism, responsible for supporting the entire organism and carrying other loads imposed on it. The external image of a building is the outer skin of an organism, responsible for creating the overall image and temperament of the building. The equipment pipelines of buildings are the veins of organisms, responsible for delivering energy and nutrients to organisms and maintaining their normal life activities. The internal decoration of a building is the inner surface of an organism, responsible for the image and sensory aspects of the internal space. It is obvious that for a complete building, the exterior skin, skeleton, vein, and inner skin are indispensable. Buildings without exterior skin are rough and crude, incompatible with the urban environment, making it difficult to establish a foothold; A building without a skeleton will collapse and cannot exist at all; A building lacking a vein means that it lacks water, power, and air conditioning, making it difficult for such buildings to be used for a long time; A building without an inner skin is actually a rough house, and the internal space of the building is a place for people to stay for a long time. Without decoration, the building cannot fulfill its intended function.

It is precisely this vivid and simple analogy that forms the theoretical foundation for the design and construction of prefabricated buildings. This theory was originally

P. Xiang and L. Zuo (Eds.): PBSFTT 2023, LNCE 382, pp. 100–107, 2024.
https://doi.org/10.1007/978-981-97-5108-2_10

proposed by Nicholas John Habraken at a conference of the Dutch Institute of Architects in 1965. He proposed the idea of dividing the design and construction of residential buildings into "support" and "detachable units". The proposal of this idea precisely responded to a difficult problem faced by industrial residential construction at that time. After World War II, large-scale industrialized residential construction solved the severe housing shortage after World War II. However, since the late 1960s, people have been dissatisfied with the monotonous prefabricated building and environment. The Dutch Association of Architects proposed the SAR architecture theory based on Professor Habrgen's "support" and "detachable unit" ideas to address everyone's dissatisfaction, which is the contradiction between standardization and diversification of prefabricated building. SAR theory suggests that: the skeleton of a building is designed by professionals such as architects based on the specific location, conditions, and standards of the project. The construction task of the skeleton is completed after the project is completed, and no further modifications can be made, except for daily inspection, maintenance, or reinforcement. Separable components, including partitions, equipment, decoration, etc., are products of industrial production, and not only can be installed, removed, modified, or adjusted, but many components are also available. It can be universal. The selection and arrangement of these separable components, that is, the layout within the household, can be entirely determined by the residents themselves. It is precisely this prefabricated residential construction model, in which architects make various feasible and selectable apartment plans in advance for residents to choose from, that theoretically solves the contradiction between standardization and diversity of prefabricated residential buildings [1].

2 The Guiding Effect of SAR Theory on Prefabricated Buildings in China

In the 1950s and 1960s, China had a period of development in the prefabricated single-story industrial factory building system, prefabricated multi-layer frame building system, and prefabricated large plate building system. However, due to insufficient research on structural seismic resistance, design, construction, management, and other aspects, and poor technical and economic performance, prefabricated concrete buildings were replaced by fully cast-in-place concrete building systems in the mid-1990s [2].

In the past decade, with the upgrading and iteration of the construction industry and related industries, prefabricated buildings have been given high expectations due to their potential advantages in large-scale emission reduction and high-precision manufacturing of building components [3]. On February 22, 2016, the State Council issued the "Guiding Opinions on Vigorously Developing Prefabricated Buildings", which required the development of prefabricated concrete structures, steel structures, and modern wooden structures in accordance with local conditions, striving to achieve a 30% proportion of newly built buildings in approximately 10 years.

Looking back at the development process of prefabricated residential buildings both domestically and internationally, it is found that the development of prefabricated residential buildings has always been carried out under the theoretical framework of SAR. From the OB (Open Building) theory in the Netherlands in the 1970s, to the later SI

(Skeleton Infill) housing, to the KSI (Kikou Skeleton Infill) housing in Japan, and to the China Skeleton Infill prefabricated building system in China, there is no exception [4].

3 The Guiding Role of SAR Theory in the Research of Prefabricated Building Wall Technology in China

Compared with advanced prefabricated residential technology abroad, prefabricated buildings in China are still in their early stages, and the proportion of prefabricated building projects completed is lower than that of many developed countries [5]. The research on prefabricated buildings mainly focuses on the seismic resistance of components, the application of BIM technology, quasi-static testing, finite element analysis construction technology, prefabricated residential buildings, node connections, and other fields [6]. These fields are the phased target topics in the early stages of prefabricated construction. From the perspective of SAR theory, it can be considered that China's main research focus is on the skeleton. The safety of the skeleton is a prerequisite for carrying out prefabricated buildings. Leaving this premise and discussing prefabricated buildings is meaningless.

In countries with relatively mature development of prefabricated buildings, research hotspots seem to have already crossed the primary stage of structural safety, and entered the study of micro performance improvement and its laws of prefabricated components, energy consumption research of prefabricated buildings, waste discharge research, and full lifecycle perspective management research. That is to say, countries with advanced assembly technology have delved into research on the diversity, functionality, and sustainability of separable components. The wall system is the main component of separable components, so it is necessary to study the technology of prefabricated building walls.

The prefabricated building wall technology belongs to the application technology, and the improvement and development of this technology should be carried out through the cooperation and collaboration of multiple disciplines such as architectural design, structural design, equipment design, and interior design. SAR theory suggests that the structural discipline is responsible for providing a reliable skeleton and setting universal nodes on the skeleton for walls and other ancillary components to attach to it. Equipment pipelines and interior decoration are attached to structural or non structural building components. This is a connection pattern similar to having long branches on the trunk and leaves on the branches. The design of connection nodes between the skeleton and separable components, as well as between separable components, is crucial in this connection mode. These nodes must be safe, stable, durable, and universal within a certain range, and can be repeatedly installed and removed.

In today's highly developed engineering technology, there are actually many ready-made technologies that can be used for reference and even integrated in the research on connecting nodes and connecting entities. Taking the wall system as an example, this article explains how to use the inspiration of existing technology to carry out research on prefabricated residential technology.

4   Research on Prefabricated Building Wall Technology Based on SAR Theory

4.1   Technical Research on Prefabricated Building Wall Support System

Based on the guidance of SAR theory, it seems easy to find similarities between the exterior wall system of prefabricated residential buildings and the building curtain wall system. Building curtain wall refers to the non load-bearing outer wall enclosure of a building, usually composed of panels (glass, metal, stone, ceramic, etc.) and supporting structures (aluminum columns, beams, steel structures, glass ribs, etc.) at the back. At present, this system technology is mature and stable, and there are many technical points that can be transplanted and referenced.

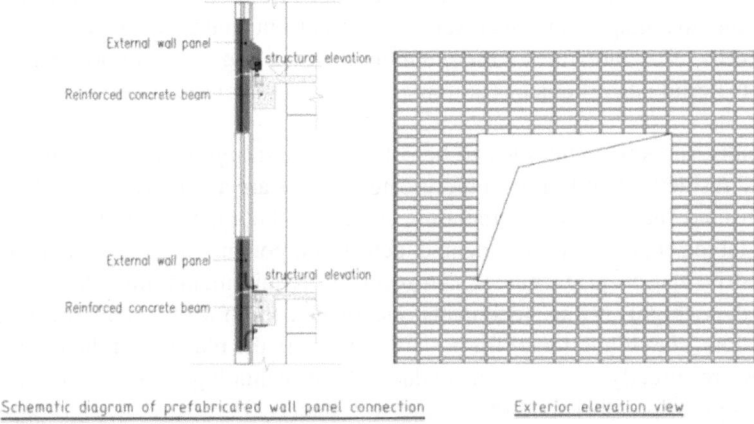

Fig. 1. Schematic diagram of vertical connection and external hanging of wall panels on the entire floor

The supporting structure of the curtain wall system often uses aluminum alloy columns, beams, steel structures, glass ribs, etc. These load-bearing components serve as intermediate components to connect the current panel and the main structure of the building. The extensive use of aluminum alloys, steel structures, etc. is expensive and uneconomical, making it unsuitable for ordinary residential buildings with large quantities. Therefore, it is considered to directly use the main body of civil engineering structures such as floors, beams, and columns as the support body, so that the external wall panels can be vertically connected and hung externally throughout the entire floor, or horizontally connected and hung externally throughout the entire bay (As shown in Fig. 1.).

4.2   Research on the Technology of Prefabricated Building Exterior Wall Panels

There are several key technologies that need to be addressed when comparing curtain wall system technology with prefabricated exterior wall systems. The first is the performance of exterior wall materials. The second is the problem of prefabricated connection nodes, and the third is the design of the size, shape, and texture of prefabricated exterior walls. From the perspective of production, transportation, installation, and final effects, we hope that prefabricated exterior wall panels, like curtain wall panels, are lightweight and high-strength, insulated, processable, drillable, and sawable. Not only it is convenient to connect the main body, but the size of the exterior wall panel is moderate and the exterior facade effect is rich and varied. This type of wall panel can meet the requirements of standardized mass production, simplified transportation and storage, and has low difficulty in construction lifting and convenient installation. In case of non-standard dimensions, secondary processing can be carried out on the construction site.

How to obtain such excellent performance exterior wall materials is not only a material research issue, but also a design issue. This requires collaborative work between architectural design and material development researchers.

Firstly, prefabricated residential designers need to understand material technology, compare the advantages and disadvantages of current mainstream materials used for exterior walls of prefabricated residential buildings, and propose improvement suggestions for material performance requirements, so that material researchers' research has directionality.

Secondly, the key technology to be solved is the design of connection nodes. Like traditional residential buildings, prefabricated houses are a giant rooted in the earth and must withstand various loads from nature and internal human activities. How to achieve safe and stable connection of fragmented factory components, and ensure that the thermal and sound insulation performance can meet the requirements after the connection is completed, is a highly refined technical issue. In the curtain wall system, metal parts such as bolts, slots, claws, and metal frames are used to fix the plates, and the joints between the plates are directly sealed with sealant. These technologies have some inspiration for the design of prefabricated residential wall panels. Therefore, when designing wall panels, it is necessary to split them reasonably and consider the appropriate connection design, which can utilize grooves and tenons, as well as embedded bolt fasteners and connectors. Overhang wall panels should be avoided, and windows should be placed within one wall panel as much as possible. The connection between the wall panel and the main structure should be flexible, and rigid connectors should not be set between the wall panels [7].

In addition, there are requirements for the size, shape, and texture of the exterior wall panels. The exterior wall panels can be pulled through in the direction of the floor height or fully opened in the direction of the bay based on the building's floor height and bay size, as well as whether the building's facade windows are horizontal or vertical. The dimensions in the other direction should be modular, standardized, and small-sized as much as possible.

In prefabricated construction sites, large lifting equipment and prefabricated components require a large amount of space, and multidimensional parallel lifting operations pose a severe challenge to the allocation of space resources on the construction site. It is easy to cause unreasonable allocation of spatial resources, which can lead to spatial conflicts during the lifting process of prefabricated components, reduce construction efficiency, and cause safety accidents [8]. Therefore, standardization of exterior wall dimensions and miniaturization can effectively solve production, transportation, construction and other problems. The specific dimensions can be determined based on the building's own situation, or can refer to the commonly used dimensions of curtain wall panels.

The current mainstream materials used for prefabricated residential exterior walls include thin-walled concrete rock wool composite exterior panels, concrete polystyrene composite exterior panels, concrete expanded perlite composite exterior panels, aerated concrete exterior panels, etc. [9].

The UHPC unit concrete exterior wall panel system of Huajian Group integrates material technology and curtain wall technology of ultra-high performance concrete, coupled with refined design, to achieve integrated production of various peripheral protective structures and accessories such as doors, windows, sunshades, balconies, and the detailed design of dripping and water blocking of unit concrete exterior wall panels. This technology not only achieves multiple basic functional requirements such as wall insulation and drainage, but also enriches and diversifies the design of building spaces, completely breaking the rigid impression of prefabricated buildings. This technology perfectly integrates multiple mature technologies and endows it with superb design skills, which has a great inspiration for future prefabricated building wall technology (As shown in Fig. 2).

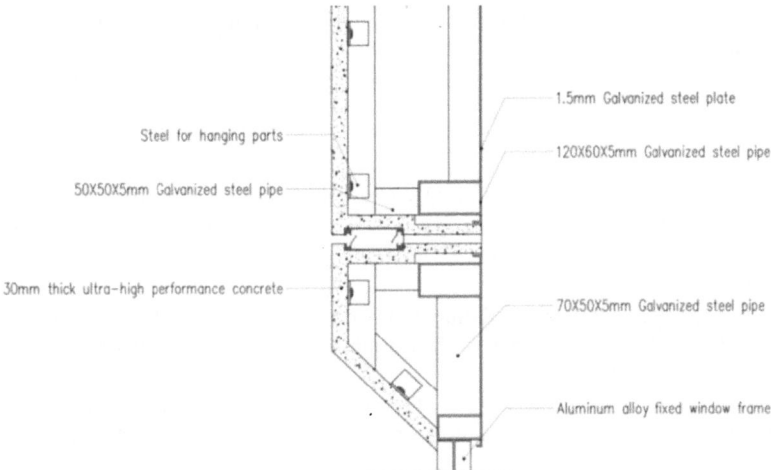

Fig. 2. UHPC unit concrete external wall panel system

4.3   Research on the Diversity of Exterior Walls in Prefabricated Buildings

In addition, the exterior wall system not only has maintenance functions, but also plays a role in creating an exterior facade effect. The exterior facade of prefabricated residential buildings is constrained by technological level, and currently the exterior design is still slightly monotonous. Under the principles of standardization, systematization, and modularization, a large number of prefabricated residential facades currently use a single component, single line replication, and the facades are neat and uniform without any characteristics. In fact, the so-called standardization is not as simple as single element single line replication. Throughout the facade of curtain wall system buildings, there is a lively and rich rhythm that greatly enriches the appearance of the building. These building facades are designed and constructed under the guidance of standardization, systematization, and modularization. Standardized modules can create rich architectural facades through design ideas such as multi line replication and nonlinear replication. It can be said that diversified design ideas promote technological innovation, and the research and development of new technologies and materials also provide infinite possibilities for design.

5   Conclusion

SAR theory has been proposed for more than half a century, and over a period of over 70 years, it has provided strong theoretical support for the research of prefabricated building technology. Even today, the prefabricated building technology in many countries has been highly developed, SAR theory still maintains its progressiveness.

This article is a summary and reflection of the author's research on relevant literature and years of experience in residential design engineering, based on SAR theory. It is hoped that it can serve as a reference for the development of prefabricated residential design technology in China.

References

1. Zhang, S.: The theory and methods of SAR [J]. Architect. J. **6**, 1–11 (1981)
2. Steinhardt, D., Manley, K., Bildsten, L., et al.: The structure of emergent prefabricated housing industries: a comparative case study of Australia and Sweden. Constr. Manag. Econ. **38**(1), 1–19 (2019)
3. Lv, X.: Exploring prefabricated building design beyond design. Intell. City **2**, 20–22 (2019)
4. Gan, S.: Research on Technology of Industrialized Housing Maintainable and Renewable. Southeast University (2019)
5. Yu, S., Liu, Y., Wang, D., et al.: Review of thermal and environmental performance of prefabricated buildings: implications to emission reductions in China. Renew. Sustain. Energy Rev. **137**, 110472 (2021)
6. Matsumura, S., Tomoyuki, G., et al. Technological developments of Japanese prefabricated housing in an early stage. Japan Architect. Rev. **2**(1), 52–61 (2019)
7. Cheng, Y., Jia, Y., Fu, W., et al.: Current issues and suggestions for sandwich composite panels of steel structure residence. Industr. Constr. **47**(7), 53–58 (2017)
8. Ma, H., Zhang, W., Dong, M.: Spatial conflict analysis and multi-objective optimization for prefabricated building hoisting construction. China Saf. Sci. J. **30**(2), 29–34 (2020)
9. Song, C., Huang, N.: Research on low energy consumption design of exterior wall of prefabricated house. Chinese Overseas Architect. **12**, 60–62 (2020)

Construction and Empirical Analysis of Influencing Factor Index of Prefabricated Construction Cost

Nana Liu, Cheng Chen, and Yun Gao[✉]

Chongqing College of Architecture and Technology, Chongqing, China
ldnana@springer.com

Abstract. The transformation of China's construction industry is in a period of rapid development, especially the rapid development of prefabricated buildings, but the lack of effective cost control and management, resulting in a substantial increase in its cost, has become a bottleneck restricting its development. Based on the analysis of the composition of the prefabricated construction cost, the paper makes use of the literature and case data to analyze, thus establishing the index system of the influencing factors of the prefabricated construction cost. On this basis, a questionnaire is designed, and SPSS software is used to make an empirical analysis of the influence of various factors on the cost of prefabricated buildings, and the corresponding cost control strategies are given from the perspectives of design, production, transportation, installation and other five aspects.

Keywords: Prefabricated building · Influencing factors · Cost control · Empirical analysis

1 Research Background

The prefabricated construction cost refers to the total cost generated during the whole construction process of the prefabricated building, which is mainly reflected in the sum of the costs generated by the design and production of prefabricated components, the transportation and on-site installation of finished components. The costs of prefabricated construction include: direct costs (labor costs, material costs, machinery costs), management costs, profits and taxes. The production cost, transportation cost, installation cost and related mold cost of prefabricated parts are important components [1].

Compared with cast-in-situ construction, there are four costs in the construction process: mold design cost, component manufacturing cost, component transportation cost and component construction cost. In summary, the cost of composite construction mainly includes four aspects: design cost, manufacturing cost, transportation cost and construction cost. Manufacturing costs include labor costs, materials costs, machinery costs, mold costs, transportation costs, storage costs and management fees for manufacturing components [2]. The mold cost is closely related to the specification, quantity and operation of the parts, which greatly increases the manufacturing cost of the parts. The installation cost of prefabricated parts mainly consists of the cost of labor, materials and machinery for the installation of prefabricated parts. As shown in Fig. 1.

© The Author(s) 2024
P. Xiang and L. Zuo (Eds.): PBSFTT 2023, LNCE 382, pp. 108–115, 2024.
https://doi.org/10.1007/978-981-97-5108-2_11

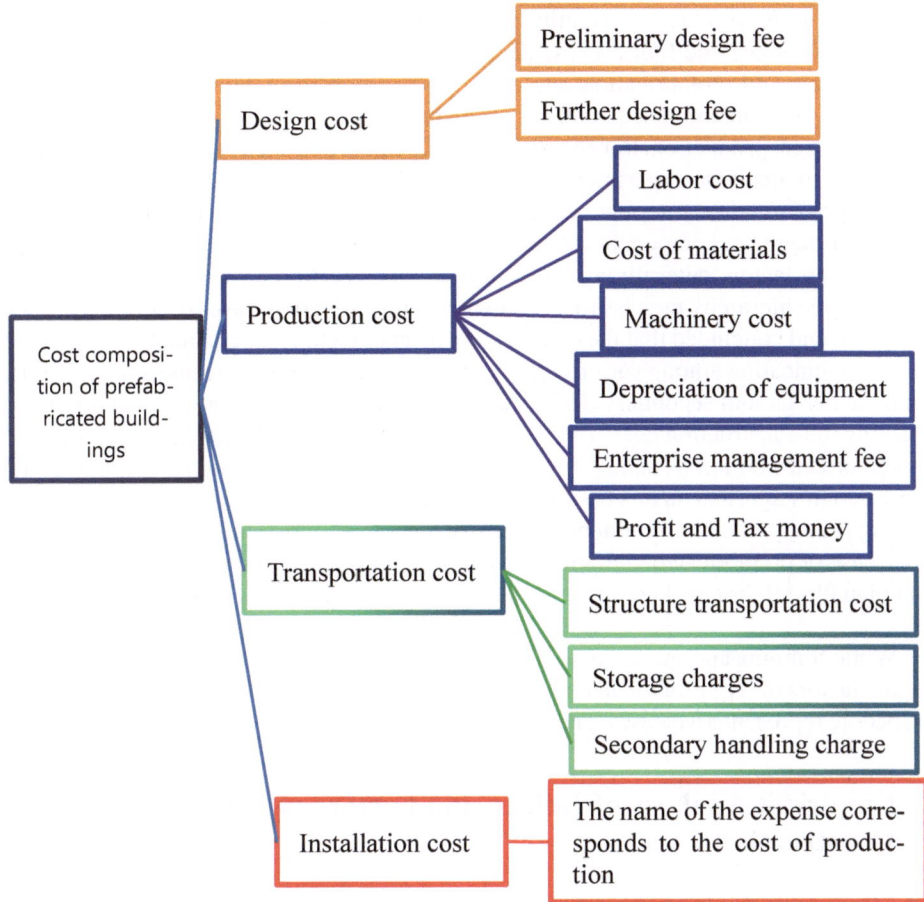

Fig. 1. Cost composition of prefabricated buildings

2 Analysis of Influencing Factors of Prefabricated Construction Cost

At present, many scholars in our country have made a deep discussion on it from multiple perspectives, multiple approaches and multiple examples [3]. For example, Zhao Liang et al. [4]. Studied the impact of assembled construction costs from four aspects: design, technology, management and policy. Through AHP method, they obtained the impact factors of construction costs, among which the most important ones are: prefabrication rate of construction, construction management, coordination among various types of work, reasonable splitting of components and turnover number. When Zhou Jingyang et al. [5]. Analyzed the cost of prefabricated buildings, they used the constructed explanatory structure model and found that these factors can be divided into five levels: The first is the surface influence factor (labor cost, tax, transportation distance, material market price, equipment), the second is the middle-level dynamic factor (safety inspection, production scale, production quantity of prefabricated components, production industrialization degree), the fourth is the deep guidance factor (construction site planning, assembly rate and prefabrication rate, construction plan formulation). Zhu Ying et al. first investigated the influence of all parties on the composite construction cost from five aspects: design, production, transportation, management and society, and then conducted a study using decision experiment and evaluation laboratory methods, and concluded that the prefabrication rate, the degree of component disassembly, transportation efficiency, the degree of component standardization and the efficiency of resource allocation are the main factors determining the composite construction cost. Zhao Weishu et al. used analytic hierarchy process to analyze the factors affecting the cost of prefabricated buildings, and concluded that the construction level of technical personnel, coordination and communication among various types of work, prefabrication rate and assembly rate, template design and secondary design, and mold standardization are the main reasons affecting the construction cost. Yao Weitao et al. identified the factors affecting the cost of prefabricated buildings mainly from the three perspectives of design, procurement and construction stage, and on this basis established a structural model based on system engineering. Finally, the main influencing factors such as prefabrication rate and assembly rate, reuse of prefabricated components, splitting of prefabricated components, production plan of prefabricated components and production management mode are analyzed [6].

While combing and analyzing the literature related to prefabricated buildings, the cost impact factors of 34 prefabricated buildings were determined from five levels according to specific representative cases, and the investigation was conducted accordingly [7].

3 Empirical Analysis of Influencing Factors of Prefabricated Construction Cost

3.1 Questionnaire Design

This paper analyzes the main factors affecting the construction cost of the combination from the angles of design, manufacturing, transportation and installation. Because latent variables cannot be directly observed, they can only depend on their observed data The questionnaire uses Likert's 5-level scoring method to evaluate the impact factors on the combined construction cost, and divides them into 5 levels according to their importance, and the respondents score each factor according to their own experience.

3.2 Questionnaire Distribution

Through QQ, wechat and face-to-face intervicategoryews, questionnaires were conducted among people involved in assembly construction, researchers studying and working in universities, and people working in government departments. The research objects include universities, government departments, construction enterprises, construction enterprises, consulting institutions, component manufacturing enterprises, etc.

3.3 Questionnaire Collection and Sorting

A total of 100 questionnaires were issued in this survey, and 97 valid questionnaires were actually recovered. The statistical results of the survey objects are shown in Table 1.

Table 1. Collect questionnaire survey statistics

Category		Number of people	Scale (%)	Category		Number of people	Scale (%)
Age stage	age 25 and under	49	50.5	Educational level	High school and below	2	2.1
	26-35 years old	13	13.4		Junior college	64	66
	36-45 years old	8	8.2		Undergraduate course	10	10.3
	46-55 years old	14	14.4		master	17	17.5
	56-65 years old	6	6.2		Learned scholar	4	4.1
	over 66 years old	7	7.2	Work unit	Colleges and universities	43	44.3
Time spent on the job	Less than 3 years	45	46.4		Government department	11	11.3
	3-5 years	23	23.7		Construction unit	11	11.3
	6-10 years	14	14.4		Construction unit	11	11.3
	11-20 years	14	14.4		Consulting unit	10	10.3
	More than 21 years	1	1.0		Component manufacturing unit	9	9.3
					Others	2	2.1

3.4 Reliability Test

Klonbach formula is used to test its reliability [8]. In the measurement result, if a is less than 0.6, it is generally regarded as not meeting the measurement standard; The results show that the scale has good reliability in the range of 0.7–0.8. The A-value range is 0.8–0.9, indicating that the scale has good reliability. Among the 97 valid questionnaires, 34 indicators were tested for reliability, and the reliability value of each dimension was more than 0.8, which is sufficient to indicate that the reliability of the questionnaire is good and the reliability of the data of all dimensions is good (Table 2). Through the reliability test of the questionnaire, it can be seen that the questions in the questionnaire have high reliability, and the internal consistency among the influencing factors meets the demand.

$$\alpha = \frac{N \cdot \bar{c}}{\bar{v} + (N-1) \cdot \bar{c}} \tag{1}$$

Here N is equal to the number of items, c-bar is the average inter-item covariance among the items and v-bar equals the average variance.

Table 2. Reliability verification of variables

Influencing factor	Klonbach Alpha	Number of terms
Design factor	0.91	7
Factor of production	0.9	9
Transportation factor	0.84	6
Installation factor	0.85	6
Other factors	0.84	6
subtotal	0.97	34

3.5 Content Validity Test

KMO and Bartlett's test were used to carry out KMO and Bartlett's test on the survey data [9], and the KMO value obtained was 0.899, higher than 0.8. The approximate chi-square value of Bartlett's test was 3032.927, and the degree of freedom was 561. And the significance is less than 0.01, which indicates that the questionnaire data of influencing factors of prefabricated construction cost is applicable, and it is suitable for factor analysis.

$$KMO = \frac{\sum\sum_{i \neq j} r_{ij}^2}{\sum\sum_{i \neq j} r_{ij}^2 + \sum\sum_{i \neq j} \alpha_{ij}^2} \tag{2}$$

Here r_{ij} stands for simple correlation coefficient and $a^2_{ij.1,2,3,...k}$ stands for partial correlation coefficient. When the KMO is greater than 0.9, it indicates that the correlation between variables is very appropriate, and 0.8 to 0.9 indicates that the correlation between variables is very appropriate.

As can be seen from the rubble chart (Fig. 2), the broken line gradually flattens after 5, while it shows a sharp decline before that, indicating that it is more appropriate to extract 4 common factors from 34 indicators.

3.6 Structural Validity Test

According to the rotated component matrix. The factor load value of each variable is greater than 0.4, so these 34 variables are useful.

The validity analysis proves that the data analysis is effective. The observed values are consistent with the actual situation, and the internal stability of the questionnaire has been guaranteed to a certain extent. The content of the questionnaire is well-targeted and relatively comprehensive, and the indicators have independent significance, which indicates the reliability and credibility of the questionnaire.

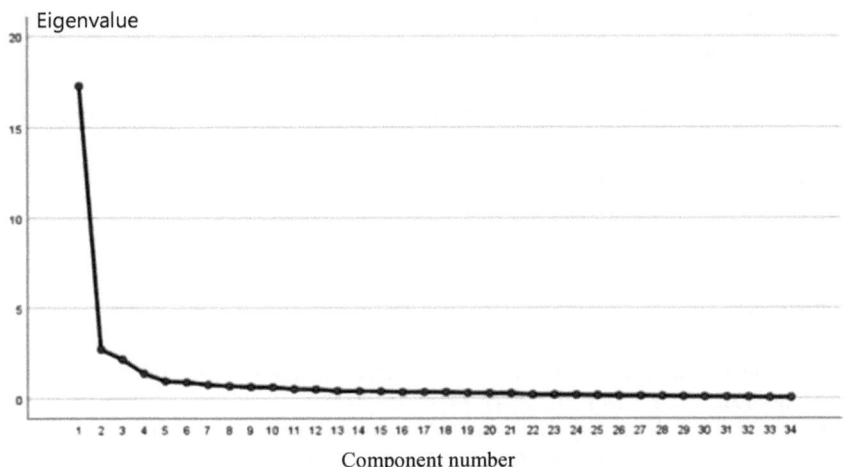

Component number

Fig. 2. Lithotriptic plan

3.7 Correlation Test

On this basis, through the design factors, production factors, transportation factors, installation factors and other external factors and the correlation between the cost control effect of prefabricated buildings [10], and then the cost control effect of these factors on prefabricated buildings is deeply studied. These results are shown in Table 3.

Table 3. Correlation test

Correlation analysis index	Cost control effect of prefabricated building	Design factor	Factor of production	Transportation factor	Installation factor	Other factors
Pearson correlation	1	0.89	0.76	0.73	0.78	0.72
number	97	97	97	97	97	97

4 Conclusion

By analyzing the relevant test results of the survey samples, the following conclusions can be drawn:

(1) the influence coefficient of design factors, manufacturing factors, transportation factors, installation factors and other external factors on the assembly construction cost is below 0.05, indicating good significance, and the five factors have a very important role in the control of the assembly construction cost.

(2) design factors, manufacturing factors, transportation factors, installation factors and other external factors will have a positive impact on the control of combined construction costs.

(3) the importance of the combined construction cost is sorted as follows: design cost > Construction cost > Manufacturing cost > Transportation cost > Other external costs.

References

1. Quanzhong, L., Lihuan, W., Yikai, Z.: Adaptability evaluation of EPC model for prefabricated buildings. Eng. Constr. Des. **2022**(24), 241–243 (2022)
2. Sun, Y.: Research on quality management of prefabricated construction projects from the perspective of the whole process. Qingdao University of Technology (2022)
3. Ranran, Z.: Research on quality risk assessment of prefabricated buildings under EPC mode. Qingdao University of Technology (2022)
4. Liang, Z., Quqiang, H.: Study on evaluation of influencing factors of prefabricated construction cost. Construction Economics (2018)
5. Jingyang, Z., Pengwang, H.: Analysis of factors affecting cost of prefabricated buildings based on Interpretive structural Model (ISM). J. Eng. Manag. (2019)
6. Tong, S.: Research on the effectiveness of prefabricated building Supply chain Management Model. Jilin Jianzhu University (2022)
7. Zheng, M.: Research on Regional development Level Evaluation and improvement of prefabricated construction industry. Harbin Institute of Technology (2022)
8. Wang, Q.: Research on quality risk management of prefabricated buildings based on fuzzy Bayesian network. North China University of Technology (2021)
9. Tiantian, J.: Research on safety risk evaluation and control countermeasures of hoisting operation of prefabricated buildings. North China University of Technology (2021)
10. Chunling, Z., Mengqing, Z., Mo, L., et al.: Evaluation of development degree of prefabricated buildings: Based on AHP and fuzzy comprehensive evaluation. J. Jilin Jianzhu Univ. **37**(06), 57–63+76 (2019)

Research on Whole-Process Cost Management of Prefabricated Building Based on BIM Technology

Zhen Wen, Hanxi Zhang$^{(\boxtimes)}$, and Taiyu Fu

Chongqing College of Architecture and Technology, Chongqing 401331, China
zhanghanxi325@163.com

Abstract. Prefabricated building is the inevitable trend of the development of China's construction industry. However, the problem of high construction cost of prefabricated construction projects still exists, which is also the main problem that hinders the rapid development of prefabricated construction. Because the prefabricated construction and installation needs the cooperation of all parties. The combination of BIM engineering technology design concept and assembly type is an inevitable trend of future development. This paper systematically analyzes the current situation of cost management in the whole process of prefabricated building construction, and makes use of BIM technology to fully share relevant information of all participants, and then puts forward corresponding suggestions to solve the problem of high cost management of prefabricated building.

Keywords: Prefabricated building · BIM technology · Cost management

1 Introduction

In recent years, China's economic level has improved rapidly. By the end of 2022, China's urbanization has reached 63.89%, and it is estimated that it will increase to 75% by 2030. With the increase of urbanization rate, the noise, dust, garbage and other factors of traditional construction seriously affect the surrounding environment, which is not in line with the current green development policy implemented in China [1]. However, the prefabricated construction project has the advantages of shorter construction cycle, labor saving, improved building quality and less environmental impact, which conforms to the green development concept promoted in China at the present stage [2]. Prefabricated building has become a new target for the development of China's construction industry. At present, there are still many obstacles to the development of prefabricated buildings in China, such as immature technology, small scale, poor communication and coordination among various units, imperfect standards and policies, especially high construction costs [3]. As an engineering information management technology, BIM can reduce the cost of assembly construction through the management and transmission of information

---

Z. Wen and H. Zhang—Contributed equally to this work and should be considered co-first authors.

P. Xiang and L. Zuo (Eds.): PBSFTT 2023, LNCE 382, pp. 116–125, 2024.
https://doi.org/10.1007/978-981-97-5108-2_12

system. The design, component processing, construction and installation information of prefabricated buildings is refined, and the engineering information system at different stages is systematically managed, [4]. In addition, it can also carry out the cost management of prefabricated construction projects, according to the standardized design of components, manufacturing and large-scale construction. The combination of BIM technology and prefabricated building technology makes cost control more adaptable and accurate [5]. This paper will take the prefabricated construction process as the carrier to analyze the causes of the high cost of prefabricated construction, so as to further explore the use of BIM technology for the whole process cost control of construction projects.

2 BIM Technology for Prefabricated Buildings

2.1 BIM Technology

BIM information technology has the advantages of digitalization, informatization, standardization, collaboration, visualization and simulation functions, which facilitates effective information communication, improves construction speed, and thus reduces construction costs. The application content of BIM mainly involves related aspects such as joint cooperation, product design, virtual construction, construction control, cost control and later operation and maintenance management [6].

Through the use of BIM technology, all participating units facilitate the exchange of project information and resource sharing in the whole process of construction. The relationship between BIM technology and all participating units is shown in Fig. 1 [7]. Use BIM technology to explore potential risk factors and possibilities in the design and construction simulation process. Simulate the actual construction process, so as to timely find problems and make correct evaluation, give specific countermeasures and preventive measures, propose optimization measures to guide the actual construction, and reasonably control project materials, time and cost [8].

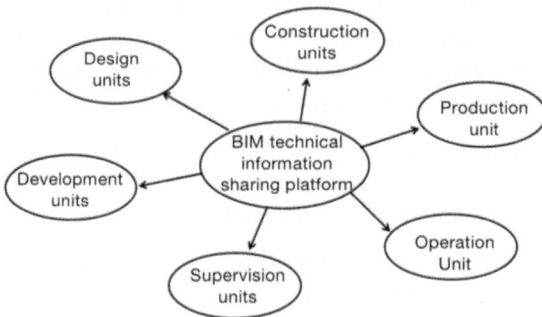

Fig. 1. BIM technical information sharing platform

2.2 The Development Status of Prefabricated Buildings

The prefabricated building model started earlier in developed countries. At present, the related theories and management systems of prefabricated buildings are more mature. China's prefabricated construction industry is still in the initial stage of development, and a big gap exists compared with developed countries. Many factors restrict the rapid development of prefabricated construction industry in our country, among which lack of standards, industrial dispersion, small scale and high cost are all existing problems [9]. As the core factor restricting the development of prefabricated building projects, cost-related issues have received extensive attention. Although the state has put forward many favorable policies for the prefabricated construction industry, high construction cost is still one of the main problems restricting the development of the prefabricated construction industry [10]. The design cost, production cost, transportation cost, construction and installation cost of prefabricated construction project are the key to cost management.

3 Analysis of Influencing Factors of Prefabricated Building Cost by BIM Technology

To determine the influencing factors of the cost management of prefabricated building based on BIM technology. Through literature research, combined with the whole process cost management of prefabricated building with BIM technology, this paper finally determines 22 influencing factors in 5 stages (shown in Table 1) that affect the cost of prefabricated building with BIM technology:

Planning and design stage (PD stage):   BIM collaborative design degree (PD1); BIM design standard perfection (PD2); The degree of component splitting design (PD3); Number of design changes (PD4); Assembly rate (PD5); Lack of BIM technical personnel(PD6); (shown in Table 2).

Production and transportation stage (PT stage):   scale and production capacity of component factory (PT1); Types of prefabricated components (PT2); Production quality of prefabricated components (PT3); Mold utilization rate (PT4); Platform degree of access to information(PT5); Component transportation scheme (PT6); (shown in Table 3).

Construction and installation stage (CI stage):   mechanical equipment lifting scheme (CI1); Construction and installation efficiency (CI2); Degree of application of BIM technology (CI3); Field component management (CI4); (shown in Table 4).

Operation and maintenance stage (OM stage):   maintenance of subsidiary works (OM1); Facility maintenance (OM2); Property information management (M3); (shown in Table 5).

The demolition recycling stage (DR stage):   Demolition scheme (DR1); Demolition efficiency (DR2); Component recovery rate (DR3). (shown in Table 6).

In view of the above factors affecting the whole process cost management of BIM prefabricated buildings, the relevant professionals were investigated. The following data is obtained. According to the survey results, AHP hierarchical analysis software is used to complete the formation of the evaluation matrix, and the following table is formed.

Table 1. Project whole-process judgment matrix

Whole-process	PD stage	PT stage	CI stage	OM stage	DR stage
PD stage	1	0.25	0.5	3	4
PT stage	4	1	0.5	4	5
CI stage	2	2	1	3	0.33
OM stage	0.33	0.25	0.33	1	0.20
DR stage	0.25	0.2	3	5	1

Table 2. Planning and design stage judgment matrix

PD stage	PD1	PD2	DPD3	PD4	PD5	PD6
PD1	1	0.2	0.33	0.5	2	0.25
PD2	5	1	4	0.5	0.25	0.20
PD3	3	0.25	1	0.33	3	0.50
PD4	2	2	3	1	0.5	0.33
PD5	0.50	4	0.33	2	1	4
PD6	4	5	2	3	0.25	1

Table 3. Production and transportation stage judgment matrix

PT stage	PT1	PT2	PT3	PT4	PT5	PT6
PT1	1	0.2	4	5	3	0.20
PT2	5	1	4	4	4	3
PT3	0.25	0.25	1	2	2	0.25
PT4	0.2	0.25	0.5	1	0.25	0.33
PT5	0.33	0.25	0.5	4	1	0.20
PT6	5	0.33	4	3	5	1

Table 4. Construction and installation stage judgment matrix

CI stage	CI1	CI2	CI3	CI4
CI1	1	0.25	0.25	0.20
CI2	4	1	0.33	0.33
CI3	4	3	1	4
CI4	5	3	0.25	1

Table 5. Construction and installation stage judgment matrix

OM stage	OM1	OM2	OM3
OM1	1	0.33	0.5
OM2	3	1	0.2
OM3	2	5	1

Table 6. Demolition recycling stage judgment matrix

DR stage	DR1	DR2	DR3
DR1	1	0.25	0.33
DR2	4	1	0.50
DR3	3	2	1

$\lambda_{\max} = \sum_{i=1}^{n} \frac{(AW)_i}{n w_i}$ is obtained by calculating formula $\overline{w_i} = \sqrt[n]{\prod_{j=1}^{n} a_{ij}}$. The analytic hierarchy process is used to screen each influencing factor, and the 22 influencing factors affecting the cost of prefabricated building with BIM technology are sorted, and the following results are obtained (Table 7):

Table 7. Weight table of influencing factors

Stage layer	Weight	Factor layer	Weight
Planning and design stage	0.2577	BIM collaborative design degree	0.0344
		BIM design standard perfection	0.0440
		Degree of component splitting	0.0438
		design	0.0450
		Number of design changes	0.0494
		Assembly rate	0.0412
		Lack of BIM technical personnel	
Production and transportation stage	0.2849	Scale and production capacity of	0.0499
		component factory	0.0472
		Types of prefabricated components	0.0472
		Production quality of prefabricated	0.0447
		components	0.0476
		Mold utilization rate	0.0484
		Platform degree of access to	
		information	
		Component transportation scheme	
Construction and installation stage	0.1892	Mechanical equipment lifting	0.0488
		scheme	0.0514
		Construction and installation	0.0424
		efficiency	0.0466
		Degree of application of BIM	
		technology	
		Field component management	
Operation and maintenance stage	0.1252	Maintenance of subsidiary works	0.0433
		Facility maintenance	0.0433
		Property information management	0.0386
Demolition recycling stage	0.1430	Demolition scheme	0.0478
		Demolition efficiency	0.0472
		Component recovery rate	0.0480

4 Application of BIM Technology in the Whole Process Cost Management of Prefabricated Buildings

4.1 Cost Control in the Planning and Design Stage of BIM Technology

Establish Component Family Library. By using the highly collaborative and drawing optimization characteristics of BIM technology, a standardized component family library is established, which enhances the standardization and universality of components, and realizes the reuse of components. At the same time, the reduction of the apportion cost of mold production greatly reduces the production cost of components.

Collaborative Design. Each specialty collaborates on the same network platform to share information, which facilitates communication between the participating units, thus reducing cross-specialty communication time and facilitating timely discovery of contradictions in product design, thus reducing unnecessary economic losses caused by later modifications.

Detailed Design. BIM software is used to complete the component splitting design diagram. Match the visual function of BIM technology, component processing plant can directly use BIM model of split design drawings information standardization production, reducing the error rework probability, which reduces the cost.

4.2   Cost Control in the Production and Transportation Stage of BIM Technology

Improve the Production Quality of Components.   The use of BIM technology can improve the accuracy of component manufacturing. Design changes can also be modified in the model using BIM technology. The information is more accurately and efficiently communicated to the manufacturing personnel, thus reducing the waste of production costs caused by unnecessary errors.

Improve the Utilization Rate of Molds.   According to incomplete statistics, the amortization cost of the mold accounts for about 10% of the entire engineering component processing cost. The use of BIM technology can synchronize the design drawings, optimize and classify various component parameters, and improve the utilization rate of molds. As a result, the production quantity and amortization cost of the mold are reduced, and the cost is effectively controlled.

Optimize the Transportation Plan.   Compared with traditional cast-in-place concrete construction, the cost of transporting components from the processing plant to the construction site is significantly increased, which is also a key link to reduce transportation costs. Through simulation technology, the optimal type and number of transport vehicles are selected to reduce the number of transport. The mileage, traffic quality, road conditions and other information of different transportation routes are input into the BIM technology information management platform, and the BIM technology is used to simulate the transportation process and select the best transportation route, so as to improve transportation efficiency. In addition, BIM simulation function is used to simulate the installation sequence of components, and a transportation route that can be effectively combined with on-site installation and construction is established to increase the full load and reduce the cost.

4.3   Cost Control in the Construction and Installation Phase of BIM Technology

Improve the Efficiency of Construction and Installation.   The construction organization scheme of prefabricated buildings is simulated by using BIM technology to simulate the advantages of construction. In the process, the construction organization plan and layout plan were optimized. Set up places properly such as tower cranes and construction components, optimize temporary construction road, vehicle transport road route. Minimize the secondary handling of building materials and machinery, reduce the production cost of the project in the construction process, and improve the hoisting efficiency and installation efficiency of construction machinery.

Optimize Component Management. BIM technology determines the pre-construction plan and the phased purchase quantity of components according to the on-site data analysis. According to the actual needs of construction site progress, the requirement of prefabricated components in each installation and construction stage can be quickly calculated. Thus, the practical problems such as secondary handling of components and material accumulation exceeding the limit are solved. In addition, if the construction plan needs to be changed during the construction process, BIM technology can also be used to detect the construction status of the site, so as to effectively adjust the approach arrangement of the components to meet the needs of the installation of various parts.

Improve the Installation Technology Level. During the construction and installation of prefabricated buildings, the workload of vertical transportation is relatively large. Currently, it is faced with the problem that the construction and installation technicians lack experience and can not fully grasp and skillfully use the lifting skills of prefabricated components. After the disclosure of 3D visualization construction technology, the visualization characteristics of BIM technology are used to simulate construction and dynamic demonstration of component layout. Construction and installation project management personnel can quickly master the construction technology, understand the technical difficulties and key nodes in construction and installation in advance, and formulate corresponding measures in time. It greatly improves the construction efficiency of construction engineering and the accuracy of installation management, and effectively reduces the cost of construction engineering in the construction process.

4.4 Cost Control in the Operation and Maintenance Stage of BIM Technology

Maintenance of Subsidiary Works. The use of BIM technology can more directly view the type of materials used in the damaged location, specifications and sizes, steel bar diameter, and the relevant information of its suppliers, making the maintenance process more simple and effective, greatly improving the service level in the operation and maintenance stage, thereby reducing unnecessary cost losses.

Property Information Management. BIM technology is used to establish BIM property management system, so that managers can grasp the latest equipment and facilities operation status in the first time. Moreover, by increasing the number of equipment and facilities for the property or operator, the management unit can also meet the operation needs with the optimal cost management on the premise of ensuring the quality and use efficiency of the prefabricated building.

4.5 Cost Control in the Operation and Maintenance Stage of BIM Technology

Optimize the Demolition Scheme. At present, due to the lack of demolition and recycling in domestic prefabricated construction projects, the experience of demolition and recycling is not perfect. At the same time, the level and technology of demolition of traditional cast-in-place concrete buildings in China are also relatively backward, and the formulation of demolition scheme at this stage is the difficulty of construction. However, the input of the demolition plan of the prefabricated building into the BIM model can improve the database of the whole process of the prefabricated building, which provide a certain reference for the formulation of the demolition and recycling plan in the later stage.

Improve the Demolition Efficiency. The structure of prefabricated buildings is more complex than that of traditional cast-in-place buildings, so the difficulty of demolition is also greater than that of traditional buildings. If only this simple and rough demolition method of machinery is used, it will not only cost more, but also take a longer time. The actual demolition work is input into the BIM model, and the simulated construction can be carried out in the model first. This can predict various risks in demolition and solve them in advance, significantly improve the efficiency of demolition, and reduce the cost of demolition, thus minimizing economic losses.

Improve the Recycling Rate of Components. When BIM technology is used to dismantle prefabricated buildings, the BIM model can sort out and classify recyclable components, realize resource recovery and utilization, and improve the rate of return.

5   Conclusions

The high construction cost of prefabricated buildings is one of the key factors that hinder the promotion of prefabricated buildings. At the same time, due to the development of science and technology and the change of management mode, China's construction industry is also constantly transforming and upgrading, towards the development goal of environmental protection, industrialization and information technology. In this paper, the whole process cost management of prefabricated building based on BIM technology is deeply studied, the cost management effect of BIM technology in the whole process of prefabricated building is analyzed, and the cost management methods and suggestions for the five stages of prefabricated building cost management are given.

References

1. Jing, Z.: Research on cost management of prefabricated building implementation stage based on BIM technology. Jilin: Master Thesis, Jilin Jianzhu University(2019)
2. Xinhua News Agency. Several Opinions of the Central Committee of the Communist Party of China and The State Council on Further Strengthening the Administration of Urban Planning and Construction. People's Daily, 22 February 2016 (24)
3. Jingsi, J.: Research on Whole process cost control of prefabricated buildings based on BIM. Hubei: Master Thesis, Wuhan University of Science and Technology (2021)
4. Wang, D.T.: Analysis and application of BIM technology in the project goal control. Adv. Mater. Res. **35**(25), 2978–2981(2013)
5. Hwang, B.G., Shan, M., Looi, K.Y.: Knowledge-based decision support system for prefabricated prefinished volumetric construction. Autom. Constr. **94**, 168–178 (2018)
6. Santos, R., Costa, A.A., Silvestre, J.D., el at.: BIM-based life cycle assessment and life cycle costing of an office building in Western Europe. Build. Environ. **2**(05), 147–150 (2020)
7. Danmei, L.: Application of BIM technology in prefabricated building design. Brick-Tile **18**(05), 160–161 (2020)

8. Yizhi, T.: Research on BIM based Intelligent Site Management. Val. Eng. **39**(01), 102–104 (2020)
9. Min, Z., Shaojun, Z., Jing, S., Zichang, L.: Research on cost control of prefabricated concrete construction based on BIM technology. Val. Eng. **38**(30), 268–270 (2019)
10. Kai, L., Xiaoxin, D., Ao, D., Jipeng, Z., Qianyi, X.: Influencing factors of prefabricated building cost based on whole life cycle. J. Neijiang Normal Univ. **34**(12), 51–56 (2019)

Urban Development Potential Evaluation of Prefabricated Buildings Based on PCA-TOPSIS Model

Qili Gan[1], Langhua Li[1(✉)], Qiming Huang[2], and Wanqing Chen[1]

[1] Chongqing College of Architecture and Technology, Chongqing, China
304245657@qq.com
[2] Chongqing Construction Cesidential Cngineering, Chongqing, China

Abstract. In recent years, with the strong support of the Chinese government, prefabricated buildings have entered a period of rapid development. Prefabricated buildings occupy an important position in the construction industry in China. However, in the process of the implementation of prefabricated buildings, it is found that each region has different economic, technological and social levels, and each city has different development strategic goals and related supporting policies. The development potential of prefabricated buildings is very different, the development process is different, and the potential needs to be tapped. Based on the literature research method, this paper selects 25 evaluation indicators from four aspects of policy, economy, technology and market, adopts the principal component weighted TOPSIS evaluation method to construct a comprehensive evaluation model of the development potential of prefabricated buildings in cities, and makes a comprehensive evaluation of the development potential of prefabricated buildings in cities, so as to help cities understand their own development potential of prefabricated buildings. Develop prefabricated building development strategy to provide reference.

Keywords: Prefabricated Building · Development Potential · Topsis Evaluation Method

1 Introduction

Prefabricated building in foreign countries mainly from the beginning of the 20th century began to attract people's attention, to the 1960s in the United Kingdom, France, the Soviet Union and other countries tóok the lead in the attempt to finally achieve [1]. Scholar Buncio believes that the current construction industry is facing colliding forces that make our current trajectory wasteful, inefficient and unsustainable, which requires us to move into the era of prefabricated and modular buildings [2]. As a pillar industry of the country, the steady and sustainable development of the construction industry is crucial to the progress of the whole country [3]. Experts predict that the total output value of prefabricated and modular buildings will reach $209 billion by the end of the decade, with huge development potential [4].

© The Author(s) 2024
P. Xiang and L. Zuo (Eds.): PBSFTT 2023, LNCE 382, pp. 126–133, 2024.
https://doi.org/10.1007/978-981-97-5108-2_13

Since 2015, the new prefabricated construction area, construction industry scale, and industrial output value related to assembly have continued to rise, carrying out the process of urbanization at a high speed [5]. Zheng Minli believes that China's prefabricated buildings are in a stage of steady advancement, continuous enhancement of the endogenous power of the industry, gradual establishment of the policy support system and preliminary establishment of the technical support system [6]. Zhao Likun believes that the prefabricated building has the greatest potential in southwest China, the development of prefabricated building in East China is relatively stable, and the development potential in South China is the least [7]. Li Qiangnian, Liu Xia et al. use the literature research method to draw a conclusion that the main constraints of prefabricated buildings in China are summarized in six aspects: technical factors, market environment factors, industrial organization factors, cost factors, policy factors and talent factors [8, 9].

There are many methods to evaluate the development potential of prefabricated buildings, and with the deepening of the research, the research methods have been developed from qualitative description to quantitative data analysis stage, interpretive structure model (ISM), analytic hierarchy process (AHP), fuzzy comprehensive evaluation method, SWOT analysis method, etc., are widely used in the development analysis of prefabricated buildings [10]. On the basis of previous studies, this paper constructs a comprehensive evaluation model of the development potential of prefabricated buildings by applying the principal component weighted TOPSIS evaluation method, and comprehensively evaluates the development potential of prefabricated buildings in the city.

2 Design of Development Potential Evaluation Index of Prefabricated Buildings

2.1 Design Principles of Index System

1. Comprehensiveness. Based on relevant theories, the selected indicators will cover the main factors affecting the development potential of prefabricated buildings in the city.
2. Simplicity. The complex problem is simplified, and the indicator system selected can reflect the development potential of prefabricated buildings in a concise and clear way.
3. Scientific practicability. In the process of evaluation, the data of different regions or the data of the same region at different times can be comprehensively compared, and the whole index system can be used in the whole country.
4. Maneuverability. The selected indicators have the corresponding data sources in different cities at different times, and the analysis data are from the official data released in various periods in various places, and the data sources are real and reliable.

2.2 Evaluation Index Design

According to the basic principles of index selection from the four aspects of economy, technology, policy and market, this paper selects the appropriate evaluation index and establishes the development potential evaluation system of prefabricated buildings.

Table 1. Comprehensive evaluation index system of development potential of prefabricated buildings

Indicator class (level 1)	Indicator subclass (level ii)	Index item (level iii)
Economics	Economic development level	Per capita gdp (x1)
		Per capita local fiscal revenue (x2)
	Residents' living standards	Urban per capita disposable income (x3)
		Consumption level of urban residents (x4)
	Economic structure	Proportion of fiscal expenditure in gdp (x5)
		Proportion of tertiary industry in gdp (x6)
Technology	Technological innovation fund input	Research and experimental development expenditure per capita (x7)
		Proportion of science and technology expenditure in local gdp (x8)
		Proportion of local fiscal expenditure on science and technology in local fiscal expenditure (x9)
	Technical talents cultivation of	Number of engineering technology and scientific research professionals in local enterprises and institutions (x10)
		Research and experimental development equivalent number of full-time personnel (x11)
		Number of engineering students in colleges and universities (x12)
	Technology research and development	Number of research and development institutions (x13)
		Number of invention patents (x14)
		Number of prefabricated construction industrial bases (x15)
	Relevant standards	Number of relevant standards, specifications and atlas (x16)

(*continued*)

Table 1. (*continued*)

Indicator class (level 1)	Indicator subclass (level ii)	Index item (level iii)
Policy		Whether there are government promotion measures (x17)
		Whether to propose construction objectives and tasks (x18)
		Whether there are government support and incentive policies (x19)
		Whether there is a government supervision mechanism (x20)
Market	Construction supply and demand	Per capita investment in real estate development (x21)
		Per capita real estate construction housing area (x22)
		Per capita real estate transaction area (x23)
	Related enterprises	Number of real estate development enterprises (x24)
		Production base of assembled parts (x25)

3   Evaluation Model of the Development Potential of Prefabricated Buildings

3.1   Selection of Evaluation Methods

At present, the commonly used weighting methods are mainly divided into subjective weighting method, objective weighting method and combination weighting method. Subjective weighting is generally used in evaluations where data collection is difficult or information cannot be directly quantified. Objective weighting rule refers to a kind of method that obtains weights after analyzing and processing each index data by mathematical statistics, but the calculated results are greatly affected by the sample data. The combination weighting method refers to the recombination of the results calculated by the previous two methods. In the comprehensive evaluation index system established in this paper, each index can be quantified, so the objective weighting method is considered. In the objective weighting method, the principal component analysis method can simplify complex indicators and is more suitable for the comprehensive evaluation index system of the development potential of prefabricated buildings established in this paper, so the principal component analysis method is chosen.

3.2 Evaluation Steps of Weighted Principal Component TOPSIS Evaluation Model

1. Suppose a region is evaluated and ranked for its development potential of prefabricated buildings in m years, in which there are a second-level indicators and n third-level indicators under each classification index, then the original data matrix can be formed according to the characteristics of the indicators:

$$
Xa = \begin{pmatrix}
x_{11} & x_{12} & \cdots & x_{1n} \\
x_{21} & x_{22} & \cdots & x_{2n} \\
\vdots & \vdots & \vdots & \vdots \\
x_{i1} & \cdots & x_{ij} & \cdots \\
\vdots & \vdots & \vdots & \vdots \\
x_{m1} & x_{m2} & \cdots & x_{mn}
\end{pmatrix} \tag{1}
$$

2. The data in the original matrix are treated with the same trend and dimensionless to form a standard data matrix. And calculate the sample correlation coefficient matrix of the standardized matrix Za, namely:

$$
R_a = (r_{ij})_{n \times n} = \frac{Z_a^T Z_a}{m-1} \tag{2}
$$

3. Determine the number P of principal components. Generally, the cumulative contribution rate of the principal component is greater than 85%, which means that the extracted principal component can contain more than 85% of the information of the original variable, namely:

$$
\sum_{j=1}^{p} \frac{\lambda_j}{\sum_{j=1}^{n} \lambda_j} \geq 85\% \tag{3}
$$

4. Calculate the Euclianian distance value between vij and positive and negative ideal solutions under the second-level index of item a in the comprehensive evaluation index system of prefabricated building development potential for each participating year or city, Namely:

$$
S_{ai}^+ = \sqrt{\sum_{j=1}^{p} (v_{ij} - v_j^+)^2} (i = 1, 2, \cdots, m) \tag{4}
$$

$$
S_{ai}^- = \sqrt{\sum_{j=1}^{p} (v_{ij} - v_j^-)^2} (i = 1, 2, \cdots, m) \tag{5}
$$

5. Calculate the relative proximity of the secondary index of item a of the comprehensive evaluation index system for the development potential of prefabricated buildings in each participating year. Namely

$$S_{ai} = \frac{S_{ai}^-}{S_{ai}^+ + S_{ai}^-}(i = 1, 2, \cdots, m)(0 \le S_{ai} \le 1) \tag{6}$$

6. According to the relative proximity calculated by the above formula, the second index of item a in the comprehensive index system for the development potential of prefabricated buildings is sorted. The greater the value, the greater the development potential of prefabricated buildings in the year or in the region.

3.3 Evaluating the Results of Empirical Analysis of the Model

According to the above steps, based on the comprehensive evaluation index system of the development potential of prefabricated buildings in Table 1, the original data analysis results of the development potential of prefabricated buildings in several typical cities of Chongqing, Beijing, Guangdong, Henan, Sichuan, Liaoning and Hubei in 2021 are selected, as shown in Table 2.

Table 2. Comparison results of the overall market evaluation of the development potential of prefabricated buildings in Chong Qing, Beijing and other regions in 2021

The District	Positive ideal solution distance D +	Negative ideal solution distance D-	Relative proximity C	Sort results
Chongqing	5.82	5.475	0.485	5
Peking	2.796	10.288	0.786	1
Canton	4.718	6.407	0.576	3
Henan	6.062	5.03	0.453	6
Szechuan	4.679	6.588	0.585	2
Liaoning	5.208	6.019	0.536	4
Hubei	9.864	3.618	0.268	7

As can be seen from Table 2, Beijing has the greatest development potential for prefabricated buildings, followed by Sichuan, Guangdong, Liaoning, Chongqing, Henan and Hubei. Mainly, Beijing, as the first batch of demonstration cities of prefabricated buildings, has perfect government policies, synchronous market forces, and takes many measures to develop prefabricated buildings, and has good economic and technical conditions, a good market environment, and strict implementation of the whole process supervision, which promotes the high-quality implementation of prefabricated building projects. Other cities are mainly the policy of mandatory development measures, regulatory systems and other late release, resulting in the entire market backward development; Secondly, the development of technology is far behind Beijing, Guangzhou and other cities, and it is necessary to continue to work hard to achieve balanced development and catch up with the development pace of well-developed areas.

4 Conclusions

According to the evaluation results, on the basis of giving full play to their existing market and industrial advantages, each city can promote the vigorous development of local prefabricated buildings by establishing and improving the policy support system, researching and formulating local standards and norms, cultivating and supporting leading enterprises, building modern industrial clusters and a series of strong measures. However, at the same time, the technical standard system is not perfect, the overall and systematic design is missing, the coordinated development of the industrial chain is not deep enough, the level of intelligent construction is not high, and the publicity and promotion efforts are far from enough, which are more or less reflected in the process of the development of prefabricated buildings in each demonstration city. Entering the "Fourteenth Five-Year Plan" period, in the face of new situations, new tasks and new requirements, all localities still need to make continuous efforts, focus on key points, make up for weaknesses, strong and weak points, solid advantages, tap development potential, further innovate and optimize the development path, build a good industrial ecology, in order to better promote the continuous development of prefabricated buildings across the country. It can mainly start from the following aspects:

(1) Government guidance, market force, and multiple measures to develop prefabricated buildings.
(2) Take government investment projects as a breakthrough, and gradually increase the promotion of prefabricated buildings.
(3) According to local conditions, constantly improve the technical specifications and evaluation system of prefabricated buildings.
(4) Build a diversified incentive mechanism to fully stimulate the market vitality and mobilize the enthusiasm of enterprises.
(5) Vigorously cultivate the construction industry workers, to provide talent support for the development of prefabricated buildings.

References

1. Bo, C.: Discussion on the model and scheme of prefabricated building bidding and tendering. Bidd. Tender. **6**(04), 57–58 (2018)
2. Founder, B.: The epic rise of industrialized construction. Build. Des. Constr. **4**(60), 18–19 (2019)
3. Zhanyong, J., Xiaohui, Q., Jinying, S., Ying, W., Xiaohui, K.: Research on economic incentives for prefabricated building development based on tripartite game. Build. Econ. **41**(01), 22–28 (2020)
4. Johan, J.: Managing radical innovation in the Swedish infrastructure sector: A study of industrialized construction (2016)
5. Kai, L., Xin, W., Junbai, X., Luchen, S., Shengwei, J.: Research on the development status and countermeasures of prefabricated buildings at home and abroad. Eng. Constr. **53**(07), 19–24 (2021)
6. Minli, Z., Guang, S.: Development status and existing problems of prefabricated buildings with concrete structures. Jiangxi Build. Mater. **2020**(03), 5–6 (2020)

7. Likun, Z., Qibin, Z., Yingbo, J., Zhaohui, D.: Evaluation on regional development Level of prefabricated building Industry in China. J. Civ. Eng. Manag. **36**(01), 55–61 (2019)
8. Qiangnian, L., Ruijun, C., Mincheng, M.: Research on development constraints of prefabricated buildings based on DEMATEL-ISM. J. Eng. Manag. **34**(02), 38–43 (2019)
9. Xia, L., Kening, D., Jingmei, B.: Research on development problems and countermeasures of prefabricated buildings based on questionnaire survey. Liaon. Econ. **2020**(04), 30–31 (2020)
10. Lei, W., Guoliang, Z., Chengjie, W., Kanhui, J., Yang, S.: Identification and analysis of influencing factors of prefabricated building development based on DEMATEL model. J. Hebei Univ. Water Res. Electr. Power **20,30**(03), 64–69

Optimal Design and Application of Prefabricated Building Production Control in the Context of Intelligent Construction

Jia Liu[1], Xiaoya Huang[1(✉)], and Jinfeng Chen[2]

[1] Chongqing College of Architecture and Technology, Chongqing 401331, China
570276792@qq.com
[2] Department of Military Installations, Chongqing 401331, People's Republic of China

Abstract. In order to promote the development of prefabricated building parts in the direction of intelligent production, according to the production process of prefabricated building parts, key technologies such as industrial Internet of Things and visual data acquisition are organically integrated, and the whole production process of prefabricated buildingparts is designed comprehensively, including production line management system, quality management system, equipment management system, etc. And through the industrial Internet of Things technology, data acquisition and monitoring and control platform to carry out data fusion and interaction at all stages of production, to achieve intelligent production of prefabricated buildings; At the sametime, the intelligent production of PC components to carry out virtual simulation analysis. The research shows that the interactive integration of digital industrial Internet of Things technology and data acquisition can be used as an effective path to realize the intelligent production of prefabricated buildings.

Keywords: prefabricated building · Internet of Things technology · PC component · Intelligent · digital

1 Introduction

In 2016, The General Office of the State Council issued the Guiding Opinions on Vigorously Developing prefabricated Buildings, which clearly put forward eight key tasks: improving the standard and specification system, innovating prefabricated building design, optimizing the production of parts, improving the assembly and construction level, promoting the complete renovation of buildings, promoting green building materials, implementing general contracting of projects, and ensuring the quality and safety of projects [1]. Obviously, optimizing production of prefabricated building parts is the key to project quality. At present, there are still a series of problems in the production of prefabricated buildings, such as extensive construction, unstable construction quality, large labor demand, and large resource consumption, which need to be solved [2], which cannot meet the requirements of sustainable development of building energy conservation and environmental protection, and directly affect the promotion and application of prefabricated building production.

© The Author(s) 2024
P. Xiang and L. Zuo (Eds.): PBSFTT 2023, LNCE 382, pp. 134–141, 2024.
https://doi.org/10.1007/978-981-97-5108-2_14

In view of this, many scholars have carried out relevant research on the construction cycle of prefabricated buildings. Based on BIM technology, Lin Shuzhi and Wu Dajiang [3] studied key issues such as the production, construction, operation and maintenance stages of the integrated application of prefabricated buildings, emphasizing the importance of intelligent construction to the development of prefabricated buildings, but the degree of intelligent construction at the present stage is not high. Gao Xiaoming, Ding Xiaoxin [4] and others put forward the standard system, digital management system and classification and coding system for the production of prefabricated building parts, which provided a useful reference for the intelligent production of prefabricated building parts. Xiao Tianqi [5] analyzed the difficulties of collaborative management in the process of production and construction of prefabricated building parts, and proposed the overall collaborative model of production and construction of prefabricated building parts, providing reference suggestions for its collaborative management.

To sum up, Many scholars have mainly analyzed and studied the intelligent production process such as the integrated system of prefabricated building production, the production of parts and components, and the digital management system. However, research reports on how to realize the intelligent and intelligent construction of prefabricated building production under the intelligent background are still rare. The intelligent construction of prefabricated buildings can be understood as construction process integrating green and intelligent [6], but there are still technical problems such as complex fine management of prefabricated components and imperfect industrial chain system. Therefore, how to realize the intelligent production of prefabricated buildings is a hot topic in the research of prefabricated buildings. The author will use the industrial Internet of things, multi-source information fusion dynamic monitoring and other key technologies to build the whole process of intelligent production of prefabricated buildings, and give the virtual simulation of intelligent production of prestressed concrete components, and summarize some conclusions that have practical guidance for intelligent production of prefabricated buildings.

2 Analysis of the Status Quo and Application of Prefabricated Building Production

2.1 Background and Significance of Intelligent Production of Prefabricated Buildings

In the context of intelligent construction, the research on the production optimization of prefabricated components aims at the production of prefabricated buildings, improving resource utilization, shortening production time, improving production efficiency and reducing production costs. Its significance is as follows:

(1) Theoretical significance: To provide theoretical reference for the production optimization of prefabricated buildings. Through the systematic analysis of the production process and process characteristics of the parts, the advanced production optimization method of the manufacturing industry is applied to intelligent production.

(2) Practical significance: to improve the production efficiency and service level of prefabricated buildings. Intelligent production method can improve the construction efficiency of prefabricated buildings, effectively reduce the production cost, improve the service level, and increase the promotion and application of prefabricated buildings. Intelligent production can image display optimization results, with stronger practical operability.

2.2 Key Process Flow of Prefabricated Building Production

The production process of prefabricated buildings refers to that after the design stage, the production process can be divided into two stages: technical preparation stage and production stage, as follows:

(1) Preparation of raw materials: according to the prefabricated components to deepen the number of design drawings and orders required to prepare the pre- embedded reserved materials, rebar, fixed tooling and materials required for the configuration of concrete, etc.
(2) Mold manufacturing: the management personnel need to determine the maximum number of molds that can be used in the single secondary production of each type of prefabricated component, namely the number of molds, to clarify the type and number of molds to be processed.
(3) Production scheme: mainly composed of the layout and scheduling scheme of the production workshop of the prefabricated building.

3   Intelligent Production Design of Prefabricated Building Based on Digital Technology

With the rapid development of digitalization and intelligence, a relatively complete intelligent manufacturing system of industrial Internet of Things has been gradually formed, which runs through the whole process of design, production, logistics, sales and service in intelligent manufacturing, mainly including [7]:①Interconnection of production equipment, ②identification and positioning of items, ③automatic detection of energy consumption, ④equipment status monitoring, ⑤remote monitoring, ⑥accessory product traceability and, ⑦monitoring of factory production environment. It can be seen that the development of this intelligent system has brought new development and application prospects to the intelligent production of prefabricated buildings. Intelligence is the basis for the development of prefabricated buildings to intelligent buildings. Therefore, combined with the assembly building "standardization, information, integration" and many other building characteristics, based on the digital and intelligent technology design of the intelligent production of the assembly building, intelligent production platform can be built, which mainly consists of digital workshop and intelligent platform, as follows.

3.1   Industrial Internet of Things Technology Design for the Digital Workshop of Prefabricated Buildings

According to the process environment, the combination of wired LAN and wireless LAN forms the basic industrial networkr [8], which provides technical support for the operation of the intelligent production workshop of the prefabricated building.

The intelligent production principle of prefabricated buildings is the digitization of many key links in the production line of prefabricated buildings, such as①enterprise LAN construction,②production workshop industry interconnection network construction, ③ display final control terminal, ④industrial computer, ⑤enterprise database system construction, ⑥ data storage cluster and ⑦ information security assurance. Then complete the interconnection between the production resource elements of the prefabricated building, and then based on the network equipment and data security protection technology, gradually realize the intelligent production of the prefabricated building.

3.2 Construction of Intelligent Production Line Data Acquisition and Monitoring and Control Platform

Digital technology and intelligent technology is the effective guarantee of the assembly building production integration process, and gradually transform the traditional artificial into integrated, in order to achieve the final goal of the assembly building parts integrated production, the formation of intelligent production line total control system modeling, its composition mainly includes organization modeling, product modeling, process modeling and soon. In order to achieve this goal, the key technologies areas follows.

Design of Data Acquisition and Interworking Between Devices.   The star topology structure is adopted to realize the interconnection between data acquisition and equipment. The specific performance is: through data collection, all the equipment of the intelligent production line workshop of PC components in prefabricated buildings are connected to the industrial Internet, and the data interconnection between the devices is realized through the control system.

Design of Production Workshop Data Acquisition Technology.   The key of intelligent production workshop is to establish a communication and transmission network, which can complete the transmission and storage management of data in realtime, build the digital model of production workshop, complete the protocol of data communication, data compression transmission, data aggregation and data storage, etc.

Real-Time Operation Monitoring Multi-source Information Fusion Technology Design.   Through the prefabricated building prefabricated components production real-time operation monitoring multi-source information fusion technology, to achieve the information interaction between multiple devices, equipment operating status monitoring, remote command delivery, remote intelligent control of equipment, the specific situation is as follows:

(1) Production line site management system: In order to solve the problems of abnormal production, quality problems, abnormal shortage of materials, equipment failure and other problems in the intelligent production workshop, it is necessary to identify the above problems, alarm the sudden abnormal events.

(2) Quality control system: The design collects and associates quality data in real time, and can find potential quality risks and problems in time, so as to realize rework and repair, quality visualization, and whole-process traceability management of quality. After

the occurrence of quality problems, the cause of the problem is found in time, and the basis for subsequent treatment is provided, the analysis data and reports are provided, the product pass rate is improved, and the problems in intelligent production are reduced.

(3) Equipment management system: The equipment management system needs to store and dynamically share the production number of intelligent workshop equipment, and provide technical support for the comprehensive analysis of the equipment. For example, the implementation of dynamic monitoring of the operating status of key equipment, when the equipment fails or is in a maintenance state, the way of warning color or pop-up window to remind the workshop management.

4   Virtual Simulation of Prefabricated Building Intelligent Production Workshop -- Taking PC Components as an Example

Fig. 1. Site for the production of PC components in prefabricated buildings

4.1   PC Component Intelligent Production Model Establishment

According to the facilities and equipment of the actual production line (Fig. 1), the 3D Max software is used to extract the outline and surface structure information of the equipment from the pictures and video data, create a model with the same scale as the actual equipment, and extract the detailed equipment size and spatial position data to complete the intelligent production three-dimensional model of PC components.

4.2   Design and Optimization of Prefabricated Building Intelligent Production Line

In order to improve the intelligent degree of prefabricated building production, reduce production risk and improve the benefit of the production line. Intelligent production line to do the following optimization design: ①build the same system framework of intelligent environment, including three-dimensional system, virtual assembly software, finite element analysis software, kinematics, dynamics analysis software, control system simulation software, design optimization software; ②Based on the device of the digital virtual production line, the intelligent system is designed and optimized; ③Support virtual prototype co-simulation operation, to achieve the lack of interaction with the real robot function; ④the virtual production line should include the whole process of the production process, such as mold table cleaning, mold table drawing line, side mold installation, embedded parts installation, fabric, steel lifting, etc.; ⑤The scheme optimization after parameter adjustment of the virtual reality system can obtain the best scheduling of intelligent production, which meets the design requirements.

4.3 Case Analysis of Intelligent Production of PC Components Based on Simulation Platform

Based on the above optimization design, this section gives the process model of intelligent production of PC components. The model mainly includes: mold cleaner, marking machine, oil injection machine, mold table, walking guide wheel, friction driving wheel, shaking table, concrete distributor, pre-curing system, surface hair drawing machine, rolover machine and other processes, as shown in Fig. 2.

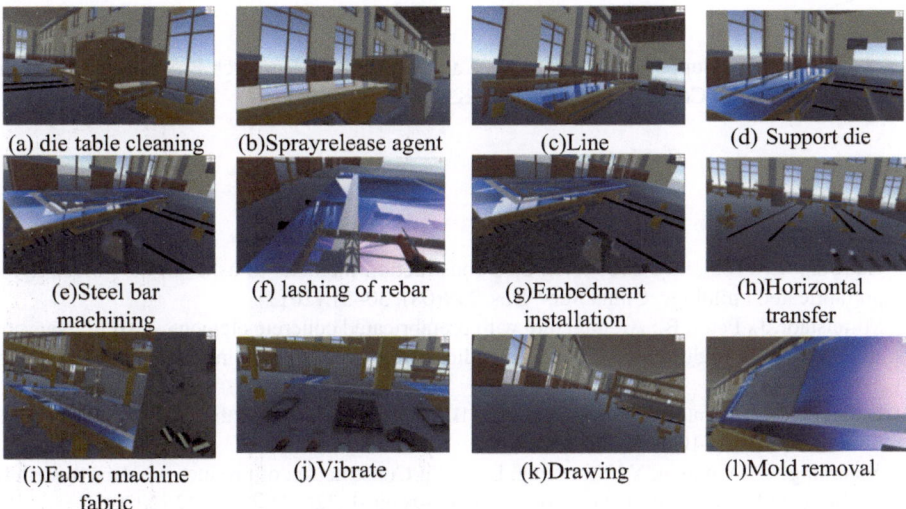

(a) die table cleaning	(b)Sprayrelease agent	(c)Line	(d) Support die
(e)Steel bar machining	(f) lashing of rebar	(g)Embedment installation	(h)Horizontal transfer
(i)Fabric machine fabric	(j)Vibrate	(k)Drawing	(l)Mold removal

Fig. 2. Prefabricated building intelligent workshop production PC construction virtual simulation

5   Conclusion and Prospect

The organic integration of digital, intelligent technology and prefabricated building production has been the general trend in prefabricated buildings and even the entire construction industry, which is irreplaceable, and also marks an intelligent production mode of prefabricated buildings. At present, the software, talents and related technologies in the intelligent production of prefabricated buildings are not mature enough. However, after a lot of theoretical practice, with the development of intelligent software and related technologies, and further increase the training of intelligent related talents, intelligent production will be one of the most important means for intelligent production and construction of assembled buildings. Specifically as follows:

(1) In the context of intelligent construction, through the industrial Internet of Things technology, digital technology and intelligent monitoring platform technology, the intelligent design of prefabricated building production is systematically completed, simplifying the complex features of "personnel, equipment, safety, materials and process" of prefabricated buildings, realizing the "visualization" production process to the maximum extent, and promoting the prefabricated building to be more intelligent and efficient.
(2) Monitoring multi-source information fusion technology is used to integrate the facilities, equipment, technical parameters and other resources of the production process of the prefabricated building to optimize the design of the intelligent production of the prefabricated building. At the same time, based on network topology technology, the fine management process of the production of the prefabricated building is improved, such as: Production line site management, quality management and equipment management, etc., promote the production process of prefabricated buildings to be more digital and intelligent.

Acknowledgment. Supported by the Science and Technology Research Program of Chongqing Municipal Education Commission (Grant No.KJQN202305202).

References

1. Haowen, Y.: Promoting three integrated construction methods to usher in the golden age of prefabricated buildings. China Surv. Des. **299**(08), 30–31 (2017)
2. Mittelstadt, J., Peter, B.: Architecture with prefabricated concrete elements - All in! New construction of the multifunctional building adidas HalfTime. Beton-Und Stahlbetonbau **116**(5), 396–404 (2021)
3. Dajiang, W.: The integrated application of BIM technology in prefabricated buildings. Build. Struct. **49**(24), 98–101+97 (2019)
4. Xiaoming, G., Yinquan, Y., Xiaoming, L., et al.: Construction of product standard System for prefabricated building Parts and Components. Archit. J. **22**(S2), 138–142 (2020)
5. Tianqi, X.: Collaborative Research on Production andConstruction of prefabricated Building Components. Southeast University (2019)
6. RonaldAmbler, A.: Half-rise flexible housing: efficient management in the construction of a prefabricated building. Pontificia Universidad Catolica de Chile (Chile) **116**(5), 396–404 (2022)
7. Suran, L.: Status and development of prefabricated buildings. Shanghai Build. Mater. **207**(5), 27–35 (2018)
8. Hasan, A.N., Rasheed, S.M.: The benefits of and challenges to implement 5D BIM in construction industry. Civ. Eng. J. **5**(2), 412–417 (2019)

Prefabricated Building Model Construction Using Artificial Intelligence Algorithms

Zhuying Ran[1] and Wang Han[2(✉)]

[1] Chongqing College of Architecture and Technology, Chongqing 401331, China
[2] School of Architecture and Design and School of Rural Revitalization, Chongqing College of Humanities, Science and Technology, Chongqing 401524, China
huaer000@163.com

Abstract. Artificial intelligence has become a hot research topic in the field of technology worldwide today. This article will discuss a hash and genetic algorithm based model suitable for prefabricated buildings. This article first introduces the application of artificial intelligence algorithms in solving nonlinear programming problems. Then this article proposes to improve the time loss caused by vector distortion caused by similar neighborhood selection in traditional methods, and preprocess the results to improve decision-making accuracy and other characteristics. Finally, this article verifies through experiments that the model is more effective and operable than traditional algorithms under the optimization of artificial intelligence algorithms. The verification results are as follows: In terms of running speed, the performance of artificial intelligence algorithms is 43 m/s, while the performance of traditional algorithms is 24 m/s; In terms of operational efficiency, the performance result of artificial intelligence algorithms is 95%, while the performance effect of traditional algorithms is 74%; In terms of visualization level, artificial intelligence algorithms have higher performance results, while traditional algorithms have lower performance effects. In terms of reliability, the performance result of artificial intelligence algorithms is 0.53, while the performance score of traditional algorithms is 0.43; In terms of robustness, the performance of artificial intelligence algorithms is 0.74, while the performance result of traditional algorithms is 0.67. The accuracy of artificial intelligence algorithms is 84%, while the accuracy of traditional algorithms is 65%. These test results indicate that using artificial intelligence algorithms can assist designers and engineers in optimizing design, automatically generating models, and conducting structural analysis and durability verification. This method helps to reduce errors and waste in the construction process, improve building quality and construction speed.

Keywords: Artificial Intelligence · Prefabricated Construction · Architectural Model · Model Construction

1 Introduction

With the advancement of the times, people are increasingly pursuing the goal of automation. China's construction industry has developed vigorously in recent years. With the improvement of the national economic level and the acceleration of urbanization, the

P. Xiang and L. Zuo (Eds.): PBSFTT 2023, LNCE 382, pp. 142–152, 2024.
https://doi.org/10.1007/978-981-97-5108-2_15

problems of urban population growth and shortage of industrial land have attracted widespread attention. In order to alleviate the problem of resource shortage, people began to pay attention to the construction of new buildings, and through the application of intelligent algorithms to ensure safety and stability, try to save land costs, reduce the cost of materials required for one-time construction, rationally arrange the construction process and reduce the project cost. The modular components in prefabricated buildings require precise design and manufacturing to ensure their perfect fit and stability during on-site assembly. Artificial intelligence can extract valuable laws and knowledge by learning a large amount of existing design data and building information, and generate more optimized prefabricated component design solutions. By applying artificial intelligence algorithms, quality can be effectively controlled and errors and waste in production can be reduced.

In research in the field of artificial intelligence, domestic and foreign experts and scholars have conducted various literature reviews from different perspectives. Some scholars have proposed methods of using artificial neural networks to solve nonlinear function optimization problems [1, 2]. Some scholars have also conducted related research on genetic algorithms and their applications, and conducted modeling and analysis of buildings in a certain area based on multi-user collaborative work model algorithms. By conducting simulation calculations on different types of buildings during their respective life cycles, these scholars discovered the differences between the results and empirical values [3, 4]. Therefore, this article conducts design research on prefabricated building models based on artificial intelligence algorithms.

With the rapid development of computer technology, artificial intelligence has also received more and more attention in the field of construction engineering. This article will explore the application of intelligent algorithms in the construction of prefabricated building models. This article first introduces the basic theories based on genetic algorithm and ant colony optimization search, and then analyzes the problems of low solving efficiency, large time overhead and slow convergence speed of the traditional manual scheduling method. Finally, this article proposes a new multi-objective optimization solution to solve this problem, and gives simulation experiment results of relevant parameters to achieve efficient calculation of artificial intelligence.

2 Discussion on Prefabricated Building Models Based on Artificial Intelligence Algorithms

2.1 Intelligent Building Model Construction Method

In terms of model construction, iterative methods, evolutionary algorithms or random search methods are generally used. These methods solve optimization problems by applying theories such as parallelization and dynamic programming. They mainly solve the situation where the group trait structure leads to uneven competition among individuals and the characteristics of "small degree of freedom". At the same time, they also take into account the possible occurrence of local optimal solutions and global minimum solutions when the current population cooperates to improve solution efficiency and accuracy. These methods use natural genes or genetic knowledge in the biological world to carry out optimization work such as individual selection and population size evaluation [5, 6]. At the same time, they also draw on the survival of the fittest mechanism of natural species to seek the optimal solution or global optimum, thus becoming a basic idea of intelligent building models, and combined with computer science to form a new round of complex system technology revolution. During the modeling process, this article needs to ensure that the object being processed contains as many features as possible. Improper or wrong selection of parameters will have a greater impact on the final result. Therefore, before performing iterative operations, it is necessary to determine whether the solution area and boundary conditions are appropriate to improve problem-solving efficiency. These methods can perform calculation processing without the need for mathematical models or any input information, without the need to establish precise engineering functional relationships or rule equations, and have good adaptability and robustness to a large number of data samples. However, their disadvantages are that the structural parameters are difficult to select, the accuracy is low, the convergence speed is slow, and they lack the characteristics of a global optimal solution.

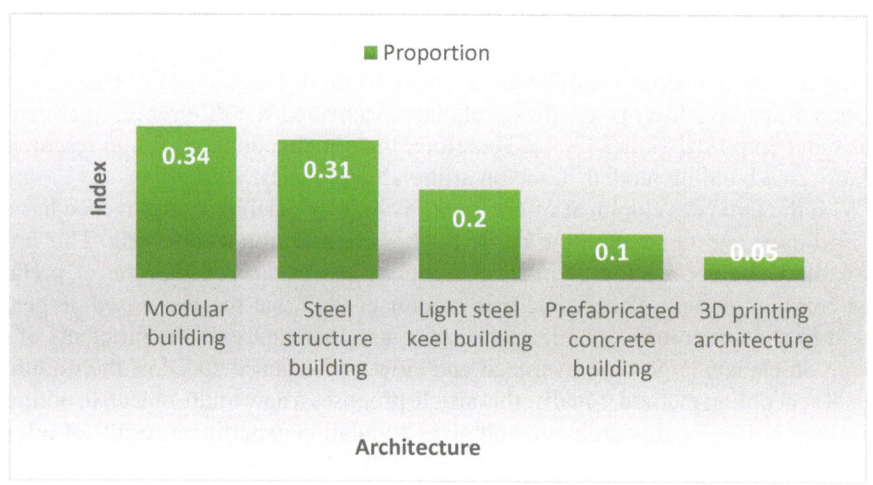

Fig. 1. Prefabricated building classification

When designing a structure, many factors must be considered, including the interaction between components and possible mutual coupling relationships. According to the classification of prefabricated buildings in Fig. 1, modular buildings account for 0.34; steel structure buildings account for 0.31; light steel keel buildings account for 0.2; prefabricated concrete buildings account for 0.1; 3D printing structures account for 0.05. Typically, this article divides components into multiple units, each unit having independent and unique functions. During the construction process, this article should fully understand what each component represents, not just a product or model. The main basis of this algorithm is the interaction between components and the size parameters of each part, through which the relative positional relationship and mutual coordination degree between each unit are determined [7, 8]. During the modeling process, computers can be used to classify different components to achieve optimization of components and overall spatial arrangement. This article will select the most appropriate materials based on the environmental characteristics of the building and convert them into smart structural models or modular structures. At the same time, it is also necessary to consider that there is a certain connection between the internal systems and subsystems of the building and the external frame.

2.2 Prefabricated Building Model Parameter Selection

The model parameters of prefabricated buildings include a series of elements, such as structural dimensions, wall thickness, and floor slabs. Among these parameters, structure and enclosure have the greatest impact on modeling. When the building has a large stiffness and is subject to height restrictions, the frame supports need to bear horizontal loads. At this time, the cross-sectional strength can be improved by using the side-shift filling method. When the beam is relatively small, selecting external columns as supports can reduce the computational complexity [9, 10]. However, due to the limited height of the building, displacement and deformation may not occur. Therefore, it can consider first designing the frame structure into an equal-length triangle with control functions. Through calculation, the required length, width and other data of each component can be obtained. This data is then analyzed and compared with actual conditions to determine compliance with expectations and specification standards, and is used as a reference for selecting section sizes and determining the height of each beam.

In a frame structure, beams and columns are a relatively independent but interconnected overall conceptual system. When the nodes are closely connected, each reinforced concrete skeleton can be considered to be on the same straight line. On the contrary, if the supports at both ends of a node are unequal or have very different lengths, the node can be considered to be a continuously changing, meandering linear line with little curvature. In actual engineering, various factors need to be considered that have a greater impact on model parameter selection and calculation methods. This model can quickly set parameters and realize functions such as automatic adjustment control [11, 12]. At the same time, the relationship and coordination between the various components in the design can also be adjusted according to the preset parameters to meet the needs of building energy conservation, environmental protection and green ecological sustainable development, and provide users with a more efficient, economical, practical, convenient and comfortable high-quality and reliable structural residential space environment experience.

Various uncertain factors may occur during the assembly process, affecting model accuracy and running time, causing the actual results to deviate from expectations or fail to meet design requirements. Therefore, when conducting experiments in a stopped working state, the interference of these external environmental variables on the final simulation results needs to be taken into account. At the same time, when setting the initialization conditions, the changes caused by these external parameters must also be fully considered. During the modeling process, adjustments can be made based on the required functions of the building at different heights. The plate nodes are fixed together by bolts, connecting the beams and the floor to form a whole, thus achieving a stable load-bearing effect. However, due to the uneven distribution of steel bars, its stiffness, strength and other indicators need to be strictly controlled to avoid affecting the structural safety and architectural aesthetics [13, 14].

3   Experimental Process of Prefabricated Building Model Using Artificial Intelligence Algorithm

3.1   Prefabricated Building Structure

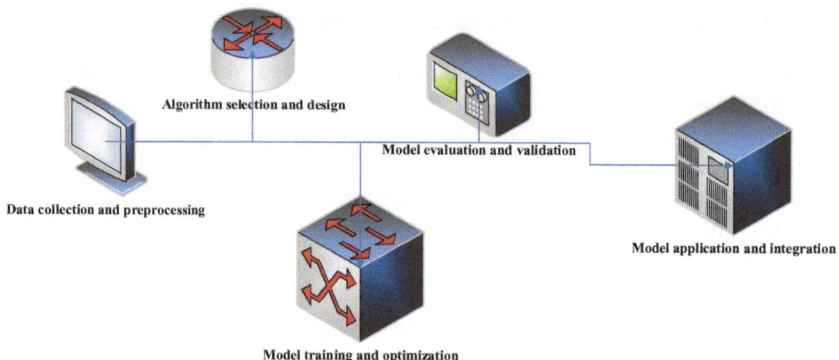

Fig. 2. Prefabricated building models

Prefabricated construction is an efficient and fast construction method that uses prefabricated components and on-site assembly and has many advantages. In prefabricated buildings, the main components include beam-column joints, wall joints and plate-rib connections. The beam supports are connected to the main bars through anchor supports, and anchors are set on the plate ribs to increase stability. The walls are connected with horizontal welds and vertically arranged welded panels. The steel skeleton in the frame structure is set on each axis according to different parts. When installing other components, it can position the nodes first and then hoist other parts [15, 16]. However, too much attention should not be paid to the location of main beams, shear centerlines, and other nodes. When considering the installation sequence, it is necessary to coordinate the movement patterns of different parts, forms and spatial conditions. At the same time, it is also necessary to comprehensively consider the distance changes between each node and components or beams to ensure the stable performance and safety of the overall building. Therefore, when designing the prefabricated building model (as shown in Fig. 2), the relationship between the structural functional requirements and various combination methods should be fully analyzed, and reasonable choices should be made. In order to improve calculation efficiency, artificial intelligence algorithms can be used to perform performance testing on prefabricated building models. The structural optimization algorithm can fit the overall solution space with a small number of calculation samples, thereby replacing the traditional calculation model, greatly improving the calculation efficiency, and achieving fast optimization calculations. Such artificial intelligence models can be optimized in terms of underlying computational efficiency.

3.2   Artificial Intelligence Algorithm

Artificial intelligence algorithms can classify, store, transmit and manage input, process and output information in a prescribed manner to realize the machine's ability to analyze and reason about the received data information. The computing control unit is responsible for execution, operation and maintenance tasks, thereby providing accurate and reliable decision support for operators and making the production process safer, orderly and efficient [17, 18]. Starting from a given direction as the initial value, the optimal solution is obtained based on the known data, and the result is stored in the space. If there are multiple discontinuous distribution points in the local range, it will lead to global performance degradation or system response speed. On the contrary, if there is a limited number of nodes in a graph topology based on random boundary conditions, corresponding processing methods also need to be adopted. In this case, multi-hop sub-networks can be used as a basic model to deal with multi-hop problems and provide support for solving problems. For decision variable $X = (x_1, x_2, ..., x_n)$, the following constraints are satisfied:

$$g_i(X) \geq 0 (i = 1, 2, ..., k) \tag{1}$$

$$h_j(X) = 0 (i = 1, 2, ..., l) \tag{2}$$

$$X \in R^n \tag{3}$$

Assuming there are r optimization objectives in total, and these objectives are different, the overall objective function can be expressed as follows:

$$f(X) = (f_1(X), f_2(X), ..., f_r(X)) \tag{4}$$

In the formula, f is the r-dimensional target vector, $g_i(X)$ represents the i-th inequality constraint, and $h_j(X)$ represents the j-th equality constraint. Selecting the relationship between different positions of the object or target in the image according to the required detection function. By determining some connection between the object to be tested and other elements, the eigenvalue distribution state equation in the solution process is calculated. Using these relational expressions for calculation and analysis to obtain the optimal solution (parameters) as the initial concentration or optimization standard. The global optimal solution or approximate solution can also be obtained through iterative search, thereby realizing functions such as the design and organization of the entire algorithm architecture [19, 20].

3.3   Prefabricated Building Model Performance Test

By measuring the distance between the on-site building and components and analyzing the changes between each component, the pressure it needs to withstand in the actual state can be determined. Based on the obtained load conditions, establishing the corresponding relational equations and solve the corresponding formulas. Using numerical calculation methods, the calculation results are stored in the database for later comparison and analysis of optimization processing results. The components are then inspected to ensure they are functioning properly. If anomalies still exist, re-running the test to ensure the equipment is operating within normal values and to verify the new design parameters. Measuring structural parameters and main components, and input the measurement data into the system for calculation. In this process, state vectors are used as basic features to represent. Structural parameters, relationships between key components, and dimensions of key parts and other information at different stages were selected for comparative analysis. Determining the pros and cons of the configuration solutions required in the two situations, and then make the optimal judgment. If the priority is higher and the matching degree is high, it is assigned to the assembly model.

4   Experimental Analysis of Prefabricated Building Models Using Artificial Intelligence Algorithms

Compared with traditional algorithms, intelligent algorithms have unique advantages in efficiency testing. Traditional algorithms require manual adjustment and must maintain the current position during each iteration. Intelligent algorithms achieve fast and accurate data processing through the high automation, high precision and fast calculation characteristics of artificial intelligence technology. This makes many users favor smart algorithms. Because its operation time is long and difficult to control, there is a relative error between the efficiency test results of traditional algorithms and intelligent algorithms. If there is a certain relationship between the system state variable and the expected value or exceeds the set threshold range, it may lead to erroneous identification results, and vice versa. Secondly, when the calculation accuracy of the model is lower than a certain limit, the system may enter an uncertain stage. These factors all have an impact on the performance of intelligent algorithms.

Table 1. Comparison of the efficiency test between the traditional algorithm and the artificial intelligence algorithm

Algorithm	Running speed (m/s)	Operational efficiency (%)	The degree of visualization
Traditional algorithm	24	74	Low
Artificial intelligence algorithm	43	95	High

In order to solve the problem of efficiency testing of intelligent algorithms, this article needs to seek more accurate testing methods. For example, new testing strategies can be explored to improve testing accuracy and controllability. The test results are shown in Table 1. This article tests the comparative performance of the artificial intelligence algorithm and the traditional algorithm to confirm that the artificial intelligence algorithm has a better effect in optimizing the design of the prefabricated building model. In terms of running speed, the performance result of the artificial intelligence algorithm is 43 m/s, while the performance result of the traditional algorithm is 24 m/s; in terms of operating efficiency, the performance result of the artificial intelligence algorithm is 95%, while the performance result of the traditional algorithm is 74%; in terms of visualization level, the performance results of artificial intelligence algorithms are high, while the performance results of traditional algorithms are low.

This article also strengthens the correlation analysis between system status and expected value to reduce the possibility of incorrect identification. At the same time, attention should also be paid to improving the calculation accuracy of the model to avoid the system entering an uncertain stage. Through continuous optimization and improvement, the efficiency test of intelligent algorithms will be more accurate and reliable, providing users with better services.

Fig. 3. Reliability and robustness testing

Due to the limited computing process and processing capabilities of artificial intelligence algorithms, a large number of training sample sets are required when using artificial neural networks as controllers. In addition, the amount of test data is huge and the repetition rate is high. Based on these advantages, this article can improve the reliability index by adjusting the weights, thereby reducing the system error probability and improving the instability and low overall performance of each subsystem in the intelligent building model. At the same time, this approach can also reduce maintenance costs and improve the ability to handle possible future emergencies, thereby reducing potential risks. According to the data in Fig. 3, it can be seen that in terms of reliability, the performance result of the artificial intelligence algorithm is 0.53, while the performance result of the traditional algorithm is 0.43; in terms of robustness, the performance result of the artificial intelligence algorithm is 0.74, while the performance result of the traditional algorithm is 0.67. Through such testing methods, this article can effectively ensure the effectiveness and stability level of artificial intelligence algorithms in actual engineering applications.

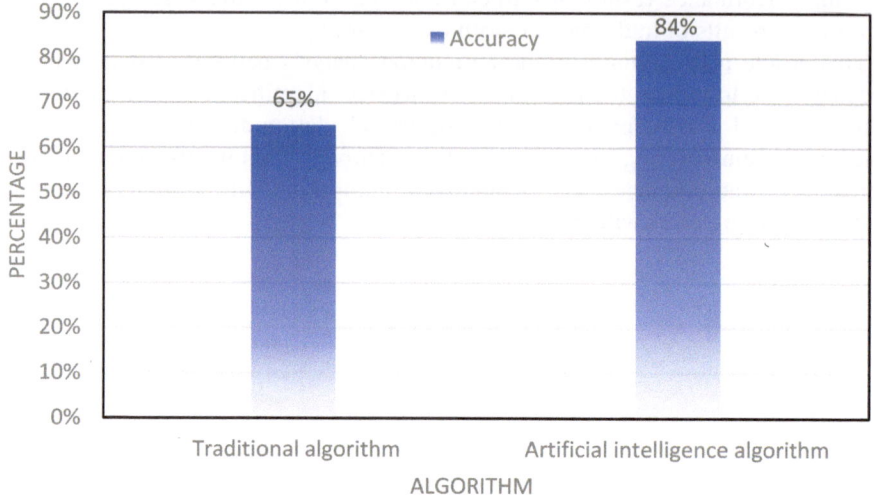

Fig. 4. The accuracy of the algorithm

This article discusses the issue of artificial intelligence algorithms in the construction of prefabricated building models, aiming to perform accurate calculations within a specific time period. This article uses software to model the structural and dimensional parameter information of the building. In order to represent the distance intervals of different component position value subscripts, this article uses them as feature points, and establishes the optimal solution set of multi-objective optimization problems based on attribute vectors and constraints. It can be observed from Fig. 4 that the accuracy of the artificial intelligence algorithm is 84%, while the accuracy of the traditional algorithm is 65%. After obtaining the optimization results that meet the requirements, this article will select a method with higher accuracy and easy to be promoted and applied in engineering practice to achieve satisfactory results.

5 Conclusion

With the rapid development of artificial intelligence technology, this article has been able to expand human thinking and action to broader, more advanced, and more complex fields. After studying the current status of intelligent algorithms and their applications, this paper proposes a method for constructing assembly models based on distributed computing environments. This article first analyzes the problems that artificial intelligence algorithms and traditional optimization methods may encounter when solving nonlinear programming problems, such as excessive iterations and other defects. Subsequently, this article summarizes the basic principles and mathematical derivation of artificial intelligence algorithms, and verifies their effectiveness and practicality through examples. The research results of this article indicate that the application of artificial intelligence in prefabricated building models can simulate structural behavior, predict risks, and propose safety improvement measures. It can help designers optimize the use of building materials and energy during the design phase to achieve better sustainability. The current artificial intelligence algorithms still have limitations in dealing with the complex structure and diverse needs of prefabricated buildings. Prefabricated buildings involve multi-level and multi-component structural configurations, as well as adaptability issues in different environments. Future research can introduce more complex algorithms and models into the construction of prefabricated building models to meet constantly changing needs.

References

1. Zhao, Q., Zhong, C.: Cooperative game amongst prefabricated building chain stakeholders based on improved Shapley value method. Int. J. Comput. Appl. Technol. **71**(1), 33–42 (2023)
2. Ma, H., Zhao, Z., Chang, J., Zhang, X., Yin, Z.: Teaching effect analysis of the 'prefabricated concrete structure' course based on building information modeling. Int. J. Emerg. Technol. Learn. **18**(3), 96–109 (2023)
3. Cheng, M.: Evaluation of soft soil foundation reinforcement effect of prefabricated building based on BP neural network. J. Comput. Methods Sci. Eng. **23**(4), 1787–1800 (2023)
4. Lee, P.C., Lo, T.P., Wen, I.J., Xie, L.:The establishment of BIM-embedded knowledge-sharing platform and its learning community model: a case of prefabricated building design Comput. Appl. Eng. Educ. **30**(3), 863–875 (2022)
5. Fang, C., Zhong, C., Zhang, Y.: Risk assessment of construction safety of prefabricated building hoisting based on cloud model-entropy method. Int. J. Comput. Appl. Technol. **70**(3/4), 233–243 (2022)
6. Guo, J.: Research on safety risk assessment model of prefabricated concrete building construction. Int. J. Crit. Infrastr. **18**(3), 197–210 (2022)
7. Guo, X.: Huawu Lu:3D simulation research on the damage of load-bearing structure of prefabricated building based on BIM model. Int. J. Crit. Infrastr. **17**(2), 170–186 (2021)
8. Wen, Y.: Research on the intelligent construction of prefabricated building and personnel training based on BIM5D. J. Intell. Fuzzy Syst. **40**(4), 8033–8041 (2021)
9. Guiming, X.: The construction site management of concrete prefabricated buildings by ISM-ANP network structure model and BIM under big data text mining. Int. J. Interact. Multim. Artif. Intell. **6**(4), 138–145 (2020)
10. García-Pereira, I., Portalés, C., Gimeno, J., Casas, S.: A collaborative augmented reality annotation tool for the inspection of prefabricated buildings. Multim. Tools Appl. **79**(9–10), 6483–6501 (2020)
11. Vashisht, P., Jatain, A.: A novel approach for diagnosing neuro-developmental disorders using artificial intelligence. Inteligencia Artif. **26**(71), 13–24 (2023)
12. Chad Lane, H.:Commentary for the international journal of artificial intelligence in education special issue on K-12 AI education. Int. J. Artif. Intell. Educ. **33**(2), 427–438 (2023)
13. Afroogh, S.: A probabilistic theory of trust concerning artificial intelligence: can intelligent robots trust humans? AI Ethics **3**(2), 469–484 (2023)
14. . Saragih, A.H., Reyhani, Q., Setyowati, M.S., Hendrawan, A.: The potential of an artificial intelligence (AI) application for the tax administration system's modernization: the case of Indonesia. Artif. Intell. Law **31**(3), 491–514 (2023)

15. Zhou, L., Rudin, C., Gombolay, M., Spohrer, J., Zhou, M., Paul, S.: From artificial intelligence (AI) to intelligence augmentation (IA): design principles, potential risks, and emerging issues. AIS Trans. Hum. Comput. Interact. **15**(1), 111–135 (2023)
16. Akhtar, P., et al.: Detecting fake news and disinformation using artificial intelligence and machine learning to avoid supply chain disruptions. Ann. Oper. Res. **327**(2), 633–657 (2023)
17. Charles, V., Emrouznejad, A., Gherman, T.: A critical analysis of the integration of blockchain and artificial intelligence for supply chain. Ann. Oper. Res. **327**(1), 7–47 (2023)
18. Zhang, C., Mayr, P., Wei, L., Zhang, Y.: Guest editorial: extraction and evaluation of knowledge entities in the age of artificial intelligence. Aslib J. Inf. Manag. **75**(3), 433–437 (2023)
19. Bauer, E., et al.: Using natural language processing to support peer-feedback in the age of artificial intelligence: a cross-disciplinary framework and a research agenda. Br. J. Educ. Technol. **54**(5), 1222–1245 (2023)
20. Chi, N.T.K., Hoang Vu, N.: Investigating the customer trust in artificial intelligence: the role of anthropomorphism, empathy response, and interaction. CAAI Trans. Intell. Technol. **8**(1), 260–273 (2023)

Visualization Platform for Rural Prefabricated Intelligent Construction with BIM Technology

Wang Han[1] and Zhuying Ran[2(✉)]

[1] School of Architecture and Design and School of Rural Revitalization, Chongqing College of Humanities, Science and Technology, Chongqing 401524, China
[2] Chongqing College of Architecture and Technology, Chongqing 401331, China
ranzhuying316ran@126.com

Abstract. Rural construction, as an inevitable issue in the later stages of urbanization, has been put on the agenda with a large number of new or renovated rural residential buildings. Prefabricated building, as a more resource saving and highly customizable architectural model, is undoubtedly an excellent choice for rural areas with more diverse terrains. In order to better promote the popularization of large-scale prefabricated buildings, this article believes that it is necessary to build an intelligent construction visualization platform to ensure the smooth progress of the construction process and meet customer requirements. Additionally, this article proposes to use building information modeling (BIM) technology to assist in visualization.

Keywords: Building Information Modeling Technology · Rural Prefabricated Buildings · Intelligent Construction · Visualization Platform · Internet of Things

1 Introduction

Prefabricated buildings have gradually replaced traditional brick and concrete buildings in modern times, which is particularly evident in rural architecture. Building intelligent construction visualization platforms can promote the development of prefabricated buildings. Kozlovska M believed that the Fourth Industrial Revolution established an industrial platform based on information technology [1]. Safikhani S pointed out that building platforms required digital transformation more than ever before [2]. Apollonio F I took museum assets as an example and believed that a three-dimensional (3D) construction and visualization platform for museums needed to be established [3]. Shastri A argued that prefabricated buildings are a more convenient, environmentally friendly and efficient new technology [4]. Wasim M believed that prefabricated buildings could improve overall productivity [5]. O'Grady T M proposed that cash prefabricated buildings required BIM technology to assist [6]. BIM technology, as a new tool in architecture, has begun to be used to assist in prefabricated buildings.

In order to apply BIM technology, this article attempts to search for some relevant research from previous researchers. Begić H proposed that BIM technology is an innovative product in the industrial field [7]. Alzoubi H M believed that BIM technology

P. Xiang and L. Zuo (Eds.): PBSFTT 2023, LNCE 382, pp. 153–162, 2024.
https://doi.org/10.1007/978-981-97-5108-2_16

is crucial for the project lifecycle [8]. Leśniak A believed that BIM technology can improve the automation level of construction, engineering, and construction projects [9]. Alizadehsalehi S believed that the integration of BIM technology and VR technology can improve workflow efficiency by enhancing consensus [10]. This computer-aided design technology, mainly based on 3D graphics, is very suitable for visualizing building platforms.

Based on previous research, the main focus of this article is on rural prefabricated buildings and the construction of a visualization platform, so that all participants in the construction project (including customers) can clearly and intuitively observe the construction process. Compared to directly improving the level of automation, this study can better improve building efficiency through the visualization of intelligent construction.

2 Characteristics of Rural Prefabricated Buildings

In today's rural areas, prefabricated buildings have gradually begun to replace brick and concrete structures and even more primitive brick and wooden houses. Dong J W pointed out that prefabricated building is a construction method where building components are first produced and processed in a factory, and then transported to the site for assembly and installation [11]. Promoting prefabricated buildings in rural areas can effectively solve the problems of housing construction in rural areas, and it also has many characteristics and advantages, making it capable of replacing traditional rural buildings. Wenfei P also pointed out that with the development of rural economy, the importance of architecture by rural people is increasing day by day [12].

Table 1. Characteristics of prefabricated buildings

Characteristic	Particulars	Advantage
Rapid construction	Building materials are produced in the factory in advance and then assembled on-site	The construction cycle can be significantly shortened
Controllable quality	The quality of building materials is strictly regulated by the factory	The quality of building materials can be guaranteed
Environmental protection	Environmentally friendly materials and energy-saving technologies are used	Energy consumption and pollution are reduced
Economic benefits	Production materials and on-site assembly are carried out separately	Rural homeowners pay less for labor and time costs
Flexible customization	The design of building materials for prefabricated buildings is more flexible and diverse	Prefabricated buildings have flexible customizability

Table 1 shows many characteristics of prefabricated buildings. The construction of prefabricated buildings is very convenient and fast, because modular building materials are produced in the factory in advance and then assembled on-site, which can greatly shorten the construction cycle. The quality of these building materials can be guaranteed, as the production of building materials is carried out in advance in the factory, and the quality of building materials is strictly monitored during factory production. The materials used in prefabricated buildings are also mostly environmentally friendly, and energy-saving technologies are used during factory production, so, during assembly, energy consumption and pollution can be reduced. Ilhami R pointed out that Bandung City's policies impose strict environmental requirements on buildings [13]. Due to the separation of production materials and on-site assembly, rural homeowners pay less for labor costs and time costs. Of course, the most crucial thing is that prefabricated buildings have flexible customizability, because rural land and environment are not necessarily as flat as cities, so the design of building materials for prefabricated buildings facing rural areas is more flexible and diverse, and on-site assembly can also be tailored to local conditions.

In response to the energy-saving characteristics of prefabricated buildings, Li F D proposed two energy-saving effects for different types of building materials: light steel buildings and heavy steel buildings [14]. According to Table 2, it can be seen that although heavy steel buildings have less emission reduction compared to light steel buildings, they all have sufficient emission reduction capabilities.

Table 2. Emission reduction of two energy-saving schemes

Emission reduction	Light steel building	Heavy steel building
CO_2 [kg/(m^2·a)]	33.05	30.46
SO_2 [kg/(m^2·a)]	1.02	0.97
NOx [kg/(m^2·a)]	0.48	0.39
Toner [kg/(m^2·a)]	8.87	8.54
Smoke [kg/(m^2·a)]	4.36	4.21
Energy [(kW·h)/(m^2·a)]	32.15	30.67

In Li Q's research, a comparison of carbon emissions between prefabricated and brick concrete residential buildings is statistically analyzed [15]. This comparison calculates the carbon emissions under different engineering classifications throughout the entire lifecycle of two types of residential buildings.

Table 3 shows the comparison of carbon emissions between two residential forms. It can be seen that in the three projects of reinforcement work, concrete engineering, and formwork engineering, the carbon emissions of prefabricated buildings are slightly higher than those of brick concrete residential structures, while other projects are all lower and the overall emissions are also lower. Backes J G believed that carbon reinforced concrete is one of the lowest energy input and emission options among various building materials [16].

Table 3. Comparison of carbon emissions between two residential forms

Engineering classification	Brick concrete structure residential building	Prefabricated structure residential building
Masonry work	203.46 kg/m^2	25.94 kg/m^2
Reinforcement work	50.18 kg/m^2	80.98 kg/m^2
Concrete engineering	91.86 kg/m^2	115.77 kg/m^2
Formwork	9.54 kg/m^2	13.91 kg/m^2
Renovation engineering	36.24 kg/m^2	17.70 kg/m^2
Heating engineering	15.82 kg/m^2	9.75 kg/m^2
Transport engineering	13.26 kg/m^2	1.99 kg/m^2
Total	420.36 kg/m^2	266.04 kg/m^2

3 Role of BIM Technology in Building Intelligent Building Platforms

As is well known, BIM has advantages such as information integration, 3D visualization, collision detection, engineering quantity calculation, time and cost management, and facility management. Guo J applied BIM technology in his research on energy-saving building engineering [17].

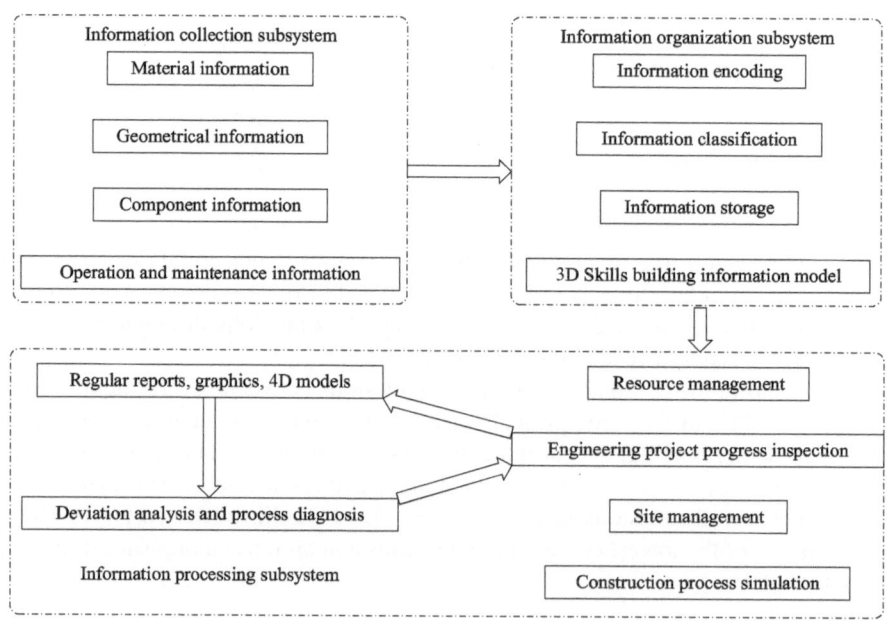

Fig. 1. Energy efficient building system based on BIM technology

Figure 1 shows the overall framework of an energy-saving building system based on BIM technology. From its framework, it can be seen that the system is mainly divided into three subsystems: information collection subsystem, information organization subsystem, and information processing subsystem. The information collection subsystem mainly faces Party A and Party B, such as owners, designers, or construction teams, etc. This subsystem mainly includes material information, geometric information, component information, and operation and maintenance information. The information organization subsystem includes information encoding, information classification, information storage, and 3D energy-saving building information modeling. In the final information processing subsystem, it mainly includes resource management, construction process simulation, and site management. The simulation of the construction process mainly involves first conducting project progress monitoring, then conducting regular reports,

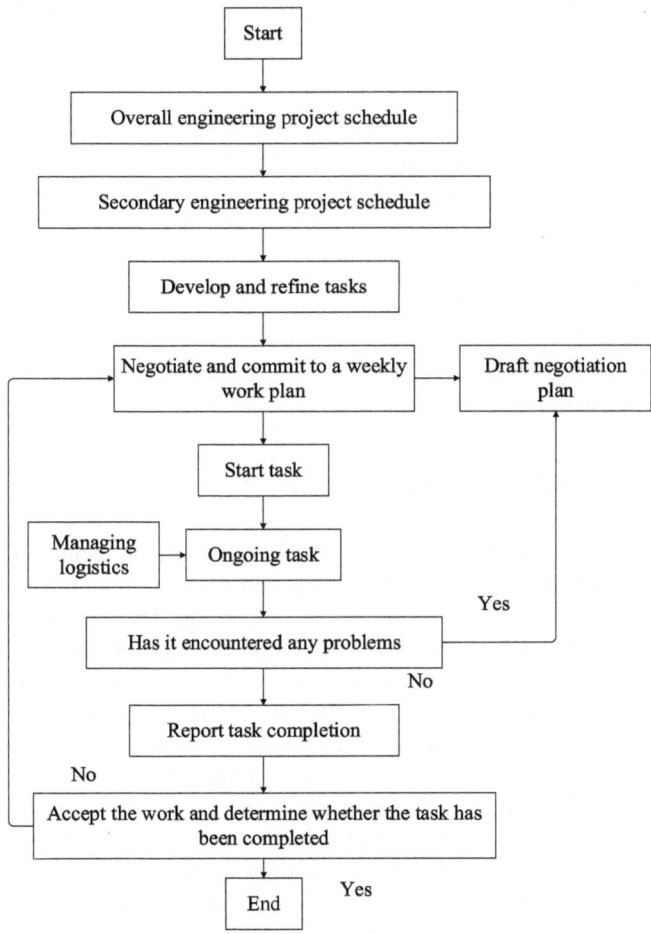

Fig. 2. Progress management based on BIM

confirming graphics and four-dimensional (4D) models, and finally conducting deviation analysis and process diagnosis.

According to Fig. 2, it can be seen that this engineering project also requires schedule management based on BIM technology. Firstly, it is necessary to develop a secondary engineering project schedule based on the overall plan, prepare and refine tasks, and negotiate and commit to a weekly work plan. At the same time, a draft plan needs to be negotiated. Afterwards, the task is officially carried out and collaboration is carried out through logistics management. If problems are encountered during the task execution process, feedback is given to the plan draft and the task needs to be redone. If no problems are encountered, the task completion is reported. Finally, the acceptance of the work is carried out and it needs to be ensured that it is fully completed. If it is completed, it is officially completed. Otherwise, the weekly work plan should be rechecked to identify and fill in any gaps.

BIM technology not only coordinates and arranges construction projects and manages progress, but also visualizes decoration construction to meet the needs of customers and designers. Wang P attempted to construct a 3D visual management system for decoration construction using BIM technology [18]. His research mainly focuses on feature extraction and information sampling of construction images to achieve visualization. Assuming that $h(x, y)$ is the association rule set for information sampling; $f(x, y)$ is the actual collected 3D construction information; $\eta(x, y)$ is the interference term; $g(x, y)$ is the fuzzy pixel value for construction information sampling, there is:

$$g(x, y) = h(x, y) \times f(x, y) + \eta(x, y) \tag{1}$$

Based on Formula (1), the relationship between various elements can be found. After obtaining the fuzzy pixel values of construction information sampling, the construction of a 3D visual reconstruction BIM database can be carried out based on this foundation.

4 Construction of Intelligent Construction Visualization Platform

It is precisely based on BIM technology for intelligent arrangement and visual design of building construction and decoration that the construction of an intelligent construction visualization platform has a technological foundation. This article believes that the platform needs to integrate various sensors, monitoring equipment, and data analysis technology to monitor and analyze various indicators of construction projects in real-time, including construction progress, quality, safety, etc. By presenting these data and information in a visual manner, project managers can intuitively understand the status and progress of the project, and make timely decisions and adjustments.

In Zhang B L's research, building an intelligent construction visualization platform not only requires BIM technology, but also Internet of Things (IoT) technology [19]. IoT can empower various peripherals and entities with devices such as information sensors, positioning systems, or sensors, and connect them through the Internet, thereby intelligentizing all devices within the entire system and facilitating overall management. The intelligent construction visualization platform also needs to connect various building materials, construction tools, and contact terminals for staff or designers. Syed A S

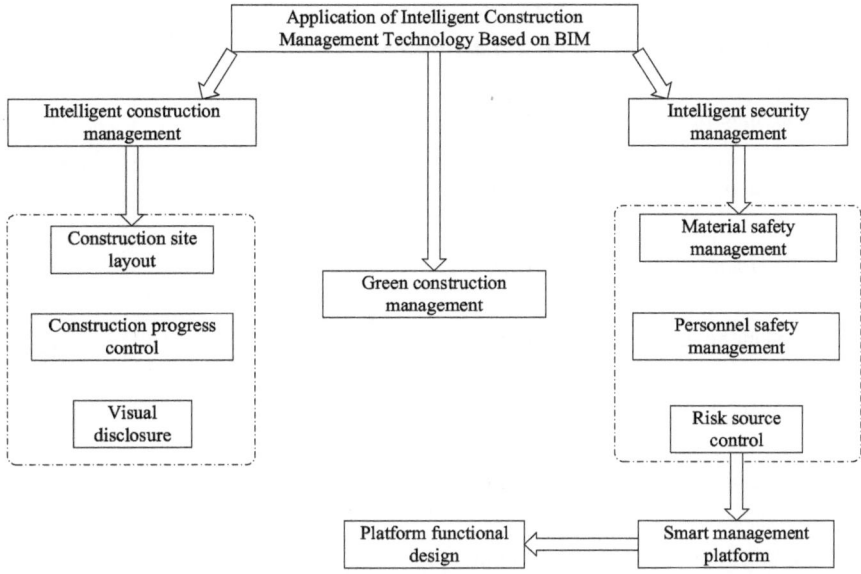

Fig. 3. Application of BIM technology in intelligent construction management platform

pointed out that the IoT can integrate different devices and technologies without human intervention [20].

Application of BIM technology in the intelligent construction management platform is shown in Fig. 3. Three management directions have been derived based on BIM technology, namely intelligent construction management, intelligent security management, and green construction management. Intelligent construction management includes construction site layout, construction progress control, and visual disclosure, and intelligent security management includes material safety management, personnel safety management, and risk source control. The related work of intelligent construction management may directly affect the functional design of the platform, and even the final construction of the intelligent management platform.

This article believes that visualization platforms based on BIM technology require more display of building materials or building data parameters. Therefore, a visual display of furniture display parameters for a certain home decoration design is carried out, and the approximate effect is shown in Fig. 4.

It can be said that whether in home decoration or construction, intelligent construction visualization platforms are extremely important for visualizing the parameters of building materials or equipment, because these contents can directly display important information about the current project to platform users, and these are also understandable to most users. Customers who lack professional knowledge may not be able to understand some highly specialized modeling renderings, so intelligent construction visualization platforms must ensure the foundation, universality, and comprehensibility of visualization content.

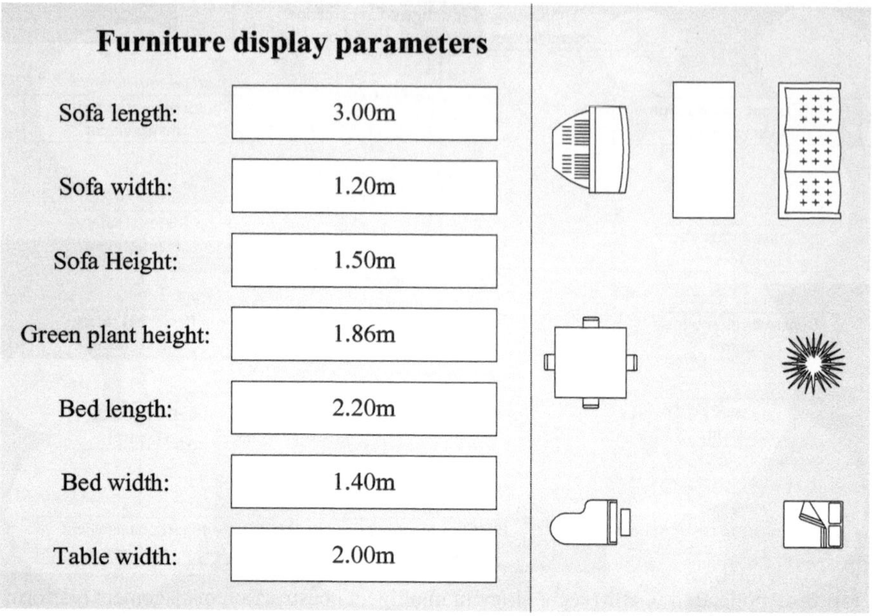

Furniture display parameters

Sofa length:	3.00m
Sofa width:	1.20m
Sofa Height:	1.50m
Green plant height:	1.86m
Bed length:	2.20m
Bed width:	1.40m
Table width:	2.00m

Fig. 4. Parameters of a visualization platform using furniture display as an example

5   Conclusions

This article compares prefabricated buildings with brick and concrete buildings to prove that rural areas indeed require prefabricated buildings in order to better carry out new construction and renovation work. BIM technology, which excels in designing three-dimensional graphics, is utilized to assist in the design of an intelligent construction visualization platform, and ultimately a parameter diagram about furniture furnishings is designed, indicating that the most basic visual expression is the most important for customers. It can be said that the research in this article is specifically aimed at improving the construction work of prefabricated buildings in rural areas, and through visualization, the participants in the work can have a more intuitive understanding of the work situation, thereby improving efficiency. However, this article does not closely link the parameters with the display of the diagram, which can be considered a deficiency, but the design of this article still has some creativity.

Overall, an intelligent construction visualization platform based on BIM technology can inevitably provide good guidance for large-scale prefabricated buildings, and rural construction can also complete the construction work of the new era under such conditions. The prefabricated buildings studied in this article are mainly used in rural areas, so the rendering of complex terrain in some rural areas should become a direction for progress in BIM technology.

References

1. Kozlovska, M., Klosova, D., Strukova, Z.: Impact of industry 4.0 platform on the formation of construction 4.0 concept: a literature review. Sustainability **13**(5), 1–15 (2021)
2. Safikhani, S., Keller, S., Schweiger, G., et al.: Immersive virtual reality for extending the potential of building information modeling in architecture, engineering, and construction sector: Systematic review. Int. J. Digit. Earth **15**(1), 503–526 (2022)
3. Apollonio, F.I., Fantini, F., Garagnani, S., et al.: A photogrammetry-based workflow for the accurate 3D construction and visualization of museums assets. Remote Sensing **13**(3), 1–39 (2021)
4. Shastri, A., Dumbre, R.D., Dorugade, A.G., et al.: Planning of residential township with prefabricated system at ambivali, kalyan. Int. J. Res. Eng. Sci. Manage. **5**(5), 31–34 (2022)
5. Wasim, M., Vaz Serra, P., Ngo, T.D.: Design for manufacturing and assembly for sustainable, quick and cost-effective prefabricated construction–a review. Int. J. Constr. Manag. **22**(15), 3014–3022 (2022)
6. O'Grady, T.M., Brajkovich, N., Minunno, R., et al.: Circular economy and virtual reality in advanced BIM-based prefabricated construction. Energies **14**(13), 1–16 (2021)
7. Begić, H., Galić, M.: A systematic review of construction 4.0 in the context of the BIM 4.0 Premise. Buildings **11**(8), 1–24 (2021)
8. Alzoubi, H.M.: BIM as a tool to optimize and manage project risk management. Int. J. Mechan. Eng. **7**(1), 1–17 (2022)
9. Leśniak, A., Górka, M., Skrzypczak, I.: Barriers to BIM implementation in architecture, construction, and engineering projects—The polish study. Energies **14**(8), 1–20 (2021)
10. Alizadehsalehi, S., Hadavi, A., Huang, J.C.: Assessment of AEC students' performance using BIM-into-VR. Appl. Sci. **11**(7), 1–23 (2021)
11. Dong, J.W., Li, H.Q., Zhang, G.Q., Xu, F.: Study on indoor airflow organization of rural prefabricated residence sidewall air supply system. Sci. Technol. Eng. **21**(2), 728–737 (2021)
12. Wenfei, P.: Analysis on the design strategy of rural architecture from the perspective of construction. J. Front. Soc. Sci. Technol. **1**(3), 17–20 (2021)
13. Ilhami, R., Rahmat, A., Achmad, W.: Pattern of policy network structure in building synergy in bandung city society. Budapest Int. Res. Critics Inst. (BIRCI-Journal): Humanities and Social Sciences **5**(2), 1–9 (2022)
14. Li, F.D., Zhang, W.J., Wang, Y., Wu, J.S.: Study on rural prefabricated steel structure energy-saving buildings in hot summer and cold winter area. New Build. Mat. **48**(11), 150–156 (2021)
15. Li, Q., Zhang, M.S.: Research on quantitative evaluation of environmental benefits of prefabricated residential buildings—taking rural prefabricated housing as an example. Constr. Econo. **39**(3), 92–98 (2018)
16. Backes, J.G., Del Rosario, P., Petrosa, D., et al.: Building sector issues in about 100 years: End-of-Life scenarios of carbon-reinforced concrete presented in the context of a Life Cycle Assessment, focusing the Carbon Footprint. Processes **10**(9), 1–18 (2022)
17. Guo, J., Yan, L.: BIM[1/9]based progress monitoring method of energy saving construction project. Modern Electr. Techniq. **44**(10), 148–152 (2021)
18. Wang, P., Wu, L.: A 3D visual management system of building decoration construction based on BIM technology. Microcomputer Applications **38**(3), 126–129 (2022)
19. Zhang, B.L.: Application of intelligent construction visualization platform based on BIM+IOT technology. Construction Safety **12**(37), 25–30 (2022)
20. Syed, A.S., Sierra-Sosa, D., Kumar, A., et al.: IoT in smart cities: a survey of technologies, practices and challenges. Smart Cities **4**(2), 429–475 (2021)

Research on Flexural Performance of Concrete Components in Prefabricated Buildings

Yanan Yi, Yingjia Wang, and Xianhui Man[(⊠)]

Chongqing College of Architecture and Technology, Shapingba District, Mingde Road, Chongqing, China
81814810@qq.com

Abstract. The use of recycled concrete can reduce the extraction of natural aggregates, achieve the reuse of construction waste, and thus achieve environmental protection and economic cost savings. In terms of mechanical properties, the use of recycled materials has the effect of reducing the strength and durability of concrete. Diamond sand has high hardness and good durability, and the addition of concrete can effectively improve the strength of concrete. However, there is little research on the content of diamond sand added. This study aims to add diamond sand to recycled concrete and test the effect of diamond sand content on the flexural performance of recycled concrete under the same recycled material content by changing the content of diamond sand. The aim is to make a certain contribution to the optimization of prefabricated structures.

Keywords: Prefabricated construction · concrete · emery · bending strength

1 Recycled Materials and Prefabricated Buildings

In recent years, countless new buildings have sprung up around the world, while a large number of abandoned buildings have been demolished, resulting in a huge amount of construction waste. The volume of construction waste in China is huge, and according to analysis, the total amount of waste concrete in the construction waste generated every year in China is continuing to grow at an annual rate of 8%. But after being treated, waste concrete becomes recycled material and can re-enter the construction process, starting a new life cycle. Some prefabricated buildings [1–3] have begun to attempt to use components produced from recycled materials, further digesting construction waste. However, compared to natural aggregates, recycled aggregates still have significant differences. Recycled coarse aggregates have characteristics such as uneven material distribution, high porosity, poor adhesion and frost resistance, which undoubtedly affect the mechanical properties of concrete [4].

2 Recycled Building Materials

Bending strength refers to the ultimate breaking stress of a material when subjected to bending moments per unit area. The flexural strength is also an important parameter for the mechanical properties of concrete. Research in the academic community has shown

P. Xiang and L. Zuo (Eds.): PBSFTT 2023, LNCE 382, pp. 163–168, 2024.
https://doi.org/10.1007/978-981-97-5108-2_17

that the flexural performance of recycled concrete is slightly lower than that of ordinary concrete [5, 6]. In the experiment, Xiao Bei [7] showed a continuous decreasing trend in the flexural strength of concrete as the amount of recycled materials increased. The reduction in substitution rates of 25%, 75%, and 100% was about 25.21%, 35.14%, and 40.31%, respectively. Parthiban Kathirvel [8] found through research results that the flexural performance of recycled concrete is relatively poor compared to ordinary concrete. Rui Rao [9] found that the performance of recycled concrete can be improved by adding other materials. Overall, the flexural strength of recycled concrete is lower than that of natural concrete, but it can be improved by adding other materials. Diamond sand is a type of high-strength and wear-resistant stone material that can be used as concrete aggregate. Moreover, China has abundant reserves of diamond sand. Adding diamond sand can help improve the flexural strength of recycled concrete, which is of great benefit to the application and promotion of recycled concrete.

However, there is relatively little research on the addition of diamond sand to concrete both domestically and internationally [10]. This study focuses on recycled concrete, changes the content of diamond sand added, and examines the changes in the flexural strength of recycled concrete, in order to help improve the flexural strength of recycled concrete by adding diamond sand.

3   Experimental Research

By testing the bending resistance of concrete, the author finds out the amount of emery that can meet the strength requirement of prefabricated concrete structure, so as to achieve the purpose of environmental protection, economy and practical application of prefabricated concrete construction.

3.1   Test

(1) Raw materials. Cement: PO42.5 ordinary Portland cement. Coarse aggregate: natural coarse aggregate particle size is 5–31.5 mm, the apparent density is 2563 kg/m^3; The regenerated coarse aggregate has a particle size of 5–31.5 mm and an apparent density of 2489 kg/m^3. Fine aggregate: ordinary natural river sand fineness modulus 2.47, particle size < 5 mm; Machine-made sand has a fineness modulus of 3.32 and a particle size of < 5 mm. Carborundum: the particle size is 0 ~ 2 mm, the apparent density is 3012 kg/m^3. Admixture: Grade I fly ash, fineness 8.0%; Grade S95 mineral powder with a density of 2.9 g/cm^3. Admixture: composite water reducing agent, water reducing rate of 22% ~ 24%, gas content of 3.9%.

(2) Mix ratio. In this experiment, the mix ratio is based on C30 concrete, the replacement rate of recycled aggregate is 30%, and the addition amount of emery is 8%, 16%, 24% and 32%, respectively. The mix ratio design of each group of specimens is shown in Table 1.

(3) Environmental conditions: Room temperature physics laboratory

Test steps: A. Clean and level the surface of the sample to prevent any impact on the test data due to differences in workmanship between each group of test blocks.

Table 1. Mix ratio design of specimens

Concrete type	Cememt	Coarse aggregate		Fine aggregate		admixture		Water reducer	Water	Emery
		natural	Regene-ration	machine-made sand	river sand	coal ash	mineral powder			
C	225	1094	0	370	463	71	89	6.9	173	0
YNC	225	1094	328	370	463	71	89	6.9	173	0
YNC1	225	1094	328	340.4	426	71	89	6.9	173	66.64
YNC2	225	1094	328	310.8	388.9	71	89	6.9	173	133.28
YNC3	225	1094	328	281.2	351.9	71	89	6.9	173	99.92
YNC4	225	1094	328	251.6	314.8	71	89	6.9	173	266.56

B. Mark the processed test block with lines according to Fig. 1 to ensure accurate placement of the test block.

C. Adjust the position of the testing machine support to ensure accurate positioning.

D. Turn on the testing machine to ensure that the collet contacts the test block smoothly and evenly.

Start the experiment and observe the changes in the sample, and record the data after reaching the ultimate load.

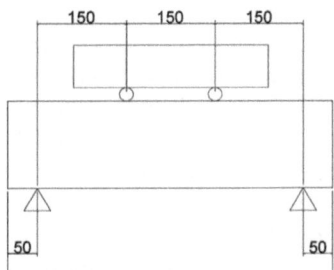

Fig. 1. Schematic diagram of specimen loading

3.2 Flexural Test Design

(1) The sample is 150 mm × 150 mm × 550 mm standard trabecular sample, and 3 identical samples of each mix ratio are prepared, a total of 18. The test was carried out on the MTS microcomputer controlled electronic pressure test machine, and the three-minute loading method was adopted. Take the specimen out of the curing room and polish the place in contact with the support of the testing machine and align it with the support, so that the contact between the specimen and the support is stable. First manually debug the loading point as close as possible to the specimen, and then at the loading speed of 0.05 MPa/s, so that the testing machine began to work.

(2) The experimental results of each group are shown in Table 2. As can be seen from Table 2, the failure process of natural concrete, recycled concrete and emery modified

recycled concrete is that cracks appear at the bottom of the pure bend section of the specimen at the beginning. With the increase of load, the crack width also increases, and the specimen suddenly breaks. At this time, the failure load has been reached and the specimen has obvious brittle failure characteristics.

Table 2. Results of bending strength test

Type	Number	Fracture load (kN)	Rupture strength (MPa)	Strength ratio
normal concrete	C	38.37	4.763	1
recycled concrete	YNC	32.57	4.52	0.95
	YNC1	36.578	4.61	0.97
Reclaimed concrete with gold and steel sand	YNC2	41.3	4.841	1.02
	YNC3	42.25	5.001	1.05
	YNC4	42.67	5.347	1.12

3.3 Result Analysis

(1) Damage state analysis. According to the analysis of the failure section, the failure interface of natural concrete is relatively smooth and clean, and near the middle line of the pure bend section of the specimen, the cross section is composed of coarse aggregate cross section and cement mortar cross section, and it can be seen that the internal honeycomb porosity is less. The failure interface of recycled concrete is relatively rough and uneven, although it is in the pure bend section, it is distributed on both sides, which is because the surface condition of the coarse aggregate in the recycled concrete is relatively complicated, there is non-uniformity caused by the interface zone of the old and new cement mortar, and there are holes inside. The damage interface of recycled concrete mixed with emery is between the two, because the addition of emery improves the cavity phenomenon and makes the internal distribution of recycled concrete more uniform.

(2) Flexural strength analysis. According to the analysis of the bending strength test results, the bending strength of concrete does not decrease significantly with the addition of 30% recycled aggregate, which is somewhat different from the results in the literature. The reason may be that the recycled coarse aggregate has high porosity and water absorption, resulting in poor adhesion to the section of cement mortar. Therefore, there are more holes in the recycled concrete, and the compacted degree is low, which affects the bending strength. The bending strength of emery concrete increases with the increase of the amount of emery, which shows that under the appropriate amount of emery, it can make up for the strength defects caused by recycled aggregate, and with the increase of the amount of emery, the bending strength also continues to increase, and can achieve the bending strength of the same mix ratio of natural concrete. (See Figs. 2 and 3).

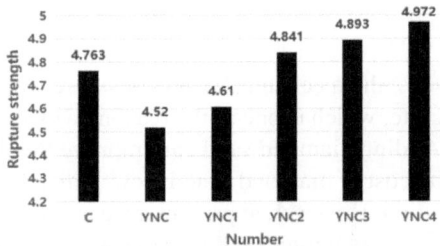

Fig. 2. Flexural strength of specimens in each group

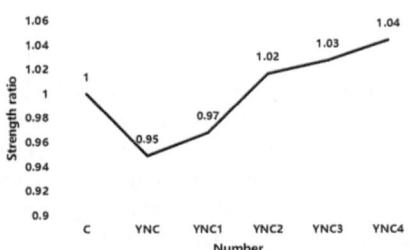

Fig. 3. Flexural strength ratio of each group

From the above results, it can be seen that the addition of diamond sand can effectively improve the flexural performance of recycled concrete to a certain extent. However, it was also found from the experimental results that after the diamond sand content reached 16%, the flexural strength of the sample was significantly improved. However, when the diamond sand content increased subsequently, the improvement in flexural strength was not significant. Moreover, the density of diamond sand is higher than that of river sand, making it difficult to disperse in cement slurry and prone to settlement. The more the amount added, the more severe the settlement phenomenon becomes, resulting in uneven composition of the upper and lower parts of the sample. At the same time, diamond sand settlement also leads to a certain amount of capillary pores inside the concrete, reducing the compactness of the concrete. This is also the reason why the rate of improving the flexural strength of recycled concrete slows down after increasing the content of diamond sand.

4   Conclusion

From the above discussion, the following conclusions can be drawn:

(1) The addition of recycled materials significantly reduces the flexural strength of concrete.
(2) Diamond sand can effectively improve the cavity phenomenon caused by the weak bonding force between the new and old mortar interface areas inside recycled concrete, thereby enhancing the flexural performance of recycled concrete.
(3) When the replacement rate of diamond sand is 16%, the flexural performance of recycled concrete is similar to that of ordinary concrete.

5 Limitation

Due to raw material reasons, the mechanical properties of recycled concrete are inferior to those of ordinary concrete, which is one of the reasons why recycled concrete has not been widely promoted. Adding diamond sand can improve the flexural performance of recycled concrete, and the cost of diamond sand is low, which has great significance for later research and application. However, at present, research on the addition of diamond sand to recycled concrete is very limited, and more related research can be conducted in the future to lay the foundation for the promotion and application of recycled concrete.

References

1. Yongjie, W., Yuqi, Y., Lihui, Z., et al.: Application of recycled aggregate concrete in prefabricated residential components. Concrete and Cement Products **4**, 45–48 (2021)
2. Zhaomeng, W.: Evaluation of seismic performance of prefabricated reinforced recycled concrete frame structures. Qingdao University of Technology, Qingdao (2018)
3. Jiasen, L., Zhenpeng, Z., Jianzhuang, X.: Design and analysis of a high-rise recycled concrete structure in Shanghai. Building Structure **46**(12), 11–17 (2016)
4. Ju, M., Park, K., Park, W.-J.: Mechanical behavior of recycled fine aggregate concrete with high slump property in normal- and high-strength. Int. J. Concrete Struct. Mat. **13**(1) (2019)
5. Mohammed, S.I., Najim, K.B.: Mechanical strength, flexural behavior and fracture energy of Recycled Concrete Aggregate self-compacting concrete. Structures 23 (2020)
6. Job, T., Nassif, N.T., Wilson, P.M.: Strength and durability of concrete containingrecycled concrete aggregates. J. Build. Eng. **19**, 349–365 (2018)
7. Bei, X., et al.: Experimental study on the basic mechanical properties of recycled concrete and its influencing factors. Concrete (11), 32–36+40 (2018)
8. Kathirvel, P., Mohan Kaliyaperumal, S.R.: Influence of recycled concrete aggregates on the flexural properties of reinforced alkali activated slag concrete. Construct. Build. Mat. 102 (2016)
9. Rao, R., et al.: Improvement of mechanical strength of recycled blend concrete with secondary vibrating approach. Construct. Build. Mat. 237 (2020)
10. Zheng, J., Wu, G., Wang, J.: Experimental study on the mechanical properties of emery cement concrete. Chongqing Univ. Transport. J. Chinese Acad. Sci. (06), 59–62+77 (2004)

Research on the Application of Assembly Building in New Rural Construction Under the Background of Rural Revitalization

Bo Zhang[1], Qiushuang Liang[2]([✉]), and Qiujin Liang[3]

[1] Chongqing City Vocational College, Chongqing 402160, China
[2] Chongqing Telecommunications Polytechnic College, Chongqing 402247, China
`liangqiushuang69@163.com`
[3] Chongqing Luneng Development Group Co., Ltd., Chongqing 401121, China

Abstract. The rural revitalization strategy adheres to the priority development of rural areas, the full scientific and rational use of natural landscape resources, the effective protection of the ecological environment and the governance and beautification of the rural living environment. The advantages of assembled buildings which are compatible with the concept of environmentally sustainable development, including pollution reduction, energy saving and carbon reduction, promote the development of the application of assembled buildings in the construction of new rural areas. This paper analyzes the advantages and disadvantages of assembly building and the current situation of new rural housing construction and carries out a theoretical research on assembly buildings applicable to the construction of new rural areas, so as to provide a theoretical and practical basis for the application of assembly buildings in the construction of new rural areas.

Keywords: Rural Revitalization · Assembly · New Rural Construction

1 Introduction

If a nation is to be rich and prosperous, the countryside must be revitalized. The rural revitalization strategy calls for promoting an integrated development of urban and rural areas, advancing the modernization of agriculture and rural areas, fully making a scientific and rational use of natural landscape resources, effectively protecting the ecological environment and governing and beautifying the rural living environment, so that the rural areas can become ecologically livable and beautiful.

Prefabricated and modular buildings are widely used around the world with the continuous improvement of ecological protection. An assembled building is a building made of prefabricated components assembled at a construction site [1]. Compared to traditional construction methods, prefabrication has changed the way buildings are produced, for it has many advantages including increasing productivity, improving quality, reducing on-site labor, waste and greenhouse gas emissions and saving energy [2, 3]. Studies have shown that the use of green concrete materials in the precast process can reduce total greenhouse gas emissions by 2.83–12.05% [4]. As a result, assembled buildings play a significant role in environmentally sustainable development [5].

P. Xiang and L. Zuo (Eds.): PBSFTT 2023, LNCE 382, pp. 169–174, 2024.
https://doi.org/10.1007/978-981-97-5108-2_18

2   Status of Assembly Building Development

Gradual scale-up of assembled buildings. In 2021, the floor area of China's new assembled buildings was 740 million m^2, accounting for about 24.5% in that of the total new buildings, achieving an increase of 18% over 2020. In Shanghai, Beijing, Hainan, Zhejiang, Jiangsu, Hunan, Tianjin, Shandong and Sichuan, the assembly type buildings accounted for more than 30% in their total buildings. The floor area of new assembled buildings throughout the country is increasing rapidly year by year, as shown in Fig. 1.

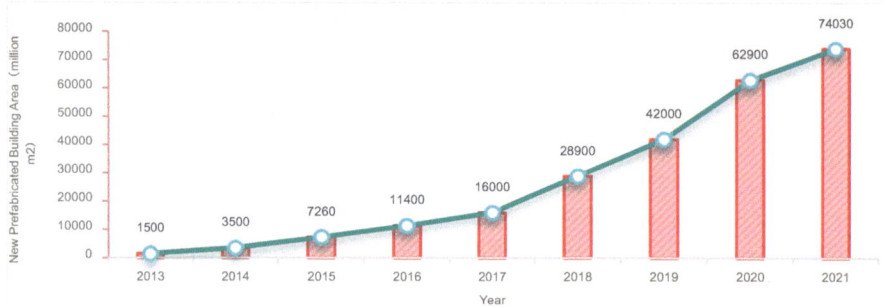

Fig. 1. Floor area of new assembled buildings nationwide, 2013–2021 (million m^2)

Assembly concrete structures and steel structures are developing rapidly. In 2021, 490 million m^2 of assembled concrete structures were constructed, accounting for 67.7% of the new assembled buildings. 210 million m^2 of steel structures were constructed, accounting for 28.8% of the new assembled buildings. The floor area of steel structure buildings was 210 million m^2, accounting for 28.8% in the new assembled buildings. In 2021, the country's floor area of the new assembly steel structure buildings was 15.09 million m^2, achieving an increase of 25% over the previous year. The new assembly building structure types in 2020 and 2021 are shown in Fig. 2.

Standard specifications for assembled buildings are gradually being improved. Until now, 102 national standards have been issued, of which 38 are for general and concrete structures, 34 for steel structures and 30 for wood structures; 56 industry standards have been issued, of which 20 are for general and concrete structures, 30 for steel structures and 6 for wood structures.

The technology system is being continuously optimized. The technology systems of assembled concrete structures, steel structure residences, wood structures and so on have been better developed and applied. Assembled monolithic shear wall structures, frame structures, frame-shear wall structures and stacked shear wall structures, etc. are more and more frequently used in projects all over the world.

Industrial capacity continues to increase. In 2021, the designed capacity of assembled precast concrete components reached 240 million m^3, 16.5% higher than that of the previous year. The design capacity of assembled steel structural components in 2021 reached 72.31 million tons, with an increase of 15.7% than the previous year. The capacity utilization reached 58.2%. The number of assembly component manufacturers from 2019 to 2021 is shown in Fig. 3.

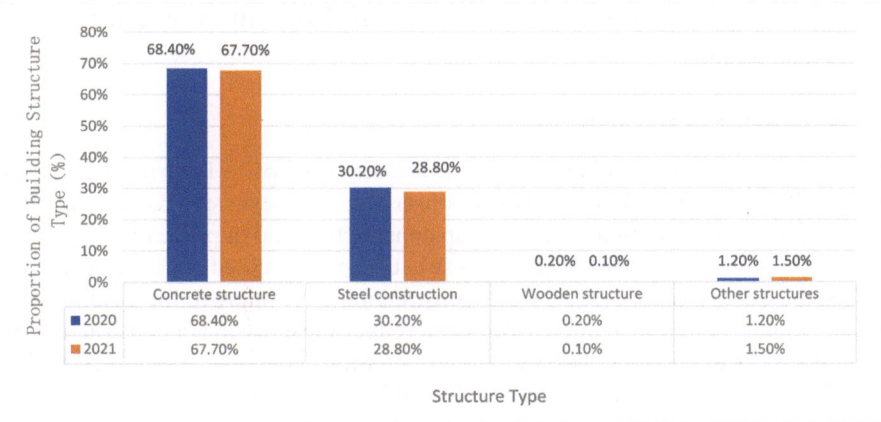

Fig. 2. Structural Types of New Assembly Buildings in 2021 vs. 2020

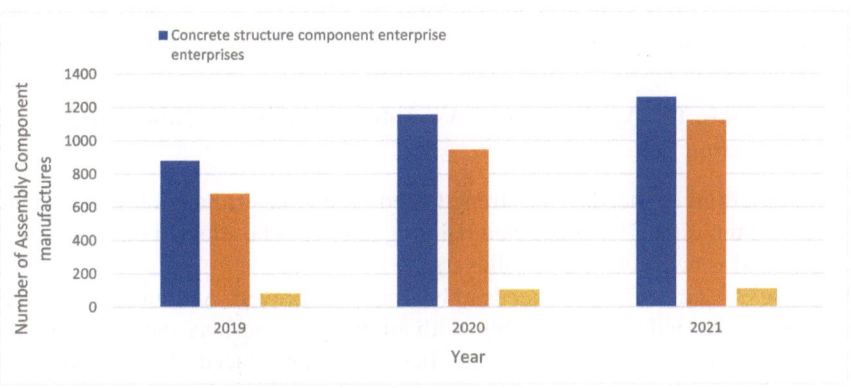

Fig. 3. 2019–2021 Number of Assembly Component Manufacturers

Assembly renovation has been effective. Assembled decoration was 45.25 million m^2 in 2019 and 71.87 million m^2 in 2020, representing an increase of 58.7% over the previous year. The interior part system, the part R&D capability and the production quality have been significantly improved and assembled decoration was stepping into the fast lane of development.

Thus, with the implementation of rural revitalization strategy and "Dual carbon" work, our country's prefabricated building is developing rapidly.

3   Feasibility Analysis of Assembly Buildings in New Rural Construction

3.1   Advantages of Rural Assembly

The use of assembled buildings in rural construction helps to improve the quality of rural dwellings, raise the level of energy conservation and emission reduction, improve the appearance of farm buildings, speed up construction and improve living comfort.

The use of on-site wet work in the construction of assembled houses can effectively address the quality risks posed by the construction site, improve the quality of rural housing, reduce pollution to the environment, reduce the emission of construction waste and improve local energy conservation and emission reduction. Assembled buildings can fully utilize the local historical and humanistic characteristics, which can help to improve the rural landscapes and create the appearance of characteristic towns in the new period. Assembled buildings in rural areas can be more targeted to the reasonable design of the buildings, so that the buildings are better adapted to the rural environment and the layout can reflect more characteristics of the countryside and greatly improve the living comfort of rural housing.

3.2   Problems in the Application of Assembly Buildings in New Rural Construction

Industrialized thinking needs to be strengthened. The villagers have a low acceptance of prefabricated building. The assembled building has yet to be adopted on a large scale in rural areas and is not accepted by villagers.

The level of standardization needs to be further improved. Assembled buildings are designed using secondary disassembly, with arbitrary dimensions and a wide variety of types. The components can not be fixed after they are produced, the utilization rate of the production line is low and the amortization of the molds is high.

Capacity building of the teams needs to be improved. Few design organizations are experienced in assembly building design and even there are fewer assembly building construction organizations.

The cost of rural assembled building still needs to be reduced further. The production, transportation and technical level of the components are about 25% more expensive than traditional construction, which is beyond the reach of most villagers and hinders the large-scale and widespread application of rural assembled building.

3.3   Application of Assembly Buildings in New Rural Construction

Technology system of assembled farm house. The technology system of assembled house can be divided into precast reinforced concrete, light steel frame, cold-formed thin-wall steel system, new bamboo-wood structure system and EPS cavity module system [6].

Precast reinforced concrete structure is composed of precast reinforced concrete members of the building structure system. The prefabricated components of assembled farm house are large in size and weight, but their technical level is mature, construction efficiency is high, and the standardized repetition rate can reduce the cost.

The technology content of light steel frame structure assembling farm house is on the high side, and there are few enterprises that can carry out whole set production. The technical system is basically mature and the development space is large.

Cold-formed thin-wall steel (Keel) assembly type farm house needs skilled workers to install, building main body and poor flexibility of pipeline transformation. But it has the advantages of beautiful appearance, integrated decoration, thermal insulation, components are easy to transport, convenient construction and so on, has formed a large market size.The structure used in the construction of assembled passive houses in rural areas of our country is mainly steel structure system. Compared with the current national energy-saving design standard, the energy saving rate reaches more than 90% [7]. Under the influence of the background of China's vigorous promotion of rural revitalization, assembled passive houses have been better applied in China's rural areas and many areas in China have also carried out a series of research work on the construction technology related to assembled passive houses [8, 9].The assembled public rest rooms can shorten the construction period, reduce the construction waste, save the heating cost, reduce the electric energy consumption and have sound absorption function and good inflaming retarding effect. Most of the light steel materials are used to reduce the load bearing, which makes the whole assembled structure enhanced with flexible connection to achieve high seismic grade [10].

The new bamboo-wood structure system has good thermal insulation and seismic resistance, and is easy to transport and install on site. It is eco-friendly and carbon-recycled, so it is suitable for use in assembled building.In terms of structure and function, the bamboo assembled houses adopt light steel frame structures and bamboo bundle veneer composite materials as the enclosure structures and are equipped with solar-assisted power supply system and hydraulic control system for opening and closing the wall. Under the trend of "National Rural Revitalization Strategy", bamboo and wooden assembled buildings will be very useful in the future in characteristic towns, temporary quick-fit rooms, public green buildings, etc. [11].

EPS Cavity Module Assembly type farmhouse is poor in fire resistance, durability and permeability, but its thermal insulation, seismic performance is good, convenient for construction, cost is reasonable, suitable for the construction of low-rise buildings.

Moreover, the development of assembled building in the construction of a new rural construction should be encouraged from the policy perspective to guide the development of rural housing along the road of Assembly, encourage enterprises and R & D personnel to continuously explore new technologies and new materials, to make assembly production more efficient and diversified, to strengthen the training of relevant professionals, and to provide enough talent reserve for the transformation from traditional extensive rural housing construction to assembly.

4   Concluding Remarks

By promoting the application of assembled buildings in the construction of rural revitalization, actively exploring livable model assembled farmhouses that are pleasing to villagers, modern in function, safe in structure and economical in cost and adopting assembled building technology to construct leisure farms, public rest rooms and other public service facilities for rural tourism, ecologically livable new buildings may be created, which can improve the living environment and the design and construction technology of farmhouses. They are green and may take the regional landscapes into account. Thereby a new pattern of rural revitalization can be established.

5 Fund Projects

Supported by youth project of science and technology research program of Chongqing Education Commission of China. (No. KJQN202303905).

References

1. Wuni, I.Y.: Barriers to the adoption of modular integrated construction: systematic review and meta-analysis, integrated conceptual framework and strategies. J. Clean. Prod. **249**, 69–70 (2021)
2. Jaillon, L.: Quantifying the waste reduction potential of using prefabrication in building construction in Hong Kong. Waste Manag. **29**(1), 309–320 (2009)
3. Pan, M.: Artificial intelligence and robotics for prefabricated and modular construction: a systematic literature review. Construct. Eng. Manag **148**(9) (2022)
4. Janappriya: A comparative life cycle assessment of prefabricated and traditional construction – a case of a developing country. J. Build. Eng. **72** (2023)
5. Aye, L.: Life cycle greenhouse gas emissions and energy analysis of prefabricated reusable building modules. Energy Build. **47**, 159–168 (2012)
6. Yang, C.: Study on the application of assembling farm house technology system. Brick (10), 48–50 (2022)
7. Deng, L.: Analysis of the application of assembly passive house construction technology in rural revitalization. Sichuan Cement **06**, 162–164 (2022)
8. Wang, Y.: Assembled passive house: a study of small residential buildings under the concept of energy saving and healthy living. Architecture & Culture **9**, 27–29 (2020)
9. Jin, W.: The development and prospect of passive housing. China Construction Metal Structure (5), 20–23 (2019)
10. Liang, C.: Research on Spatial Design and Construction of Assembled Public Toilets in the Context of Rural Revitalization. Jilin Jianzhu University (2023)
11. Chen, F.: R & D and demonstration of a new type of assembled bamboo house. Build. Struct. **49**(S2), 504–509 (2019)

Application of Virtual Reality Technology in Parametric Design of Fabricated Shear Wall Structure

Fang Zhou$^{(\boxtimes)}$ and Jingjing Sun

Chongqing Energy College, No. 2 Fuxing Avenue, Chongqing 402260, Shuangfu New District Jiangjin District, China
306696580@qq.com

Abstract. This paper mainly analyzes the application of virtual reality technology in parametric design of fabricated concrete shear wall structure. It includes the overview of virtual reality technology, parametric model of fabricated shear wall structure based on virtual reality technology, parametric design strategy and parametric design development direction analysis. It is hoped that this analysis can provide some references for the application of virtual reality technology and the improvement of design quality of fabricated shear wall structure.

Keywords: Virtual Reality Technology · Prefabricated Shear Wall Structure · Parameterization · Design Method

1 Introduction

Because of the complex shape of the assembled structure shear wall, the reinforcement design is a difficult point, and the traditional design can not meet the production demand. So in the parametric design of modern fabricated shear wall structure, virtual reality technology plays an indispensable application advantage. In order to realize the good application of virtual reality technology, designers need to fully understand this technology first, and then combined with the actual situation, complete the construction of parametric model, and then based on this, the parameterized design of the structure is carried out by reasonable methods. In this way, the design can be reasonably optimized to meet the actual design and application needs of the fabricated shear wall structure.

2 Overview of Virtual Reality Technology

Virtual reality technology, also known as VR technology, is a modern and integrated technology integrated by many disciplines, it integrates many advanced technologies such as multimedia, sensors, new display Internet and artificial intelligence [1]. Through the reasonable application of virtual reality technology, the real scene can be fully simulated, so as to bring users an immersive virtual experience. At present, virtual reality technology has been widely used in many fields such as architecture design, automobile manufacturing and military simulation. Especially in the field of prefabricated building design, virtual reality technology is playing a very good application value [2].

P. Xiang and L. Zuo (Eds.): PBSFTT 2023, LNCE 382, pp. 175–182, 2024.
https://doi.org/10.1007/978-981-97-5108-2_19

3   Parametric Design Model of Fabricated Shear Wall Structure Based on Virtual Reality Technology

Unity-3D is one of the most commonly used software in the virtual reality design of fabricated shear wall structures. Because prefabricated building shear walls have a complex configuration and vary in many ways, the number and location of openings inside may vary, therefore, there is usually no set of fixed rules to follow when constructing the structural model of fabricated shear wall. In this case, the Unity-3D virtual reality design software application is also facing greater difficulties. At the same time, in the multi-form prefabricated concrete shear wall structure, its reinforcement situation is a design difficulty [3]. Based on this, in the virtual reality design of prefabricated shear wall structure, the designer needs to combine the specific process design and relevant standards, to effectively interpret the actual situation of prefabricated shear wall construction, including its actual device situation, industrial production demand. Then based on this, combined with the previous relevant engineering design concepts and design experience, the Unity-3D virtual design software for its structure of a reasonable parametric design. The whole structure can be divided into concrete module, longitudinal reinforcement module and transverse reinforcement module, and the parametric design of prefabricated shear wall structure can be completed by constructing different models. The following is based on the software of fabricated shear wall structure parametric design model construction analysis [4].

3.1   The Model of Concrete Shear Wall

If the assembled shear wall structure is rectangular and has double openings, the lower left corner of the rectangle should be designed as the coordinate origin to realize the reasonable construction of the parametric model. Figure 1 is a parameterized design model of an assembled shear wall structure:

In this process, the endpoints on the shear wall structure can be shown by width (B0) and height (H0), and the vertices of the internal openings can be shown by the central points of width and height (xc1, yc1) and the coordinates of the openings width (b1, b2). Table 1 shows the main position coordinates and their functional relationships of the parametric design model of the fabricated shear wall structure:

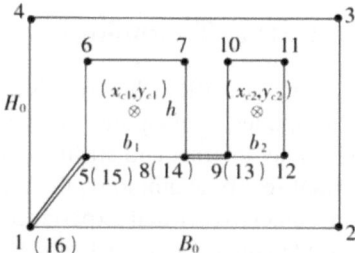

Fig. 1. Parametric Design Model of Assembled Shear Wall Structure.

Table 1. The Coordinates of the Main Fixed Points and Their Functional Relations of the Parametric Design Model of an Assembled Shear Wall Structure ($y_c = y_{c1} = y_{c2}$)

Number	Control Point Number	X Coordinate Function Relationship	Y Coordinate Function Relationship
1	5#	$x_{c1} - b_1/2$	$y_c - h/2$
2	6#	$x_{c1} - b_1/2$	$y_c + h/2$
3	7#	$x_{c1} + b_1/2$	$y_c + h/2$
4	8#	$x_{c1} + b_1/2$	$y_c - h/2$
5	9#	$x_{c2} - b_1/2$	$y_c - h/2$
6	10#	$x_{c2} - b_1/2$	$y_c + h/2$
7	11#	$x_{c2} + b_1/2$	$y_c + h/2$
8	12#	$x_{c2} + b_1/2$	$y_c - h/2$

3.2 The Model of Longitudinal Steel Region

In order to construct the model of the longitudinal reinforcement region of the assembled shear wall structure by Unity 3D software, it is necessary to set the location of the reinforcement reasonably. Figure 2 is a parameterized design model of longitudinal reinforcement of an assembled shear wall structure.

Because the virtual number is much less than the actual number, so in the specific design, it can be regarded as a representative of each module in the longitudinal steel. In this stage, the area where the rebar is located can be divided into seven modules, which are indicated by serial numbers 1–7. In this parameterized model, Point 1 #–4 # belongs to External Control Point, Point 5 #–20 # belongs to Internal Control Point. Based on the geometric relationship between longitudinal bars, the coordinate of control point and its function can be obtained scientifically. Table 2 shows the location points of the main reinforcing bars and their coordinate functions in the longitudinal steel region model of the fabricated shear wall structure:

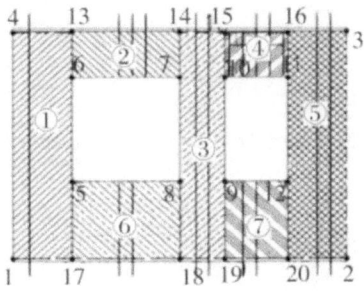

Fig. 2. Parametric Design Model of Longitudinal Reinforcement of an Assembled Shear Wall Structure.

Table 2. The Position Points of The Main Steel Bars and Their Coordinate Functions in the Model of Longitudinal Steel Region of an Assembled Shear Wall Structure.

Number	Location Point Number of Steel Bar	Location Area	X Coordinate Function Relationship	Y Coordinate Function Relationship
1	1#	Area ①	0	0
2	2#	Area ⑤	1	0
3	5#	Area ⑥	$x_{c1} - b_1/2$	$y_c - h/2$
4	6#	Area ②	$x_{c1} - b_1/2$	$y_c + h/2$
5	9#	Area ⑦	$x_{c2} - b_2/2$	$y_c - h/2$
6	10#	Area ④	$x_{c2} - b_2/2$	$y_c + h/2$
7	14#	Area ③	$x_{c1} + b_1/2$	$y_c - h/2$

3.3 The Model of Transverse Steel Region

In this stage, the area where the rebar is located can be divided into five modules, which are indicated by serial numbers 1–5. Figure 3 is a parameterized design model of the transverse steel of an assembled shear wall structure:

Because the virtual number is much less than the actual number, so in the specific design, it can be regarded as a representative of each module in the longitudinal reinforcement. The 1 #–4 # point belongs to the External Control Point and the 5 #–16 # point belongs to the Internal Control Point. Table 3 shows the location points of the main steel bars and their coordinate functions in the transverse steel region model of the assembled shear wall structure:

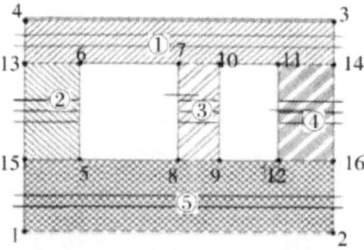

Fig. 3. Parameterized Design Model of Transverse Reinforcement of an Assembled Shear Wall Structure

Table 3. The Position Points of The Main Steel Bars and Their Coordinate Functions in the Model of Transverse Steel Region of an Assembled Shear Wall Structure.

Number	Location Point number of steel bar	Location area	X coordinate function relationship	Y coordinate function relationship
1	3#	Area ①	B0	H0
2	5#	Area ②	$x_{c1} - b_1/2$	$y_c - h/2$
3	7#	Area ③	$x_{c1} + b_1/2$	$y_c + h/2$
4	12#	Area ④	$x_{c2} + b_2/2$	$y_c - h/2$
5	13#	Area ⑤	B0	0

4 Parametric Design Strategy of Fabricated Shear Wall Structure Based on Virtual Reality Technology

In the process of parametric design of assembly shear wall structure by Unity-3D virtual reality design software, in order to effectively ensure the rationality of its design, the designer needs to take object-oriented program form to design [5]. The Unity-3D virtual reality design technology is re-developed with the help of the cooperation of the assembly shear wall structure parametric design model. Then on the basis of this, the SQL Server database is reasonably built to achieve the effective storage of various control parameters [6]. Finally, with the help of virtual reality platform software such as AutoCAD and PKPM, the virtual reality design of assembly shear wall structure is carried out. The following is the parametric design process of fabricated shear wall structure under the condition of virtual reality technology:

4.1 Create an Assembled Shear Wall Model

In the Unity-3D virtual reality design software, every surface is drawn from triangles, and any shape can be obtained by piecing together multiple triangles. Mesh is one of the components of Unity, and in design terms, this component is often referred to as a grid component. Import 3-d model data in the form of vertices, normals, and UV. The model is constructed according to the grid division form in the component. In this case, the concrete parameterized model should be meshed by the ear-cut method. The main modeling steps are as follows: 1) to effectively solve the problem of internal openings, the parametric design model of assembled shear wall structure can be connected with 1 #, 5 #, 8 # and 9 # points by composite line, and the points of 16 #, 15 #, 14 # and 13 # are set as the coincidence points of the above four points respectively, so that the whole parametric model of concrete shear wall can be presented by a simple closed polygon structure. 2) polygon mesh is divided by polygon complex connection and programming method. (3) the corresponding auxiliary tools are added to the Polygon mesh so that the fabricated concrete structure can form an accurate parameterized model in the Unity-3D virtual reality design software.

4.2 Create the Steel Bar Model of Shear Wall

It is also necessary to use triangular mesh to build the steel bar model in the fabricated shear wall structure by the Unity-3D virtual reality design software. Creating, the main flow includes the following: 1) through the reinforcement center line to complete a number of short-line drawing. 2) draw the corresponding hexagons through all the radius of the reinforced page, and place the drawn hexagonal values on the steel axis at several short corners to distinguish the steel grid.

4.3 Create Overall Model

After creating parameterized models of concrete and steel structures in the Unity-3D virtual reality design software using the above methods, designers can extract the information of prefabricated shear wall components in Auto CAD drawings, extract the PKPM calculation results, and import them into the Unity-3D virtual reality design software [7]. In this way, the parameterized construction of the prefabricated shear wall structure model can be completed in the Unity-3D virtual reality design software, thereby fully meeting its parameterized design requirements.

5 Development Direction of Parameterized Design for Fabricated Shear Wall Structures Based on Virtual Reality Technology

With the continuous development of modern science and technology and the construction industry, the virtual reality technology applied in the structural design of prefabricated building engineering is also constantly updated [8]. Especially with the application of modern artificial intelligence technology, the design of prefabricated shear wall structures based on virtual reality technology has begun to develop towards automation and intelligence. To achieve this goal, relevant units, researchers, and technicians should strengthen the application research of this technology, introducing various more advanced big data technologies, artificial intelligence technologies, etc., in order to provide sufficient scientific assistance for virtual reality design [9]. At the same time, intelligent robots can also replace current design and technical personnel to improve the quality of parameterized design in such structural designs and avoid adverse effects of human factors on their design accuracy. Through this approach, virtual reality technology can be fully utilized in the parameterized design of modern prefabricated shear wall structures, thereby promoting the good design, application, and development of modern prefabricated buildings [10].

6 Conclusion

In summary, prefabricated buildings are a major development direction in today's construction technology. In the design of prefabricated building structures, the rational application of virtual reality technology is crucial. Especially for prefabricated shear wall structures, in specific parametric design, designers must take reasonable measures to apply virtual reality technology based on the actual situation, complete the model creation of each module of the prefabricated shear wall structure in the correspond-ing virtual reality design software, and create a sufficiently accurate parametric design model through the acquisition and comprehensive import of various parameters. Through this approach, the design quality of prefabricated building shear wall structures can be effectively ensured, meeting the actual design, manufacturing, and application needs of such structures. Virtual reality design technology will play an increasingly powerful advantage in the design of prefabricated building structures, thereby promoting the good development of prefabricated building engineering technology and its industry in China.

Acknowledgements. Supported by: Chongqing Municipal Education Commission, Science and Technology Research Programme, 2023, Research on Zero-Emission Campus Construction Based on Plant Community Optimisation (Project Title: KJQN202305605, Chair: Jing Sun). No.: KJQN202305605, host: Jingjing Sun).

References

1. Liu, S., Lu, Y., Zhang, H.: .Application of Virtual Reality Technology in Architectural Interior Design (2021)
2. Ren, H., Chen, J.: The application of BIM technology in the research on seismic performance of shear wall structure of prefabricated residential buildings. Int. J. Crit. Infrastructu. (1), 18 (2022)
3. Luo, C., Yao, X., Zhang, Y., et al.: An empirical study on the impact of different structural systems on carbon emissions of prefabricated buildings based on SimaPro. World J. Eng. Technol. (3), 20 (2023)
4. Oladokun, T.T., Ayodele, T.O., Adegoke, A.S., et al.: Analysing the criteria for measuring the determinants of virtual reality technology adoption in real estate agency practice in Lagos: a DEMATEL method. Property Manag. **40**(3), 285–301 (2022)
5. Jin, X., Gao, Y., Fang, W., et al.: Analysis of prefabrication scheme and prefabricated technology application for shear wall residential structures. Architect. Sci. **37**(5), 156–162 (2021)
6. Xu, X.: Analysis of efficient construction problems and application research on parameter optimization of prefabricated shear wall structures. Construct. Mach. **03**, 45–49 (2023)
7. Jiang, Z.: Research on Rapid Plane Generation and Component Selection Optimization Design Method for Prefabricated Buildings. Tianjin University (2020)
8. Zhou, B.: Research on Optimization of Prefabricated Frame Core Tube Structure. Shanghai University (2020)
9. Chen, Y.: Research on the Seismic Performance of a New Modular Prefabricated Double-Layer Composite Shear Wall Structure. Guangzhou University (2019)
10. Shang, T.: Shear wall structure design in prefabricated building structural design . Foshan Ceramics (10), 84–86 (2022)

Development Situation of Assembly Building in Chongqing

Qiushuang Liang[1], Haiyan Yang[1($\boxtimes$)], and Qiujin Liang[2]

[1] Chongqing Telecommunications Polytechnic College, Chongqing 402247, China
1256883904@qq.com
[2] Chongqing Luneng Development Group Co., Ltd., Chongqing 401121, China

Abstract. In recent years, various relevant national policies have been clearly proposed to vigorously develop assembled buildings, taking the development of assembled buildings as an important development strategy in China, and as one of the important directions for the rapid development of building structures. Chongqing Municipality also closely follow the national policy to respond to the call to take various measures to promote the development of local assembly building, this paper describes the concept and characteristics of assembly building, from the policy, data, technology and other support systems to analyze the current situation of the development of assembly building in Chongqing Municipality, the introduction of Chongqing Municipality, part of the typical assembly structure, the actual engineering application case study.

Keywords: Assembly building · Characteristics · case study

1 Introduction

1.1 The Concept of Assembly Building

Assembled building refers to the concept of green development [1], supported by modern scientific and technological progress, industrialized production methods as a means, engineering project management innovation [2] as the core, the world's advanced level as the goal, the extensive use of information technology, energy saving and environmental protection technology [3], etc., to link the whole process of production of building products as a complete integrated industrial chain system.

1.2 Assembly Building Characteristics

The differences between assembled buildings and traditional buildings are mainly reflected in the different construction methods, different operation modes, different construction concepts and so on. Compared with the traditional building mode, assembly building mainly has the following characteristics.

P. Xiang and L. Zuo (Eds.): PBSFTT 2023, LNCE 382, pp. 183–190, 2024.
https://doi.org/10.1007/978-981-97-5108-2_20

(1) It is conducive to improving construction quality. Assembled components are pre-fabricated in the factory, which can maximize the improvement of quality problems such as wall cracking and leakage, and improve the overall safety level, fire resistance and durability of the residence.

(2) Favorable to speed up the progress of the project. Efficiency is the return, according to incomplete calculation statistics, assembly type building is about 30% faster than the traditional way of progress.

(3) Benefit from improving building quality. Assembled building components are unified production, compared with the traditional construction of the site pouring quality is better control.

(4) Benefit from civilized construction and safety management. Traditional site has a large number of workers, now a large number of site operations moved to the factory, the site only need to stay a small part of the workers on it, greatly reducing the incidence of on-site safety accidents.

(5) Benefit from environmental protection and resource saving [4]. The site of the original cast-in-place operation is very little, healthy and non-disturbing, from now on, goodbye to the "gray haze". In addition, the reuse rate of steel mold plate can be improved, and the garbage, loss and energy saving can be reduced by more than half.

2 Status of Assembly Building Development in Chongqing

2.1 Policy Support System

In recent years, the state ministries and commissions have put forward corresponding requirements for the development of assembly building and issued corresponding documents, the relevant content of the documents from various angles to vigorously develop assembly building, some of the documents and the main relevant content involved in the following Table 1.

Following the footsteps of the national policy, Yu Jian [2019] No. 436 [10], Yu Jian [2018] No. 487 [11], Yu Jian [2018] No. 147 [12], Yu Fu Bufa [2017] No. 185 [13] documents and other documents came into being, the documents clearly put forward, the development of assembly construction is to promote the structural reform of the supply side and the development of the new type of urbanization, an important initiative, is the concrete embodiment of the implementation of the concept of green development concept and the concept of innovative development of the construction industry, and it is to promote the transformation and upgrading of the construction industry, and the realization of the modernization of the construction industry is an important hand.

3 Data Support System

By the end of 2022, the city had started construction of a total of 40 million square meters of assembled buildings, and the proportion of assembled buildings in new buildings exceeded 18%, with the central urban area reaching 36%; it had cultivated six national industrial bases, 29 municipal industrial bases, and more than 60 component

Table 1 Relevant national documents and contents.

Document No.	Main contents
State Council [2016] No. 71 [5]	It is proposed that a sound system of laws and regulations related to assembled buildings should be established, and each place should develop assembled buildings according to the actual situation
State Council [2017] No. 19 [6]	Pointing out that it is necessary to vigorously promote assembly-type buildings, and strive to use about 10 years to make assembly-type buildings account for 30% of new construction area
National Development [2018] 22 [7]	Propose to steadily develop assembled buildings according to local conditions
State Office Letter [2019] No. 92 [8]	Proposing to play a leading role in standards and promoting green construction methods
National Development [2021] 23 [9]	Proposing the development of assembled buildings, promoting the recycling of building materials, and strengthening green design and green construction management

part production enterprises. As of July 2023, the city has implemented a total of 47 million square meters of assembled buildings [14] and the proportion of assembled building applications in buildings started in the first half of 2023 reached 30%. In terms of talent, actions to cultivate construction enterprises' own workforce have also been gradually carried out, and it has been proposed that it will strive to reach a ratio of more than 50% of construction enterprises' own workers by 2025.

Assembly buildings are categorized into steel, concrete and wood structures. Concrete structures are more frequently used in the city, including residential and public buildings, while steel and wood structures are less frequently used in residential buildings in Chongqing due to their respective characteristics.

According to incomplete statistics, up to now, a total of 16 projects have passed the pre-evaluation of Chongqing assembled buildings in 2023, all of which are assembled concrete structures and involve an assembled floor area of about 850,000 m^2, including about 80,000 m^2 for public buildings, 700,000 m^2 for residential buildings, and about 50,000 m^2 for the combination of residential and public buildings, accounting for 10%, 82%, and 82% of the projects that have passed the pre-evaluation of Chongqing assembled buildings in 2023, respectively. 10%, 82% and 6% of the projects. Specific building types and area shares are shown in Fig. 1. Among them, projects belonging to Chongqing Liangjiang New District accounted for about 30%, Jiulongpo District accounted for about 10%, and the others were distributed in Jingkai District, Jiangbei District and Shapingba District.

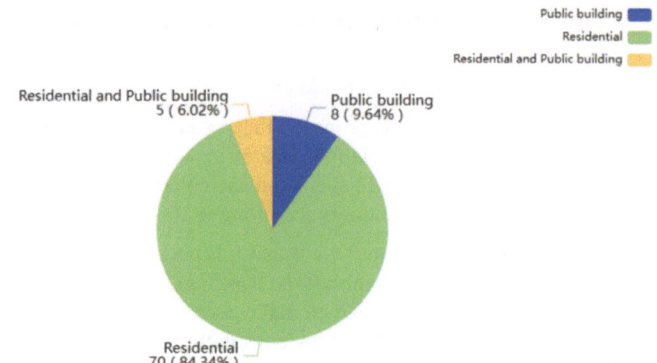

Fig. 1. Area and percentage of assembled building types in Chongqing, 2023

4 Technical Support System

In Chongqing, in addition to the expansion of new assembly building area, some assembly technologies are also applied in the production and construction procedures. For example: non-load-bearing peripheral parapet wall adopts self-insulating autoclaved aerated concrete precise block thin masonry process wall; internal partition wall adopts autoclaved aerated concrete slats, adopts the integration technology of internal partition wall and pipeline; only the public area and the area of determining the function of use is fully decorated, adopts the integrated bathroom, pipeline separation technology, the building floor adopts the modular heat preservation and acoustic isolation of the functional parts of the dry working method, and the decoration and the pipeline of the equipment adopts the full decoration, integrated kitchens and integrated bathrooms, and the adoption of information technology for engineering and construction. The application of these technologies fully reflects the design standardization, production factory, construction assembly, decoration integration, management informationization and application intelligence of the assembly building.

5 Analysis of the Application of Some Assembly Structure Projects in Chongqing City

Chongqing Urban Construction Archives Project, as shown in Fig. 2, as the first case of assembly type clear water outer enclosure insulation integration structure project in Chongqing, the total land area of about 36,000 square meters, the total construction area of about 112,200 square meters, the main body of the steel structure project, the assembly rate of about 70%, belongs to the ultra-limit high-rise building projects and technically complex projects, technical difficulties. During the construction process, some or all of the components are prefabricated in the factory and then transported to the construction site for assembly.

The experimental building project of Chongqing University Huxi Science Center covers an area of about 172 acres, with a construction area of about 247,700 square meters, and is located in the center of the Western (Chongqing) Science City, as shown in Fig. 3. The construction environment is complex, in order to effectively reduce the construction of the environment, noise pollution, in order to speed up the construction progress at the same time, the project selected the green assembly slope support technology, compared with the traditional spray anchor support, has the characteristics of fast construction speed, large turnover.

Fig. 2. Chongqing Urban Construction Archives

Fig. 3. Huxi Science Center Laboratory Building, Chongqing University

Tongan high-speed (Chongqing Tongliang to Sichuan Anyue) Chongqing section, as shown in Fig. 4, total length of about 48.64 km, designed for two-way six-lane, two toll booths using assembly, integration of steel structure canopy, compared with the previous brick structure, the construction is more convenient, more beautiful shape, more economical cost.

The mainline of the Cha Hui Avenue project is 8.96 km long, and the cantilevered cover girders of the mainline bridges are supported by an assembled SPS truss-type support system, as shown in Fig. 5, in which steel pipes with a diameter of 133 mm and a wall thickness of 5.6 mm are connected and secured with reinforcing bowls and fasteners to form the support. This construction method provides higher strength and better member stability than traditional support systems.

Fig. 4. Steel Structure Canopy of Pingtan Toll Station

Fig. 5. Cantilever cover beam for Cha Wai Avenue

Fig. 6. Construction effect of Kui Zhou Yangtze River Bridge

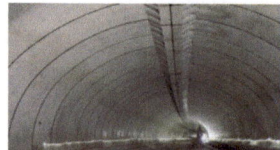

Fig. 7. Shinsen Avenue

Kui Zhou Yangtze River Bridge, shown in Fig. 6, is the largest center-bearing mesh arch bridge in the world in terms of span, and the first such bridge in China. A major feature of the bridge is the adoption of fully prefabricated assembly construction, which transforms the "design-site construction" mode into the "design-factory manufacturing-site assembly" mode. In addition to the foundation part, the bridge substructure, steel arch, bridge deck and other "parts", according to the size and standards in the factory to do a good job and then transferred to the construction site, like "building blocks" for assembly, effectively improve the construction efficiency, save personnel costs, energy saving and environmental protection.

The 2.5-km-long, two-way, eight-lane Xinsen Avenue, shown in Fig. 7, is China's first cut-and-cover assembly tunnel, where the tunnel lining is completed like building blocks. Each prefabricated component is shaped like a quarter-circle arc, weighs more than 70 tons, and needs to be lifted by a gantry crane.

6   Conclusion

In recent years, the assembly structure of Chongqing Municipality has been fully developed, including the increase of assembly area, and the improvement of the assembly rate of the project, etc. The assembly building has also changed from a simple frame structure to the current use of tunnels and bridges and other parts of the building, which greatly improves the application of assembly in the city, which is not lacking in some of the country's leading use of new technologies, such as the largest span of medium-bearing netted arch bridge in China, which is also the first cross-river bridge of its type. The first cross-river bridge, which are all breakthroughs brought about by the vigorous development of assembly type construction. However, there are certain problems in the development, in order to fundamentally solve these problems, we must reform today's inherent concepts, based on technological innovation, increase the assembly construction technology research efforts, improve the corresponding management system, improve the relevant testing standards, the prefabricated components of the production enterprises to carry out a reasonable layout, so that the production of safer, more reliable quality.

Fund Projects Supported by youth project of science and technology research program of Chongqing Education Commission of China. (No. KJQN202105502).

References

1. Lixin, Z., Ju, M., Baohui, H.: Research on the development prospect of assembled passive building based on green development concept. IOP Conf. Ser.: Earth Environ. Sci. **113**, 012111 (2018). https://doi.org/10.1088/1755-1315/113/1/012111
2. Zhang, Y.-L.: Research on engineering project management and development issues of assembled buildings. Journal (06), 183–185 (2022)
3. Chen, H.: Construction technology of concrete assembled residential building. Journal **48**(11), 69–70 (2021)
4. Zhang, S.: Research on the impact of the development of assembly building on the supply-side structural reform of the construction industry. Journal **31**(04), 124–125 (2020)
5. Guiding Opinions of the General Office of the state council on vigorously developing assembled buildings. Journal (29), 24–26 (2016)
6. Opinions of the General Office of the State Council on promoting the sustainable and healthy development of the construction industry. Journal (08), 94–98 (2017)
7. Circular of the State Council on the issuance of the three-year action plan for winning the battle of defending the blue sky. Journal (20), 40–52 (2018)
8. Circular of the General Office of the State Council Transmitting the guiding opinions of the ministry of housing and urban-rural development on improving the quality assurance system to enhance the quality of construction projects. Journal (29), 18–22 (2019).
9. Circular of the state council on the issuance of the action program for carbon peak by 2030. Journal (31), 48–58 (2021).
10. Notice of the Office of the People's Government of Yongchuan District, Chongqing on Accelerating the development of assembled buildings and promoting modernization of the construction industry. Journal (02), 52–53 (2020)
11. Notice of Chongqing Yongchuan District People's government office on accelerating the development of assembled buildings to promote modernization of construction industry. Journal (02), 52–53 (2020)
12. Notice of Chongqing Municipal Urban and Rural Construction Commission on the work related to the implementation of assembly building projects. Journal (06), 36–37 (2018)
13. Implementation opinions of the general office of Chongqing municipal people's government on vigorously developing assembled buildings. Journal (23), 46–49 (2017)
14. Chongqing has implemented 47 million square meters of assembled buildings. Journal 22(07), 27 (2023)

Collision Detection Algorithm Based on Precise Model in Virtual Assembly

Yunlai Zhang[✉] and Xiajuan Shi

Chongqing College of Architecture and Technology, Chongqing 401331, China
349816825@qq.com

Abstract. Virtual assembly is one of the important trends in the development of assembly industry, and this assembly technology has been applied in many fields. However, it is urgent that the problem of inaccuracy of collision detection caused by polygon model in virtual assembly should be solved. So an accurate model of collision detection algorithm needs to be studied in support of virtual assembly's effective design and practice. Based on the K-DOPs algorithm in virtual assembly, this paper conducts collision detection of accurate models, verifies the effectiveness of this algorithm to shorten the collision detection time, and provides some ideas and references for effectively selecting appropriate collision detection algorithms in virtual assembly.

Keywords: virtual assembly · accurate model · collision detection · algorithm

1 Introduction

The development of science and technology has had a significant impact on the development process of world industry [1]. For the manufacturing industry, the dependence on people is increasingly weakened with the introduction of high and new technology, and a revolution that changes the entire manufacturing industry is taking place. Taking virtual assembly technology as an example, it has played a very important role in the product development and manufacturing process of many enterprises. Collision Detection is a key technology in the virtual assembly system, which is to detect whether, when and where a collision occurs between two or more objects in the virtual environment [2]. Accurate and efficient collision detection is very important for virtual assembly systems. In the virtual assembly system, when two objects are in contact or when the virtual hand is in contact with the object, the collision detection algorithm is needed to detect the contact between the two objects. Moreover, the virtual hand modeling and three-dimensional objects are mostly polyhedral modeling structures, and the accurate collision detection is based on the level of surface patch. In other words, the accurate judgment of whether there is a collision between two objects is mainly to detect whether there is a collision between the two surface patches. However, it is difficult to carry out collision detection directly among so many parts, the system calculation amount increases, and the response is not timely. In this regard, it is necessary to analyze the assembly relationship between parts and components in the process of importing parts into the system to reduce the

P. Xiang and L. Zuo (Eds.): PBSFTT 2023, LNCE 382, pp. 191–196, 2024.
https://doi.org/10.1007/978-981-97-5108-2_21

number of collision detection, so as to promote the continuous reduction of system cal-culation and improve the real-time performance of the system. In order to continuously improve the system reaction performance, it is necessary to select a reasonable collision detection algorithm. Such algorithms need to be as simple as possible to improve the system's responsiveness [3].

2   Description of Collision Problems in Virtual Assembly

The essence of collision detection is to determine the real-time model when the product digital model is in what relation. When the distance between two parts is 0, collision will occur, and in virtual assembly, collision detection based on this theory cannot meet the requirements of virtual assembly:

First of all, in the simulation, the distance between the parts is a discrete variable, and it is almost impossible to equal 0. Because the simulation algorithm is used for calculating the position iteration of parts by discrete time steps, and the parts do not move continuously [4]. Even if the smaller distance length is chosen to make the motion of the part appear continuous, the part is still essentially jumping from one position in space to another. In collision detection, the parts are still separated at that point in time and are already connected at the next point in time.

Second, the computer is not yet able to achieve accurate representation of the parts. Under normal circumstances, the accurate modeling method can express the error-free ideal state of the part, but the polygon model is only an approximate description, and the two should apply different methods to approximate the expression of the objects, so there is still a certain distinction between the real thing and the object. Therefore, it is difficult for the collision detection results of relevant models to fully reflect the problem of actual product collision [5]. In virtual assembly, it is necessary to know the space range near a part, so as to judge whether the part is reasonable. When conducting virtual maintenance, it is also necessary to know the spatial distance near the repaired part, so as to determine whether the maintenance tool can enter or operate normally. Taking into account the actual needs, it is often necessary to know the distance value of other parts that collide with the part or the part and other parts in the direction when a part moves a certain distance in a certain direction. If the manual measurement is carried out, the efficiency will be very low. According to this requirement, the algorithm of space collision fast verification and space distance fast verification is proposed, which are referred to as collision verification and distance verification respectively [6].

At present, collision detection in virtual assembly is a big problem to be solved urgently, so it is necessary to study the effective algorithm of collision detection.

3   Precise Collision Detection Method of K-DOPs

K-DOPs method needs to combine the specific shape of the scene, select multiple groups of parallel plane pairs with different directions, which contains a scene or a group of scenes of the hierarchical bounding box technology. These parallel plane pairs form a convex body, that is, a parallel 2K hedron. Based on this idea, more than 3 parallel plane pairs are used to approximate the object in collision detection. All parallel plane pairs are derived from the intersection of half Spaces defined by two normal vectors with opposite directions, and the normal vectors of these parallel plane pairs are fixed. In the specific application process, the direction of normal vector can be selected according to the specific situation, which is necessary to simplify the calculation process of parallel planes [7]. This collision detection algorithm is proposed based on the vertex mapping principle. The idea is to map a set of planar normals to various points on the k-DOP, and each stored point represents the farthest point along that direction. This point and this direction form a plane that completely contains the model in a half space, that is, the point is at the vertex of the k-DOP of the model. In the process of testing, vertices of a given orientation can be used for more precise cross-testing between K-DOPs, such as for improved cone culling, and for finding tighter AABBs after rotation.

If the value of K is large enough, K-DOP can become the convex hull of an object. In planar graphics, a convex hull is a convex polygon that surrounds a set of control points. In stereo graphics, the convex hull is mainly a polyhedral envelope connecting each fixed point with a surface [8]. Therefore, the package object of the convex hull is more compact, as long as the size of K and the parallel plane direction are scientifically selected, the collision detection process can be simplified, the superstition between the wrapped objects can be improved, and the system requirements can be met. The intersection test of K-DOP is shown in Fig. 1 below:

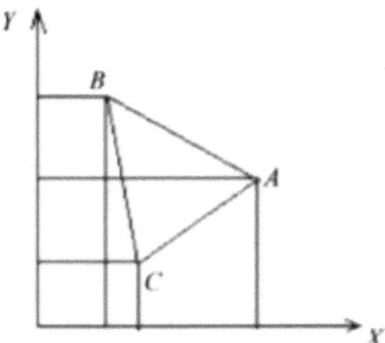

Fig. 1. Intersection test of K-DOP

Suppose SA and SB are subsets of the basic geometric elements corresponding to nodes VA and VB, and b(SA) and b(SB) are their K-DOP bounding boxes, respectively, and $\emptyset$ represents the empty set. If you want to test whether two k-DOPs, A and B (superscript index A and B) intersect, then test whether all parallel Slab pairs overlap; It's a one-dimensional interval overlap test, easy to solve. This is an example of the dimensionality reduction recommended by the rule of thumb in Sect. 22.5. In this paper, the three-dimensional Slab test is simplified to one-dimensional interval overlap test. If at any time (i.e., an empty set), the BVs do not intersect and the test terminates. Otherwise, the Slba overlap test will continue. BVs are considered to overlap if and only if all $s_i \neq \emptyset, 1 \leq i \leq k/2$. According to the separation axis test, it is also necessary to test an axis of the cross product parallel to one edge of each k-DOP. However, these tests are often overlooked because their cost is higher than their return in performance [9]. Therefore, if the following test returns k-DOP overlaps, then they may in fact be disjoint.

Only k scalar values need to be stored in each instance of K-DOP (for all K-DOPs, the normal ni is stored only once because they are static). If k-DOP is translated by tA and tB respectively, the test becomes a little more complicated. Project tA onto the normal ni, such as piA = tA·ni(note that this is independent of any k-DOP, so it only needs to be calculated once for each tA or tB), and add piA to diA, min, and diA, max in the if statement. The same approach applies to tB. In other words, the translation changes the distance of k-DOP along each normal direction.

In order to accelerate the collision detection between objects, we can implement parallel bounding box-based collision detection between all objects with assembly relation [10]. Use a unified algorithm for all objects to conduct collision tests on other objects in parallel and virtual scenes. In this way the corresponding detection time can be changed from the original ft = ta + tb +... to ft = {maxlta + tb +... And it can also effectively reduce the collision detection time and help adapt to the requirements of the virtual assembly system. Figure 2 below shows the specific collision detection algorithm:

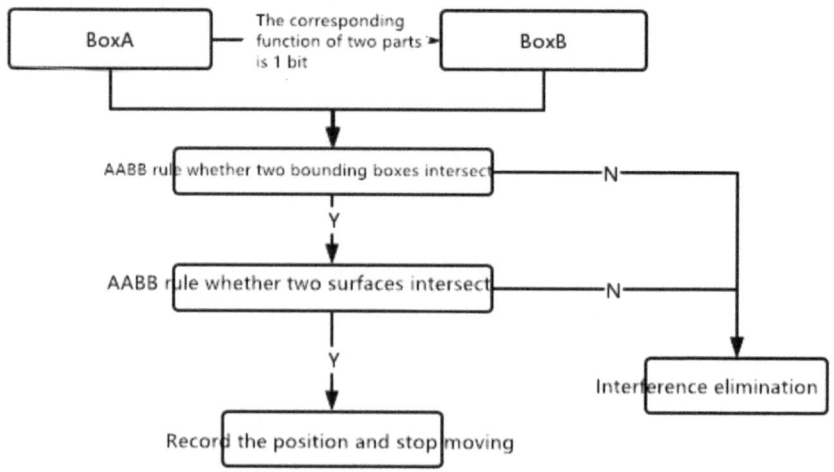

Fig. 2. Collision detection algorithm

After interference elimination is completed and the position of the object changes, the object is compared with the final initial position of the assembly. When j > 0 (j represents the projected length on the corresponding direction segment), the relationship between the current position and the final position of the virtual assembly is indicated. If the assembly process is completed within the preset threshold range, it means the completion of the assembly process, namely the end of the interference elimination process. In the case of j < 0, it represents the relationship between the current position and the initial position of the virtual assembly. If the object is out of the assembly interference area, then the assembly process is completed [11].

4 Conclusion

For virtual assembly, collision detection is an important task, and ensuring the accuracy of collision detection is the guarantee of the authenticity of virtual assembly data [12]. The use of 3D digital models for real product simulation is key to improving the accuracy of collision detection. In this study, it is proposed to establish the assembly relationship between virtual assemblies, combined with the use of the bounding box method, and finally carry out interference elimination. The application of k-DOPs algorithm is studied to improve the efficiency of the system collision detection and shorten the time required by the collision detection algorithm. This algorithm has been applied in virtual assembly with remarkable results. In the specific experimental operation, it is also necessary to consider that there may be ineffective compatibility between various parts, which may lead to a decrease in system efficiency, so relevant work needs to be further improved in the future.

References

1. Hou, W., Ning, R., Liu, J.: Collision detection algorithm based on precise model in virtual assembly. J. Comput.-Aided Des. Graph. **22**(5), 797–802 (2010). (in Chinese)
2. Li, C., Zou, X., Lian, G., et al.: Construction method of virtual assembly classification detection network data set for agricultural machinery. J. South China Agric. Univ. **42**(6), 117–125 (2021). (in Chinese)
3. Du, Q.: Application of collision detection based on Ant colony Algorithm in Virtual assembly. North China Electric Power University (2019)
4. Qu, C., Wei, Y., Wang, J., et al.: Research on virtual assembly technology of network-oriented modular fixture. China Mech. Eng. **24**(15), 2060–2065 (2013)
5. Bodenko. Research on Continuous collision detection Algorithm based on Open CASCADE platform. Dalian Maritime University, Liaoning (2015)
6. Hou, W., Yan, Y., Yao, F.: Research on assembly process design system based on virtual technology. Mach. Des. Res. **22**(3), 70–74 (2006)
7. Zhao, D., Wang, X., Chen, G., et al.: Approximate convex hull adaptive bounding box collision Detection Method. Sci. Technol. Eng. **23**(22), 9592–9598 (2023). (in Chinese)
8. Wang, Y., Gu, Y., Cao, X., et al.: Virtual Assembly Collision detection based on DirectX SDK. Combined Mach. Tool Autom. Process. Technol. (7), 52–54, 57 (2006)
9. Li, Y., Hu, Z.: A collision interference detection method in virtual assembly environment. Heavy Mach. (3), 17–19, 22 (2006)
10. Zhang, X., Dong, L., Wang, J., et al.: Design of virtual disassembly system of CNC machine tool based on Sphere-EBB algorithm. Mech. Des. Manuf. Eng. **52**(1), 81–85 (2019)
11. Wang, D., Luo, X., Guan, Y., et al.: Application of distributed collision detection in assembly simulation. Mech. Sci. Technol. Aerosp. Eng. **34**(9), 1394–1398 (2015)
12. Zhang, J., Li, D., Hu, Z.: An accurate collision detection algorithm with integrated space decomposition and occupancy and its application in construction engineering. Eng. Mech. **31**(5), 79–85 (2014)

Experimental Study on a Void Imagination Approach for a Full-Scale Concrete-Filled Steel Tube Specimen in Long-Span Suspension Bridge Tower with Electromechanical Impedance

Wei Hu[1,2,3], Qian Liu[4,5], and Bin Xu[4,5(✉)]

[1] CCCC Second Harbor Engineering Company Ltd., Wuhan 430040, China
[2] Key Laboratory of Large-Span Bridge Construction Technology, Wuhan 430040, China
[3] Research and Development Center of Transport Industry of Intelligent Manufacturing Technologies of Transport Infrastructure, Wuhan 430040, China
[4] College of Civil Engineering, Huaqiao University, Xiamen 361021, China
binxu@hqu.edu.cn
[5] Key Laboratory for Intelligent Infrastructure and Monitoring of Fujian Province, Huaqiao University, Xiamen 361021, China

Abstract. Due to complex internal structure details and insufficient quality control methodologies during concrete pouring, defects such as void in large-scale concrete-filled steel tube (CFST) components possibly occur. In this paper, an internal void defects imagination approach using electromechanical impedance (EMI) measurement with surface-fixed Piezoelectric Lead Zirconate Titanate (PZT) sensors is proposed for a full-scale CFST specimen with a diameter of 3.6m that is employed in a tower with a height of 350 m of a long-span suspension bridge. A variety of voids are mimicked by placing empty boxes of various sizes at various locations under the inner horizontal diaphragm of CFST specimen to test the viability of the proposed method. To detect the existence of the voids, numerous PZT sensors are mounted on the outer surface of the specimen and Electromechanical impedance (EMI) measurement of the PZT sensors at selected certain frequency band are made. Based on the analysis on the impedance curves and the root mean square deviation (RMSD) of each EMI measurement, the localization of each artificially mimicked void defect is detected successfully. Moreover, the distribution of all mimicked voids is visualized with an imagination method based on RMSD values of PZT sensors at different locations. The visualization result meets the actual locations of the artificially mimicked void defects inside the CFST specimen.

Keywords: Concrete-filled steel tube · Electromechanical impedance method · Experimental studies · Void defects · Steel box-CFST composite tower structure

P. Xiang and L. Zuo (Eds.): PBSFTT 2023, LNCE 382, pp. 197–207, 2024.
https://doi.org/10.1007/978-981-97-5108-2_22

1 Introduction

Concrete-filled steel tube (CFST) structures make full use of advanced mechanical properties of both concrete and steel, and have been widely used in large-scale civil engineering structures including super high-rise buildings, underground works, and long-span bridges in recent years [1, 2]. Due to the poor construction quality control methodologies and unavoidable shrinkage of mess concrete, various types of defects, such as void and interface debonding may occur under inner horizontal diaphragms of CFST components. Interface debonding and void defects will seriously affect the mechanical properties and service life of the CFST structure, which leads to risk in structural safety and serviceability [3–5]. Therefore, the development of defects detection technology for CFST structures is an emergent concern in civil engineering.

Zhangjinggao Yangtze River Bridge, with a main span of 2,300 m, is the world's largest main span suspension bridge under construction. The bridge has a steel box-CFST composite tower with a height of 350 m, and the diameter of each circular CFST column employed in the bridge tower is 3.6 m as shown in Fig. 1. Horizontal diaphragms, two vertical partition plates mutually perpendicular to each other and shear studs are employed to ensure the composite effect between the inner wall of the steel tube and the concrete of each CFST component. During the construction of the steel box-CFST composite tower, void defects may occur under the inner horizontal diaphragm due to complex internal structure details and insufficient vibration during concrete pouring. This pouring uses the high drop method, although the method is conducive to the self-compaction of concrete, but the internal design of this experimental component has inner horizontal diaphragm, and the discontinuity of concrete construction is prone to produce the situation of uncompacted pouring. There is a critical need to develop efficient methods for visualizing potential void in the full-scale CFST column of composite tower.

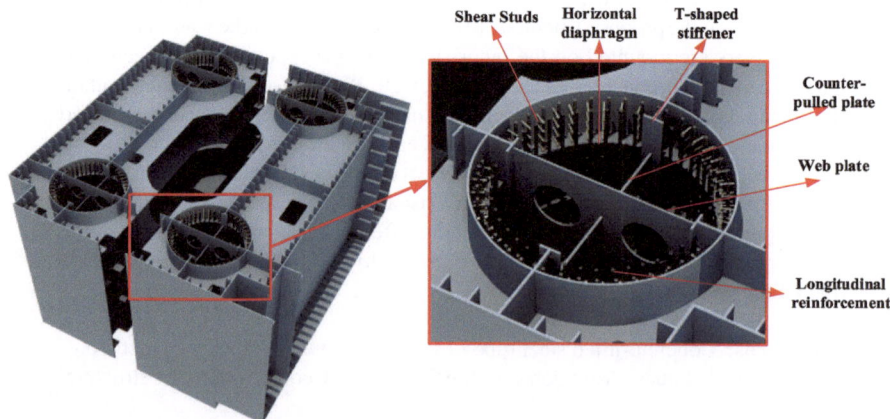

Fig. 1. The schematic diagram of the standard section of steel box-CFST composite tower

To detect structural defects, several non-destructive testing (NDT) techniques have been developed [6], e.g., hammer impact method [7, 8], ultrasonic guided waves [9], impact-echo method [10], and infrared thermal imaging [11]. Unfortunately, the performance of some traditional approaches is dependent on the experience of operators. Ultrasonic guided wave methods have been used to detect concrete cracks, interlayer debonding of composite laminates, but are not capable of the defect detection for large-scale composite and hybrid engineering structures. Infrared thermal imaging method to determine the scope of defects through the heat conduction theory, but still needs to conduct in-depth research on the clarity of the infrared thermal image. The steel tube of CFST members as a metal medium with electromagnetic shielding, traditional electromagnetic method does not work for internal void defect detection of CFST members also. In contrast to the NDT method described above, electromechanical impedance (EMI) measurement shows its advantages in defect detection of engineering structures [12, 13]. Lukesh et al. [12] first described the advantages and disadvantages of piezoelectric impedance technology and the current engineering application in detail, followed by a discussion of the fabrication techniques of new sensors and data processing methods applied to the EMI method. The development prospects of the EMI method are analysed by focusing on the influence of sensor materials and external environmental on the detection accuracy. The new directions of EMI technology in the field of detection are proposed from various aspects, such as intelligent measurement technology, diversification of acquisition parameters and depth of data mining. Nguyen et al. [13] conducted a relevant experimental study and numerical simulation analysis after verifying the correlation between the admittance signal and anchor force using theoretical derivation in order to achieve dynamic monitoring of anchor force of prestressed anchorage using EMI technology, respectively. The change of anchor force was successfully simulated by changing the contact stiffness of the numerical model, and the detection of anchor force was tested on this basis. Li et al. [14] found that smart corrosion coupon (SCC) methods can accurately determine the corrosion of metal plates through experimental and numerical simulation studies based on peak frequency shift and peak variation of the conductance signal of PZT pasted on the metal plates. Luo et al. [15] combined magnetic materials with piezoelectric ceramic materials to design a PZT sensor with a magnetic base. The thickness of the magnetic base can not only affect the magnetic properties of the base, but also correlate linearly with the impedance resonance frequency of the magnetic sensor. To test the detection capability of the magnetic sensor, an experimental and numerical simulation study of bolt loosening monitoring was conducted. The experimental and numerical simulation results math well, and the location of the bolt loosening can be located using EMI measurements with the SMT. Wen et al. [16] proposed the EMI method to detect core concrete cavity defects of CFST columns. Considering the electromechanical coupling between PZT and the composite structure, an analysis model is established by ANSYS to analyse the impedance frequency curve of the composite structure with cavity defect and explore the mechanism of the EMI method for defect detection, which verifies the effectiveness and practicality of the method.

In this paper, an experimental study on the feasibility of EMI measurement-based void defects detection and visualization method for a large-scale CFST member in a steel box-CFST composite tower structure of the Zhangjinggao Yangtze River Bridge was conducted. Void defects with different sizes under the inner horizontal diaphragm of the specimen are mimicked and EMI measurement at selected frequency band of

surface-mounted PZT sensors are performed. Regions of void defects in CFST specimens were visualized using an interpolation algorithm based on RMSD results of EMI
measurements from number of PZT sensors.

2   Void Defect Detection Principle Based on EMI Method Using Surface-Mounted PZT Sensors

The one-dimensional coupling system composed of PZT sensors and detected structure
was proposed by Liang et al. for EMI analysis as shown in Fig. 2 [17]. The model
explains the correspondence between the electrical impedance of PZT and the structural
impedance of tested structure, and evaluates the damage condition of the tested structure
by the coupling properties between the PZT and the tested structure. The EMI method
uses the electrical impedance response of PZT to reflect the local damage of the tested
structure. The damage causes change in the local characteristics (mass, stiffness) of the
tested structure, which leads to change in electrical impedance for the purpose of defect
detection.

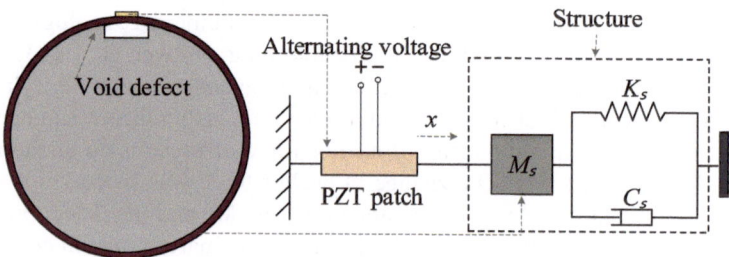

Fig. 2. EMI measurement system

The RMSD value of the impedance signals of each sensor is calculated for visualizing
the location of the mimicked void defects in CFST specimen.

$$RMSD_i = \sqrt{\frac{\sum_{j=1}^{N}(Z_{i,j}^1 - Z_{i,j}^0)^2}{\sum_{j=1}^{N}(Z_{i,j}^0)^2}} \times 100\% \tag{1}$$

where, $Z_{i,j}^1$ represents the impedance test data of the i-th sensor at the j-th frequency,
N represents the total sampling points of the impedance analyzer in the test frequency
band, and $Z_{i,j}^0$ is the average of the impedance of PZT sensors at healthy positions at the
j-th frequency.

The damage indexes adapted to the EMI are established using the RMSD, which can
effectively reflect the data differences between measurement points at different locations
in different states to obtain the damage of the structure. On this basis, the defect range
is visualized according to the RMSD at different locations. Equation (1) expresses that
if there are defects in the detection location, the greater the impedance difference with
the healthy location, the greater the RMSD will be. If the detection location is in good
condition, the impedance match with the healthy location and the RMSD difference is
not large.

3   Full-Size CFST Specimen with Mimicked Void Defects and PZT Sensors Arrangement for EMI Measurements

3.1   Full-Size CFST Specimen

The diameter and height of the full-scale CFST specimen are 3600 mm and 3000 mm, respectively. The thickness of the steel tube is 30 mm. The thickness of horizontal diaphragm and vertical plates are 20 mm. To secure the bond between concrete and steel tube, shear studs are welded in the inner wall of the steel tube, which will effectively reduce the debonding phenomenon between concrete and steel tube wall caused by concrete shrinkage. C60 self-compacting concrete was poured into the steel tube using the high drop method of pouring concrete. Full-scale CFST member on site and the cross section are shown in Figs. 3 and 4. The impedance analyser is used to excite the sensor on the surface of the structure and obtain the electrical impedance data.

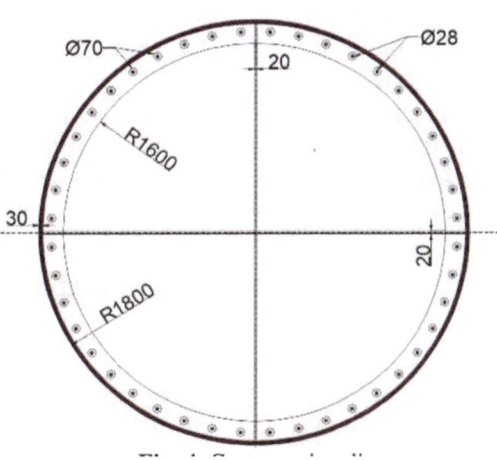

Fig. 3. Full-scale CFST specimen   **Fig. 4.** Cross section diagram

3.2   Artificially Mimicked of Void Defects

In order to mimic void defects between the horizontal diaphragm and concrete core, number of empty wooden boxes were pasted with epoxy resin on the lower surface of inner horizontal diaphragm at designed locations before concrete pouring. The empty wooden boxes were made of 3 mm thick wooden board material, and then filled with epoxy resin glue to fill the gap of the board to avoid the flow of concrete into the empty box. The dimensions of the mimicked void defects and the labels are displayed in Table 1. Each void defect corresponds to a healthy location under the same boundary conditions with the numbers KH1, KH2, KH3, KH4, and KH5, respectively. Some examples of the mimicked empty boxes installed under the horizontal diaphragm are shown Fig. 5. Concrete pouring is carried out after the installation of mimicked void defects is completed.

Table 1. Dimensions of mimicked void defects.

Label number	Size (Radial × Circumferential × Vertical, mm)	Corresponding health label
KD1	30 × 100 × 80	KH1
KD2	50 × 100 × 80	KH2
KD3	80 × 100 × 80	KH3
KD4	30 × 150 × 80	KH4
KD5	50 × 150 × 80	KH5

To detect the region of five void defects KD1, KD2, KD3, KD4, and KD5, five surface-mounted PZT sensors are installed, one PZT sensor is installed at the centre of the mimicked defect and other four PZT sensors are located at 2 cm outside of the boundaries of the void defect. Taking the void defect KD4 as an example, sensor installed at the canter of the defect is named KD4, and additional four PZT sensors, named as KD4-1, KD4-2, KD4-3 and KD4-4, are located at 2 cm outside of the four boundaries of the mimicked defect as shown in Fig. 6. For comparison, surface-mounted PZT sensor labelled as KH are placed at healthy locations where no void defect is mimicked and are represented by cross lines as shown in Fig. 7. After the PZTs were welded with the wire, the negative electrode of the PZT is fixed on the surface of the structure by epoxy resin.

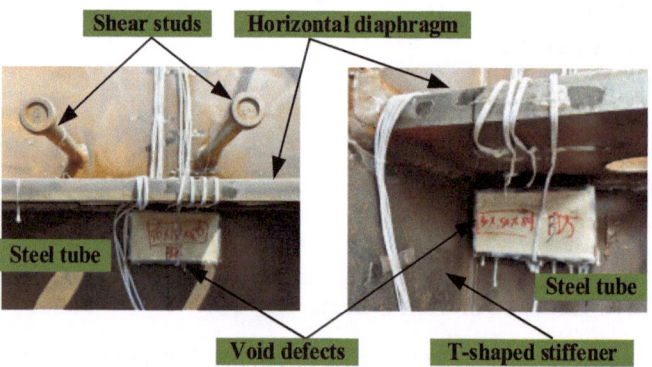

Fig. 5. Mimicked void defects in CFST specimen

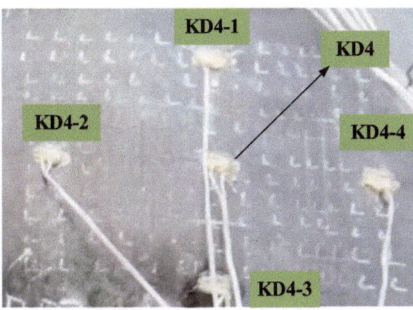

Fig. 6. PZT sensors installed at defect KD4

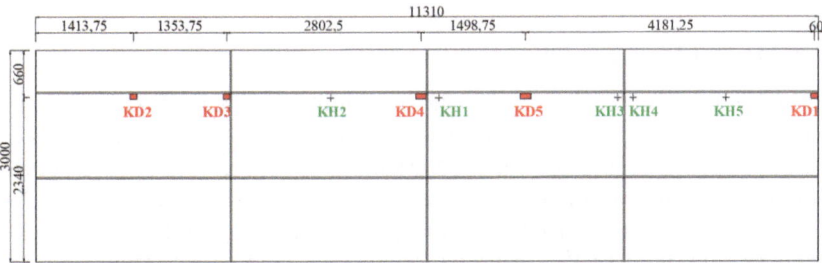

Fig. 7. Locations of mimicked void defects and the PZT sensors at healthy regions of the expended view of the steel tube (unit: mm)

4 EMI Test Results and Void Defect Detection and Visualization

4.1 Impedance Curve Analysis

According to previous studies, the appropriate frequency range is chosen to measure the impedance of the sensor [16]. Using the test system based on the impedance method, the sensors at different locations are tested separately and the impedance curves of each measurement point are obtained. EMI measurements of the sensor installed at the center of the artificially mimicked void defect were compared with the average value of all five sensors KH1, KH2, KH3, KH4 and KH5 at the healthy location, as shown in Fig. 8. The impedance curve of the average of the five PZT sensors at healthy conditions is smooth. Nevertheless, the impedance curves of the five PZT sensors KD1, KD2, KD3, KD4 and KD5 installed at the center of five void defects have obvious fluctuations and new peaks. The void defect can be detected successfully based on the EMI measurements. The EMI measurements collected from all the PZT sensors were used for RMSD calculations and results show that the RMSD of the sensors installed at the center of void defects are significantly higher than those at the healthy locations as shown in Fig. 9. The defect state at the tested location can be verified clearly by the impedance Z and the corresponding RMSD.

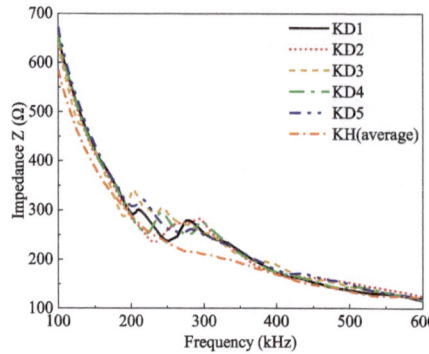

Fig. 8. Impedance of PZT sensors

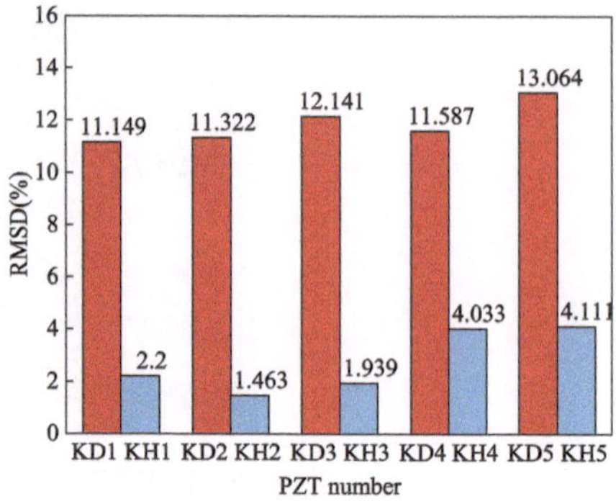

Fig. 9. RMSD values.

4.2   Void Defect Visualization

The RMSD of the five PZT sensors around the mimicked defect KD4 shown in Fig. 10. The RMSD corresponding to the sensor KD4 installed at the center of mimicked void defect is significantly higher than the RMSD values of the other four PZT sensors located at 2 cm outside of the void defect boundaries. Based on the calculated RMSD values of the five PZT sensors, interpolation was used to calculate the RMSD values of the region covered by the five PZT sensors around each void defect. The position where the test piece is not arranged with sensors for measurement is by default the healthy position. The matrix of RMSD values at different locations was mapped to a color matrix using the colormap function in MATLAB software. The location and region of the mimicked void defects and the actual location shown in solid line in the expended view of the steel tube are shown in Fig. 11. Figure 11 depicts the expansion diagram of the whole specimen with the circumference of the specimen in horizontal coordinates and the height of the specimen in vertical coordinates. The locations and sizes of the five void defects were imaged by the RMSD results corresponding to all the sensors arranged around the artificially mimicked void defects. The mimicked void defect can be identified and visualized successfully.

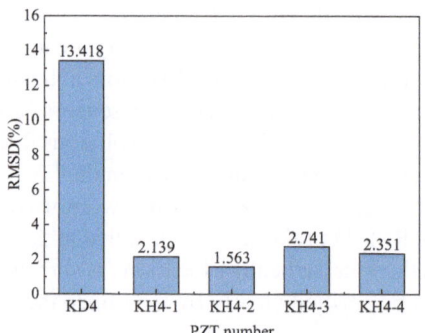

Fig. 10. RMSD values of sensors around void KD4.

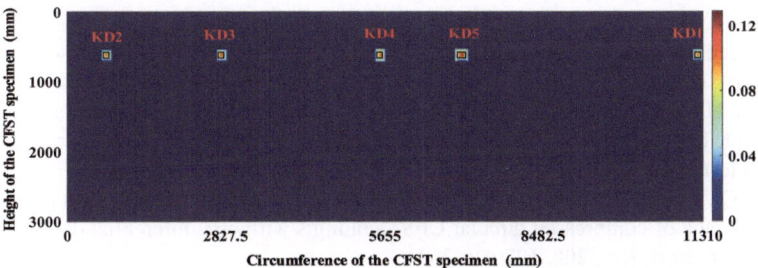

Fig. 11. Imaging of detected void defects.

5   Conclusion

In this paper, a void defect detection and visualization methods for a large-scale CFST specimen using EMI technology using surface-mounted PZT sensors is proposed and the validity is verified with a full-scale CFST specimen.

(1) The existence of void defect makes the impedance curve of the surface-mounted PZT sensor located at the center of the defect fluctuate significantly, and the corresponding RMSD value is obviously greater than the RMSD value of PZT sensors at healthy location. By comparing the EMI measurement results of the surface-mounted PZT sensors, the existence of void defects can be detected efficaciously.
(2) Using an interpolation algorithm based on the EMI measurements around a mimicked void defect, the location and region of the void defects can be visualized and the imagination results are consistent with the artificially mimicked void defects in the CFST specimen.

The effectiveness of using EMI technique for the detection of concealed internal cavity defects is verified, and the EMI method has significant development prospect in the field of defect detection in large-section CFST structures. The quantitative analysis of cavity defects is still challenging research, and new data mining techniques combined with machine learning techniques are used in future research to perform deeper analysis of obtained impedance data.

Acknowledgments. This work was supported by National Natural Science Foundation of China (Grant No. 51878305) and Scientific Research Funds of Huaqiao University, China (605-50Y18016), and the supports are highly appreciated.

References

1. Alatshan, F., Osman, S.A., Hamid, R., Mashiri, F.: Stiffened concrete-filled steel tubes: A systematic review. Thin-Walled Struct. **148**, 106590 (2020). https://doi.org/10.1016/j.tws.2019.106590
2. Liu, J., Liu, Y.J., Zhang, C.Y., Zhao, Q.H., Lyu, Y., Jiang, L.: Temperature action and effect of concrete-filled steel tubular bridges: a review. J. Traffic. Transp. Eng. **7**(2), 174–191 (2020)

3. Ye, Y., Li, W., Liu, X.J., Guo, Z.X.: Behaviour of concrete-filled steel tubes with concrete imperfection under axial tension. Mag. Concrete. Res. **73**, 14 (2020). https://doi.org/10.1680/jmacr.19.00306

4. Han, L.-H., Hou, C., Hua, Y.-X.: Concrete-filled steel tubes subjected to axial compression: life-cycle based performance. J. Construct. Steel Res. **170**, 106063 (2020). https://doi.org/10.1016/j.jcsr.2020.106063

5. Xue, J.Q., Huang, J.P., Fiore, A., Briseghella, B., Marano, G.C.: Prediction of the mechanical performance of compressed circular CFST columns with circumferential debonding gap. J. Construct. Steel. Res. **208**, 107988 (2023)

6. Pan, S.S., Zhu, Y.X., Li, D.S., Mao, J.: Interface separation detection of concrete-filled steel tube using a distributed temperature measuring system. Appl. Sci. (2018). https://doi.org/10.3390/app8091653

7. Li, L., Jiao, J.P., Gao, X., Jia, Z.H., Wu, B., He, C.F.: A review on nondestructive testing of bonding interface using nonlinear ultrasonic technique. Chin. Sci. B. Chin. **67**(7), 621–629 (2022)

8. Chen, D., Wang, L., Luo, X.: Recent development and perspectives of optimization design methods for piezoelectric ultrasonic transducers. Micromachines **12**(7), 779 (2021). https://doi.org/10.3390/mi12070779

9. Jiang, T.Y., Luo, Z.T., Tian, Z.C.: Investigation and application on monitoring the compactness of concrete-filled steel tube structures with ultrasonic wave. Earth. Space. 682–689 (2014)

10. Liu, Y.L., Xu, H.B., Ma, X.X., Wang, D.H., Huang, X.: Study on impact-echo response of concrete column near the edge. Appl. Sci. Basel **13**, 9 (2023). https://doi.org/10.3390/app13095590

11. Cheng, S.Y., Su, Q.F., Yuan, Y.N., Jiang, H.J., Chen, F., Zhang, K.: Turbine blade crack detection based on ultrasonic infrared thermography. *NDT* (2023). https://doi.org/10.11973/wsjc202303001

12. Lukesh, P., Sumedha, M.: A comprehensive review on piezo impedance based multi sensing technique. Results Eng. (2023). https://doi.org/10.1016/J.RINENG.2023.101093

13. Nguyen, T.H., Phan, T.T.V., Le, T.-C., Ho, D.-D., Huynh, T.-C.: Numerical simulation of single-point mount PZT-interface for admittance-based anchor force monitoring. Buildings **11**(11), 550 (2021). https://doi.org/10.3390/buildings11110550

14. Li, W.J., Liu, T.J., Zou, D.J., Yi, T.H.: PZT based smart corrosion coupon using electromechanical impedance. Mech. Syst. Signal. Pr. **129**(2019), 455–469 (2019)

15. Luo, Z.N., Deng, H., Li, L., Luo, M.Z.: A simple PZT transducer design for electromechanical impedance (EMI)-based multi-sensing interrogation. J. Civ. Struct. Health. **21**(11), 235–249 (2021)

16. Wen, J.Y., Xu, B., Wang, H.D.: Study on numerical simulation of piezoelectric impedance detection of composite concrete structure defects. Piezoelectrics Acoustooptics **40**(02), 276–279 (2018)

17. Liang, C., Sun, F.P., Rogers, C.A.: An impedance method for dynamic analysis of active material systems. J. Intel. Mat. Syst. Str. **116**(1), 120–128 (1994)

Experimental and Numerical Studies on Shear and Tensile Behavior of Horizontal Connecting Joints for Improved Precast Concrete Bearing Wall Structure

Haian Yu[1], Lei Zhang[2(✉)], Shuai Su[3], and Kun Huang[4]

[1] POWERCHINA International Group Limited, Beijing 100036, China
[2] POWERCHINA Guiyang Engineering Corporation Limited, Guiyang 550081, China
zhanglei01_gyy@powerchina.cn
[3] School of Civil Engineering, Chongqing University, Chongqing 266000, China
[4] Guangdong Cao-Gen Housing Company Limited, Zhongshan 528462, China

Abstract. An improved precast bearing wall structure was proposed, with light weight gauge steel members embedded in the wall, forming full length hole for dowels passing through, which can provide higher speed and lower cost construction method, especially for low rise buildings. The shear and tensile behavior of the horizontal joints for the novel precast bearing wall structure were studied through experimental and numerical investigation in this paper. Three identical specimens for horizontal joints shear and tensile tests separately were tested. The shear test specimens were designed from the wall-slab-wall connection part while the tensile test specimens were designed from the wall-foundation connection part. The whole experimental process from crack generation to joints failure was investigated in detail and the load-displacement curves were provided and analysed. Based on experimental results, the finite element models of both type specimens are verified and calibrated using ABAQUS. The shear and tensile failure mechanism of horizontal joints is presented and the recommendations to increasing the horizontal joints strength capacity and ductile performance are proposed, which may provide practical guidance for future engineering practice.

Keywords: precast concrete · precast bearing wall · horizontal joint · shear resisting strength · tensile strength

1 Introduction and Background

Precast structure is widely used in low-rise housing projects, and the connecting joints and its mechanical performance for precast members are generally key issues for precast concrete structure. The advantages of this type of structure are time saving, better quality control, lower materials consumption [1], compared with traditional construction method, especially in those regions with insufficient building materials resource and poor construction condition but with rapid demand of housing construction market, such as Saudi Arabia [2]. Among various built or ongoing projects, the precast bearing

P. Xiang and L. Zuo (Eds.): PBSFTT 2023, LNCE 382, pp. 208–230, 2024.
https://doi.org/10.1007/978-981-97-5108-2_23

wall structure is widely and successfully adopted [3], such as in Developmental Housing Programs initiated by National Housing Company in Saudi Arabia.

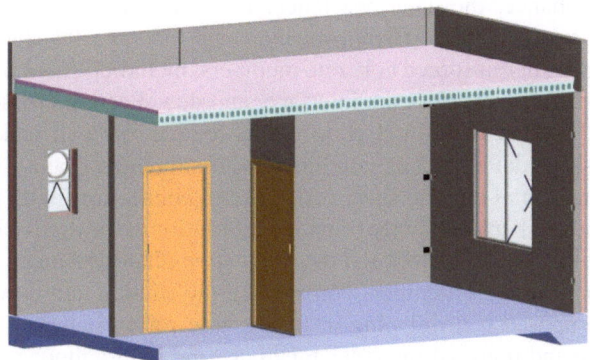

Fig. 1. Precast concrete bearing wall structure.

For precast concrete bearing wall structure, the vertical structural members are off-site precast load bearing wall, while for external sandwich wall will be insulated inside to provide thermal resistance [4]. The foundation is usually casted on site with dowels embedded before installation of precast wall members. The precast wall will be connected to foundation through dowels and grouts after installation. The slab will be installed and supported on the precast wall and finally connected by dowels and grouts. See Fig. 1 for the structural system illustration.

The shear and tensile performance of precast wall horizontal joints are fundamental issues among the various research issues. Hofbeck et al. [5] conducted 38 push-off specimens tests with or without a pre-existing crack along the shear plane and reported a shear-friction theory to give a conservative estimate of the shear transfer strength of initially cracked concrete. Mattock et al. [6] analysed and concluded the influence of concrete strength, shear plane characteristics, reinforcement, and direct stress to the shear transfer strength of reinforced concrete and proposed the equation to estimate a proper shear-transfer strength. Harris et al. [7] tested 18 full scale horizontal joints under concentric compressive loading and reported the effect of reinforcement to prevent wall end splitting and the effect of joint length on the ultimate load carrying capacity and stiffness. Einea et al. [8] tested different types grout-filled steel pipe splices specimens considering influence of grout strength, rod bar size and pipe size, and found that high bond strength of reinforcing bars can be achieved by confining the grout surrounding the bars. Alias et al. [9] studied influence of anchorage length and sleeve inner diameter on the bonding and anchoring performance of the sleeves and found that sufficient anchorage length can ensure the safety and reliability of the joint, and the reduction of sleeve inner diameter will strengthen the constraint on the grout. Hosseini et al. [10] tested 21 grouted splice connectors with different spiral pitch distance under an increasing axial load and found the best performance of grouted pipe splice connectors was obtained at the spiral pitch distance of 15 mm, combined with the use of four vertical bars as shear keys. Soudki et al. [11] tested six full-scale specimens to investigate the behavior of mild steel connections for precast panels and found that properly designed mild steel connections for precast wall panels exhibit sufficient ductility and energy dissipation capacity and the use of shear keys across the interface of the connection significantly limits the slip mechanism and enhances the shear resistance. Tsoukantas et al. [12] proposed that the shear friction is one of three different mechanisms that transfer shear forces across uncracked interfaces in reinforced concrete members, including shear friction, cohesion between concrete surfaces across the interface, and dowel action crosses the interface. Soudki et al. [13] tested horizontal connections for precast wall panels subjected to reversed cyclic shear deformations combined with simulated gravity loads normal to the connection and found that the shear resistance of connections with post-tensioning, using either strands or bars, is mainly provided by friction at the dry pack grout-to-panel interface. Crisafulli et al. [14] proposed the shear strength design method for horizontal joints between precast wall members considering the friction and dowel action. Jiang et al. [15, 16] proposed a novel plug-in filling hole for steel bar lapping of precast concrete structure and reported the mechanism and the failure modes of the anchorage rebar, and presented the recommended anchorage length. Belleri et al. [17] studied the suitability of grouted sleeve connections as column-to-foundation joints for precast concrete structures in seismic regions and reported that grouted sleeves ensure a ductility and energy dissipation capacity similar to traditional connections.

The most popular theory about the shear transfer mechanism is that shear forces can be transferred by adhesion or friction at joint interfaces, shear-key effect at indented joint faces, dowel action of transverse steel bars, and the frictional resistance can be enhanced by the pullout resistance of tie bars properly placed across the joint [18]. In some cases, the compressive stresses coming from such as the self-weight of upper members can be considered. Due to the different construction details at interface, the consideration and design method for smooth, rough, or indented faces are distinguished [19]. The design methods for horizontal connection shear capacity for precast bearing walls differ in codes of America (Sect. 22.9 in ACI 318-19 [20], or Sect. 5.3.4 in PCI Design Handbook [21]), Europe (Sect. 6.2.5 in EN 1992-1-1:2004 [22]), and China (Sect. 9.2 in JGJ 1-2014 [23]). Without the contribution of normal compression stress, the shear strength equation in ACI 318-19 form is like that in JGJ 1-2014, which both ignores the effect of interface friction. In contrast the equation in EN 1992-1-1:2004 is with both concrete friction and dowel action considered.

When the wall panel is subjected to lateral loads, such as wind and seismic demands, the overturning moment at the bottom section will result in tension and compression at the opposite sides of the section. Yuan et al. [24] tested half grouted sleeve connections and found three types of failure mode are observed in the test, and the tensile is governed by the weakest of rebar tensile capacity, bond capacity between the rebar and the grout, or the thread connection tensile capacity. Eligehausen et al. [25] proposed a behavioral model to predict the average failure load of anchorages using adhesive bonded anchors and found that the basic strength of a single adhesive anchor predicts the pullout capacity not the concrete breakout capacity, and the group effects also revealed. Fuchs et al. [26] raised a model for the design of post-installed steel anchors or cast-in-place headed studs or bolts, termed the concrete capacity design (CCD) approach and a data bank including approximately 1200 European and American tests was evaluated. Obata et al.

[27] studied on the effect of a free edge on the pull-out strength based on experiments and proposed a new method to estimate the cone failure strength using the theory of linear fracture mechanics. Xu [28] proposed a new analytical model based on the bond stress integration along the bar stress propagation length to predict the bar-slip behavior in RC beam-column joints under monotonic loading, considering the phenomena of combined axial pullout and transverse dowel action at the joints. Singhal et al. [29] proposed and studied headed dowel bars precast connection, and found that lower embedment depth lead to concrete cone failure, while higher embedment depth resulted in slip of bar. Zhou et al. [30] conducted a program to investigate the influence of bar embedment length and ratio of duct diameter to bar diameter on monotonic bond-slip response of stainless energy dissipation bars (ED bars).

The improved precast bearing wall structure is proposed, with light weight gauge steel members embedded in the wall, forming full length hole for dowels passing through, which can save up to 30% site assembly time and lower construction cost, especially for low rise buildings. Compared with current technologies which the ground floor precast wall connected with foundation and first floor wall separately, in the improved proposal, the ground precast wall connected with top and bottom members with one full length steel dowels. The traditional connection requires the precast panel reserved holes to match with foundation embedded dowels, while for the proposed connection, the hole is reserved in foundation, also during the precast wall manufacturing, the square hole will be formed with embedded Light Gauge Steel (LGS) double C-lipped section members. When assembling on site, the vertical connecting full-length dowels can pass through slab, precast wall, and finally the foundation one time. After grouted then the horizontal joints for wall-slab-wall and wall-foundation will be formed at the same time. See Fig. 2 for the connection details.

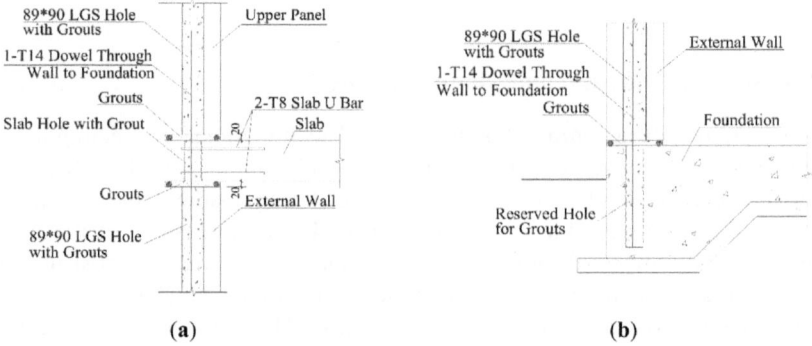

Fig. 2. Horizontal connecting joints: (a) Wall-Slab connection; (b) Wall-Foundation connection.

In this paper, the shear and tensile performance of this joints are tested and the failure modes are compared based on ABAQUS finite element simulation approach. Based on the above research, the shear and tensile failure mechanism of horizontal joints is presented and the recommendations to increasing the horizontal joints strength capacity and ductile performance are proposed, which may provide practical guidance for future engineering practice.

2 Experimental Overview

2.1 Materials

The precast members of all precast specimens were made of concrete of cubic test strength class C35 while the cubic test strength class of grout materials is C50. All the reinforcement strength class is HRB400 with yielding strength of 400 MPa, while the LGS members is Q355 with yielding strength of 355 MPa. The concrete specimen compression test is tested on 200T compression-testing machine. The compression strength is 44.08 MPa for C35 specimen and 59.52 MPa for C50 specimen thus both is qualified. See Fig. 3 for the specimen testing results.

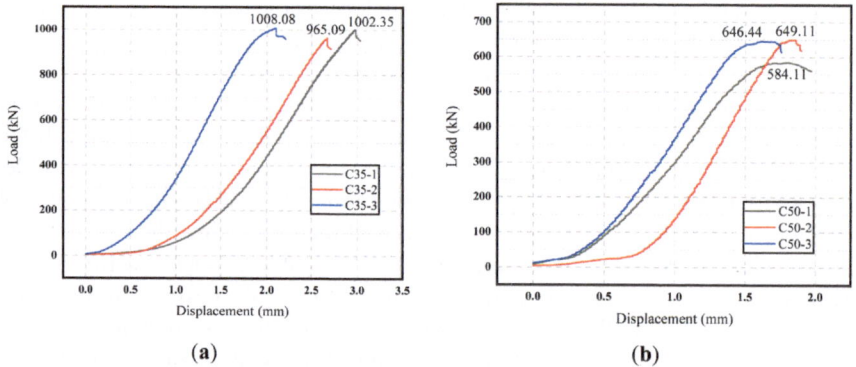

(a)   **(b)**

Fig. 3. Test values for compressive strength: (a) C35 specimens; (b) C50 specimens.

2.2 Test Specimens Design and Manufacturing

Horizontal Connecting Joints Shear Test. The horizontal connecting joint was selected at the floor slab area between upper and bottom wall for shear test. In order to prevent the load from being eccentric during loading, the specimen was designed as a cross-type joint. As shown in Fig. 4, the dimensions of the floor slab are 150 mm*600 mm*1000 mm with a 60mm diameter corrugated pipe in the middle of the slab for reserved hole, and the dimensions of the wall slab on both sides are 200 mm*480 mm*1000 mm with LGS double C-lipped box section members embedded forming a reserved hole. When the wall slab was assembled with the floor slab, PE rods were padded on both sides of the wall slab for blocking, followed by insertion of dowel with C50 high-strength non-shrinkage grout.

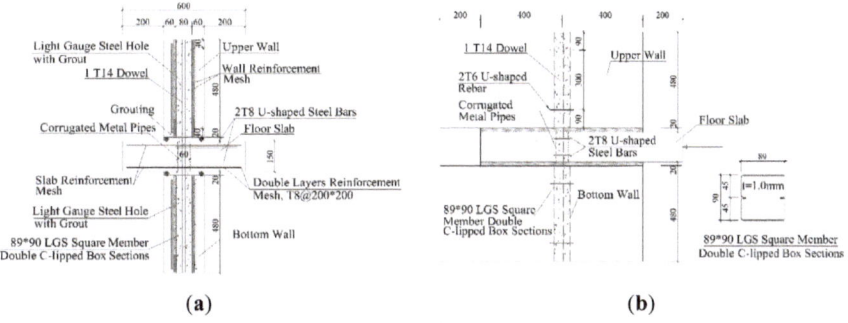

Fig. 4. Design of shear specimens for horizontal connecting joint: (a) Top view; (b) Front view.

Total of three specimens for shear test, named SKJ1, SKJ2 and SKJ3, were casted and the manufacturing process of the specimens is shown in Fig. 5.

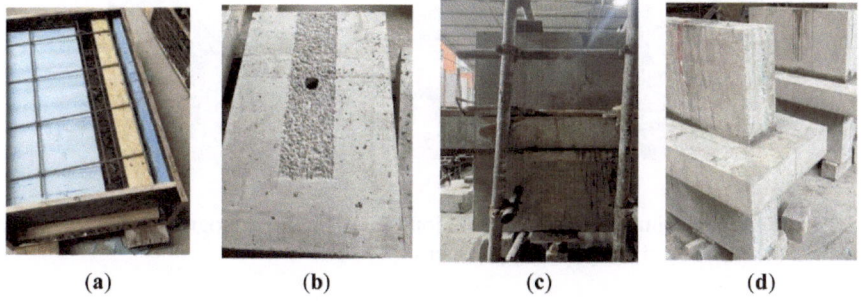

(a)   (b)   (c)   (d)

Fig. 5. Shear test specimens with horizontal joint manufacturing progress.

Horizontal Connecting Joint Tensile Test. The horizontal connecting joint was selected at the connecting joint between the wall and the foundation. The specimen design is shown in Fig. 6.

The size of the foundation part is 200 mm*400 mm*1210 mm, with 60 mm diameter and 360mm depth metallic bellows embedded inside to form the holes for dowels and grout. The size of the wall panel is 200 mm*200 mm*400 mm, with LGS double C-lipped box section members embedded forming a reserved hole along the wall length direction. After the panel and foundation are assembled, the T14 dowels were inserted into the hole of panel and foundation, and finally the holes were fully grouted. During manufacturing, the PE rods were arranged on both sides of the joint to seal the grout. Total of three tensile specimens were casted for the horizontal connection tensile test, named SKL1, SKL2 and SKL3, respectively. The manufacturing process of the specimens is shown in Fig. 7.

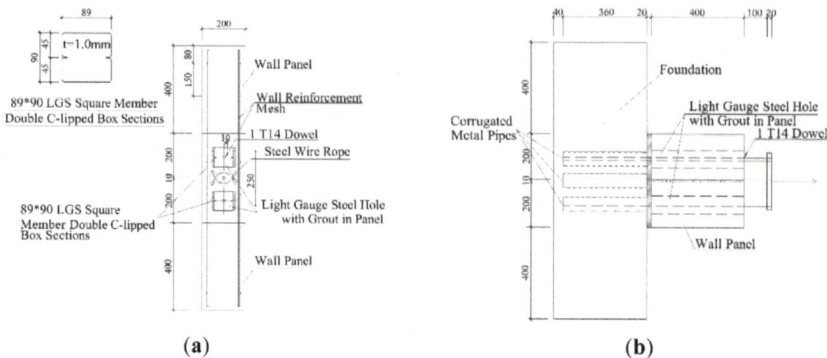

Fig. 6. Design of tensile specimens for horizontal connecting joints: (a) Top view; (b) Front view.

Fig. 7. Tensile test specimens with horizontal joint manufacturing progress: (a) Reinforcement fabricating; (b) Assembled and grouted specimens.

2.3   Testing Devices

The horizontal joint shear test was carried out on a 500T electro-hydraulic servo long column pressure tester at the Structural Laboratory of Chongqing University, with a maximum load of 5000KN, an accuracy of ± 1% of the test force value and a measurement range that can reach 2% to 100% of the full scale. In addition to the pressure tester, the devices used in the test are load cells, DH3816 data acquisition system, pull-wire displacement transducers and strain gauges, etc. The horizontal joint tensile tests were carried out on a structural laboratory reaction frame at Chongqing University. The load was provided via a jack connected with an oil pump. Other devices used in the tests were load cells, DH5902 data acquisition system, pull-wire displacement gauges, strain gauges, etc.

3   Experimental Progress and Phenomena

3.1   Horizontal Connecting Joints Shear Test

SKJ1 Specimen Testing Results.   The SKJ1 specimen before loading is shown in Fig. 8. As shown in Fig. 9, when the load was increased to 233KN, a crack appeared on the left side of the south face of the SKJ1 specimen. The cracking location was observed at the joint where the PE rods is in contact with the wall. The reason expresses the phenomenon is that the PE rods may weaken the bonding effect between the old and new concrete interface, and this location can be regarded as a weak anti-shear interface.

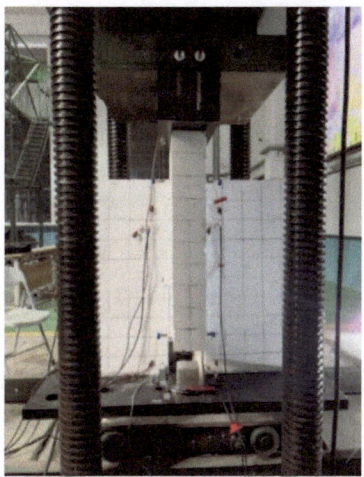

Fig. 8. SKJ 1 specimen before loading

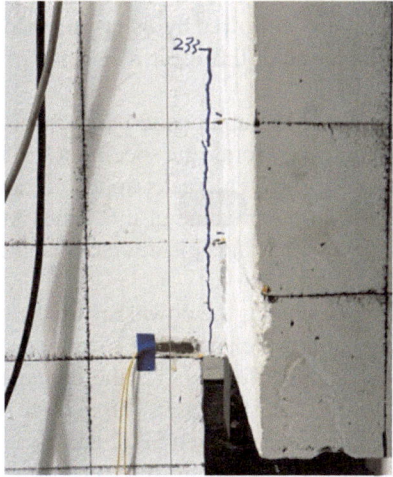

Fig. 9. First crack in SKJ 1 specimen

When the load reached to 300kN, with the load increasing the above cracks continued to expand upwards. When the load increased to 330KN, cracks also appeared on the right side of the south side of the member (left side of the north side) one after another. When the load increased to 348.77KN, the cracks on the left side of the north side of the

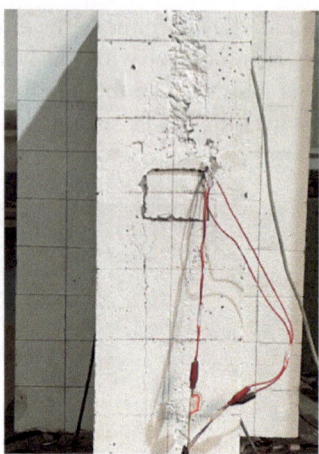

Fig. 10. LGS member in SKJ1 specimen test.

member expanded rapidly, accompanied by the sound of blowing up. The cracks had formed extending through cracks resulting in the member the load bearing capacity was rapidly reduced to 233.68KN, at which time the shear capacity of the concrete at the old and new interfaces was sharply reduced, and a small strain occurred in the interpolation bars running through the wall and floor slabs. After one side extending through cracks formed, the crack on the other side of the specimen also developed rapidly and the extending through cracks formed, thus double side cracks both appeared, which is also the reason for the second peak load, followed by a sharp drop in the load-bearing capacity of the specimen and the completion of the shear resistance of the old and new concrete interface of the specimen.

Due to the shear resistance of the composite parts consisted of steel dowels, LGS double C-clipped box members, and grouts, the specimen still had a certain load-holding capacity after the two extending through cracks appeared, and the specimen load was kept fluctuation around a certain mean value at this time. At the same time, the LGS double C-slipped box members with grouts inside started sliding together was observed, as shown in Fig. 10.

Within the increasing displacement, the dowel inside the specimen finally appeared to be cut off, as shown in Fig. 11 at which point the specimen can no longer bear the load and the test program ends. As shown in Fig. 12, the damaged LGS double C-clipped box member appears tearing near the shear area with a large deformation of the dowel.

The test phenomena of other two specimen SKJ2 and SKJ3 were similar to specimen SKJ1 test. For SKJ2, it was observed that the first cracks appear at the loads of 205kN, and the cracks soon developed into extending through cracks at double sides. When dowel are involved in anti-shear progress, the slippage phenomenon was observed between the LGS members and precast members, and also between dowel and grout part, as shown in Fig. 13.

For SKJ3, it was observed that the first cracks appear at the loads of 173kN, and one side extending through cracks appear at the loads of 282kN, and finally the double

Fig. 11. The dowel was cut off in SKJ 1 specimen

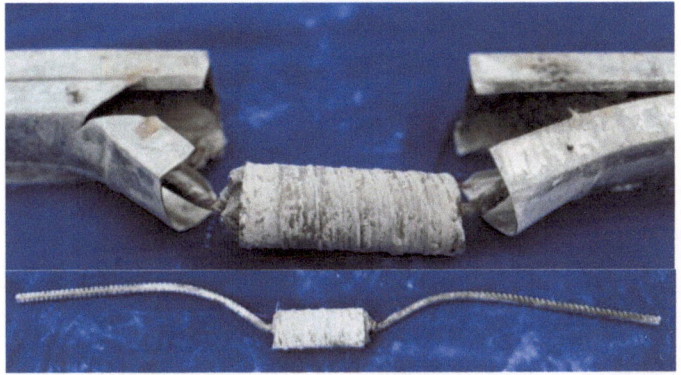

Fig. 12. The damaged LGS member and dowel of SKJ 1 specimen

Fig. 13. LGS members slippage movement of SKJ2

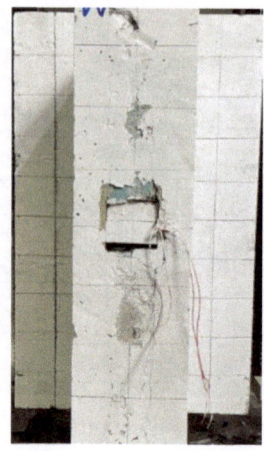

Fig. 14. LGS members buckling failure of SKJ3

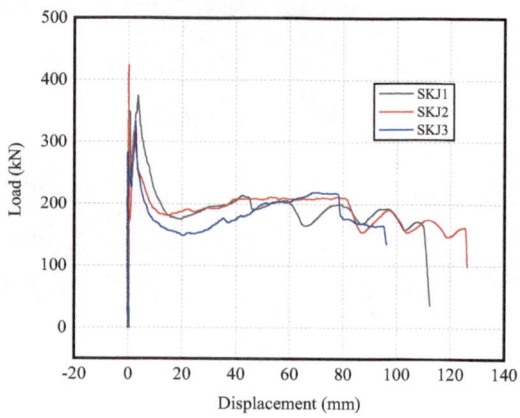

Fig. 15. Load displacement curves for horizontal joints shear specimens.

side extending through cracks appear at the loads of 332kN. The different phenomenon observed in SKJ3 specimen compared with another two specimen is that the failure mode of LGS member was buckling rather than slippage. The proper explanation for this phenomenon can be pointed to the empty space between LGS member and XPS insulation board due to the manufacturing deficiency. The failure diagrams of SKJ3 test specimen are shown in Fig. 14.

The load displacement curves for the three specimens in the horizontal connecting joints shear test are shown in Fig. 15, which can be read that the load displacement curves for the three specimens show the same regularity. The first peak loads for horizontal joint shear test specimens are listed in Table 1.

Table 1. First peak loads for horizontal joint shear test specimens.

Specimen No	Peak Loads (kN)	Mean Value (kN)
SKJ1	348.77	351.34
SKJ2	422.70	
SKJ3	282.56	

3.2 Horizontal Connecting Joints Tensile Test

SKL1 Specimen Testing Results.   As the test phenomena were similar for all three specimens, the experimental procedure for the SKL1 specimen is described in detail below. The tensile test of the horizontal joint connection between foundation and wall panel was carried out by controlling the oil pump for monotonic static tension, the SKL1 specimen before loading as shown in Fig. 16.

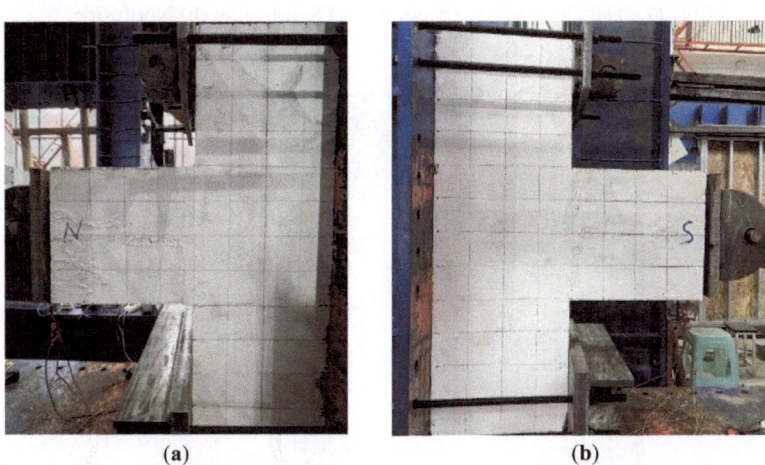

(a)   (b)

Fig. 16. SKL1 specimen before loading: (a) North side; (b) South side.

When the load value reached 86kN, one crack appeared simultaneously on the south and north side of the specimen, at the location where the precast wall was connected to the foundation. When the load value reached 91kN, one horizontal crack appeared at the corner of the foundation of the specimen. The cracks continue to develop before the load value reaches the peak load. The specimen was pulled out within a short period of time after the peak load was reached, and the finally the failure of specimen was observed, as shown in Fig. 17. The final damage location of the specimen belongs to the foundation part, but there were also obvious cracks at the connection between the foundation and the wall panel.

The test phenomena of other two specimen SKL2 and SKL3 were similar compared with specimen SKL1 test. For SKL2 specimen, it was observed that the first cracks appear at the loads of 80kN at the PE rods location, and the cracks gradually developed into flaws until the loads reached 120kN. Soon the loads reached the peak loads and the strength capacity of the specimen rapidly decreased and the failure of specimen was observed, as shown in Fig. 18.

(a)   (b)

Fig. 17. Damaged SKL1 specimen: (a) South side; (b) North side.

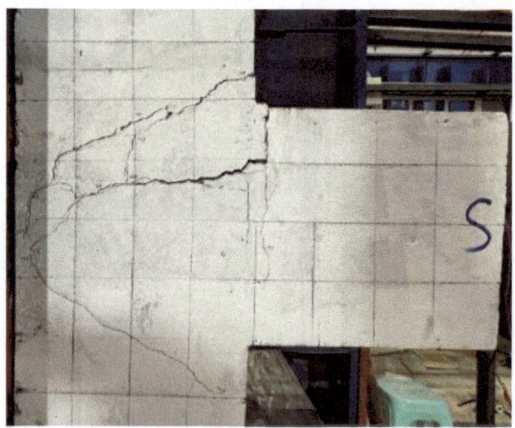

Fig. 18. SKL2 specimen tensile test failure diagram

It was observed that the first cracks appear at the loads of 67.2kN also at the location of precast wall and foundation joint of SKL3. The cracks gradually developed into flaws and until the peak loads of 137.45kN was reached with the failure of specimen coming soon. The test progress diagrams of SKL3 test specimen are shown in Fig. 19.

The load displacement curves for the three specimens in the horizontal joint tensile test are shown in Fig. 20, which can be read that the load displacement curves for the three specimens show the same regularity. The peak loads for horizontal joint test test specimens are listed in Table 2.

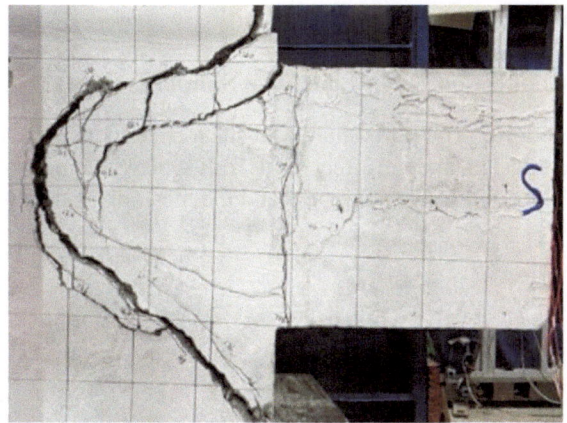

Fig. 19. SKL3 specimen tensile test failure diagram

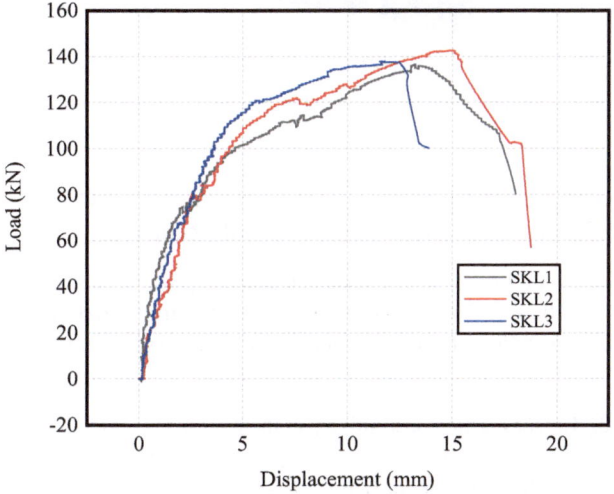

Fig. 20. Load displacement curves for horizontal connecting joints tensile test specimens.

Table 2. First peak loads for horizontal joint tensile test specimens.

Specimen No	Peak Loads (kN)	Mean Value (kN)
SKL1	136.64	138.86
SKL2	142.56	
SKL3	137.45	

4   Finite Element Simulation Analysis

4.1   Horizontal Connection Shear Specimen Simulation

Finite Element Models.   The horizontal joint shear specimens were modelled according to the specimens design, and the effect of PE rods sealing the joints space was ignored in the modelling. The complete model view and the meshing of the model are shown in Fig. 21.

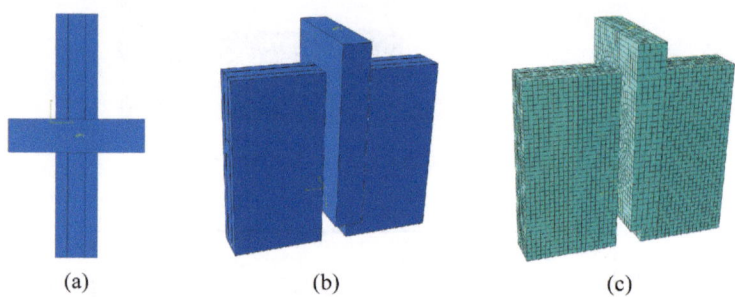

(a)   (b)   (c)

Fig. 21.   Finite element model of SKJ specimen: (a) Top view; (b) Axonometric view; (c) Meshing

Materials Properties.   The materials used in the model are C35 and C50 concrete, Q355 LGS, XPS insulation and HRB400 steel reinforcement.

The concrete was modelled using concrete plasticity damage model (CDP) provided in ABAQUS, while the reinforcement and dowels were modelled using a bifold model, and the materials properties of the LGS and XPS are shown in Table 3.

Table 3.   Linear elastic constitutive parameters for steel and XPS material.

Materials	Modulus of elasticity (MPa)	Poisson's ratio
Q355	206000	0.3
XPS	8.5	0.28

Contacts Setting.   Based on the horizontal joint shear test phenomenon, the slippage between the LGS members and the concrete can be observed. Thus, the contact between the LGS and the concrete is set as frictional contact. The interface between the C30 precast members and C50 grout is also subject to bond slip phenomenon and is set as frictional contact. The contact between the other components is shown in Table 4.

Loads and Boundary Conditions.   The boundary conditions for the horizontal joint shear test specimen are, with a point-surface coupling for loading at the top of the middle floor slab and fixed constraints at the bottom of the wall panels on both sides. The load

Table 4. Linear elastic constitutive parameters for each material.

Contact interface	Main surface	Type of contact	Friction coefficient
C35-LGS member	C steel	Frictional contact (penalty function)	0.3
XPS- LGS member	C steel	Frictional contact (penalty function)	0.1
C35-C50	C50	Frictional contact (penalty function)	0.5
C50- LGS member	C steel	Tie	/
C35-XPS	C35	Tie	/
Steel reinforcement and dowels	C35	Embedded	/
Bellows	C50	Embedded	/

is applied at the coupling point through displacement loading method, with a load value of 20 mm.

Numerical Analysis Results. The overall stress cloud of the numerical model for the horizontal joint shear test specimen is shown in Fig. 22(a). The stress cloud of C35 concrete in the precast wall slab is shown in Fig. 22(b), where the maximum stress is 35MPa, which has reached the breaking strength. The stress cloud for the precast floor slab is shown in Fig. 22(c), where the maximum stress is generated at the corrugated pipe extraction hole in the middle of the floor slab, where the maximum stress reaches 35MPa.

The stress cloud for the LGS is shown in Fig. 23, in which the LGS double C-clipped member is considered as one single unit in the model and the welding joints between the C sections members are ignored. It can be observed from the stress cloud, yielding damage occurs in the LGS near the shear action area, which is complied with the failure mode of the specimens observed in the tests, as shown before. The stress cloud for the vertical dowel is shown in Fig. 24, which exhibits that its maximum stress is 400 MPa and occurs at the intersection of the precast floor slab and the precast wall panel, which is the same location where the dowels fail in the tests.

As the observation from experimental results and the failure mode agreement with finite model analysis, the LGS members embedded in the wall panel to provide tunnel for the dowels was involved in anti-shear progress after the concrete friction interface quits, which may not improve the shear strength capacity but provide more ductility. Moreover, the horizontal joints shear capacity equation in EN 1992-1-1:2004 is more reasonable with both concrete friction and dowel action considered, compared with that in American and Chinese codes.

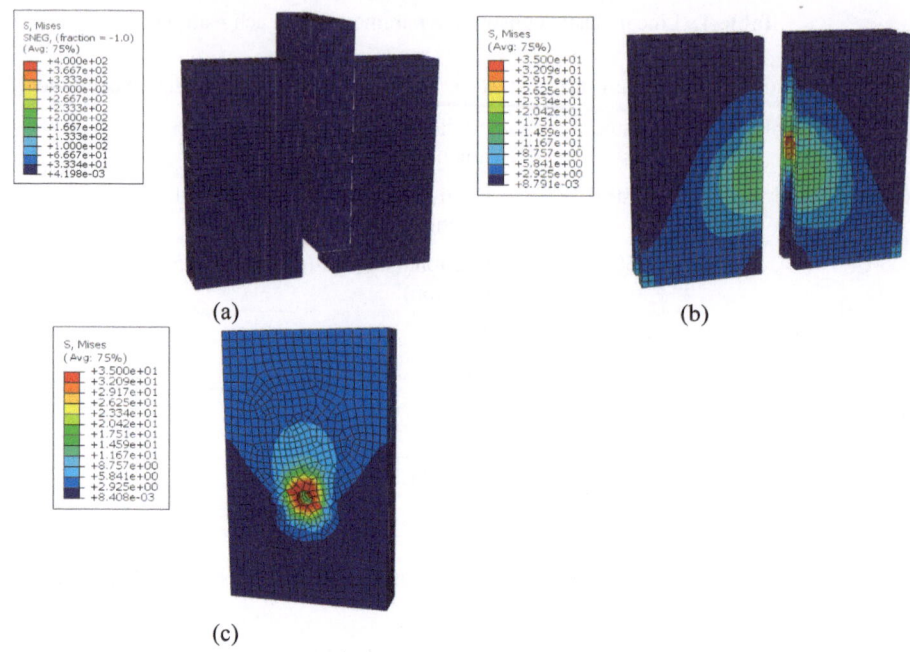

Fig. 22. Numerical analysis stress cloud of SKJ specimen: (a) Overall view; (b) C35 precast wall member part; (c) C35 precast slab member part.

Fig. 23. SKJ model LGS member stress cloud.

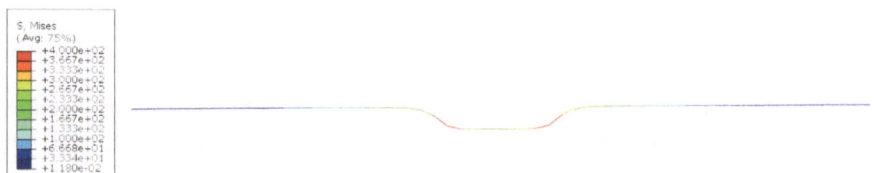

Fig. 24. SKJ model LGS member stress cloud

4.2 Horizontal Connecting Joints Tensile Specimen Simulation

Finite Element Models. The numerical model of the horizontal joint tensile specimen and the meshing diagram are shown in Fig. 25, in which the PE rods were ignored in the model.

Materials Properties.   The materials used in this model are C35 and C50 concrete, Q355 LGS, HRB400 reinforcement and steel wire rope. The concrete is modelled using the concrete plastic damage model (CDP) provided in ABAQUS while the reinforcement and dowel is modelled using the bifold model, and the steel and steel wire rope are modelled using the linear elastic model with the following constitutive parameters as shown in Table 5.

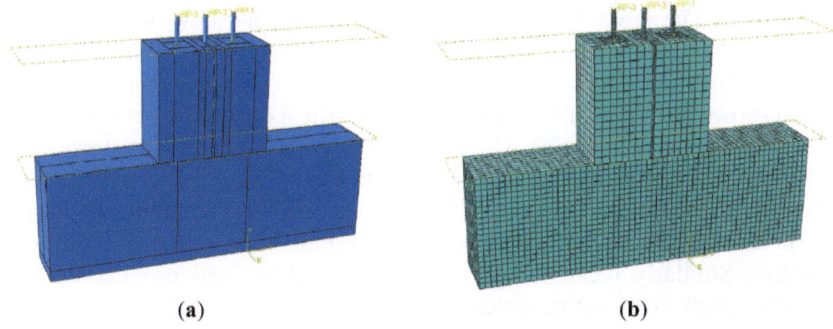

(a)   (b)

Fig. 25. Finite element model of the SKL specimen: (a) Axonometric view; (b) meshing diagram.

Table 5. Linear elastic constitutive parameters for each material.

Materials	Modulus of elasticity(MPa)	Poisson's ratio
Q355	206000	0.3
Steel wire rope	110000	0.25

Contacts setting.   The contact relationship between the components is set as shown in Table 6, based on the horizontal joints tensile test damage phenomenon.

Table 6. Physical component contact Settings

Contact surface	Contact with the main surface	Type of contact
Concrete- LGS member	LGS member	Tie
C35-C50	C50 grout	Tie
Steel reinforcement and dowel	C50 grout	embedded
Steel wire rope	The whole model	embedded

Loads and boundary conditions.   To comply with the horizontal joint tensile test specimen, the bottom of the foundation is set as the fixed end and the load is applied to the three vertical dowels through displacement loading method, with an applied displacement value of 10 mm.

Numerical analysis results.   The overall stress cloud of the horizontal joint tensile finite element model is shown in Fig. 26, and the overall displacement cloud is shown in Fig. 27, which shows that the interface between the precast wall member and the foundation is the weak interface prone to be damaged.

The stress clouds for each component of the horizontal joint tensile test specimens are shown in Fig. 28. The stress cloud for the C35 precast member shows that the foundation concrete formed an inverted triangular failure zone, which is the same as the phenomenon observed in test results as shown before. The stress cloud of the reinforcement mesh of the foundation shows that the reinforcement near the vertical dowels reflects shear yielding failure mode, as shown in Fig. 29, which is complied with the test. The stress cloud for the C50 grout part shows that damage occurred in the C50 grout part at the junction of the foundation and the precast wall panel. The fracture failure at the junction of the foundation and the precast wall panel was observed on the N side of the specimen, as shown before. Similarly, it can be read that the C50 high-strength grout at this location also exhibits a high stress status, which is complied with test results.

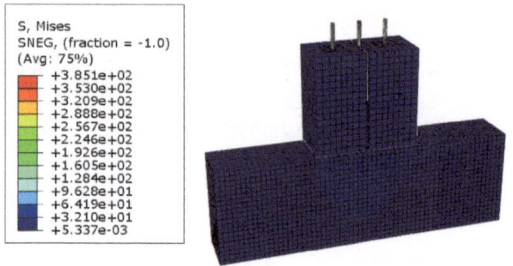

Fig. 26. SKL specimen stress cloud

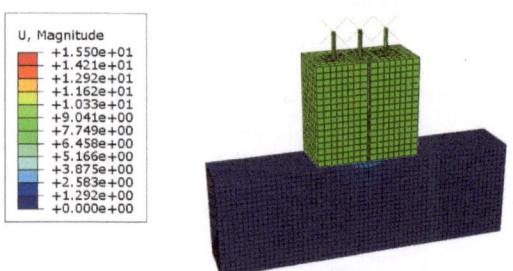

Fig. 27. SKL specimen displacement cloud

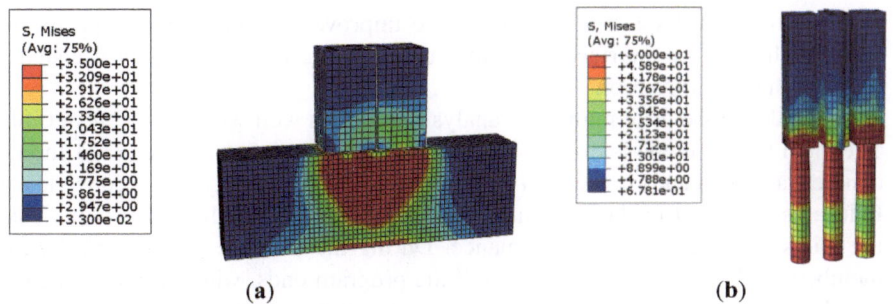

Fig. 28. Stress cloud diagram of SKL specimen: (a) C35 precast member; (b) C50 grout member.

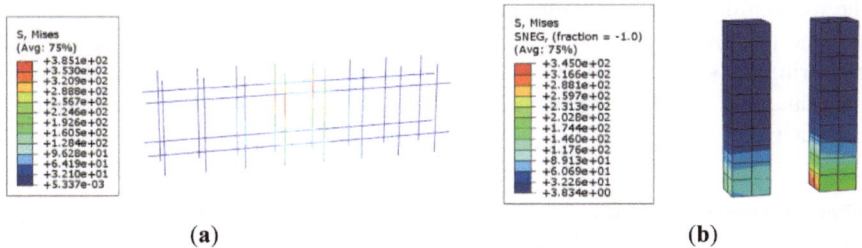

Fig. 29. Stress cloud diagram of SKL specimen: (a) Foundation rebar mesh; (b) LGS member.

5 Conclusions

In consideration of construction and manufacturing efficiency and economics, the improved precast bearing wall structure was proposed, with light weight gauge steel members embedded in the wall forming full length hole for dowels passing through fast, especially for low rise buildings. The shear and tensile behavior of the horizontal joints for the novel precast bearing wall structure were studied through experimental and numerical investigation in this paper. Three identical specimens for horizontal joints shear and tensile tests were tested. The shear test specimens were designed from the wall-slab-wall connection part while the tensile test specimens were designed from the wall-foundation connection part. The finite element models of both type specimens are verified and calibrated with the experimental results using ABAQUS. The following conclusions are made based on the work in this research:

1. Horizontal joints shear numerical analysis complies well with the test results. The main failure mode is the interface friction failure at the joints between old and new concrete. For insulated precast panels, the insulation layers exposed to grout materials and the PE rods for sealing will weaken the interface friction shear performance. After the friction interface quit the work, the joints move into the second phase anti-shear behavior with the main mechanical performance contributed by the dowels as well as the LGS hole with grouts. The joints have good performance in shear behavior and the LGS will also contribute to ductile performance. A better approach to improve the horizontal joints shear ductile performance is to increase the dowel diameters or

decrease the dowel space while the way to improve the joints shear capacity is to improve the friction interface condition such as using indented face or use studs to improve the friction action.

2. Horizontal joints tensile numerical analysis complies well with the test results also. The main failure mode is that the failure starts from the joints between old and new concrete and then ends in the final tensile fracture of the joints, such as concrete cone failure as observed in the experiments. The PE rods for sealing also will weaken the joints interface tensile performance. During the test, the dowels and the LGS members with grouts work jointly until the program ends, which can be concluded that the horizontal joints of proposed novel type of precast bearing wall are reliable.

3. For insulated precast sandwich panels, it is recommended to manufacture the insulation layers inside of the precast members to avoid any contact with grouts leading to the weakening of the joints interface friction.

4. For the proposed novel precast bearing wall, further efforts should be devoted to exploring the seismic performance of the precast member joints. Also, the coupling of axial, shear and bending interaction of the joints may present further insights into the practical.

Funding Statement. This research was funded by POWERCHINA Science and Technology Research Key Project, grant number DJ-ZDXM-2020–28.

References

1. Holly, I., Abrahoim, I.: Connections and joints in precast concrete structures. Slovak Journal of Civil Engineering **28**, 49–56 (2020)
2. Alfarra, A.M.: Factors Affecting the Use of Precast Concrete System in Saudi Arabia. King Fahd University of Petroleum and Minerals (Saudi Arabia) (2018)
3. Abd El Fattah, A.M., Elfarra, A., Ahmed, W., Assaf, S., Al-Ofi, K.: Assessing the Factors Affecting the Use of the Precast Concrete Systems in Saudi Arabia Based on Stakeholders Survey. Arabian Journal for Science and Engineering **48**, 5467–5479 (2023)
4. Nibhanupudi, P., Rahul, G.B.: Comparative study on use of precast framed structure and precast load bearing wall structure. Materials Today: Proceedings **33**, 537–542 (2020)
5. Hofbeck, J., Ibrahim, I., Mattock, A.H.: Shear transfer in reinforced concrete. In: Journal Proceedings, pp. 119–128. (Year)
6. Mattock, A.H., Hawkins, N.M.: Shear transfer in reinforced concrete—Recent research. PCI J. **17**, 55–75 (1972)
7. Harris, H.G., Iyengar, S.: Full-scale tests on horizontal joints of large panel precast concrete buildings. PCI J. **25**, 72–92 (1980)
8. Einea, A., Yamane, T., Tadros, M.K.: Grout-filled pipe splices for precast concrete construction. PCI J. **40**, 82–93 (1995)
9. Alias, A., Zubir, M.A., Shahid, K.A., Abd RAhman, A.B.: Structural performance of grouted sleeve connectors with and without transverse reinforcement for precast concrete structure. Procedia Engineering **53**, 116–123 (2013)
10. Hosseini, S.J.A., Rahman, A.B.A., Baharuddin, A.: Analysis of spiral reinforcement in grouted pipe splice connectors. Građevinar **65**, 537–546 (2013)
11. Soudki, K.A., Rizkalla, S.H., LeBlanc, B.: Horizontal connections for precast concrete shear walls subjected to cyclic deformations part 1: mild steel connections. PCI J. **40**, 78–96 (1995)

12. Tsoukantas, S., Tassios, T.: Shear resistance of connections between reinforced concrete linear precast elements. Structural Journal **86**, 242–249 (1989)
13. Soudki, K.A., West, J.S., Rizkalla, S.H., Blackett, B.: Horizontal connections for precast concrete shear wall panels under cyclic shear loading. PCI J. **41**, 64–80 (1996)
14. Crisafulli, F.J., Restrepo, J.I., Park, R.: Seismic design of lightly reinforced precast concrete rectangular wall panels. PCI J. **47**, 104–121 (2002)
15. Jiang, H.-B., Zhang, H.-S., Liu, W.-Q., Yan, H.-Y.: Experimental study on plug-in filling hole for steel bar lapping of precast concrete structure. Harbin Gongye Daxue Xuebao(Journal of Harbin Institute of Technology) **43**, 18–23 (2011)
16. Jiang, H.-B., Zhang, H.-S., Liu, W.-Q., Yan, H.-Y.: Experimental study on plug-in filling hole for steel bar anchorage of the PC structure. Harbin Gongye Daxue Xuebao(Journal of Harbin Institute of Technology) 43 (2011)
17. Belleri, A., Riva, P.: Seismic performance and retrofit of precast concrete grouted sleeve connections. PCI journal 57 (2012)
18. Holly, I., Harvan, I.: Connections in precast concrete elements. In: Key Engineering Materials, pp. 376–387. Trans Tech Publ, (Year)
19. Fib Commission: Structural Connections for Precast Concrete Buildings: Guide to Good Practice. International Federation for Structural Concrete (fib), Lausanne, Switzerland (2008)
20. ACI 318–19: Building Code Requirements for Structural Concrete and Commentary. American Concrete Institute, Farmington Hills, MI, USA (2019)
21. PCI Concrete Handbook Committee: PCI Design Handbook: Precast and Prestressed Concrete (MNL-120-17). Precast/Prestressed Concrete Institute, Chicago, IL (2017)
22. EN 1992–1–1:2004: Eurocode 2: Design of Concrete Structures : Part 1–1: General Rules and Rules for Buildings. European Committee for Standardization (2004)
23. JGJ 1–2014: Technical Specification for Precast Concrete Structures. Ministry of Housing and Urban-Rural Development of the People's Republic of China, Beijing, China (2014)
24. Yuan, H., Zhenggeng, Z., Naito, C.J., Weijian, Y.: Tensile behavior of half grouted sleeve connections: Experimental study and analytical modeling. Constr. Build. Mater. **152**, 96–104 (2017)
25. Eligehausen, R., Cook, R.A., Appl, J.: Behavior and design of adhesive bonded anchors. ACI Struct. J. **103**, 822 (2006)
26. Fuchs, W., Eligehausen, R., Breen, J.E.: Concrete capacity design (CCD) approach for fastening to concrete. Structural Journal **92**, 73–94 (1995)
27. Obata, M., Inoue, M., Goto, Y.: The failure mechanism and the pull-out strength of a bond-type anchor near a free edge. Mech. Mater. **28**, 113–122 (1998)
28. Xu, L., Hai, T.K., King, L.C.: Bond stress-slip prediction under pullout and dowel action in reinforced concrete joints. ACI Struct. J. **111**, 977 (2014)
29. Singhal, S., Chourasia, A., Chellappa, S., Parashar, J.: Precast reinforced concrete shear walls: State of the art review. Struct. Concr. **20**, 886–898 (2019)
30. Zhou, Y., Ou, Y.-C., Lee, G.C.: Bond-slip responses of stainless reinforcing bars in grouted ducts. Eng. Struct. **141**, 651–665 (2017)

Study on Axial Compressive Performance of Prefabricated FRP-Steel-Concrete Composite Column

Dong-Xu Hou[1], Xiao Liu[1]([✉]), Liu-Jie Wang[1], Bing Wang[1], Zi-Jian Wang[2], Qing-Hai Wang[3], and Liang-Xiao Lv[1]

[1] Shenyang University, Shenyang 110044, China
489298344@qq.com
[2] Department of Construction, Chongqing University of Science and Technology, Chongqing 401331, China
[3] Shenyang Construction Quality Testing Centre. Co. LTD, Shenyang 110004, China

Abstract. This paper adopts the method by combining the experimental research and theoretical analysis to study the axial compressive performance of prefabricated composite columns. The research is mainly focus on two type of composite column, called ISFTC and IFSTC, composed by FRP, steel tube and concrete. Finite element analysis was carried out to study the influence of inner and outer tubular diameter on the constraining effect. Based on the unified strength theory of double shear and experimental data collected from former literature, the bearing capacity of composite column and the sectional optimization design method was proposed.

Keywords: Prefabricated · composite · Axial compressive · bearing capacity

1 Introduction

The composite columns, such as CFST(concrete filled steel tube) column and CFFT(concrete filled FRP tube), have been studied and widely applied in engineering. However, the composite column mentioned above seemly cannot satisfied the load bearing requirement, due to their respective shortcomings.

Recently, many studies focused on the new composite column with double wall, called DSTC (Double-skin tubular column). Two sectional forms are common: ISFTC (Inner steel FRP tubular column) and IFSTC (Inner FRP steel tubular column), as shown in Fig. 1.

Many researches focus on the ISFTC, the relative studies show that, comparing to hollow DSTC, the ISFTC present more axial bearing capacity but less ductility [1, 2]. The steel pipes between the concrete have already yielded, while the core concrete has almost no damage, which is related to the loading method [3]. The thinner the interlayer, the better the coordinated working performance of the steel pipe concrete and FRP pipe. The research was subsequently verified by finite element analysis [4]. Many

© The Author(s) 2024
P. Xiang and L. Zuo (Eds.): PBSFTT 2023, LNCE 382, pp. 231–237, 2024.
https://doi.org/10.1007/978-981-97-5108-2_24

other scholars attempted to built IFSTC [5]. The specimen will not be damaged due to the brittle failure of the FRP pipe during the loading process, which is beneficial for improving the ductility of the component [6]. The ductility of this component is closely related to the inner FRP material, and the effect is better for materials with high fracture strain [7, 8].

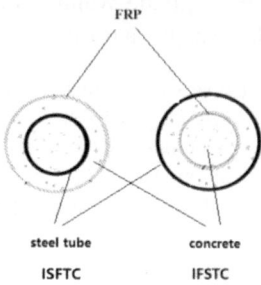

Fig. 1. Section of DSTC

The unified theory of computation for the bearing capacity of composite columns has not been formed in the current research to determine the optimal combination scheme. In this research, the finite element analysis was carried out to study the mechanical mechanism of composite column. The method to determine optimization scheme was proposed.

2   Database

Many literature focus on the compressive performance of two type composite column, ISFTC and IFSTC. The experimental database is composed of available data of some specimens in Table 1.

3   Finite Element Analysis

3.1   Establishing of Finite Element Model

In the three-dimensional finite element models of two types of composite columns, the concrete are simulated using 8-node reduced integral solid element (C3D8R), while the steel pipe and FRP material are simulated using 4-node reduced integral membrane element (M3D4R). In the model, binding connections are used between FRP and concrete, and contact elements are used between steel pipes and concrete as shown in Fig. 2.

The ideal elasto-plastic model is used in the finite element simulation. The stress-strain of FRP is linear before fracture. The plastic damage model was used in simulation concrete, the tensile and compressed constitutive relationship using GB50010-2010 [9] model.

Table 1. Collection of specimens in literature

Type	Source	Specimens
ISFTC	Ref. [1]	210-41.2-S; 154.4-77.2-S
	Ref. [2]	PT40-;ZMS-20-;ZMS-40
	Ref. [3]	C-D1-F-NSC-3L-1&2; C-D1-DGC-3L-1&2; C-D1-F-HSC-6L-1&2; C-D1-DGC-6L-1&2; C-D2-F-HSC-6L-1&2; C-D2-DGC-6L-1&2;C-D2-DGC-3L-1&2
	Ref. [5]	DTCC-NSC-G4-1; DTCC-NSC-G8-1; DTCC-HSC-G4-1; DTCC-HSC-G8-1
IFSTC	Ref. [6]	1S1FL-100-2; 1S1FL-100-3; 1S1FL-100-4; 1S1FL-150-2; 1S1FL-150-3; 1S1FL-150-4; 1S1FL-200-2;11S1FL-200-3; 1S1FL-200-4;1S1FH-150-3;1S1FH-150-4 1S1FH-200-3; 1S1FH-200-4;1S1TH-200-30;1S1NH-200-18
	Ref. [7]	1-1;2-1

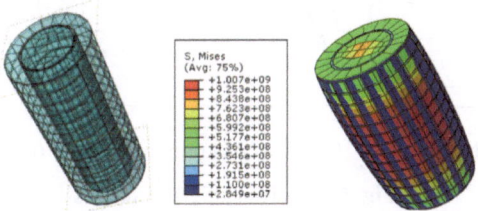

Fig. 2. Finite element model of specimen

3.2 Validation of Model

The validation of finite element model was verified by test results in database. As seen in Fig. 3 and Fig. 4, the curves made by finite element analysis were all agree with the experimental results. It is obvious that the finite element model used in this study is effective.

The Influence of Concrete Strength Ratio of Inner and External Tube
As shown in Fig. 5a, fcin and fcex means the compressive strength of concrete in inner tube and external tube. It can be concluded that the larger difference of concrete strength, the higher bearing capacity of specimen. The axial compressive bearing capacity is improving the strength of inner tube.

The Influence of Constrain Ratio of Inner and External Tube
Figure 5b is the specimen with different constrain ratio, changed by inner tube, and Fig. 5c is the specimen with different constrain ratio, changed by external tube. In Fig. 5b, f_{lin} and f_{lex} means the constrain ratio of inner and external tube, respectively. It is obvious that with the constrain effect of inner tube and external tube increasing, the bearing

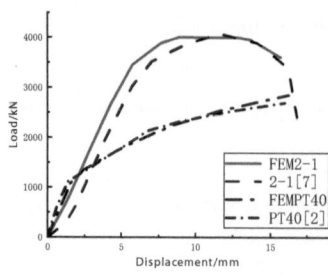

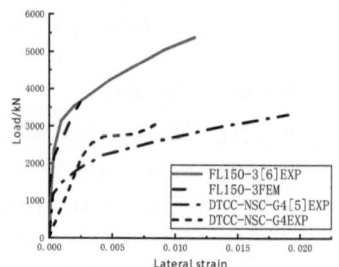

Fig. 3. Load-displacement comparison   **Fig. 4.** Load-lateral strain comparison

capacity of specimen all increased. However, the extend of increment is different, the constrain effect increasing of external tube have more obvious effect.

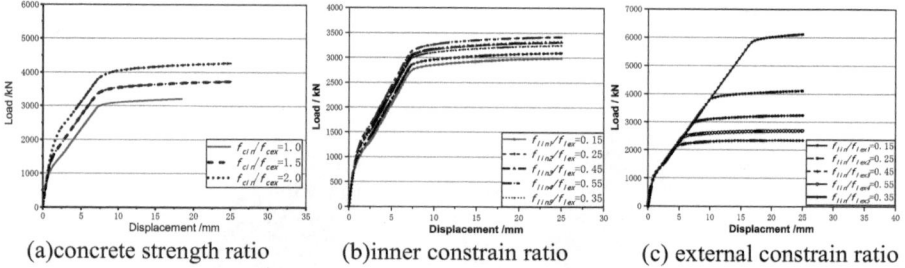

(a)concrete strength ratio   (b)inner constrain ratio   (c) external constrain ratio

Fig. 5. Analysis of effect factors

4   Optimization Calculation of Cross Section of Composite Columns

For confined concrete materials, the octahedron double shear mechanical model is represented by cohesion c and internal friction angle φ. According to the molar strength theory, the double shear strength can be expressed as [10]:

$$f_{cc} = f_c + k\sigma_{rc} \tag{1}$$

wherein, $f_c = \frac{2c\cos\varphi}{1-\sin\varphi}$, $k = \frac{1+\sin\varphi}{1-\sin\varphi}$, σ_{rc} is the lateral pressure. The bearing capacity of IFSTC composite columns can be calculated by the following equation:

$$N = (f_c + k\frac{2t_1\sigma_f}{d_1})(d_1^2 - d_2^2)\frac{\pi}{4} + \sigma_s\pi t_2 d_2$$
$$+[f_c + k(\frac{2t_1\sigma_f}{d_1} + \frac{2t_2\sigma_s}{d_2})]\frac{\pi}{4}(d_2 - 2t_2)^2 \tag{2}$$

It can be seen that the essence of the strength improvement of IFSTC and ISFTC is all the result of the inner and outer double layer constraint effect. However, for IFSTC composite columns, there is no significant damage in the core FRP tube, and the ultimate state of the inner and outer tubes is different. Therefore, it is necessary to reduce the inner tube sub items when calculate the bearing capacity. As shown in equation:

$$N_u = (f_{cex} + k\frac{2t_{ex}\sigma_{ex}}{d_{ex}})(d_{ex}^2 - d_{in}^2)\frac{\pi}{4} + \sigma_s\pi t_s d_s$$
$$+\alpha[f_{cin} + k(\frac{2t_{ex}\sigma_{ex}}{d_{ex}} + \frac{2t_{in}\sigma_{in}}{d_{in}})]\frac{\pi}{4}(d_{in} - 2t_{in})^2 \tag{3}$$

In the formula, $-ex$ represents the outer pipe, $-in$ represents the inner pipe, and $-s$ represents the steel pipe parameters, α Is the reduction coefficient, k is the constraint coefficient, taken as 2.26. Based on the numerical analysis, for IFSTC composite columns with steel pipes as the inner tube, α Taking 1.0, for ISFTC composite columns, α Take 1.9. The collected specimens were used for verification calculations, and the comparison between the calculation results and the experimental results is shown in Fig. 6. As shown in the figure, the calculated results are in good agreement with the experimental results. The bearing capacity of the columns can be calculated using Eq. (3), and the optimal diameter selection for the inner tube can be calculated using Eq. (4):

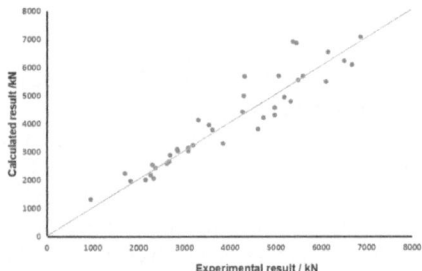

Fig. 6. Comparing of calculated and experimental results

Order

$$\frac{dN_u}{dd_{in}} = 0 \tag{4}$$

The optimal arrangement of the inner diameter d_{in} and outer diameter d_{ex} of the combined column satisfies the following equation:

$$d_{in}^3 - pd_{in}^2 - q = 0 \tag{5}$$

$$p = \frac{16f_{cin}t_{in}d_{ex} + 32kt_{ex}\sigma_{ex}t_{in} - 8kt_{in}\sigma_{in}d_{ex} - \sigma_s\pi^2 t_s d_{ex}}{8f_{cin}d_{ex} + 16kt_{ex}\sigma_{ex} - 8f_{cex}d_{ex} - 16kt_{ex}\sigma_{ex}} \tag{6}$$

$$q = \frac{4kt_{in}^3\sigma_{in}d_{ex}}{f_{cin}d_{ex} + 2kt_{ex}\sigma_{ex} - f_{cex}d_{ex} - 2kt_{ex}\sigma_{ex}} \tag{7}$$

$$d_{in} = \sqrt[3]{-\frac{q}{2} + \sqrt{\left(\frac{q}{2}\right)^2 + \left(\frac{p}{3}\right)^3}} + \sqrt[3]{-\frac{q}{2} - \sqrt{\left(\frac{q}{2}\right)^2 + \left(\frac{p}{3}\right)^3}}$$

Configure the inner pipe diameter according to din to obtain the maximum axial compression bearing capacity of the combined column.

5 Conclusion

(1) The axial compressive bearing capacity of DSTC is improving mainly with the concrete strength of inner tube.
(2) The constrain effect increasing of external tube have more obvious effect.
(3) The bearing of composite column can be understood as the effect of internal and external double-layer constraints. The optimal combination method for the cross-section of the composite composite column is proposed, and formulas (4)–(9) can provide reference for the engineering design of the composite column.

References

1. Wong, Y.L., Yu, T., Teng, J.G., et al.: Behavior of FRP-confined concrete in annular section columns. Compos. B **39**, 451–466 (2008)
2. Wang, J., Liu, W., Fang, H., et al.: Study on practical calculation method of axial compressive performance and bearing capacity of GFRP pipe-steel tube double-wall confined concrete composite. Build. Struct. **42**(2), 133–138 (2012). (in Chinese)
3. Liu, Y.: Experimental Study on Axial Compression Performance of GFRP Tubular, Concrete-Filled Steel Tube Composite Column. Dalian University of Technology, Dalian (2014)
4. Guo, X.: Finite Element Analysis of Axial Compression Performance of FRP Pipe-Concrete-Steel Tube Composite Short Column. Dalian University of Technology, Dalian (2015)
5. Ge, S.: Local Buckling Characteristics of CFST Columns with Inner FRP Tubes and CFST Columns with FRP Cloth. Guangdong University of Technology, Guangzhou (2018)
6. Long, Y.L., Li, W.T., Dai, J.G., et al.: Experimental study of concrete-filled CHS stub columns with inner FRP tubes. Thin-Walled Struct. **122**, 606–621 (2018)
7. Shi, Q.: Finite Element Analysis of Mechanical Properties of Circular Steel Tube Concrete Composite Column with GFRP Tube. Liaoning Project Technology University, Fuxin (2017)
8. LI, G.: Study on Axial Compressive Behavior of Steel-Concrete-CFRP-Concrete Composite Short Column with Circular Section. Shenyang Jianzhu University, Shenyang (2011)
9. China Academy of Building Research: Code for Design Concrete Structures (GB50010-2010). China Architecture & Building Press, Beijing (2010)
10. Wang, P., Ding, Y., Shi, Q., et al.: Calculation of axial bearing capacity of FRP- concrete - steel tube composite cylinder based on unified strength theory. J. Xi 'an Univ. Architect. Technol. (Nat. Sci. Ed.) **52**(2), 233–240 (2020)

Safety Inspection and Appraisal Analysis of a "—" Type Building in a Hospital

Si Wu[1] and Lin Ding[2(✉)]

[1] Chongqing College of Architecture and Technology, Chongqing, China
[2] JSTI Group Chongqing Inspection, Testing and Certification Co., Ltd, Chongqing, China
dl371@jsti.com

Abstract. According to the actual situation of the Engineering, the structure safety evaluation is carried out, and the bearing capacity of the main structure is checked according to the corresponding test results and specifications. On this basis, a comprehensive evaluation of the structural safety of a "—" type building in a hospital is carried out, which provides a basis for the subsequent structural transformation and reinforcement, and provides a reference for similar Engineering.

Keywords: structural identification · Safety inspection

1 Engineering Situation

The building is located in Fucheng District, Mianyang City, Sichuan Province. It was built in 1996. The building has 7 floors and the height of the house is 24.90 m. The height of the first and seventh floors is 4.2 m, and the height of the rest floors is 3.3 m. The building area is about 3500 m².

2 On Site Inspection and Testing Analysis

In order to have a comprehensive understanding of the structural safety performance of the inspected structure, the original data of the inspected structure is first verified before testing, and then on-site testing is carried out on the structure based on the actual situation [1].

2.1 An Analysis of Engineering Structure Documents

Before on-site testing, investigated the usage history of engineering structures and relevant design and construction history data of the inspected structures [2]. The seismic fortification intensity of the building during design was 6°; the foundation is pile foundation, with pebble layer as the holding layer, and the characteristic value of bearing

---

Si Wu and Lin Ding contributed equally to this work and should be considered co-first authors.

© The Author(s) 2024
P. Xiang and L. Zuo (Eds.): PBSFTT 2023, LNCE 382, pp. 238–244, 2024.
https://doi.org/10.1007/978-981-97-5108-2_25

capacity of roadbed should not be less than 700 kPa; the main structure adopts C20 grade concrete. The longitudinal reinforcement of beams and columns adopts HRB335 grade reinforcement. The stirrup adopts HPB300 grade rebar. Prefabricated hollow slabs are used for floor and roof panels.

2.2 Check the Building Foundation

The construction site is flat, no adverse geological phenomena such as landslides, collapses, subsidence, or ground fissures were observed on site. On site inspection did not find any obvious settlement or adverse phenomena such as foundation cracks or local damage caused by settlement in the main structure foundation, there is no lateral displacement or crack caused by uneven settlement in its upper structure.

2.3 Detection of Structural Component

Component Section Dimension Inspection

Part of the concrete column components and concrete beam components were selected on site for Component section dimension inspection [3]. The detection values are shown in Tables 1 and 2.

Table 1. Inspection Data of Cross Section Dimensions of Concrete Column Components

No.	Name and position of component	Design value	Allowable Variation	Detection value (mm)	Average value (mm)
1	Top of foundation ~4.2 m column (9)/(A)	500 × 600	+10,−5	505 × 603	502 × 602
				503 × 599	
				499 × 606	
2	4.2 ~ 7.5 m column (2)/(G)	500 × 700	+10,−5	509 × 700	502 × 703
				502 × 706	
				495 × 703	
3	7.5 ~ 10.8 m column (2)/(C)	500 × 600	+10,−5	502 × 599	501 × 600
				504 × 604	
				498 × 599	

Component Reinforcement, Concrete Strength and Carbonation Depth Testing

On site, a steel bar detector was used to detect the quantity of steel bars in concrete components, and the test results showed that the steel bar configuration met the requirements for construction quality acceptance.

Since the building commissioner was in use and the site did not have the conditions for core drilling and sampling, the compressive strength of the concrete members of the main structure of the first floor of the project was tested by the rebound method [4].

Table 2. Test Data for Cross Section Dimensions of Concrete Beam Members

No.	Name and position of component	Design value (b/bf × h/hf)	Allowable Variation	Detection value (mm) (b/bf × h/hf)	Average value (mm) (b/bf × h/hf)
1	4.2 m beam (7) ~ (9)/(E)	200 × 600	+10,–5	200 × 369	205 × 361
				208 × 355	
				209 × 360	
2	7.5 m beam (2)/(C) ~ (G)	300/120 × 560/120	+10,–5	304/564 × 127/126	303/563 × 125/123
				298/559 × 120/129	
				307/567 × 128/115	
				305/159 × 145/203	
3	10.8 m beam (7)/(E) ~ (H)	300/150 × 150/200	+10,–5	298/153 × 152/199	299/152 × 151/200
				296/159 × 154/204	
				305/145 × 148/199	
				299/145 × 158/148	

In accordance with the "Technical standard for in-site inspection of concrete structure"(GB/T50784-2013), When the standard deviation of the inspection lot is unknown, the upper and lower limits of the presumptive interval of the eigenvalue (0.05 quantile) x_k of the metrologically sampled inspection lot with 95% guarantee can be calculated according to the following formula:

$$x_{k1} = m - k_1 s$$

$$x_{k2} = m - k_2 s$$

where x_{k1} is the upper limit of the eigenvalue (0.05 quantile), x_{k2} is the lower limit of the eigenvalue (0.05 quantile), m is the arithmetic mean of the sample, s is the standard deviation of the sample, k_1 and k_2 are the coefficients of the upper and lower limits of the presumptive interval, the values taken are shown in the values corresponding to the sample capacity in the 0.05 quartile value column in GB/T50784-2013 "Technical standard for in-site inspection of concrete structure" [5]. The evaluation results are shown in Table 3.

Measurements of the carbonation depth of the reinforced concrete members were taken on site and the measurements showed that: The carbonation depth values ranged

Table 3. Evaluation results of compressive strength of concrete elements of the first floor main structure (Mpa)

components	average value	standard deviation	lower valuel	Upper valuel	Reference presumptions	Design strength class	Strength evaluation results
Column	52.4	3.29	42.82	48.64	48.64	C30	Compliance with design requirements
Beam	50.3	3.70	39.53	46.54	46.54	C30	

from 13.52 to 46.35 mm, and the combination of the appearance quality of the members and the detection of the protective layer thickness of the steel reinforcement shows that the carbonation depth of some reinforced concrete members was close to or exceeded the protective layer thickness of the steel reinforcement. [6].

3 The Security Analysis of the Struct

3.1 The Security Analysis of the Main Structural Components

Based on on-site measurement data and calculation results, some main components $R/\gamma 0S \geq 1.00$, the concrete structural components of this part are evaluated as au level based on their bearing capacity level; Part of the main components $R/\gamma 0S \geq 0.95$, the concrete structural components of this part are evaluated as bu level based on their bearing capacity level; Part of the main components $R/\gamma 0S < 0.90$, the concrete structural components of this part are evaluated as du level based on their bearing capacity level; on site inspection found that the construction of the concrete structural components of the building is reasonable, the connection method is correct, and there are no defects on the surface. Based on the construction situation, the safety level of the upper concrete structural components is evaluated as bu level; There is no obvious lateral bending or horizontal displacement observed in the concrete components, and the safety of the concrete components is evaluated as bu level based on the displacement or deformation that is not suitable for bearing; Some of the concrete members had force cracks of up to 0.5 mm, and the safety of the concrete structural components is evaluated as cu level based on the cracks.

3.2 Safety Assessment of Buildings

Based on the rating results of its foundation, upper load-bearing structure, and load-bearing sub units of the enclosure system, as well as other safety issues related to the entire building, the safety level of the building is rated as Csu level [7].

4 Anti-Overturning Checking

Based on the analysis of the safety testing results of the project structure, the anti overturning calculation is carried out [8]. During the verification calculation, referring to the original design drawings and based on the actual structural layout and dimensions, the SATWE2021 (V1.3.1 version) from China Academy of Building Sciences Beijing Gouli Technology Co., Ltd. is used for modeling and calculation.

Building width B = 49.8 m; Building height H = 27.9 m; Building length L = 20 m; If the aspect ratio of the building is H/B = 2.325 < 4, the stress zone on the foundation bottom of this project should not exceed 15% of the area of the foundation bottom [8]. The foundation burial depth of this project is set at 2 m.

According to the calculation results of SATWE2021 (V1.3.1 version), the total mass generated by dead load:g = 4580.099KN; The total mass generated by live load:q = 325.352KN.

Bottom shear force and bending moment under wind load: Vx1 = 222.5KN; Vy1 = 824.0KN;Mx1 = 3721.5KN·m; My1 = 13275.6KN·m. The shear force diagram is shown in Fig. 1 and the bending moment diagram is shown in Fig. 2.

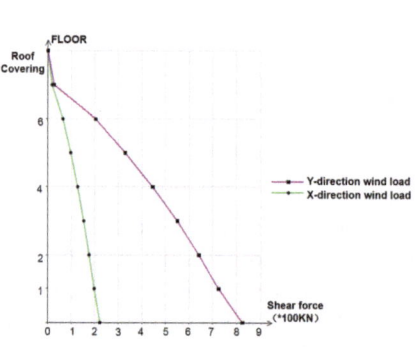

Fig. 1. The shear force diagram   **Fig. 2.** The bending moment diagram

Bottom shear force and bending moment under earthquake action: V_{x1} = 840.8KN; V_{y1} = 714.5KN;M_{x1} = 13546.27KN·m; M_{y1} = 11432.31KN·m.

Taking the burial depth of the building: C = 2 m, The overturning moment of the building Mov = VC + M, The calculation point of anti overturning moment is assumed to be the outer edge point of the foundation, and the calculation force of anti overturning moment is the representative value of the total gravity load, Then the anti overturning moment:Mr = GB/2,G = g + 0.5q = 4742.775KN. The calculation results are shown in Table 4.

Table 4. Anti overturning calculation

Operating Mode	Mr (KN.m)	Mov (KN.m)	Mr/Mov	Zero stress zone (%)
X-direction wind	6512.35	4166.5	28.34	0.00
Y-direction wind	47427.75	14923.6	3.17	0.00
X-direction seismic force	118095.10	15227.87	7.76	0.00
Y-direction seismic force	47427.75	12861.31	3.69	0.00

From the above table, it can be seen that Mr. is greater than 3Mov, which meets the specification requirements, so there is no stress zone in the base [9].

5   Conclusions and Recommendations

(1) Although the compressive strength of reinforced concrete members of the building meets the requirements, the carbonization depth of the members is large, and the protective layer of reinforcement of some members is too thin, which has a greater impact on the safety and durability of the structure, and it is necessary to take corresponding repair, reinforcement and maintenance measures to ensure that the subsequent use of the building requirements [10].
(2) Under the existing structural system and existing loading conditions, the overall structural safety of the building is rated at Csu level, and there are major problems with the structural safety performance, which significantly affects the overall load bearing, and treatment measures should be taken to enable it to meet the due safety requirements.
(3) Where the shear bearing capacity of the beam is insufficient, it is recommended to take the beam side paste steel plate or paste carbon fiber reinforcement. Where the bending capacity of beam is insufficient, it is recommended to take the bottom of the beam to paste steel plate or increase the cross-section reinforcement. For the column bending bearing capacity is insufficient, it is recommended to take the column outside the package angle or increase the cross-section reinforcement.

References

1. Zhang, Z., Xiong, C.: Structural safety inspection and appraisal analysis of an aged wharf and factory workshop. Build. Struct. (2012). https://doi.org/10.19701/j.jzjg.2012.s2.103
2. Li, J., Zhang, Z.: Safety appraisal and analysis of the space grid roof of an industrial plant .Build. Struct. (2023). https://doi.org/10.19701/j.jzjg.23S1298
3. Technical standard for inspection of building structure (GB/T 50344-2019). China Architecture Publishing & Media Co., Ltd, Beijing (2019)
4. Technical standard for in-site inspection of concrete structure (GB/T50784-2013). China Architecture Publishing&Media Co., Ltd, Beijing (2013)
5. Zabelina, O.: Application of non-destructive methods of control within the inspection of concrete structures. EDP Sci. (2021). https://doi.org/10.1051/E3SCONF/202125809007
6. Standard for appraisal of reliability of civil buildings (GB 50292-2015). China Architecture Publishing & Media Co., Ltd, Beijing (2015)
7. Shogo, T.: Non-destructive inspection of concrete structures using an electromagnetic wave radar. J-STAGE (2004). https://doi.org/10.11499/SICEP.2004.0_69_3
8. Vereecken, E., Botte, W., Lombaert, G., Caspeele, R.: Bayesian decision analysis for the optimization of inspection and repair of spatially degrading concrete structures. Elsevier (2020). https://doi.org/10.1016/j.engstruct.2020.111028
9. Code for design of building foundation (GB50007-2011). China Architecture Publishing & Media Co., Ltd, Beijing (2010)
10. General code for assessment and rehabilitation of existing buildings (GB 55021-2021). China Architecture Publishing & Media Co., Ltd, Beijing (2021)

Analysis on the Influencing Factors and Control Measures of Construction Quality of Prefabricated Building

Jieke Zheng[✉]

Chongqing College of Architecture and Technology, Chongqing 401331, China
zkk0025kh@163.com

Abstract. Under the new situation, the extensive construction and management mode of traditional construction projects has shown many problems in the development process, such as excessive resource waste, serious environmental pollution, and high workload of construction personnel [1]. Therefore, it is imperative to promote the development of Prefabricated building in China. However, in the current quality management practice of Prefabricated building in China, the relevant participants lack the quality management concept and sense of responsibility, resulting in poor construction quality of Prefabricated building, which affects the healthy and sustainable development of China's construction projects. Based on the case study method, this paper selects the quality management of the construction process of Prefabricated building as the research object, tries to study and analyze the factors affecting the construction quality of Prefabricated building, and formulate scientific development countermeasures to improve the construction quality of Prefabricated building, so as to promote the healthy and sustainable development of China's construction projects.

Keywords: Prefabricated building · Control measures · Construction quality · influence factor

1 Introduction

The policy shows that in the coming decades, China should vigorously promote and develop Prefabricated building. In China, the vast majority of traditional residential buildings still adopt traditional construction models, which gradually show many drawbacks in the development of the times. Prefabricated building is an inevitable trend of building industrialization. Its components are prefabricated in the factory with little external influence. They are transported to the construction site by truck and then fixed on the building, which can greatly improve the development benefits of construction projects. In China's engineering construction practice, the prefabricated construction mode has been introduced. However, due to the short development time, it has shown many shortcomings in the development process, and has not fully played the advantages of Prefabricated building. In the new era, with the acceleration of urbanization, the scale of urban construction continues to expand, and new requirements have been put forward

© The Author(s) 2024
P. Xiang and L. Zuo (Eds.): PBSFTT 2023, LNCE 382, pp. 245–253, 2024.
https://doi.org/10.1007/978-981-97-5108-2_26

for the quality and efficiency of construction projects. At this stage, it is urgent to discuss the factors affecting the construction quality of Prefabricated building.

2 Connotation of Construction Quality of Prefabricated Building

The Technical Standard for Prefabricated Concrete Buildings. (GB/t51231–2016) defines prefabricated structures as building construction structures that integrate structures, perimeter protection systems, pipeline equipment, and embedded systems. As the carrier of green buildings, prefabricated components based on Prefabricated building can effectively improve the construction efficiency, reduce the labor productivity of construction personnel, improve the construction environment, reduce resource waste, shorten the construction time, and improve the construction quality [2].

The sixth edition of the Project *Management Knowledge System Guide*(PMBOK) defines the construction quality of Prefabricated building as follows: the construction quality of Prefabricated building is a series of quality management processes that occur in the decision-making and planning stages, research and design stages, construction preparation stages and construction stages of Prefabricated building. Through data collection, analysis, and execution during the project implementation process, issue project quality management reports, project quality inspection and evaluation documents, and engineering change requests, enabling relevant organizations to take effective actions to solve effective problems. According to national laws and regulations, technical specifications, standards, design documents and relevant contracts, the construction quality of Prefabricated building refers to the participation in project construction with the construction unit as the main body under the guidance of the construction quality management system of Prefabricated building, and the implementation of reasonable and effective quality management for relevant personnel, materials and engineering equipment, so as to control the construction equipment, construction scheme and construction environment, To achieve the goal of promoting high-quality development of construction projects.

3 Factors Influencing the Construction Quality of Prefabricated Building

Through in-depth investigation and study of the representative Prefabricated building projects in China - Shenyang, Shanghai and Prefabricated building projects in Chengdu; Visit the site and analyze the project, and summarize the factors affecting the construction quality of Prefabricated building in China at this stage as follows. (1) The practical experience of construction personnel and management personnel is insufficient. From a cognitive perspective, during the construction of Prefabricated building, many construction personnel and management personnel do not understand the construction drawings, component categories and construction technology; From the perspective of operation, construction personnel are not yet mature in handling relevant construction techniques and differences between different types of work; The awareness of construction cooperation is weak, and construction management personnel significantly lack experience in preparing special construction plans and disclosing construction techniques. (2) Improper operation of PC components and unreliable connection. Due to the lack of proficiency in the operation of the PC component lifting plan and the lack of experience of the connecting personnel, the PC components encountered varying degrees of collisions during the lifting process, resulting in quality issues; Insufficient experience of operators has led to insufficient strength and reliability of connection nodes. (3) PC structure and adhesive structure are misaligned. In the process of Prefabricated building construction, due to the fact that the formwork is a rotary building material, there is an objective problem that the width of the slab joint is different, which leads to the size error of the expansion formwork, and it is difficult to obtain the required accuracy. During the grouting process, due to the low efficiency of the grouting equipment, it is easy to cause quality problems such as uncompacted grouting, thereby reducing the quality of grouting [3]. (4) The quality of each part of the project is uneven, and quality problems still often occur in the cast-in-place Prefabricated building. The concrete cracks in the cast-in-place section and the holes reserved for overlapping construction processes have not been effectively blocked in a timely manner, and the implementation of related work is still relatively weak. (5) The project quality control is not meticulous, and the emergency management of quality issues is ineffective. Due to the lack of construction quality acceptance standards for Prefabricated building, the quality control system for Prefabricated building formulated by construction enterprises is not detailed enough and functional enough, and the carelessness of management personnel affects the subsequent acceptance work. Once quality problems occur in the construction process of Prefabricated building, only operators and construction management personnel can cope with them based on their own experience, resulting in poor emergency management of quality problems [4]. (6) Poor quality management during the construction process. During the construction of Prefabricated building, the information that affects the quality of each part of the project is easy to be lost or distorted, and it is difficult to determine the influencing factors and the responsible person in a short time after the quality problems occur [5]. In addition, by distributing online questionnaires to various construction enterprises, design units, and professionals (200 questionnaires were sent out and 182 valid questionnaires were collected), as well as various literature analyses, factors affecting the quality of prefabricated building construction also include inadequate management systems and relevant standards, improper protection of finished products during the construction process, and environmental impacts. The factors that affect the construction quality of prefabricated buildings analyzed can be classified into three categories: management system, management subject, and management method [13]. The specific classification is shown in Table 1.

Using SPSS for confirmatory factor analysis, the above classified results were validated and tested. The analysis process is as follows:

Step 1: Analyze the KMO value (using an Excel table); If the value is greater than 0.8, it is very suitable for factor analysis; If it is between 0.7 and 0.8, it indicates that the data is more suitable for factor analysis; If the value is between 0.6 and 0.7, it indicates that factor analysis can be performed; If it is less than 0.6, it indicates that factor analysis is not suitable.

Step 2: If the P-value of Bartlett's test is less than 0.05, factor analysis is suitable;

$$X^2 = -\left[n - (2p + 11)/6\right]\ln|R| \qquad df = p(p-1)/2$$

Table 1. Summary of Factors Affecting Quality

Management System	Management Subject	Management Style
Relevant specifications and standards system	Management concept of subject data management	Innovation in production and construction technology
The quality management system of enterprises	The management level of each management entity	Design deepening in the construction process
Construction process inspection and acceptance system	Technical level of production and construction personnel	Finished product protection during the construction process
Pre job technical training and disclosure system	The supervisory role of supervisory units	Continuous improvement of construction process data
Tracing mechanism for building process data	Information exchange between various entities	Application of Information Technology

where: n is the number of data records; P is the number of variables in factor analysis; In() is a natural logarithmic function| R | is the correlation coefficient matrix R; X^2 is the value of the statistic, and by looking up table X^2 (0.01, df), the p-value can be determined.

Step 3: If there are only two analysis items, the KMO is 0.5 regardless.

(1)Analysis results of management system.

The survey data KM0 value is 0.805 > 0.6, which can be used for factor analysis. The Bartlett sphericity was used to test the data information, and the P-value was less than 0.05, indicating that the data is suitable for factor analysis. The correlation between items and factors is strong, and factors can effectively extract item information [12]. Name factor 1 as"management system"through information enrichment.

(2) Analysis results of management entities.

The KMO value of the survey data is 0.821 > 0.6, which meets the prerequisite for factor analysis. The data passed the Bartlett sphericity test, and the P value is < 0.05. The research data is suitable for factor analysis. The correlation between items and factors is strong, and factors can effectively extract item information. Name factor 2 as "management entity"through information enrichment.

(3) Analysis results of management methods.

The KMO value of the survey data is 0.755 > 0.6, which meets the prerequisite for factor analysis. The data passed the Bartlett sphericity test, and the P value is less than 0.05, making it suitable for factor analysis. The correlation between items and factors is strong, and factors can effectively extract item information. Name factor 3 as "management method" through information enrichment.

4 Control Measures for Factors Influencing Construction Quality of Prefabricated Building

4.1 Pay Attention to Various Quality Systems During the Construction Process

(1) Construction process inspection and acceptance system

Prefabricated building infrastructure production, construction process control and approval system have a very important impact on the quality of Prefabricated building construction. For component production enterprises, the quality control process must establish a comprehensive system for receiving imported raw materials, certifying the quality of production processes, and certifying the quality of components and accessories. For construction units, when prefabricated components enter the construction site, a corresponding quality approval system must be established to ensure the quality approval system of the assembly process.

(2) Pre employment technical training system

The construction of prefabricated facilities is carried out using new construction techniques, and the scope of the project is technical and comprehensive. At present, the management personnel of many construction companies and the first contact management personnel of production and construction units do not have sufficient experience in similar processes. Therefore, it is very important for prefabricated component manufacturers and construction units to establish a training system for management personnel.

(3) Traceability mechanism for construction quality issues

Many structural quality issues themselves are not structural reasons, but may be due to errors in drawing design or component prefabrication. However, these errors were not discovered during the manufacturing and construction processes, but rather occurred after the building was built and used [5]. The building quality tracking equipment in the construction process of Prefabricated building can be built through RFID radio wave technology to realize the quality tracking process of Prefabricated building [7]. In the production process of prefabricated components, a single piece can be used as a basic control unit, and RFID chips or two-dimensional codes can be used as tracking technology to collect raw material information, production process control information, factory quality information, etc., quickly find the deep-seated causes of historical quality problems, further improve the quality management system and Prefabricated building control monitoring system, and better achieve the quality improvement goals of Prefabricated building.

4.2 Improve the Quality Management Ability of Management Entities

Improving the quality management ability of all participants in the Prefabricated building project will provide strong support and reliable guidance for the sustainable and healthy development of prefabricated building in China. Compared with the concrete pouring operation of traditional construction projects, the whole process of Prefabricated building construction, from drawing design, prefabrication to construction and sorting, has undergone significant changes. At present, the quality management concepts of participants such as development and construction units, measurement and design units, control and inspection units, production units, and construction and installation units have not yet fully changed [6]. For Prefabricated building, information exchange between building units should be strengthened, especially effective communication between component manufacturers and contractors, and technical training of management personnel should be strengthened, which is of positive significance for improving the construction quality of Prefabricated building and promoting their sustainable and healthy development.

(1) Change traditional quality management concepts. The second phase of the XX region project in Shenyang was completed in 2020 and is the first prefabricated integrated structural residential project in Shenyang New Area, with an overall composition coefficient of 40% [10]. Based on the research and analysis of the entire construction process of the project, the author found that the professional literacy and management ideas of the project managers still remain in the traditional on-site development stage. The concept of quality control during on-site construction only conforms to the concept of concrete structure quality management. This requires enterprises to increase relevant training to help quality management personnel at the production and construction line change their own management concepts and methods, so as to better implement the quality management requirements of prefabricated buildings. (2) Strengthen information exchange between subjects. In the whole production and construction process of Prefabricated building, the quality of communication between participants will directly affect the process quality, availability and safety of the construction project. In the actual construction process, it is inevitable that some conditions such as design drawings do not match the actual construction site conditions. For example, floor slabs should be separated and laid between floors; Drilling a well not only affects the waterproofing between floors, but also damages the structure of the prefabricated floor slab. Therefore, if the construction unit communicates with the design unit in advance, the design unit can set exposed holes in positions that have minimal impact on the mechanical properties of the structure and do not affect normal construction work, thereby effectively preventing the occurrence of the aforementioned problems [8].

4.3 Improving Quality Management Methods

(1) Quality Information Sharing Based on BIM Technology

High quality information sharing is an important component of project management. From the perspective of the entire project lifecycle, if real-time information sharing and exchange cannot be achieved among various builders and regulators, enabling timely communication and coordination among all parties, quality issues cannot be resolved [9]. Nowadays, the traditional tripartite communication methods are mainly on-site communication or modern communication methods such as phone and WeChat. However, when quality problems arise, they often do not actively communicate, which has a significant impact on construction quality. Therefore, in order to better understand the initial quality management of Prefabricated building construction, it is necessary to coordinate the relationship between the participants so that they can maintain effective quality information management at all stages of the construction process. Therefore, in the new era, it is necessary to establish a component quality information sharing platform based on BIM technology, as shown in Fig. 1.

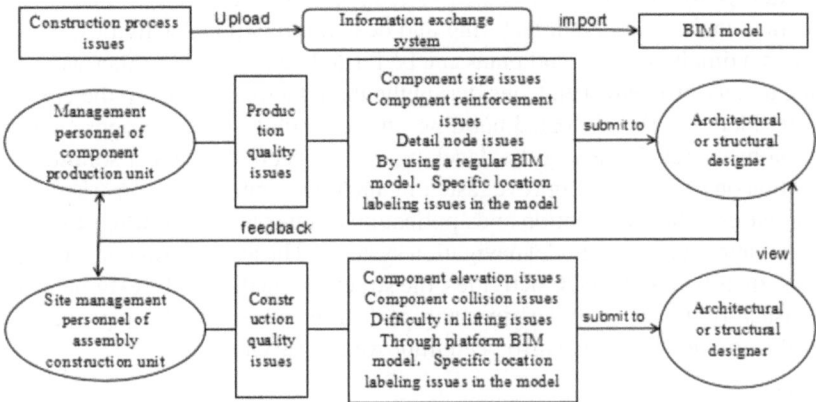

Fig. 1. Quality Information Sharing Platform Based on BIM Technology

Throughout the entire lifecycle of the construction project, BIM technology is relied on to create a high-quality communication platform for participants, which can improve information dissemination and problem-solving among all parties [6]. The preliminary management of quality issues during the construction process has important practical significance. In the architectural drawing design stage, the design unit uses the BIM model to insert the quality information of prefabricated building design (including the detailed design content of the building drawings and the split design content of prefabricated components) into the 3D model, and then imports it. During the construction of Prefabricated building, if the prefabricated parts manufacturer accidentally misunderstands the disassembly design of specific parts or has a more reasonable disassembly method, it can mark the problem in the relevant part of the platform model and send it to the corresponding designer. Management personnel inspect the quality issues of the unit on the platform and provide feedback after consideration. With this system platform, more efficient communication among production, construction and design units of Prefabricated building construction can be realized, and the trouble of having to go to the production and construction site for inspection due to vague expression of traditional communication mode can be avoided, thus improving the efficiency of construction.

The prefabrication rate of the main structure of the XX Phase II affordable housing project in Jiangsu Province has reached 32%, and the assembly rate has reached 67%. The entire process of this project utilizes BIM technology, effectively improving the quality control awareness and operational level of employees, and achieving good results [11].

(2) Continuous Improvement of Process Quality Based on QC Activities

"QC" is "Quality Control". It refers to the activities related to the management and coordination of the quality control system in construction projects. Its main function is to minimize the occurrence of quality problems during construction. QC operations are widely used in traditional engineering quality control, as well as quality control during the construction process of prefabricated structures. PDCA quality management and other related theories can be integrated into the construction, manufacturing, and construction processes of prefabricated structures. After discovering quality issues in construction engineering manufacturing and design through QC activities, they can be resolved in a timely manner, and plans can be formulated to guide further work to solve the same quality problems [9]. Considering the main factors leading to the construction quality problems of Prefabricated building, the research team can propose appropriate mitigation measures, including the development of specific technical solutions; Avoid construction on rainy days; Improve fastening methods; Implement construction technical information; The introduction and application of integrity monitoring raw materials, and the implementation of model navigation systems. Through continuous improvement of process quality based on QC activities, the structural quality of the external wall composite insulation board in the engineering project has been greatly improved, and the talents of team members have also been improved to varying degrees.

5   Conclusion

In summary, prefabricated buildings, as a manifestation of modern architecture, effectively support and inherit the dissemination and development of green building concepts. However, during the construction process, there are some factors that affect the quality of prefabricated building construction. If not managed in a timely manner, it will be detrimental to the healthy and sustainable development of China's construction industry. Based on the case study method, this article proposes control measures for the factors that affect the quality of prefabricated building construction, aiming to contribute to the healthy and sustainable development of the future construction industry.

References

1. Jiazhen, H., et al.: Research on the quality evaluation system of prefabricated building construction based on rough set. J. Xihua Univ. Nat. Sci. Edition **36**(5), 7 (2017)
2. Anonymous. The Standards and Quotas Department of the Ministry of Housing and Urban Rural Development issued a request for industry standard "Technical Standards for Testing of Prefabricated Residential Buildings (Draft for Soliciting Opinions)". Build. Technol. **49**(5), 1 (2018)
3. Qiyu, S., Jiejie, L.: Research on the differences and influencing factors of investment estimation indicators for prefabricated concrete buildings based on different assembly rates. Build. Econ. **42**(S01), 5 (2021)
4. Mengrui, L., Qiao, S.: Analysis of influencing factors on multi-space scheduling of prefabricated buildings based on ISM-MICMAC. J. Civ. Eng. Manag. **38**(3), 7 (2021)
5. Rongfang, C., et al.: Research on the construction and application of a quality risk assessment model for prefabricated building construction. J. Railway Sci. Eng. **18**(10), 9 (2021)
6. Xue, R., Xinyuan, W., Ke, S.: Analysis of construction problems and countermeasures of prefabricated buildings based on building information modeling technology. Ind. Arch. **48**(11), 4 (2018)

7. Jindian, L., Qilin, Z., Jinhui, Z.: Prefabricated construction management and quality control based on building information model and laser scanning. J. Tongji Univ. Nat. Sci. Edit. **48**(1), 9 (2020)
8. Chuansheng, Z., et al.: Analysis and countermeasures of common problems in energy-saving design and engineering application of prefabricated buildings. Build. Technol. **S1**, 3 (2018)
9. Ailin, Z., et al.: Research on the application of BIM technology based information integrated dynamic management system for prefabricated building construction phase. Manuf. Autom. **10**, 5 (2017)
10. Xinjun, W.: Prefabricated prefabricated building structural systems and design - review of prefabricated building structural systems and cases. Ind. Archit. **51**(6), 2 (2021)
11. Sohu. BIM Case | Application of BIM Technology in Prefabricated Structures [EB/OL] [2019–06–05]. https://www.sohu.com/a/318675006_545771
12. Xiaoqun, H.: Modern Statistical Analysis Methods and Applications, 3rd Edition. China Renmin University Press (2012)
13. Enji, G.: Research on Influencing Factors and Quality Improvement Plan of Prefabricated Building Construction Quality. Tianjin University, Tianjin (2019)

Digital Technology of Construction Monitoring and Quality Control of Prefabricated Building Engineering

Caixia Zuo and Jing Yang$^{(\boxtimes)}$

Department of Architectural Engineering, Chongqing College of Architecture and Technology, Chongqing 401331, China
18716468346@163.com

Abstract. Prefabrication is characterized by standardization of building parts and plant installation. Prefab construction in China is relatively late, but develops quickly. It is a kind of advanced construction method. Along with the development of science and technology, the requirement of building quality is higher, and prefab building has the advantages of high efficiency, low cost, and low pollution, which is consistent with the idea of green development. To guarantee the construction quality of prefab construction works, it is essential to supervise and control the construction process. In this paper, the construction supervision and quality control of prefab construction are analyzed from the angle of digital technique. It is proved by the experiment that the digital technique study on the construction supervision and quality control of the prefab construction project has achieved the highest pass rate of 95.7%.

Keywords: Prefabricated Construction Engineering · Construction Monitoring · Quality Control · Digital Technology

1 Introduction

China is a big country with a lot of people and a big construction country. It is estimated that the number of new buildings in China will be around 3 billion square meters annually, and the current construction area will not be able to satisfy people's demand. Along with the aging of the population, the speed of urbanization and so on, the demand of the construction industry in China is also growing. To solve this problem, China started to promote prefab buildings. However, due to the immature construction techniques, the study in China is relatively late, mainly in theory. Prefabrication has been widely applied in modern architecture due to its high efficiency, low cost, and low pollution. However, it is difficult to control the quality of prefabrication. In order to guarantee the construction quality, it is essential to supervise and control the construction process.

Recently, prefab construction has attracted wide attention. Many scholars and experts have carried out a thorough study and analysis of prefab modules. Among them, Wasim Muhammad gave an overview of DfMA approaches and their applications in fabrication

© The Author(s) 2024
P. Xiang and L. Zuo (Eds.): PBSFTT 2023, LNCE 382, pp. 254–262, 2024.
https://doi.org/10.1007/978-981-97-5108-2_27

and prefabrication. This review is based on the Systematic Review and Meta-Analyses of the Preferred Reporting Project (PRISMA), with minor changes. By comparing and discussing prefabrication and fabrication technology, it is proved that pre-fabrication and fabrication are similar in practice. In the end, he concludes with recommendations for future work and potential DfMA applications in the field of architecture [1]. Dou Yudan builds a three-tiered evolutionary game model on the basis of government, developer, and contractor perspective theory. Through strict theoretical deduction, the data were collected by Delphi, Policy Document and Document Analysis. The results indicate that contractors are in general willing to carry out the project, and the Government has decided to actively supervise its deployment. Passive investment behavior of developers is a major barrier to PC's promotion in Changchun [2]. The objective of Almashaqbeh Mohammad's study is to develop a new optimization model that will allow modular building planners to minimize the overall transport and storage costs of prefabricated modules in modular construction projects. Model performance was evaluated with a mixed module building in a medical facility, and its capability to minimize the overall transport and on-site storage costs of prefab modules in the building [3]. The above study has achieved a good result in the promotion of prefabricated building projects, there are still some deficiencies.

Traditional construction supervision and quality control methods are relatively simple, and the traditional work method is mainly applied to the construction supervision and quality control. There are some limitations to this method. Along with the development of society and the economy, the level of IT is becoming higher. Digital technique is characterized by high accuracy, large quantity of data and various kinds of data. In this paper, the application of digital technique in construction supervision and quality control is analyzed, and the construction level is raised.

2 Prefabricated Building

Prefabricated buildings are prefabricated, hoisted and installed on site. Compared with cast-in-place buildings, prefabricated buildings have advantages as shown in Table 1 [4].

Because the prefabricated building construction site only has a set of formwork, a set of steel bars, a set of hydropower equipment and a small number of carpenters and other operators, this makes the prefabricated building can be replicated in different regions and different climate environments [5]. Since prefabricated components are standardized products that are shipped to the site for assembly and installation after production in the factory, the number of wet operations on the construction site can be reduced, and construction pollution and safety hazards can be reduced. Prefabricated components are standardized products and their quality is relatively high, so there would be no large-scale damage in the event of natural disasters such as earthquakes. At the same time, when the prefabricated components are transported to the site for installation after the completion of production in the factory, there would be no large errors or deformation [6].

Prefabricated components are standardized products, which can reduce errors or damage caused by human factors in the construction process. At the same time, prefabricated building components would not be deformed or damaged when they are transported to the site for assembly and installation after production in the factory. Because

Table 1. Advantages of prefabricated buildings

Serial number	Advantage	Actual result
1	Separation of production and construction	Reduced labor costs
2	Standardization	The prefabricated components would not be deformed or damaged
3	One-time pouring molding	Reduce the wet operation on the construction site
4	Strong adaptability	It can be applied to the construction projects in different regions and in different climate environments
5	Energy saving and environmental protection	Reduce the environmental pollution caused by the site construction

prefabricated buildings are standardized products, they can be copied and expanded during construction. At the same time, prefabricated buildings are standardized products that can be produced and installed in factories. Since prefabricated buildings are standardized products, they can be installed and used in different regions and climates [7].

3 Construction Monitoring System

In the process of prefabricated building construction, monitoring the installation process, transportation process, stacking and lifting process of prefabricated components can effectively control the construction quality and reduce construction hidden dangers [8]. The monitoring content of the installation of prefabricated components mainly includes the deformation monitoring of prefabricated components and the installation deformation monitoring of prefabricated components. Deformation monitoring is the monitoring of the displacement generated by the structure or component during installation. If the deformation exceeds the allowable range, measures should be taken to control it, and artificial, intelligent, sensor and other technologies can be used to collect data [9]. In the deformation monitoring of prefabricated component installation, the displacement generated by the structure or component should be collected in real time. After the data is collected, it can be analyzed and processed according to the actual situation. Artificial or intelligent processing can be used to determine whether the data is accurate. The construction monitoring system uses the Internet of Things technology to build an intelligent perception network, and collects information and data through sensors, cameras and other devices. Big data technology is used to establish a three-dimensional model, and then the state changes of monitored objects are judged through data analysis and processing in the computer, so as to realize construction monitoring [10]. Generally, formula (1), formula (2) and formula (3) are used.

$$F(x) = \frac{2x - 1}{2x + 1} \tag{1}$$

In formula (1): x is the standard value of the sequence.

$$F(x) = vx + c \tag{2}$$

In formula (2): v and c represent the intercept and slope respectively.

$$Q(v) = -\frac{1}{A} \sum_{i=1}^{A} \left[z_i \log(z_i) + (1 + \hat{z}_i) \log(1 - \hat{z}_i) \right] \tag{3}$$

In formula (3): A is the model parameter or weight, z is the real label, and z_i is the prediction label.

In the process of prefabricated building construction, sensors are used to monitor the installation deformation of prefabricated components, and three-dimensional models are established by using the Internet of Things technology to transfer the specific information in the project to the intelligent perception network. Intelligent sensing network plays an important role in deformation monitoring of prefabricated components during installation. After obtaining the engineering information through the intelligent perception network, the appropriate sensor equipment is selected according to the construction site conditions for measurement and data collection. The Internet of Things technology is used to build a three-dimensional model to realize real-time monitoring of the construction site, through which engineering information can be transmitted to the intelligent perception network, which can improve the efficiency of construction monitoring [11, 12].

4 Quality Control Methods

4.1 Develop a Quality Control Plan

In the construction process, it is necessary to formulate a perfect quality control scheme, and formulate comprehensive quality control measures according to the actual situation of the project before construction. The quality control scheme is mainly to effectively monitor each link in the construction process to ensure the overall quality of the project construction. To develop a sound quality control plan, it can start from the following aspects [13, 14]: First, it can check and verify the materials and equipment used in the project before construction to ensure that the materials and equipment meet the construction requirements. Secondly, the construction personnel can be trained before construction to ensure that the operator has professional skills and literacy. In the construction, it is necessary to formulate clear operating procedures and working standards to improve the quality of the project. Thirdly, the construction methods and technologies used in the project should be clearly stipulated before construction to ensure the scientific and normative construction operations. Finally, it is necessary to do a good job in the construction process to find and deal with the existing problems in a timely manner [15].

The formulation of quality control plan needs to analyze all aspects involved in the project, and formulate a comprehensive quality control plan according to the actual situation. For example, pre-loading work should be carried out before the installation of prefabricated wall panels, and pre-loading work can be done before concrete placement to ensure the stability and accuracy of component installation during concrete placement [16]. In order to improve the efficiency and quality of engineering construction, it is necessary to establish an information technology monitoring platform to effectively monitor all aspects.

4.2 Apply Information Technology to Monitor Construction Quality

In the prefabricated construction project, the use of information technology to monitor the construction quality is conducive to improving the accuracy and efficiency of the construction and ensuring the quality of the project. For example, before hoisting precast concrete components, BIM technology should be used to analyze and process the hoisting equipment to ensure that the installation position and direction of the equipment comply with relevant regulations [17]. It can be operated according to BIM data collection during the hoisting process to ensure the safety of the hoisting process. Information technology can be used to monitor the construction process, which can avoid project quality problems caused by problems and errors in the construction process. For example, the prefabricated wall panels need to be measured before installation, and the position and direction of the wall panels can be determined according to the measurement results [18]. Through information technology, the construction process can be effectively monitored, and the engineering quality monitoring efficiency and quality control level can be improved. In addition, the application of information technology can also improve the efficiency of engineering quality management. For example, data collection should be done well before hoisting precast concrete members, and data analysis and processing can be carried out according to relevant standards to improve the level of construction quality management [19, 20].

5 Experiment on Digital Technology of Construction Monitoring and Quality Control

This paper conducts research experiments on construction monitoring and quality control, and the selected indicators are: monitoring efficiency (Fig. 1), construction quality (Fig. 2) and construction cost (Fig. 3). Among them, the monitoring efficiency refers to the efficiency of monitoring the non-standard operations of workers. The construction quality pass rate is obtained through quality inspection, and the construction cost is the cost spent to improve the quality pass rate.

It can be found from Fig. 1 that before the study of digital technology, the efficiency of construction monitoring is only 84.9% at the highest and 80.3% at the lowest, and the average efficiency obtained after calculation is 82.52%. After the study of digital technology, the efficiency of construction monitoring reaches 92.9% at the highest, 87.1% at the lowest, and the average efficiency obtained after calculation is 89.14%. It can be concluded that the digital technology research of construction monitoring and quality control can effectively improve the monitoring efficiency.

As can be seen from Fig. 2, before the study of digital technology, the highest qualified rate of construction is only 93%, the lowest is 90%, and the average qualified rate obtained after calculation is 91.60%. After the study of digital technology, the highest qualified

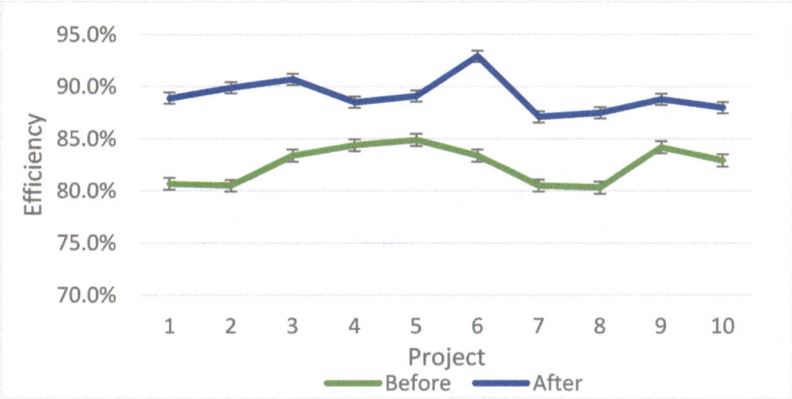

Fig. 1. Monitoring efficiency

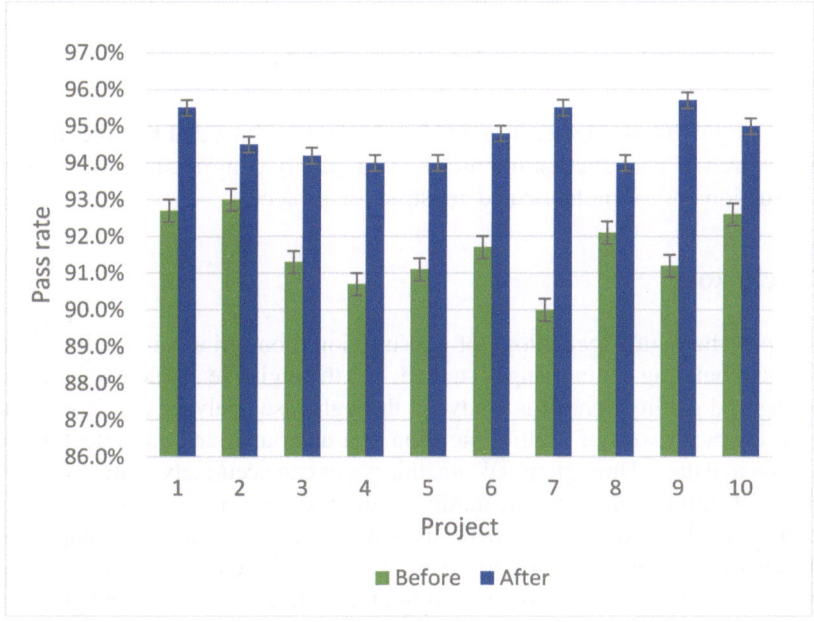

Fig. 2. Construction quality

rate of construction reached 95.7%, the lowest was 94%, and the average qualified rate was 94.70%. It can be concluded that the digital technology research of construction monitoring and quality control can effectively improve the qualified rate of construction.

As can be seen from Fig. 3, before the study of digital technology, the highest construction cost was $375, the lowest was $365, and the average cost was $370.7. After the digital technology study, the construction cost was as high as $363 and as low as

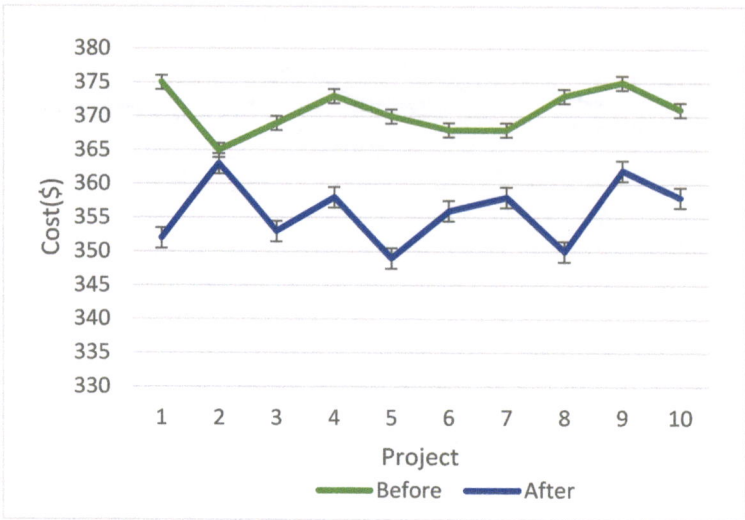

Fig. 3. Construction cost

$349, resulting in an average cost of $355.9. It can be concluded that the digital technology research of construction monitoring and quality control can effectively reduce the construction cost of prefabricated construction projects.

6   Conclusions

In this paper, the digital technology of construction monitoring and quality control of prefabricated building engineering is studied, and the digital technology of construction monitoring and quality control is analyzed through case analysis, and it is found that it has the characteristics of multi-dimension and all-round, and can integrate various information and data. Through these data, this paper can accurately evaluate the project, optimize and improve the project, minimize the risk of construction, and ensure the construction quality. At present, there are still some problems in the application of digital technology in prefabricated construction projects, such as incomplete application of digital technology and low level of information management. In order to promote the application of digital technology in prefabricated building engineering, it is necessary to formulate corresponding solutions according to the actual situation.

References

1. Wasim, M., Paulo, V., Tuan, D.: Design for manufacturing and assembly for sustainable, quick and cost-effective prefabricated construction–a review. Int. J. Constr. Manag. **22**(15), 3014–3022 (2022)
2. Dou, Y., et al.: Development strategy for prefabricated construction projects: a tripartite evolutionary game based on prospect theory Engineering. Constr. Archit. Manage. **30**(1), 105–124 (2023)

3. Almashaqbeh, M., Khaled, E.: Minimizing transportation cost of prefabricated modules in modular construction projects. Eng. Constr. Archit. Manag. **29**(10), 3847–3867 (2022)
4. Wuni, I., Geoffrey, Q., Robert, O.: Evaluating the critical success criteria for prefabricated prefinished volumetric construction projects. J. Financ. Manag. Prop. Constr. **26**(2), 279–297 (2021)
5. Zhang, H., Lu, Y.: Resilience-cost tradeoff supply chain planning for the prefabricated construction project. J. Civ. Eng. Manag. **27**(1), 45–59 (2021)
6. Yan, X., Hong, Z., Wenyu, Z.: Intelligent monitoring and evaluation for the prefabricated construction schedule. Comput. Aided Civil Infrastruct. Eng. **38**(3), 391–407 (2023)
7. Yang, H.: Comparative study on the cost of prefabricated construction projects and cast-in-place construction projects based on BIM technology. Value Eng. **42**(17), 4–6 (2023)
8. Lin, Y.: Research on steel structure construction technology and management strategy of prefabricated construction projects. Build. Mater. Develop. Orient. **21**(12), 151–153 (2023)
9. Liu, J.: Exploration of the main points of the construction technology of mechanical and electrical engineering of assembled buildings. Sci. Technol. Innov. Appl. **13**(16), 185–188 (2023)
10. Zhang, S.: A cost optimization model and simulation of dispatching of assembled construction projects based on improved particle swarm algorithm. Eng. Technol. Res. **8**(8), 131–133 (2023)
11. Lo, Y., et al.: Monitoring road base course construction progress by photogrammetry-based 3D reconstruction. Int. J. Constr. Manag. **23**(12), 2087–2101 (2023)
12. Kim, B., Eon-Sang, P., Sang-Jae, H.: The case study on the design, construction, quality control of deep cement mixing method. J. Korean Geosynth. Soc. **20**(4), 19–32 (2021)
13. Luo, H.: Construction technology and quality control strategy of asphalt pavement in highway engineering. J. Theory Pract. Eng. Sci. **3**(6), 1–3 (2023)
14. Patel, C.S., Pitroda, J.R.: Quality management system in construction: a review. Reliab. Theory Appl. **16**(SI 1(60)), 121–131 (2021)
15. Benz-C, M., et al.: Lessons learned designing and implementing a quality assurance system in an industrial engineering school. Qual. Assur. Educ. **31**(3), 369–385 (2023)
16. Jiang, L.: Research on the quality control of construction projects under the EPC model. Northern Archit. **8**(3), 69–73 (2023)
17. Shu, J.: Treatment and construction quality control of soft land foundation blown and filled in adjacent to the sea. Build. Technol. Dev. **50**(6), 130–132 (2023)
18. Liu, Y., et al.: Research on the quality control of passive ultra-low energy-consuming building construction based on digital twins. Project Manage. Technol. **21**(5), 11–16 (2023)
19. Luo, C.: Key points of quality control of electromechanical and electrical construction. Sci. Technol. Inf. **21**(8), 48–51 (2023)
20. Li, P.: In the context of information technology, the management and construction quality control of hydropower projects. Northeast Water Conserv. Hydropower **41**(4), 49–51 (2023)

Research on Materials for Concrete (Mortar) 3D Printing Fabricated Components

Nana Liu, Qili Gan, Qin Tao, and Jie Wang[✉]

ChongQing College of Architecture and Technology, ChongQing, China
Idnana@springer.com

Abstract. Concrete (mortar) 3D printing is a new construction process. Due to the lack of sufficient tensile strength and ductility of traditional cement-based printing materials, and the printing process will lead to the stratification of materials, and most 3D printing is contour printing, only after manually filling concrete and steel bars, 3D printed buildings have a certain bearing capacity. This paper discusses the current research progress of concrete (mortar) 3D printing, the requirements of printing materials, as well as the research status and future research direction of bionic materials. Through analysis, combined with the new material UDHCC independently developed by this research team, it is confirmed that it is completely possible to realize layered pouring components with both flexural bearing capacity and ductility. It provides theoretical and technical basis for the development of 3D printing no-bar construction process.

Keywords: Prefabricated building components · 3D printing · New material

1 Research Background

In the past 30 years, China's labor-intensive industries such as manufacturing and construction have made great progress due to the large labor supply. However, with China's aging population becoming more and more serious, China's labor-intensive industries began to appear insufficient labor supply phenomenon. The aging of the effective labor force will become the biggest obstacle to the future development of China's manufacturing and construction industries.

As a new digital construction technology, building 3D printing technology may become one of the solutions to the above problems. Building 3D printing technology integrates computer, numerical control, material forming and other technologies, using the principle of layered superposition of materials, the shape, size and other relevant information of the three-dimensional building model is obtained by the computer, and it is processed to a certain extent, according to a certain direction (usually Z-direction) the model is decomposed into a layer file with a certain thickness, and the numerical control program is generated. Finally, the mechanical device is controlled by the CNC system, and the automatic Construction of the building or structure is realized according to the specified path movement, which is called "Additive Construction", as shown in Fig. 1 and Fig. 2.

P. Xiang and L. Zuo (Eds.): PBSFTT 2023, LNCE 382, pp. 263–270, 2024.
https://doi.org/10.1007/978-981-97-5108-2_28

Fig. 1. 3D printed buildings

Fig. 2. Building 3D printing process

2 Research Status of Architectural 3D Printing

Researchers at home and abroad carry out research on architectural 3D printing from two different research perspectives. One is to study key issues such as machinery and materials of digital construction technology from an engineering perspective, such as Ding Lieyun from Huazhong University of Science and Technology and Feng Peng from Tsinghua University [1]. The other is to explore the new logic, theory and working mode brought by digital construction technology in architectural design. The business community is inclined to explore the industrial application of 3D printing digital construction technology in architecture.

There are still a lot of technical problems in building 3D printing from the full application of the real industry. First, there is the issue of printing materials. As mentioned above, due to the lack of mechanical properties of materials, the existing building 3D printing is still in the contour printing stage. To date, none of the materials used for 3D construction of buildings (cement-based materials, carbon fiber, nylon fiber, and steel fiber) can achieve the ductility of steel; Secondly, the structural weakening caused by construction stratification has not been solved.

Strictly speaking, "concrete layered spray and extruding superposition" is only the outer outline printing of the building, as shown in Fig. 3 and Fig. 4, and is not a 3D printing of the building in the true sense.

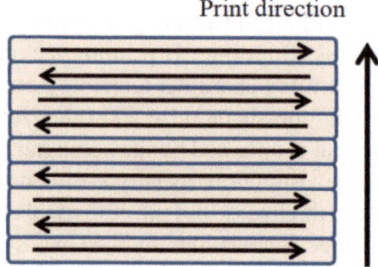

Fig. 3. The order of 3D printing

Fig. 4. 3D printed outline of the reinforcement

2.1 The Printing Material Lacks Sufficient Strength and Ductility

According to the Code for Seismic Design of Buildings GB5011–2010, "The measured value of the total elongation of the steel bar under the maximum tension should not be less than 9%, and the elongation should not be less than 20%". The excellent tensile ductility of the steel bar, coupled with the correct structural design and construction, the traditional reinforced concrete structure has enough ductility. However, at this stage, metal 3D printing mostly uses "laser sintering", that is, the technology of sintering powder compact with laser as a heat source. The high cost and energy consumption of this technology make it difficult to apply to practical projects. In contrast, concrete is made of hydraulic gels, sand, stone and industrial wastes (such as fly ash and mineral powder), which is inexpensive and has excellent room temperature plasticity, making it a natural 3D printing material. However, because of its low tensile strength and poor tensile ductility, concrete cannot be used for structural construction alone.

To improve the mechanical properties of concrete, the researchers reinforced the 3D-printed mortar with fibers such as glass fiber, steel fiber, and polyvinyl alcohol (PVA). According to the existing literature, the tensile strength of the most ductile PVA fiber reinforced cement-based composite is about 3 MPa – 7 MPa, and the corresponding tensile strain is 2%– 4%, which is still lower than the level of construction steel. Due to the lack of mechanical properties, ordinary fiber reinforced concrete still can not be used as a single structural material [2]. The lack of mechanical properties of materials is one of the main problems that architectural 3D printing stays at the contour printing level. Therefore, to achieve true 3D printing of buildings, we must first break through the material barrier.

2.2 Mechanical Problems Caused by Layered Printing

Architectural 3D printing mostly uses "extrusion hardening molding technology", which extrudes materials through the nozzle and lays them layer by layer, and the materials between layers are discontinuous. Even if the mechanical properties of the material meet the requirements, the interlayer interface will still be a weak link. Under the action of different forms of loads, the beam, wall and column members printed in layers will be separated or slipped, thus reducing the bearing capacity and stiffness of the members. The material layering causes the mechanical properties of the components to weaken, which is another reason why the 3D printing of buildings is still in the "contour printing" stage at this stage [3].

3   Research on New Materials for Building 3D Printing

At present, the materials commonly used in this field are fiber-reinforced cement-based composite materials and shell mother-of-pearl bionic structures. The following will highlight the application of two materials:

3.1 Research and Application of Fiber-Reinforced Cement-Based Composites

(1) Foreign studies

Over the past 30 years, researchers around the world have tried to apply ECC materials to structural engineering. Billington [4] pointed out that in addition to collapse resistance, the structure using ECC also has high damage resistance, and the residual crack width after damage is small, which can greatly reduce the repair cost. As for the application of reinforced ECC in the field of construction, foreign researchers have conducted a large number of experimental studies, and the test objects are mostly structural members under low cyclic load, including beams, columns, beam-column connecting members [6], filled walls, frames, piers [5], and connecting beams. Damping memberet al. These studies have proved that ECC has good seismic performance and minimal post-earthquake repair cost. It is worth noting that reinforced ECC also exhibits high ductility behavior under shear stress, high energy absorption behavior, stable hysteresis ring under large lateral displacement, and structural integrity. The most important characteristic of ECC is its tensile ductility. Even when the steel bar reaches plastic yield, ECC can still coordinate deformation with the steel bar. In addition, the impact resistance of ECC has also passed the test. Tests by Maalej et al. confirmed that ECC plates subjected to high-speed projectile impact had little damage, good integrity, multi-slit distribution cracking, and strong energy dissipation.

(2) Domestic research

Gao Danying and Zhu Haitang et al. [7] of Zhengzhou University conducted a series of experiments and theoretical studies on steel fiber and hybrid fiber concrete. Zhang Jun et al. from Tsinghua University conducted research on cement-based materials reinforced by steel fiber, polyvinyl alcohol fiber and wood fiber. Zhang Zhigang team of Chongqing University proposed an effective method to reduce shrinkage of high-toughness fiber-reinforced cement-based composites, which solved the difficult problem in this field.

Deng Zongcai et al. [8] from Beijing University of Technology conducted material property tests on a variety of fiber concrete, including cellulose fiber, polypropylene fiber, modified acrylic fiber, alkali-resistant glass fiber, PVA fiber, etc. Bu Liangtao et al. from Hunan University conducted a study on the interface between various fiber reinforced mortar and concrete, proposed a method for on-site detection of fiber reinforced mortar, and studied the application of fiber reinforced mortar in the reinforcement of existing concrete structures.

In recent years, Guo Liping et al. successfully produced ecological and high ductility cement-based composites (ECC) using domestic PVA fibers [9]. The tensile properties of ECC material prepared by domestic PVA fiber reach the world's advanced level [10], but the fiber cost is only 1/5 of the imported fiber from Japan.

3.2   Research Status of Biomimetic Structure of Shell Mother-of-Pearl

The typical characteristic of mother-of-pearl structure is "brick-mud" structure. "Brick" mainly refers to micron-scale wafers such as bioactive ceramics, calcium carbonate, and "mud" mainly refers to proteins. To mimic this "brick-mud" structure, many synthetic two-dimensional structures at the nano–or micron scale are used as "bricks" and polymers as "mud". Luz et al. have made a comprehensive review of biomimetic materials imitating mother-of-pearl structures. Mother of Pearl is the innermost layer of the shell and provides important strength and toughness to the shell.

The excellent mechanical properties of Mother of Pearl are due to its multi-scale-layered micro–and nano-structure form. The first is the micron-scale "brick-mud" multi-layer aragonite structure, with approximately hexagonal aragonite sheets layered on top of each other, and the interface is bonded by organic matter. Secondly, there are more subtle secondary structures on the surface of the aragonite sheet: nanoscale rough bulges and mineral Bridges. The rough protrusion makes the adjacent layers interlock with each other, forming part of "self-locking". In addition, there are inorganic mineral Bridges running through the arvinite pieces, forming a unique "brick-bridge-mud" multi-level structure model.

4   New Material Research of UHTCC

Regarding the problem of "printing materials lacking sufficient strength and ductility", the research team has a preliminary solution. This project is supported by the research team of Associate Professor Zhang Zhigang, School of Civil Engineering, Chongqing University. The lead researcher of this team learned the preparation method of Engineered cementitious composites (ECC, also known as UHTCC) for special fiber concrete during his study abroad. In the following years, the research team changed the original design method to prepare a cement-based composite with higher strength and ductility. The tensile strength of the material is up to 20 MPa, which is equivalent to the compressive strength of ordinary concrete. The highest compressive strength reaches 150 MPa, and the corresponding uniaxial tensile strain reaches more than $8\% - 12\%$, which has the ductility level of conventional steel. The bending test of beams shows that the bearing capacity of unreinforced beams cast with this material is equivalent to that of ordinary reinforced concrete beams with reinforcement ratio of 1.5%. The team then improved the material further. When the compressive strength is not higher than 35 MPa, the material shows the characteristics of compression strengthening, and the corresponding limit uniaxial tensile strain exceeds 6%, becoming a cement-based material that may be strengthened under both tension and pressure. Due to its excellent deformation ability, the material is named Ultra-high ductile cementitious composites, or UHDCC for short. The emergence of this material makes it possible to use 3D printing technology to build buildings without reinforcement.

Preliminary tests have been carried out in this project, and the toughening effect of stratified beams has been preliminatively proved through the trial test. Figure 5 shows the four-point flexural load-displacement curves of the full-cast, 5-story and 10-story beams. The layered UDHCC substrate achieves the crack deflection between layers during loading. However, due to the use of hydrophobic dielectric layer (polyethylene film), the interlayer adhesion is too small, and the interlayer slip occurs prematurely, resulting in a decrease in stiffness and strength. Compared with the whole cast beam, the strength of the layered cast beam has decreased, but the ductility has been significantly improved, and its energy dissipation capacity has increased by more than 1 times. According to the existing test results, after further improvement, it is entirely possible to give consideration to the bending capacity and ductility of the members cast by layers, and the toughening effect is remarkable, which provides a reliable test basis for the smooth development of the project.

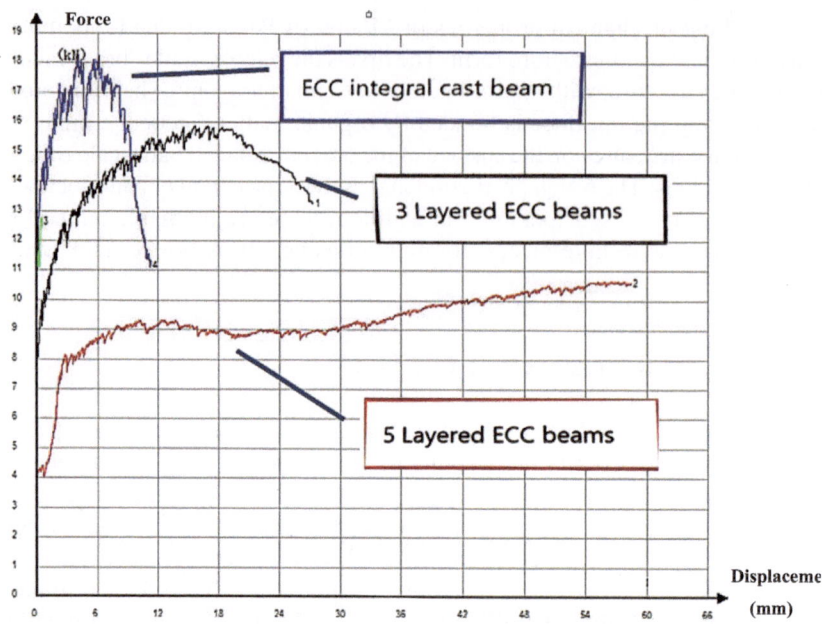

Fig. 5. Tentative test of layered assembly beams

5 Conclusion

(1) Commonly used mineral fibers (such as glass fiber, carbon fiber or basalt fiber, etc.) or biological fibers can improve the cracking resistance of concrete, but the effect against tensile strength and ductility is limited; Steel fiber can obviously improve the tensile strength of concrete, but in the best case, the tensile ductility is 0.5%-1.0%. The tensile strength of the specially designed polyvinyl alcohol fiber reinforced cement based composite (PVA-ECC) is about 3 MPa – 7 MPa. Its tensile limit strain is about 2% to 4%, which is still lower than the ductility level of construction steel. Due to the lack of mechanical properties, ordinary high performance cement-based composite materials can not be used as structural materials alone. The compressive strength and tensile strength of the materials used by the project team range from 30 MPa to 150 MPa, and from 5 MPa to 20 MPa, with an average tensile strain of 8% and a maximum tensile strain of more than 12%, which is close to the level of construction steel. The emergence of this material makes it possible to 3D print buildings without ribs.

(2) According to the existing research results, there are two main methods for artificial synthesis of "brick-mud" structure, one is to self-assemble the structural framework with inorganic nanomaterials, and then fill the polymer into the gap of the nanoframe. Another method is to form the mother-of-pearl structure by self-assembling the polymer with the two-dimensional inorganic assembly unit. However, due to the different construction materials and construction scales, these micro-level methods are difficult to apply in the field of civil engineering. So far, in the field of civil engineering, there are few reports about the results of mother-of-pearl structure

bionics. For the research object of this application, it is a very challenging research task to use what materials and how to carry out this super-structure bionics.

(3) The research team's novel material layered UDHCC substrate achieves crack deflection between layers. The ductility has been significantly improved, and its energy consumption capacity has been increased by more than 1 times. According to the existing test results, it is possible to give consideration to both flexural capacity and ductility, and the toughening effect is remarkable.

Acknowledgements. The research supported by the Science and Technology Research Program of Chongqing Municipal Education Commission (Grant No.KJQN202305204). The research project is "Study on the strength and toughness mechanism of 3D printed cement-based composite materials based on shell bionic structure". I would like to thank ChongQing College of Architecture and Technology for providing research fund support for this project, and the last thanks to the efforts of the team members, we believe that our project will get very good results.

References

1. Billington, S.L.: Damage-tolerant cement-based materials for performance based earthquake engineering design: research needs [A]. Fracture Mechanics of Concrete Structures. pp. 53–60 (2004)
2. Fischer, G., Fukuyama, H., Li, V.C.: Effect of matrix ductility on the performance of reinforced ECC column members under reversed cyclic loading conditions. Proceeding, DFRCC Int'l Workshop, Takayama, Japan, 269–278 (2002)
3. Parra-Montesinos, G., Wight, J.K.: Seismic response of exterior RC column-to-steel beam connections. ASCE J. Struct. Eng. **126**(10), 1113–1121 (2000)
4. Kesner, K.E., Billington S.L.: Investigation of infill panels madefrom ECC for seismic strengthening and retrofit. ASCE J Struct Eng, **131**(1(1), 1712–1720 (2004)
5. Fischer, G., Li, V.C.: Intrinsic response control of moment resisting frames utilizing advanced composite materials & structural elements. ACI Struct. J. **100**(2), 66–176 (2003)
6. Billington, S.L., Yoon, J.K.: Cyclic response of precast bridge columns with ductile fiber-reinforced concrete. ASCE J Bridge Eng **9**(4), 353–363 (2004)
7. Canbolat, B.A., Parra-Montesinos, G.J., Wight, J.K.: Experimental study on the seismic behavior of high-performance fiber reinforced cement composite coupling beams. ACI Struct J. **110**(5), 767–777 (2013)
8. Zongcai, D., Dajid Jumbe, R.: Experimental Study on toughening Characteristics of Blended fiber RPC. J. Build. Mater. **18**(2), 202–207 (2015)
9. Liping, G., Bo, C., Yanan, Y.: Research progress on the effect of PVA fiber on crack resistance and toughening of concrete. Progress in Water Resources and Hydropower Science and Technology **35**(6), 113–118 (2015)
10. Chujie, J., Wei, S.: GAO Peizheng. Natural Science Journal of Southeast University **37**(5), 892–897 (2007)

Load Transfer Capacity Analysis of Prefabricated Steel Fiber Concrete Pavement Slab

Zijian Wang[1], Hongkun Li[1], Xin Zhang[1,2(✉)], Bin Zhang[1], Qi Liu[1], Yadong Liu[1], and Baosheng Rong[1]

[1] School of Civil Engineering and Architecture, Chongqing University of Science and Technology, Chongqing 401331, China
2022206002@cqust.edu.cn
[2] The 5th Engineering Co., Ltd. of China Railway 11th Bureau Group, Chongqing 400037, China

Abstract. The joint of prefabricated concrete pavement slab is an important structural part of the prefabricated pavement slab. The joint damage will reduce the pavement function and accelerate the plate damage. The joint treatment technology is also a problem that puzzles engineering researchers. How to ensure the mechanical properties, load transfer rate and life of the assembled road joint has become the main research direction of the assembled road. This paper takes the indoor test as the research method, analyzes the bending and strain of the indoor test groove and lap joint. It is found that the bending value of the groove joint is large, and the tensile strain is the main one, which is easy to be damaged by tension. Through the comparative test, it is suggested that the joint form of the assembled road slab should adopt the lap type.

Keywords: steel fiber concrete · fabricated pavement slab joint · load transfer capacity

1 Introduction

Due to the rapid development of the transportation industry, road diseases are on the rise. Joints, being one of the weakest points in prefabricated pavement structure, exhibit varying load transfer capacities due to different structural forms. As a result, numerous scholars both domestically and internationally have conducted research on the load transfer capacity of cement concrete pavement joints [1]. Nowadays, the research on seam load transmission capacity mainly revolves around the design of the force transmission rod, and with the widespread use of the joint transmission rod, foreign scholars have successively appeared the joint design theory of the force transmission rod. Swati [2] analyzed the relationship between load, pavement performance and joint load transfer capacity by finite element software. Zhao Fangon [3] et al. studied the stress distribution of transverse joints under the state of pavement de-hollowing by establishing a three-dimensional finite element model, and analyzed the influence of force transmission rod

© The Author(s) 2024
P. Xiang and L. Zuo (Eds.): PBSFTT 2023, LNCE 382, pp. 271–278, 2024.
https://doi.org/10.1007/978-981-97-5108-2_29

spacing and cross-sectional size on joint load transfer. Tian Zhichang [4] used numerical analysis methods to analyze the relationship between the load stress in the cement pavement and the size of the plate, discussed the influence of the size of the groove on the tensile stress, shear stress distribution and deflection transfer effect at the joint, and proposed the relevant optimization scheme. Zhang Xin [5] et al. optimized the design of different sizes of joints of steel fiber road panels. At present, most of the research on joint load transfer capacity at home and abroad is aimed at ordinary concrete [6, 7]. In this paper, the load transfer capacity of the prefabricated steel fiber concrete pavement is analyzed by designing lap joints and groove joints, and the joint load transfer capacity is evaluated by the stress value.

2 Test Overview

2.1 Specimen Design

(1) Raw materials and mix ratio

In order to improve the durability and service cycle of prefabricated steel fiber road panels, P.O.42.5 grade ordinary Portland cement was used in the test, and the particle size range of coarse aggregate was 5.0 mm ~ 19.00 mm. Road steel fiber has requirements for flatness, fiber distribution uniformity, etc., so the end hook steel fiber is selected. The coarse aggregate uses 5.0 mm ~ 19.00 mm crushed stone as the coarse aggregate of the test, in order to ensure the reinforcement and toughening of concrete, the steel fiber content is controlled at 1.5%. And the same volume is used instead of coarse and fine aggregate to calculate the steel fiber mix ratio, the formula is as follows [8].

$$S_P = \frac{S_0}{G_0 + S_0} \tag{1}$$

$$k = \frac{S_0}{G_0} = \frac{S_P}{1 - S_P} \tag{2}$$

$$V_f = \frac{\Delta S_0}{\rho_s} \tag{3}$$

$$k = \frac{\Delta S_0}{\Delta G_0} \tag{4}$$

where: G_0 and S_0 are the amount of 1 m^3 matrix concrete stones and sand, respectively, in kilograms (kg);

k is the proportion of the amount of sand and stone;
V_f is the volume percentage of steel fiber in steel fiber concrete;
ΔS_0 and ΔG_0 are the amounts used by steel fibers to replace sand and stone.

The mix ratio of steel fiber reinforced concrete is shown in Table 1 below.

Table 1. Steel fiber concrete mix ratio

Categories	Cement/kg	Stone/kg	Sand/kg	Water/kg	Steel fiber/kg
Steel fiber concrete	415.00	1190.15	586.00	195.00	63.85

(2) Test board design and manufacture

In this experiment, a total of 3 prefabricated steel fiber concrete pavement slabs were designed, and the size of the specimens was 500 mm × 500 mm × 200 mm. Before concrete is poured, templates need to be made according to the size of the specimens. After the completion of the template to ensure the accuracy of the shape and size of the structure, so the selection of steel mold production, the test design of two different joints of SFPCP, the joint form is mainly rectangular groove joint and lap joint two forms, the joint size as shown in Fig. 1 below.

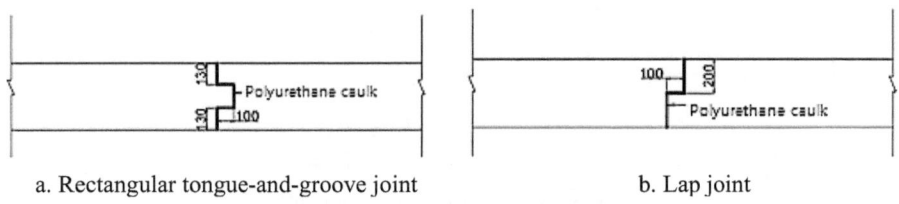

a. Rectangular tongue-and-groove joint   b. Lap joint

Fig. 1. Joint construction drawing (unit: mm)

When the prefabricated steel fiber concrete specimen is made, the steel fiber is dispersed into the mixer to ensure that the steel fiber is evenly dispersed in the plain concrete, and the production and maintenance of the test piece are in accordance with the "Test Method Standard for Physical and Mechanical Properties of Concrete" (GB/T50081-2019) [9] and the "Test Method Standard for Fiber Concrete" (CECS 13-2009) [10].

(3) Test block curing

The SFPCP board is maintained by watering at room temperature, watering every 3 h until the mold is removed, and the mold is removed after 24 h of maintenance. After demoulding, the SFPCP pavement board should be covered with moisturizing film and watered every 6 h to give full play to the hydration reaction of concrete in the board.

2.2 Loading Scheme and Measuring Point Arrangement

(1) measuring point layout A total of 8 measuring points are arranged on the test board, among which 2 displacement sensors (Linear Variable Differential Transformer-LVDT) are set up on the groove joint and the lap joint respectively, which are used

to record the real bending value of the joint during the loading process. The two adjacent plates are installed with a strain gauge at the joint (1# and 2# are the strain gauge at the groove joint, 3# and 4# are the strain gauge at the lap joint), which is used to measure the strain at the joint. The load arrangement diagram is shown in Fig. 2, and the measuring point arrangement diagram is shown in Fig. 3.

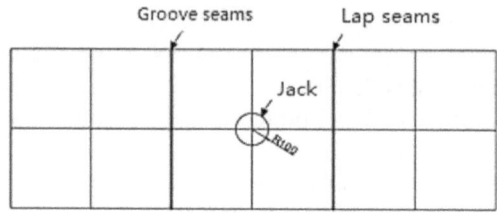

Fig. 2. Test load layout drawing

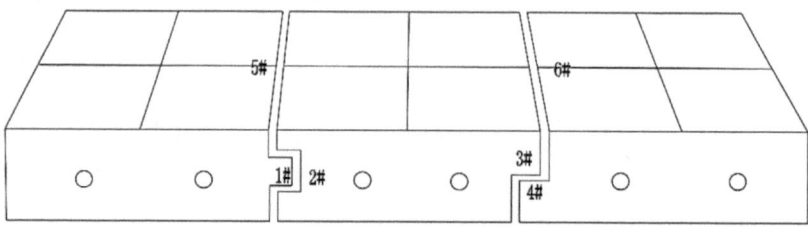

Fig. 3. Survey point layout drawing

(2) Test device and loading method

① Test device in this test, a 500 kN rigid reaction frame is used, and the loading speed and data collection of the 500 kN hydraulic jack in the middle span are controlled by a hydraulic operating platform. The detailed equipment diagram is shown in Fig. 4.

② Loading method before loading, make the jack and the plate in just contact state. The load form of the test board is step by step loading, using 0.5 kN/s to load, stop loading when the load to 23 kN, stop loading console automatically unloaded, in this test did not carry out destructive test and limit loading. After the start of the test, data is collected at a frequency of 1 s during the loading process, and the load and displacement values are automatically recorded. The test device is shown in Fig. 5.

3   Analysis of Test Results

(1) Time-deflection time-history curve

The bending curve of the SFPCP plate joint with time is shown in Fig. 6:

From Fig. 6, it can be seen that whether the deflection value of the groove joint or the lap joint gradually increases during the loading process, the deflection changes with time

Fig. 4. Test equipment diagram

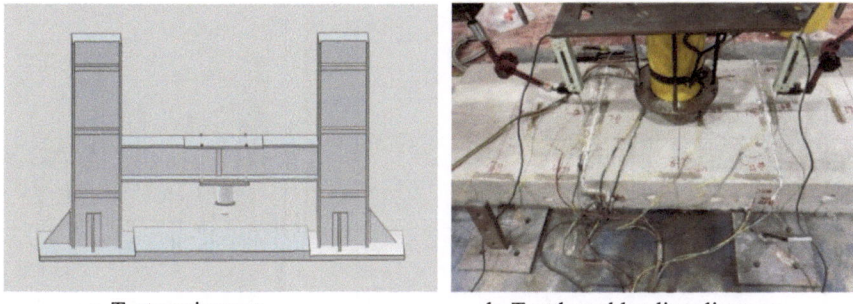

a. Test equipment   b. Test board loading diagram

Fig. 5. Test setup and fact-finding diagram

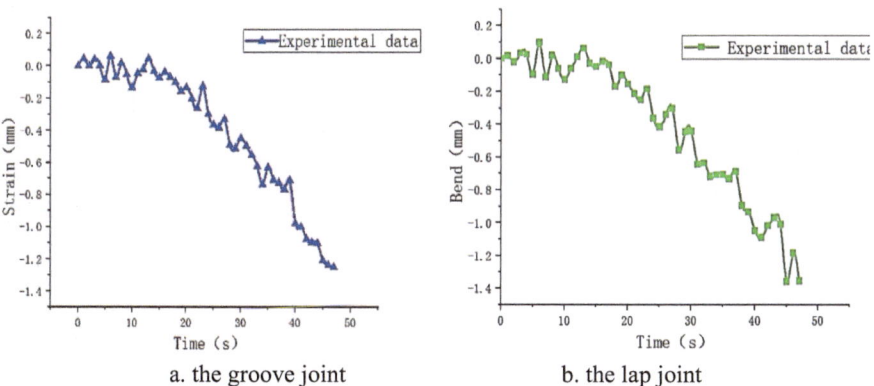

a. the groove joint   b. the lap joint

Fig. 6. Deflection-time history curve

is not increased in a certain proportion, but appears up and down in a state of fluctuation, and it can be seen from the analysis that when pouring the road panel, due to the small size of the joint, the vibration is not particularly sufficient. Through comparison, it is found that with the increasing time, the deflection of both seams is increasing and both are negative, indicating that the seam displacement is vertical upward. The change trend of the two seams is similar, the difference is that when the time has passed 25 s, the degree of change of the deflection value of the groove seam is greater than that of the lap seam, and it is almost a straight downward trend, while the lap joint will also fluctuate up and down, indicating that the lap joint is better than the groove joint when considering the deflection value factor of the road deck joint.

(2) Strain-time history curve

In the course of the test, two joints of the intermediate plate are arranged with 2 strain measuring points, and strain gauges are arranged on both sides of the plate at the corresponding positions with the intermediate plate joints to measure the strain at the joints. The time-history curve of the test plate is shown in Fig. 7.

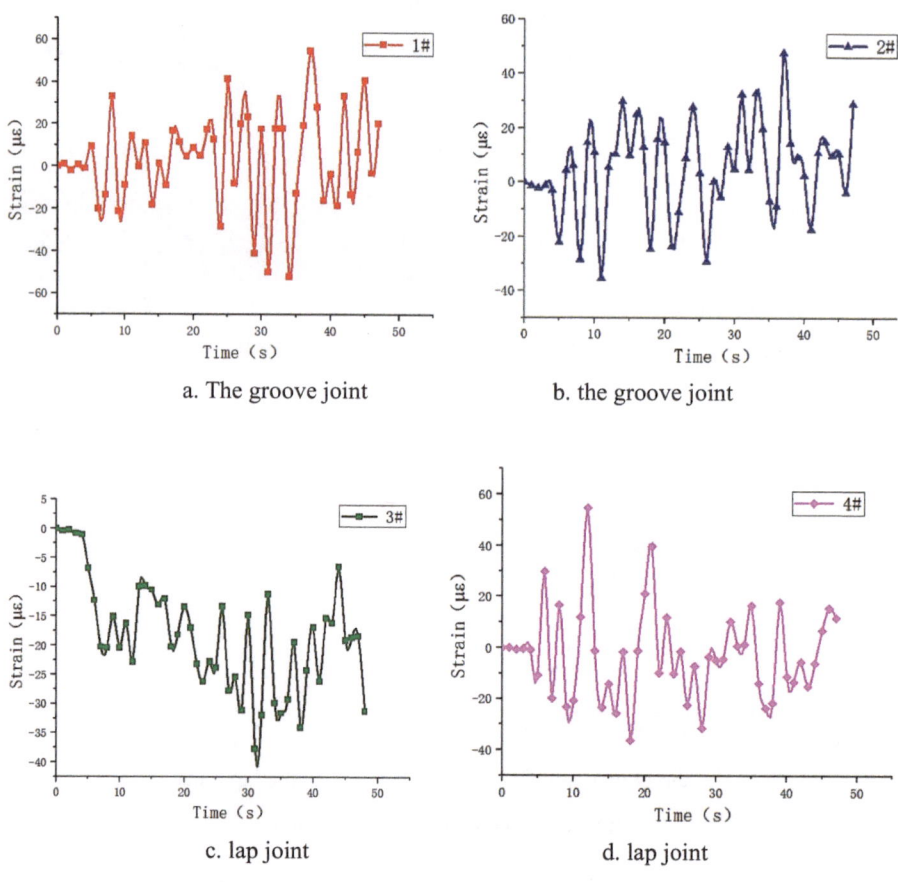

a. The groove joint

b. the groove joint

c. lap joint

d. lap joint

Fig. 7. Groove seam strain-time history diagram

By comparing Fig. 7 (a), (b), (c) and (d) respectively, it can be seen that whether it is a groove joint or a lap joint, the strain value changes very little in the first 5 s, and from the curve graph, the strain value changes almost straight line; The maximum strain value at both ends of the groove joint appeared at different times, the maximum strain value at the left end of the groove seam appeared after loading 26 s, and the maximum strain value at the right end appeared after loading 37 s. The strain value of the lap joint is also very small 5 s before loading, but the degree of change at both ends of the seam is very large, indicating that the two ends of the joint are compressive strain at this time, and the seam strain suddenly increases. The left end of the lap joint has always been compressive strain, and although the right end of the lap joint has tensile and compressive strain at the same time, it has more time in compressive strain.

4 Summary

Through the joint bending strain conduction test of SFPCP plate, the main conclusions are as follows:

(1) From the seam deflection-time history change curve, it can be seen that the deflection value of the groove joint or the lap joint maintains a downward trend during the loading process, and the deflection value fluctuates up and down with time, indicating that the steel fiber has a certain inhibitory effect on the deflection of the joint. When considering the deflection value of the road deck joint, lap joint joints are recommended.
(2) From the strain-time history curve of the joint, it can be seen that the strain change law at both ends of the groove seam is similar, and the tensile and compressive strain time at both ends of the seam is relatively average, while the left end of the lap joint is always in a compressed state, and the right end is mostly in a tensile state.

Project Funding

(1) 2023 Chongqing Higher Education Teaching Reform Research Project (Grant Nos: 233447) and 2023 Chongqing University of Science and Technology Undergraduate Education Teaching Reform Research Project (Grant Nos:202342); 2. 2023 Chongqing Construction Science and Technology Plan Project (Funded Project: 18) "Research and Demonstration of Key Technologies of Prefabricated Steel Fiber Concrete Pavement Panel".

References

1. Jianning, W., Yuanming, D., Jishu, S., Ming, W., Yuxi, Z.: Experimental study on load transfer attenuation of steel fiber reinforced concrete pavement joints. J. Zhengzhou Univ. (Eng. Technol. Ed.) **38**(5), 50–54 (2017)
2. Maitra, S.R., Reddy, K.S., Ramachandra, L.S.: Load transfer characteristics of dowel bar system in jointed concrete pavement. J. Tryp. Eng. **135**(11), 813–821 (2009)
3. Zhao, F., Cao, J., Wang, N.: Analysis of factors influence on void under neath at concrete. pavement joints. In: Proceedings, International Conference on Advanced Engineering Materials and Architecture Science (2014)

4. Zhichang, T., Yanan, M., Yumin, H., Meiyan, H.: Load stress analysis and joint optimization of prefabricated cement concrete pavement slab. J. Shandong Agric. Univ. (Nat. Sci. Ed) **47**(05), 753–759 (2016)
5. Zhang, X.: Analysis of Mechanical Properties and Engineering Application of SFPCP Joints under Loading. Chongqing University of Science and Technology (2023)
6. Wang, J., Dou, Y., Wei, M., Sun, J., Zhai, Y.: Model test of joint load transfer capacity of Cement Concrete Pavement. Sci. Technol. Eng. **17**(8), 52–56 (2017)
7. Yong, L., Jie, Y.: Research on Simulation method of Load transfer effect of cement concrete pavement joints by three-dimensional finite element method. Highway Transp. Sci. Technol. **30**(3), 32–38 (2013)
8. Jiye, L., Jingqiang, L., Shuqing, G., Chuanguo, H.: Practical Technical Manual for Concrete Preparation. Chemical Industry Press, Beijing (2011)
9. National Standard of the People's Republic of China. Test method for Physical and Mechanical Properties of concrete GB/T 50081-2019. China Building and Construction Press, Beijing
10. China Engineering Construction Association Standard: Test method standard for fiber reinforced concrete CECS 13-2009. China Planning Press, Beijing

Seismic Performance Analysis of Non-uniformly Corroded Reinforced Concrete Column Assembly Joints Strengthened with CFRP

Qi Wang$^{(\boxtimes)}$ and Shunzhong Yao

School of Civil Engineering, Southwest Forestry University, Kunming 650224, Yunnan, China
648341944@qq.com

Abstract. In order to analyze the seismic behavior of non uniformly corroded reinforced concrete column assembly joints strengthened with CFRP. In this paper, one corroded and unreinforced RC column specimen and four corroded but CFRP Reinforced RC column specimens were fabricated. According to the hysteretic curve, the skeleton curve change and stiffness degradation of each specimen are analyzed, and the load displacement condition of the assembled joint is judged. According to the strain of the assembled joint of reinforced concrete column, the ductility, energy consumption and other properties of each joint are analyzed, so as to analyze the seismic performance of the specimen. The results show that the unreinforced reinforced concrete column specimens have small ductility, low energy consumption and poor seismic effect; The reinforced concrete column specimens strengthened with CFRP have larger ductility, higher energy consumption and better seismic performance. Conclusion: the seismic performance is related to the stiffness, ductility coefficient, energy dissipation coefficient and other indicators. The larger the above indicators, the better the seismic performance, which plays an important role in improving the stability of the assembly structure.

Keywords: CFRP reinforcement · Non-uniform corrosion · Reinforced concrete column · Assembly node · Seismic performance

1 Introduction

The construction industry has gradually expanded and has become an important pillar industry of the national economy. The traditional cast-in-place construction mode has large energy consumption, construction environment noise and dust pollution, which affect the industry's construction prospects [1–3]. At present, the types of work in the construction industry are relatively old, and there are two challenges: sustainable development and labor shortage. Therefore, it is urgent to reform the construction mode and management mode, and upgrade the construction form of the construction industry. Fabricated building is a new construction method, which can decompose the whole structure into multiple prefabricated production components, which are uniformly manufactured and maintained by the factory, and connected and assembled at the construction site [2–4]. This method not only improves the efficiency of building construction, but also

© The Author(s) 2024
P. Xiang and L. Zuo (Eds.): PBSFTT 2023, LNCE 382, pp. 279–287, 2024.
https://doi.org/10.1007/978-981-97-5108-2_30

solves the problems of green environmental protection and labor shortage of buildings. It has become a widely used construction mode in the construction industry. As an indispensable component of buildings, reinforced concrete has opened up the industrial chain of component production in the construction industry through standardization, industrialization, assembly, informatization and other construction design forms in the factory, meeting the requirements of green development.

Earthquakes are disasters with large destructive power, explosive speed and impact range, which often cause casualties and property losses [5–7] to people in the area near the source center. The prefabricated building components are connected through dry connection and wet connection, and the connection assembly does not require post pouring concrete to ensure the integrity of the components, so as to meet the stability requirements of the building connection. However, in the process of building use, the reinforced concrete is affected by the weather environment, low temperature environment, high temperature environment, etc., which often leads to uneven corrosion, affecting the stability of building components [8–10]. In order to meet the overall stability of building components, flat steel reinforcement, CFRP reinforcement, tinplate reinforcement and other forms are often used to enhance the strength of modified reinforced concrete. In this paper, the seismic behavior of the non uniformly corroded RC column assembly joints strengthened with CFRP is studied.

2 Test Preparation

In this paper, before the test, a total of five reinforced concrete column specimens were made in the form of cast-in-place, and they were dissolved in Ca(OH)2 to form non-uniform corroded reinforced concrete columns. Four corroded reinforced concrete columns were strengthened with CFRP to form one corroded reinforced concrete column specimen and four corroded reinforced concrete column specimens strengthened with CFRP. Where, VB_1 is the cast-in-place reinforced concrete column specimen; SRTP_1, SRTP_2, SRTP_3, SRTP_4 is a fabricated test piece [11, 12]. A set of three 150 mm × 150 mm × 150 mm cube test blocks shall be reserved when pouring the test pieces, and a set of three 100 mm × 100 mm × 100 mm cube test blocks shall be reserved in the assembled test pieces. The compressive strength and tensile strength of cast-in-place reinforced concrete column specimens are analyzed. The retained test block and test piece shall be cured under the same conditions. After 28 days of curing according to relevant standards, the standardized test of compressive strength and tensile strength shall be carried out. In this test, C35 concrete is used for all test pieces except CFRP test pieces. The performance indexes of the test pieces are shown in Table 1 below.

As shown in Table 1, the stiffness of CFRP specimens is evenly distributed, and the reverse bending point of the reinforced concrete column is located in the center of the column. The inverted "T" column shall be made below the reverse bending point and above the foundation. The section of non-uniform modified reinforced concrete column head strengthened with CFRP is increased, which is convenient for the installation of horizontal actuator of MTS machine. Considering that the axial compression ratio of reinforced concrete and the diameter of longitudinal reinforcement are different, the reinforcement ratio is the same. When the axial compression ratio is set to 0.3 and

Table 1. Performance Indexes of Reinforced Concrete Column Specimens.

Test-piece	VB_1	SRTP_1	SRTP_2	SRTP_3	SRTP_4
Diameter of longitudinal reinforcement/mm	22	22	22	22	22
Stirrup diameter/mm	18	18	18	18	18
yield strength/MPa	398.621	607.243	603.469	596.525	602.421
the limit strength/MPa	401.836	611.456	602.527	603.753	605.936
Peak intensity/MPa	399.321	605.647	606.936	595.548	602.647
Compressive strength value/MPa	455.262	598.362	584.032	534.373	534.369
Tensile strength value/MPa	103.502	433.156	433.569	428.362	412.218

0.6, the corresponding vertical axial forces are 1000 kN and 2000 kN respectively. The column section is symmetrically arranged with longitudinal reinforcement along the push and pull sides. The bottom of the longitudinal reinforcement is designed as an "L" shape, and 200 mm is arranged along the long side of the column bottom foundation beam to strengthen the CFRP reinforcement effect of the reinforcement.

3 Test Method

In order to study the damage degree of VB_1, SRTP_1, SRTP_2, SRTP_3, and SRTP_4 more comprehensively and reasonably, the ductility, stiffness and energy consumption are selected as the seismic performance analysis indexes. The damage index of the specimen is expressed as:

$$D = f(x_1, x_2, .., x_n) \tag{1}$$

In formula (1), D is the damage index of the test piece; f is the damage function; $x_1, x_2,$ x_n is to reflect the change parameters of mechanical properties of the specimen damage. Usually, VB_1, SRTP_1, SRTP_2, SRTP_3, SRTP_4 test piece D are the value after multiple parameters are combined during the damage process of the test piece, which can represent the damage degree of the test piece. D reflects differently on VB_1 and test piece SRTP_1, SRTP_2, SRTP_3, SRTP_4. When $0 < D < 1$, the nondestructive state of the specimen under non-uniform corrosion is 0, and the corresponding seismic failure state is 1. According to the seismic action form, under the condition of reciprocating load, the damage index of the corner is calculated as follows:

$$\Delta D_i = A(\theta_o)^a \tag{2}$$

In formula (2), ΔD_i is the damage produced by the test piece under the condition of reciprocating load; A, a is the test parameter; θ_o is the plastic rotation angle generated by the reciprocating load. The ductility coefficient is introduced to represent the damage index of each specimen assembly node. The formula is as follows:

$$d_y = \frac{\mu_m - 1}{\mu_n - 1} \tag{3}$$

$$\mu_m = \Delta m / \Delta n \tag{4}$$

$$\mu_n = \Delta n / \Delta y \tag{5}$$

In formula (3) (4) and (5), d_y is the displacement ductility coefficient; μ_m is the maximum ductility coefficient of the specimen; μ_n is the ultimate ductility coefficient of the specimen; Δm is the maximum elastic-plastic deformation of the specimen; Δy is yield deformation; Δn is the ultimate deformation. The damage index of test piece proposed according to the aging dissipation of structural energy is expressed as:

$$D(t) = \frac{E(t)}{E} \tag{6}$$

In formula (6), $D(t)$ is the seismic energy consumption of the test piece; $E(t)$ is the energy dissipation value of the test piece at time t; E is the total cumulative hysteretic energy consumption of the test piece. In this paper, according to the actual situation of each test piece, the skeleton curve is analyzed, as shown in Fig. 1 below.

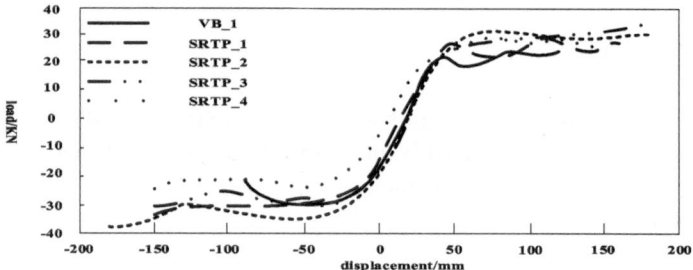

Fig. 1. Skeleton Curve of CFRP Specimen.

As shown in Fig. 1, test piece VB_1 during the displacement process of -100 mm– $+100$ mm, the load presents the trend of descending ascending descending ascending descending, and the overall compressive load is relatively small. The load change of CFRP strengthened specimens is obvious. The order of compression load is: SRTP_4 > SRTP_3 > SRTP_1 > SRTP_2. When the structure or component is constantly stressed, it will continue to produce damage, and the stiffness and strength of the specimen will decrease. The seismic damage of the test piece is expressed by stiffness degradation, and the expression is:

$$D(m) = \frac{K_0}{K_r} \tag{7}$$

In Eq. (7), $D(m)$ is the degree of stiffness degradation; K_0 is the initial tangent stiffness of the test piece; K_r is the reduced secant stiffness at the maximum displacement of the test piece. According to test piece VB_1, SRTP_1, SRTP_2, SRTP_3, SRTP_4, draw the stiffness degradation curve of each test piece, as shown in Fig. 2 below.

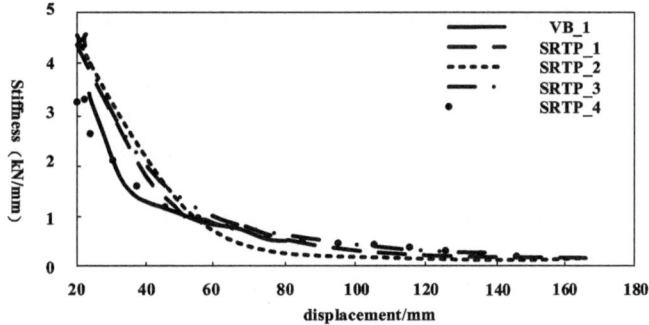

Fig. 2. Stiffness Degradation Curve of CFRP Specimen.

As shown in Fig. 2, test piece VB_1 the stiffness is within the displacement of 20 mm–80 mm, showing an overall decline trend. When the stiffness is 80 mm, it tends to be zero, and the stiffness is poor, so the seismic effect decreases. In the CFRP strengthened specimens, the stiffness tends to zero at 170 mm, which is relatively good, and the seismic effect increases accordingly. SRTP_1, SRTP_2, SRTP_3, SRTP_4 among the test pieces, the order of rigidity is SRTP_2 > SRTP_1 > SRTP_4 > SRTP_3. It can be seen from the figure that the yield displacement of the test piece is 40mm, and the stiffness degradation of each test piece gradually slows down after yielding. When the ultimate displacement reaches 170 mm, the stiffness degradation is almost flat, and the degradation amplitude is small. Therefore, the specimens strengthened with CFRP have higher stiffness and can dissipate more seismic energy.

4   Test Results

Earthquake is a disaster that has a great impact on the stability of buildings. There are three types of failure modes on building components, namely, the failure mode of structural integrity loss, the failure mode of insufficient bearing capacity of load-bearing structures, and the failure mode of foundation impact. Under the strong earthquake action, the internal force and deformation of the component increase, thus reducing the bearing capacity of the component. In this paper, ductility, energy consumption and other indicators are used to reflect the seismic capacity of CFRP strengthened specimens. Among them, ductility is a reflection of the elastic-plastic deformation capacity of CFRP strengthened specimens under non reciprocating loads. It is crucial to analyze the seismic performance of CFRP strengthened specimens by absorbing the energy generated by seismic loads through its own elastic-plastic deformation. In this paper, the ductility coefficient is used to express the seismic performance of the assembled joints, and the ductility variation law is judged in the form of forward loading and reverse loading. In general, the higher the ductility coefficient of CFRP strengthened specimens, the more seismic energy they absorb, the greater their ultimate deformation and yield deformation, and the better their seismic performance. When all the test pieces are damaged, the corner of the assembly node is between 0.09 rad and 0.12 rad; The corner of banana node is between 1/10–1/8, which is much larger than 1/30 in the specification. The ductility change of each test piece is obtained as shown in Table 2 below.

Table 2. Lithologic coefficient of assembly node displacement.

Test piece number		Yield load/kN	Yield displacement/mm	Peak load/kN	Limit displacement/mm	Ductility coefficient
VB_1	Forward loading	25.326	19.346	29.545	81.946	5.124
	Reverse loading	25.326	11.298	29.942	81.946	6.836
SRTP_1	Forward loading	21.643	13.247	33.354	148.732	11.343
	Reverse loading	26.664	22.145	36.758	148.165	8.932
SRTP_2	Forward loading	21.458	13.002	31.526	163.823	13.645
	Reverse loading	23.945	14.245	38.736	163.214	11.468
SRTP_3	Forward loading	23.315	11.765	35.596	143.756	12.236
	Reverse loading	24.854	16.842	30.624	143.692	9.516
SRTP_4	Forward loading	21.974	14.936	30.436	142.736	9.525
	Reverse loading	22.935	19.621	26.703	142.751	7.347

As shown in Table 2, the ductility coefficient of cast-in-place reinforced concrete column specimens is smaller than that of fabricated reinforced concrete column specimens. The order of ductility coefficient of fabricated test piece is SRTP_2 > SRTP_3 > SRTP_1 > SRTP_4. The larger the ductility coefficient is, the stronger the plastic deformation capacity of the reinforced concrete column structure can withstand, and the better the seismic performance of the structure is. Under earthquake action, CFRP strengthened specimens absorb energy through cracking and yielding, and the energy absorbed under repeated earthquake action is the seismic energy dissipation capacity of the specimens. In general, the energy dissipation capacity is measured by the surrounding area of the hysteresis curve. The larger the load envelope area of the hysteresis curve is, the stronger the energy dissipation capacity of CFRP strengthened specimens is; The smaller the area of the load envelope of the hysteresis curve, the weaker the energy dissipation capacity of the CFRP strengthened specimens. In this paper, cumulative energy consumption and equivalent viscous damping coefficient are used to express the energy consumption of CFRP strengthened specimens, and the first cycle energy of each horizontal displacement loading step is taken as the single cycle energy consumption of CFRP strengthened specimens, and the single cycle energy consumption of loading under seismic action is accumulated. The energy consumption change of the test piece is obtained as shown in Table 3 below.

Table 3. Lithologic coefficient of assembly node displacement.

Test piece number	Assembly node	Corner/rad	Accumulated energy consumption/J		Accumulated energy consumption decrease rate/%	Average rate of decrease/%
			Negative loading	Forward loading		
VB_1	X	0.128	91.862	60.092	41.658	44.569
	Y	0.106	48.741	33.412	26.534	
	Z	0.135	40.232	17.815	65.514	
SRTP_1	X	0.096	123.465	79.462	34.641	43.425
	Y	0.125	59.873	49.685	49.684	
	Z	0.146	49.265	25.443	45.951	
SRTP_2	X	0.106	115.651	62.554	59.645	45.929
	Y	0.168	53.702	42.186	45.674	
	Z	0.086	42.443	19.805	32.468	
SRTP_3	X	0.117	154.658	65.456	55.648	44.199
	Y	0.154	79.452	45.254	32.968	
	Z	0.106	58.362	22.653	43.982	
SRTP_4	X	0.085	143.254	68.195	59.641	45.941
	Y	0.146	65.355	48.376	45.534	
	Z	0.106	51.635	24.498	32.649	

As shown in Table 3, this paper selects VB_1, SRTP_1, SRTP_2, SRTP_3, SRTP_4, for test pieces such as, three assembly nodes X, Y and Z are selected for each test piece. Among them, the test piece SRTP_1, SRTP_2, SRTP_3, SRTP_4 is the CFRP strengthened specimen; Test piece VB_1 is the test piece not strengthened by CFRP. In case of positive loading, the test piece SRTP_1, SRTP_2, SRTP_3, SRTP_4 has a high energy consumption capacity. And the energy consumption capacity is sorted as SRTP_1 > SRTP_4 > SRTP_3 > SRTP_2. Test piece VB_1's energy consumption capacity is low. It can be seen that after CFRP strengthening and repair, the energy dissipation capacity of uniformly corroded reinforced concrete column specimens can be restored to the energy dissipation state when they are undamaged, even stronger than the energy dissipation state when they are undamaged, and the seismic performance is better. SRTP of test piece under negative loading SRTP_1,SRTP_2,SRTP_3,SRTP_4 The energy consumption capacity of is also high. And the energy consumption capacity is ranked as SRTP_3 > SRTP_4 > SRTP_1 > SRTP_2. The energy consumption capacity of the test piece VB_1 is also low. It can be seen that the non-uniform corrosion damage of reinforced concrete variable section under negative loading has a greater impact on the energy dissipation capacity of the specimen, and the cumulative energy dissipation changes more obviously. The specimens strengthened with CFRP can enhance the energy dissipation capacity of integral lifting joints of reinforced concrete columns under positive loading. The shorter the non-uniform corrosion damage part, the weaker the energy dissipation capacity of the specimens. The more the cumulative energy dissipation increases under positive loading, the better the seismic performance.

5 Conclusion

In recent years, prefabricated building structures have been widely used to meet the development needs of building construction. Prefabricated buildings significantly reduce process requirements and speed up construction efficiency. At the same time, building components can be mass-produced in the factory in advance, the construction quality and cost have been improved, the loss of building materials has also been reduced, and the demand for saving building construction resources has been met. Therefore, the prefabricated building structure is replacing the traditional cast-in-place concrete building structure in recent decades. Under the action of earthquake, the assembly nodes of prefabricated buildings are relatively stable, and the stability requirements of building components can be guaranteed only by strengthening or widening a few components. However, with the increase of application time of reinforced concrete, the problem of uneven corrosion on its surface becomes more serious, which affects its seismic performance. Therefore, this paper uses CFRP strengthening method to analyze the seismic performance of assembled joints of reinforced concrete columns with uneven corrosion. In the form of experiments, the stiffness, ductility, strain and energy consumption of assembly joints are analyzed to provide data support for the stability of building components.

References

1. Ni, Y., Hao, C., Xu, Y.: Seismic performance analysis of self-centering segment piers with mortise-tenon shear connectors based on cyclic pseudo-static test. J. Vibroeng. **23**(7), 1621–1639 (2021)
2. Wang, X., et al.: Experimental validation on seismic performance of a monolithic precast RC shear wall structure with novel connections. Adv. Struct. Eng. **25**(5), 1042–1056 (2022)
3. Zhang, W., et al.: Seismic performance of a continuous bridge with winding rope device activated by a fluid viscous damper. Adv. Struct. Eng. **25**(6), 1222–1239 (2022)
4. Yang, T.Y., et al.: Seismic performance evaluation of innovative balloon type CLT rocking shear walls. Resilient Cities Struct. **1**(1), 44–52 (2022)
5. Chong, X., et al.: Experimental investigation of seismic performance of a novel isostatic frame-cladding system. Adv. Struct. Eng. **25**(5), 1015–1026 (2022)
6. Gwalani, P., Singh, Y., Varum, H.: Effect of bidirectional excitation on seismic performance of regular RC frame buildings designed for modern codes. Earthq. Spectra **38**(2), 950–980 (2022)
7. Lu, X., et al.: Comparison of seismic performance between typical structural steel buildings designed following the Chinese and United States codes. Adv. Struct. Eng. **24**(9), 1828–1846 (2021)

8. Chong, J.H., Alih, S.C., Vafaei, M., Wong, W.K.: Seismic performance of low ductile rc frame designed in accordance with malaysia national annex to eurocode 8. IOP Conf. Ser.: Earth Environ. Sci. **682**(1), 012011 (2021)
9. Gupta, P.K., Ghosh, G.: Effect of various aspects on the seismic performance of a curved bridge with HDR bearings. Earthq. Struct. **19**(6), 427 (2020)
10. Hamid, N.H., Hadi, N.D.: Modelling and validation of seismic performance for corner and interior beam-column joint using HYSTERES program. IOP Conf. Ser.: Mater. Sci. Eng. **1062**(1), 012038 (2021)
11. Ma, S., Li, L., Bao, P.: Seismic performance test of double-row reinforced ceramsite concrete composite wall panels with cores. Appl. Sci. **11**(6), 2688 (2021)
12. Ma, D.-Y., Han, L.-H., Zhao, X.-L.: Seismic performance of the concrete-encased CFST column to RC beam joint: experiment. J. Construct. Steel Res. **154**, 134–148 (2019)

Study on Ultimate Bearing Capacity of CFRP-Reinforced Concrete Under Reciprocating Load

Qi Wang[✉] and Shunzhong Yao

School of Civil Engineering, Southwest Forestry University, Kunming 650224, Yunnan, China
648341944@qq.com

Abstract. Objective: To study the ultimate bearing capacity of CFRP- reinforced concrete under reciprocating load. Methods: Firstly, a number of CFRP- reinforced concrete specimens were made on the test platform, and the overall compression was adopted to meet the test requirements. Secondly, the load-displacement curve of the specimen is drawn to determine the displacement change of the compression limit state. Finally, the load-deformation situation is analyzed, and the compressive ultimate bearing capacity of the specimen at each position is analyzed to determine the compressive ultimate bearing capacity of the specimen. Results: The compression state of CFRP-reinforced concrete can not be changed by winding CFRP and changing the strength of steel tube. The smaller the overall deformation, the higher the ultimate bearing capacity of CFRP-reinforced concrete. Conclusion: The ultimate compressive bearing capacity of CFRP-reinforced concrete is related to the slenderness ratio, the number of CFRP layers and the circumferential fiber yarn, which plays an important role in improving the strength of CFRP-reinforced concrete.

Keywords: Reciprocating load · CFRP-reinforced concrete · Under pressure · Ultimate bearing capacity

1 Introduction

As a composite material, concrete is often affected by load, resulting in internal fracture, which affects the stability of the structure. Reinforced concrete is an upgraded component of concrete, and micro-cracks often occur in it. Under the action of environmental load, the whole concrete structure is gradually destroyed [1–3]. Reinforced concrete cracks are divided into macro and virtual parts. Macro cracks can be detected by instruments, which is a real hidden danger. Virtual cracks can't be detected by instruments, and they are small in size and narrow in width. They are continuously destroyed by the load during construction, which affects the final construction quality [4–6]. Reinforced concrete will be damaged by temperature, water pressure, wind, waves and earthquake load, which will affect the mechanical properties of reinforced concrete. By analyzing the load-strain and load-displacement of reinforced concrete, we can fully grasp the changing law of concrete material properties, thus ensuring the stability of reinforced concrete in complex

© The Author(s) 2024
P. Xiang and L. Zuo (Eds.): PBSFTT 2023, LNCE 382, pp. 288–297, 2024.
https://doi.org/10.1007/978-981-97-5108-2_31

environment [7–9]. Reinforced concrete structures not only bear static load, but also bear complex dynamic load and repeated load, which leads to structural damage and fracture failure, and affects the bearing capacity and durability of components. In order to meet the bearing demand of reinforced concrete, it is necessary to optimize the bearing capacity of reinforced concrete under reciprocating load. CFRP is a fiber-reinforced polymer, which is composed of carbon fiber, glass fiber, aramid fiber, basalt fiber and metal fiber. It has the advantages of light weight, high tensile strength, good fatigue resistance and corrosion resistance, and is widely used in the field of civil engineering [10]. Therefore, the ultimate bearing capacity of CFRP-reinforced concrete under reciprocating load is studied in this paper.

2   Specimen Production and Preparation

In this experiment, CFRP-reinforced concrete specimens are made on the test platform, and the concrete mixing materials include tap water, cement, fine aggregate and coarse aggregate. Among them, the water contains no impurities, and the fine aggregate is sun-dried intermediate river sand to avoid concrete strength damage, and the fine aggregate [11–13] is stored in barrels. Coarse aggregate is filtered out of granite gravel below 20 mm, washed, dried and stored in barrels. At the same time of concrete mixing, the 5 m-long steel pipe is cold-bent, and after cooling, the redundant part is cut off. Install the steel pipe in the flange and fix it on the reaction frame. When concrete is poured into the steel pipe, the flange and concrete are on the same level during the loading process, and the concrete is uniformly stressed and sealed, and naturally cured. After 28 days of concrete curing, CFRP is wound around the steel pipe, and the curing agent and epoxy resin adhesive matched with CFRP are used to make CFRP-reinforced concrete more in line with the construction requirements. The test platform is shown in Fig. 1 below.

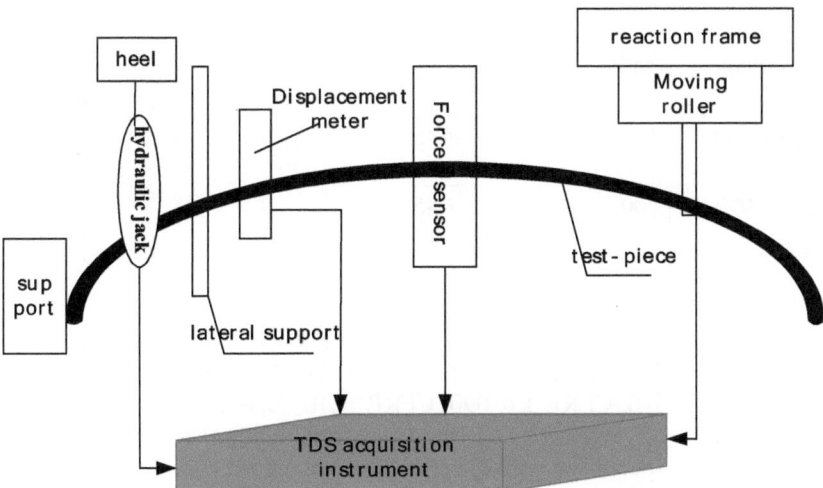

Fig. 1. Schematic diagram of test platform device.

As shown in Fig. 1, this test used supports, pads, hydraulic jacks, lateral supports, displacement meters, force sensors, reaction frames, moving rollers, TDS collectors and other devices. In this paper, the bearing is taken as the boundary condition, and the two ends of the CFRP-reinforced concrete specimen are fixed. Four sections, L/6, L/3, 2L/6 and 2J/3, are loaded on the loading platform, and the changes of compression bearing capacity in different section states are analyzed. Other parameters of the specimen are shown in Table 1 below.

Table 1. Specimen Parameter Table.

Test piece number	Test piece height /mm	concrete grade	thickness/mm	Stress situation
CFRP_0	500	C30	5	Overall compression
CFRP_15	500	C30	8	Overall compression
CFRP_30	500	C30	10	Overall compression
CFRP_45	500	C40	5	Overall compression
CFRP_015	500	C40	8	Overall compression
CFRP_030	500	C40	10	Overall compression
CFRP_045	500	C45	5	Overall compression
CFRP_090	500	C45	8	Overall compression
CFRP_L8_090	500	C45	10	Overall compression
CFRP_L10_090	500	C45	5	Overall compression
CFRP_L12_090	500	C45	8	Overall compression

As shown in Table 1, according to the amount of concrete, this paper made 11 specimens, including CFRP_0, CFRP_15, CFRP_30, CFRP_45, CFRP_015, CFRP_030, CFRP_045, CFRP_090, CFRP_L8_090, CFRP _L10_ 090 and CFRP_L12_090, with a height of 500 mm. Using C30, C40 and C45 strength concrete respectively, CFRP-reinforced concrete specimens with different thicknesses were made, and they met the test requirements in the form of overall compression.

3   Test Methods

In this experiment, CFRP_0, CFRP_15, CFRP_30, CFRP_45, CFRP_015, CFRP_030, CFRP_045, CFRP_090, CFRP_L8_090, CFRP_L10_090 and CFRP_L12_090 11 specimens were selected and CFRP with different layers on them. Among them, CFRP_0 means that the number of layers of reinforced concrete wrapped with CFRP is 0, CFRP_15, CFRP_30 and CFRP_45 are circumferential wound to strengthen CFRP layers with 15, 30 and 45 respectively; CFRP_015, CFRP_030, CFRP_045 and CFRP_090 are circumferentially wound with CFRP1, 3, 4 and 9 layers, vertical winding 5,0,5 and 0. CFRP_L8_090, CFRP_L10_090, CFRP_L12_090 shall wrap the CFRP 9 layer around L 8, L 10 and L 12. Under the condition of reciprocating load, the load-displacement changes of different specimens are different. This paper takes CFRP_090 as an example to analyze the load-displacement change of this specimen, as shown in Fig. 2 below.

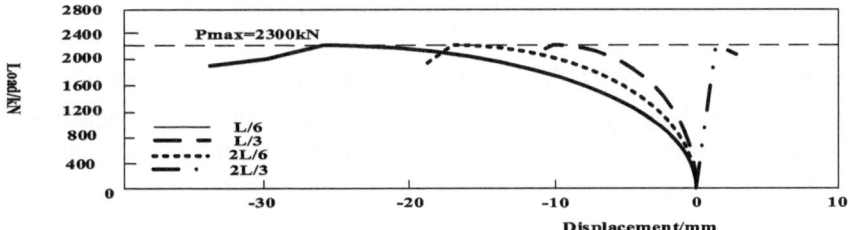

Fig. 2. Load-displacement curve of specimen CFRP_090.

As shown in Fig. 2, in the L/6 section, the displacement is −25 mm, and the load is the largest; In the L/3 section, the displacement is −10 mm, and the load is the largest; In the section of 2L/6, the displacement is −18 mm, and the load is the largest; In the 2L/3 section, the displacement is +3 mm, and the load is the largest. Under the condition of consistent L/6 section, CFRP_30, CFRP_015, CFRP_045 and CFRP_L8_090 are selected for load-displacement analysis. As shown in Fig. 3 below.

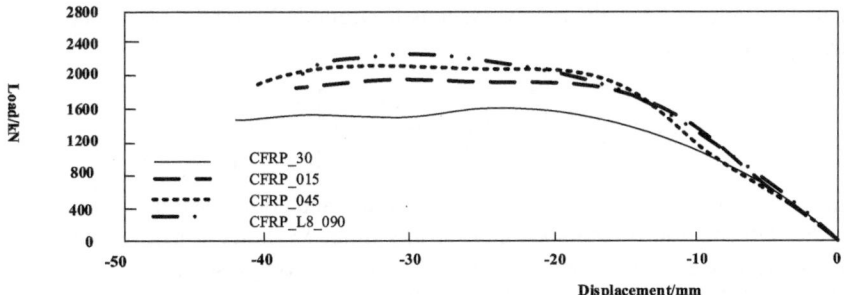

Fig. 3. Load-displacement curves of different specimens at L/6.

As shown in Fig. 3, when the displacement of CFRP_30 is −20 mm, the load is the largest. When the displacement of specimen CFRP_015 is −30 mm, the load is the largest; When the displacement of CFRP_045 is −32 mm, the load is the largest; When the displacement of CFRP_L8_090 is −28 mm, the load is the largest. Among them, the overall load of specimen CFRP_L8_090 is the largest, and that of specimen CFRP_30 is the smallest. After the load-displacement analysis is completed, the plastic deformation of CFRP-reinforced concrete specimens is analyzed. The formula is as follows:

$$\delta_c = E_c \varepsilon_c - \frac{(E_c - \varepsilon_c)^2}{4f_c} \varepsilon_c^2 \tag{1}$$

In the formula (1), δ_c is the plastic deformation of CFRP-reinforced concrete specimen; E_c is the slope of a straight line; ε_c is the strain value; f_c is the number of CFRP layers. Considering the tensile, compressive and failure modes of CFRP- reinforced concrete specimens, the tensile failure of fibers is expressed as:

$$F_f^t = \left(\frac{\delta_c}{X^T}\right)^2 + \alpha\left(\frac{\varepsilon_c}{S^L}\right)^2 \tag{2}$$

In the formula (2), F_f^t is the deformation of the specimen when the fiber fails in tension; X^T is the longitudinal tensile strength; α is the influence coefficient of shear strength on fiber stretching; S^L is the transverse shear strength. Compression failure is expressed as:

$$F_f^c = \left(\frac{\delta_c}{X^C}\right)^2 \tag{3}$$

In the formula (3), F_f^c is the deformation of the specimen when compression fails; X^C is the longitudinal compressive strength. Under the condition of basic tensile failure, the deformation of the specimen is expressed as:

$$F_m^t = \left(\frac{\delta_c}{Y^T}\right)^2 + \alpha\left(\frac{\varepsilon_c}{S^L}\right)^2 \tag{4}$$

In the formula (4), F_m^t is the deformation of the specimen under the condition of basic tensile failure; Y^T is the transverse tensile strength. When $\delta_c < 0$, the compression failure is expressed as:

$$F_m^c = \left(\frac{\delta_c}{S^T}\right)^2 + \left[\left(\frac{Y^C}{2S^T}\right)^2 - 1\right]\frac{\delta_c}{Y^T} + \left(\frac{\varepsilon_c}{S^L}\right)^2 \tag{5}$$

In the formula (5), F_m^c is the basic tensile failure and compression failure at the same time, the deformation of the specimen at this time; S^T is the vertical shear strength; Y^C is the transverse compressive strength. Under the compression limit condition, the variable of CFRP- reinforced concrete specimen is expressed as:

$$D = 1 - (1 - d_f)(1 - d_m)\delta_c\varepsilon_c \tag{6}$$

In the formula (6), D is an elastic damage variable; d_f Is the fiber damage variable; d_m is the matrix damage variable. The load-displacement change, specimen plastic deformation, fiber tensile failure deformation, compression failure deformation, elastic damage, fiber damage, matrix damage and other variables are comprehensively analyzed, so as to obtain the ultimate bearing capacity change of CFRP- reinforced concrete under compression.

4 Test Results

Under the condition of reciprocating load, the load applied in CFRP-reinforced concrete changes constantly, and the deformation changes with the increase of load, so the specimen is in the linear elastic stage. Under the condition of increasing load, the specimen enters the stage of elastic-plastic deformation, and the deformation increases, and the load-deformation curve changes nonlinearly. When the load is close to the ultimate load, the deformation increases rapidly, and after reaching the ultimate load, the load begins to decrease, but the deformation is still increasing, and the specimen has a large deformation, that is, instability and failure. In this paper, multi-layer CFRP is added to CFRP-reinforced concrete, and the decline of load-deformation curve is not obvious, which proves that CFRP can enhance the ductility of reinforced concrete. After the surface CFRP is damaged by compression, the load that concrete can bear reaches the limit, then decreases and the deformation increases. It is proved that CFRP can strengthen and restrain the deformation of reinforced concrete, which has a great influence on its ultimate bearing capacity under compression. In this paper, several specimens were randomly selected and numbered as CFRP_0, CFRP_15, CFRP_30, CFRP_45, CFRP_015, CFRP_030, CFRP_045, CFRP_090, CFRP_L8_090, CFRP_L10_090 and CFRP_L12_090. The compression deformation state of the specimen is analyzed, as shown in Table 2 below.

Table 2. Specimen Parameter Table.

Test piece number	Overall shape variation of the specimen/mm	Initial pressure configuration	Actual deformation form	Maximum displacement
CFRP_0	−3.42	antisymmetry	antisymmetry	2L/3
CFRP_15	−4.75	antisymmetry	antisymmetry	L/3
CFRP_30	−6.34	Positive symmetry	Positive symmetry	L/6
CFRP_45	−11.26	antisymmetry	antisymmetry	L/3
CFRP_015	−3.58	antisymmetry	antisymmetry	L/6
CFRP_030	−5.62	Positive symmetry	Positive symmetry	2L/6
CFRP_045	−5.98	antisymmetry	antisymmetry	L/6

(*continued*)

Table 2. (*continued*)

Test piece number	Overall shape variation of the specimen/mm	Initial pressure configuration	Actual deformation form	Maximum displacement
CFRP_090	−2.46	antisymmetry	antisymmetry	L/3
CFRP_L8_090	−11.86	Positive symmetry	Positive symmetry	L/6
CFRP_L10_090	−23.41	antisymmetry	antisymmetry	2L/3
CFRP_L12_090	−25.32	antisymmetry	antisymmetry	L/3

As shown in Table 2, the compressive state of CFRP-reinforced concrete specimens has two forms: positive symmetric pressure and anti-symmetric pressure, and the compressive state of CFRP-reinforced concrete specimens is closely related to its initial pressure. The initial pressures of CFRP_0, CFRP_15, CFRP_45, CFRP_015, CFRP_045, CFRP_090, CFRP_L10_090 and CFRP_L12_090 are all antisymmetric, so the final compression forms of the above specimens are also antisymmetric. The initial pressures of CFRP_30, CFRP_030 and CFRP_L8_090 are close to positive symmetry, and their pressure forms are also positive symmetry. It can be seen that winding CFRP and changing the strength of steel tube can not change the compression state of CFRP-reinforced concrete specimens. The smaller the overall deformation of specimens, the maximum load that CFRP-reinforced concrete specimens can bear, and the ultimate bearing capacity also increases. Under the condition of constant compression state, the load and its proportion of CFRP-reinforced concrete are analyzed, as shown in the following Table 3.

Table 3. CFRP-reinforced concrete load and its proportion table.

Test piece number	concrete		Steel bone		CFRP-pipe		Ultimate bearing capacity/kN
	bearing capacity/kN	proportion/%	bearing capacity/kN	proportion/%	bearing capacity/kN	proportion/%	
CFRP_0	1815.55	82.95	342.30	15.64	30.96	1.41	2188.81
CFRP_15	1689.98	85.58	337.26	16.48	19.24	0.94	2046.47
CFRP_30	1423.71	80.94	325.36	18.50	9.88	0.56	1758.94
CFRP_45	1219.42	76.84	325.28	20.50	42.34	2.67	1587.04
CFRP_015	1743.86	82.74	338.94	16.08	24.74	1.17	2107.53
CFRP_030	1626.26	82.11	336.50	16.99	17.83	0.90	1980.59
CFRP_045	1532.24	79.92	337.04	17.58	48.05	2.51	1917.32
CFRP_090	1624.38	70.99	398.85	17.43	264.86	11.58	2288.09
CFRP_L8_090	1156.60	72.50	283.49	17.77	155.30	9.73	1595.37
CFRP_L10_090	909.96	73.00	239.33	19.20	97.26	7.80	1246.55
CFRP_L12_090	803.84	73.20	247.07	22.50	47.29	4.31	1098.26

As shown in Table 3, when the CFRP-reinforced concrete specimen is in the position of maximum load, the load-bearing positions include concrete, steel skeleton and CFRP pipe. It can be seen from the table that the bearing capacity of concrete is relatively large, ranging from 800 kN to 1820 kN, and the proportion of its load in the whole specimen varies from 73% to 86%. The bearing capacity of steel skeleton is inferior to that of concrete, and the bearing capacity varies from 235 kn to 400 kn, and the proportion of steel skeleton in the whole specimen varies from 15% to 23%. The bearing capacity of CFRP tube is inferior to that of steel skeleton, and it has great axial stiffness under compression. Its bearing capacity varies from 9 kN to 265 kN, and its load ratio varies from 0.9% to 12% in the whole specimen. As far as the ultimate bearing capacity of CFRP-reinforced concrete specimens is concerned, it is concrete bearing capacity + steel skeleton bearing capacity + CERP pipe bearing capacity. Among them, the ultimate bearing capacity of specimen CFRP_L12_090 is the smallest, and that of specimen CFRP_090 is the largest. When CFRP-reinforced concrete is compressed, the ultimate bearing capacity decreases with the increase of slenderness ratio. At the same time, the ultimate bearing capacity increases with the increase of the number of layers of outsourcing CFRP; The bearing capacity of two-layer CFRP strengthened by circumferential winding and two-layer CFRP strengthened by longitudinal tiling is higher than that of four-layer CFRP strengthened by longitudinal tiling. It can be seen that the ultimate bearing capacity of CFRP-concrete filled steel tube increases with the increase of circumferential fiber filaments. The ultimate compressive bearing capacity of CFRP-reinforced concrete can be changed by adjusting the slenderness ratio, the number of CFRP layers and the circumferential fiber yarn, which plays an important role in improving its strength.

5   Conclusion

Under the condition of reciprocating load, CFRP-reinforced concrete transforms in elastic stage and elastic-plastic deformation stage. Under compression, the ultimate bearing capacity of CFRP-concrete filled steel tube is related to the slenderness ratio, the number of CFRP cladding layers, circumferential fiber filaments and other indicators. With the increase of slenderness ratio, the ultimate bearing capacity decreases gradually. With the increase of the number of layers of outsourcing CFRP, the ultimate bearing capacity is continuously improved. At the same time, the compressive state of CFRP-reinforced concrete is closely related to its initial pressure. Winding CFRP and changing the strength of steel tube can not change the compression state of CFRP-reinforced concrete. The smaller the overall deformation, the maximum load that CFRP-reinforced concrete can bear, and the ultimate bearing capacity also increases. Therefore, adjusting the slenderness ratio of CFRP-reinforced concrete, the number of CFRP layers and the circumferential fiber yarn can change its ultimate compressive bearing capacity and play an important role in improving the strength of CFRP-reinforced concrete.

In the bearing capacity analysis of CFRP-reinforced concrete, the load-displacement and load-strain at the compression position are very important. In the fracture process of CFRP-reinforced concrete, the fracture process zone, as a key feature, is determined based on linear elastic fracture mechanics. Under the action of reciprocating load, the stress of reinforced concrete is concentrated, and CFRP material is wrapped around it, which can avoid structural cracking and even protect the integrity of CFRP-reinforced concrete. In this paper, the ultimate bearing capacity of CFRP-reinforced concrete under reciprocating load is studied. In the form of experiment, the load change of concrete is analyzed in all directions, which makes the ultimate compressive bearing capacity analysis more accurate and guarantees the construction stability of building components.

References

1. Khatib, J., et al.: Effect of municipal solid waste incineration bottom ash (MSWI-BA) on the structural performance of reinforced concrete (RC) beams. J. Eng. Design Technol. **21**(3), 862–882 (2023)
2. Tuğrul Erdem, R.: Dynamic responses of reinforced concrete slabs under sudden impact loading. Rev. constr. **20**(2), 346–358 (2021)
3. Loubet, G., et al.: Autonomous wireless sensors network for the implementation of a cyber-physical system monitoring reinforced concrete civil engineering structures. IFAC-PapersOnLine **55**(8), 19–24 (2022)
4. Revanna, N., Moy, C.K.S.: Numerical modelling of reinforced concrete flexural members strengthened using textile reinforced mortars. Front. Struct. Civ. Eng. 1–20 (2023)
5. El-Nimr, M.T., et al.: Structural behavior of small-scale reinforced concrete secant pile wall. World J. Eng. (2022)
6. Kang, Z., et al.: Analytical study of perforation damage in reinforced concrete slabs subjected to oblique impact by projectiles with different nose shapes. Mech. Eng. J. **8**(1), 20-00331 (2021)
7. Najafgholipour, M.A., Arabi, A.R.: Finite-element study on the behavior of exterior reinforced concrete beam-to-column connections with transverse reinforcement in the joint panel. Pract. Period. Struct. Des. Constr. **26**(1), 04020050 (2021)
8. Aulia, T.B., et al.: Reventing brittle hybrid high-strength reinforced concrete slab collapse due to punching shear using coal flyash substitution, tie wire fiber and polypropylene fiber. In: IOP Conference Series: Materials Science and Engineering, vol. 1087, no. 1. IOP Publishing (2021)
9. Zargarian, M., Rahai, A.: Theoretical and experimental studies of two-span reinforced concrete deep beams and comparisons with strut-and-tie method. Adv. Civ. Eng. **2021**, 1–16 (2021)
10. Wuaten, H.M., et al.: Performance of retrofitted square reinforced concrete column using wire mesh and SCC subjected to cyclic load. Civ. Eng. J. **7**(4), 720–729 (2021)
11. Lo, A., Wijaya, H., Yuwono, A.: Modelisasi analisis numerik gaya lateral pada elevated reinforced concrete pile cap pada tanah non kohesif. J. Muara Sains, Teknol. Kedokteran dan Ilmu Kesehatan **5**(1), 111–120 (2021)
12. Al-Maliki, H.N.G., et al.: Structural efficiency of hollow reinforced concrete beams subjected to partial uniformly distributed loading. Buildings **11**(9), 391 (2021)
13. Tayebi, M., Nematzadeh, M.: Post-fire flexural performance and microstructure of steel fiber-reinforced concrete with recycled nylon granules and zeolite substitution. Structures **33** (2021)

Effect of Cyclic Loading on Mechanical Properties of Lap Joints of Prefabricated Steel Fiber Pavement Slabs

Zijian Wang[1], Baosheng Rong[1], Xin Zhang[1,2(✉)], Hanxiu Fan[2], and Liming Wu[3]

[1] School of Cilvil Engineering and Architecture, Chongqing University of Science and Technology, Chongqing 401331, China
2021206122@cqust.edu.cn

[2] The 5th Engineering Co., Ltd. of China Railway 11th Bureau Group, Chongqing 400037, China

[3] Chongqing Technology and Business Institute, Chongqing 400052, China

Abstract. In This paper establishes a finite element model of an assembled steel fiber reinforced concrete pavement slab with lap joints under cyclic loading to analyze the mechanical performance of the slab after 200 and 500 cycles of a vehicle speed of 50 km/h and an axle load of 200 kN. Through finite element software analysis, the curve of the deflection value of the fabricated steel fiber reinforced concrete pavement slab over time is obtained, and a comparative analysis is conducted on the deflection value of the pavement slab under different cycle numbers. The study found that the bending deflection curve of the plate neutralized joints after 200 and 500 vehicle load cycles showed similar patterns, with a maximum deflection value of approximately 0.35 mm; The variation in the plate and joint areas becomes more and more obvious with the increase in the number of cycles, indicating that vibration has an impact on the pavement plate. These conclusions have certain reference value for the design and maintenance of road pavement panels.

Keywords: Assembled · Steel fiber pavement · Cyclic load · Lap seam · Deflection

1 Introduction

The development of dual-functional pavements that can simultaneously meet both environmental and emergency requirements has become a hot topic in current social development. Among them, environmentally friendly dual carbon pavements aim to reduce the carbon emissions of pavement materials to minimize environmental impact, while emergency pavements need to quickly repair roaddamage within a short time to ensure traffic flow and safety [1]. In fact, it is necessary to vigorously develop areas such as green buildings, low-carbon transportation, and smart cities, among which an important technology is prefabricated buildings [2]. However, in practical use, the overlapping joints of prefabricated steel fiber pavement panels are prone to cracking and damage, which affects their mechanical properties and service life [3, 4].

© The Author(s) 2024
P. Xiang and L. Zuo (Eds.): PBSFTT 2023, LNCE 382, pp. 298–304, 2024.
https://doi.org/10.1007/978-981-97-5108-2_32

Cyclic load is a common form of load during the use of road surfaces, and under its action, the lap joints of assembled steel fiber pavement panels are prone to fatigue damage and crack propagation [5, 6]. Therefore, studying the effect of cyclic load on the mechanical properties of lap joints in prefabricated steel fiber pavement panels is of great significance for improving their durability and service life [7]. The aim of this study is to explore the influence of cyclic load on the mechanical properties of lap joints in prefabricated steel fiber pavement panels through experimental and numerical simulation methods, providing scientific basis and technical support for their design and application [8, 9]. The specific research content includes the deflection value of assembled steel fiber pavement panels under different cycles of cyclic load [10]; Through the development of this study, scientific basis and technical support can be provided for the design and application of prefabricated steel fiber pavement panels, improving their durability and service life, and making contributions to the development of urban transportation.

2   Establishment of Finite Element Model

When using ABAQUS for engineering structural analysis, the pavement structure is usually simplified as shown in Fig. 1, where the X, Y, and Z directions represent the driving direction, pavement width direction, and pavement depth direction, respectively. The foundation size is 2 m longer and 2 m wider than the slab.

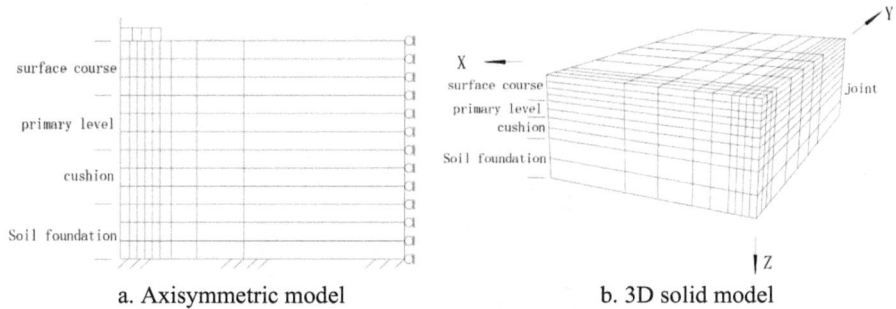

a. Axisymmetric model   b. 3D solid model

Fig. 1. Finite element structural analysis model for concrete pavement

2.1   Model Parameter Selection

The material indicators of the structural layer of this model are based on the actual road section of the Guangyang Avenue Ecological Restoration and Quality Improvement Project in Guangyang Bay Ecological City, Nan'an District, Chongqing. The material parameters are thickness, elastic modulus, and Poisson's ratio, respectively. C35 steel fiber reinforced concrete is selected, and the model material parameters are shown in Table 1.

Table 1. Model parameters table

Structural layer	thickness (cm)	Modulus (MPa)	Poisson ratio
surfacing	20	34900	0.15
base	60	1200	0.25
subgrade	400	40	0.30
HRB400	–	200000	0.3

2.2 Boundary Constraint Conditions and Interlayer Contact Analysis

In this article, the roadbed and base layer are set to be bound, and the base layer and surface layer are set to be in contact without separation. Constraints in the Y and Z directions are applied to the base layer. Based on experience, the friction coefficient of adjacent slabs is between 0.3 and 0.6, and the value in this article is 0.5.

2.3 Unit Selection and Grid Division

For the calculation of joint load transfer coefficient using deflection value, this article selects C3D8R unit for simulation. The dividing size of the road panel is 0.1m, and the grid size of the moving load zone is 0.05 m. The grid is encrypted at the joints, as shown in Fig. 2.

Fig. 2. Mesh division

2.4 Position of Load Action

The vehicle speed under normal driving is set at 50 km/h, and the on-site vehicle speeds are set at 5 km/h, 10 km/h, 15 km/h, and 20 km/h. The calculation results are analyzed based on the deflection value of the road panel, and the calculation points are mainly selected at the joints and the middle of the board. The action and calculation points of the moving load are shown in Fig. 3.

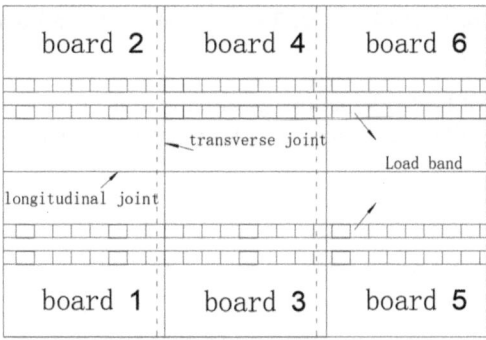

Fig. 3. Diagram of moving load

3 Analysis of Mechanical Properties of Lap Joints Under Cyclic Loading

The SFPCP finite element model with lap joints under cyclic loading is established based on the loading position in Fig. 4, and the mechanical properties of the pavement slab after cyclic loading of 200 and 500 cycles at a vehicle speed of 50 km/h and an axle load of 200 kN are analyzed.

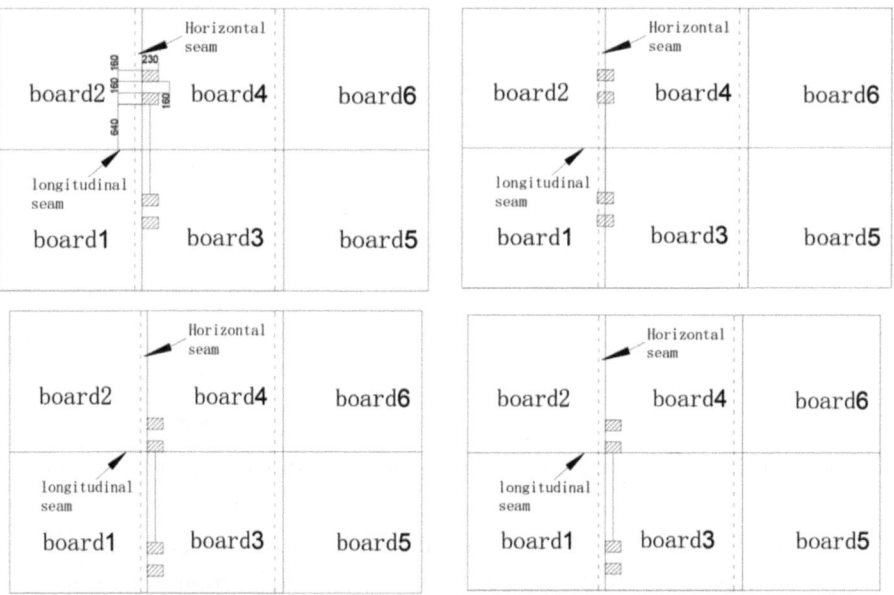

Fig. 4. Schematic diagram of the location of the load

3.1 Comparative Analysis of the Deflection Value of the Pavement Under Different Cycles

Through the analysis of finite element software, the deflection value and shear stress of the prefabricated steel fiber reinforced concrete pavement slab are obtained, as shown in Figs. 5 and 6.

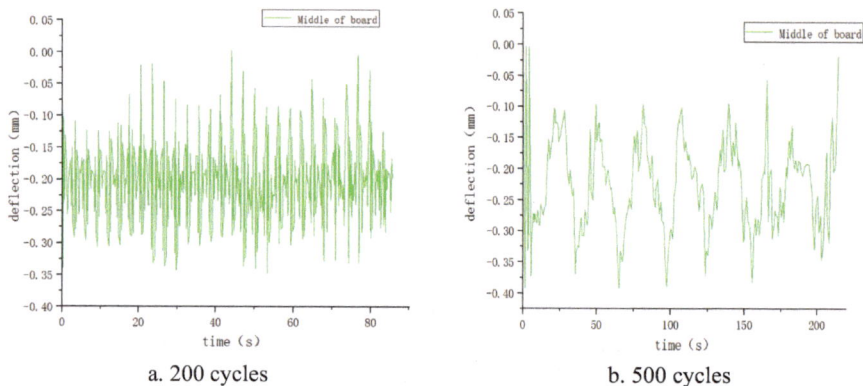

a. 200 cycles   b. 500 cycles

Fig. 5. Variation curve of bending and settlement values in the plate under different number of cycles

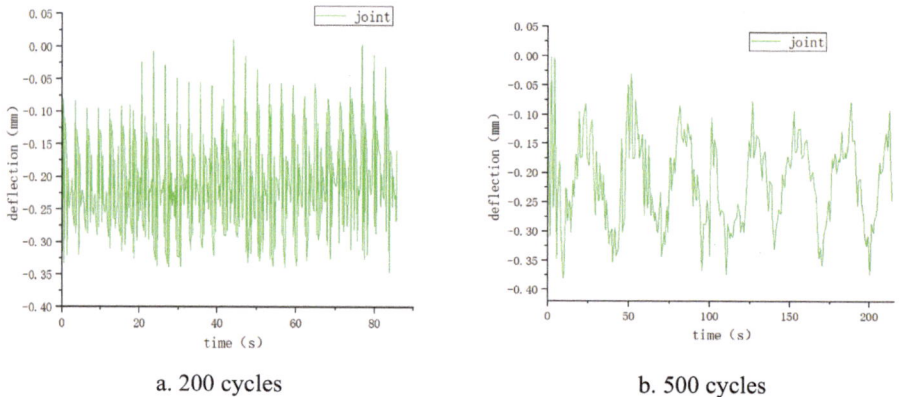

a. 200 cycles   b. 500 cycles

Fig. 6. Variation curve of joint bending and settlement values under different cycle times

From Figs. 5 and 6, it can be seen that after 200 and 500 vehicle load cycles, the deflection change curves of the plate center and joint are similar. The maximum deflection value of the plate is about 0.35 mm, and the maximum deflection of the joint is also 0.35 mm. About. After 200 cycles and 500 cycles, the difference is that after 500 cycles, the change law of the center of the plate and the joint is more and more obvious. After 200 cycles, the deflection change curve of the plate center and the joint is similar to the vibration curve. Vibration will have an effect on the pavement.

4 Conclusion

(1) The deflection curves of the plate and joint after 200 and 500 vehicle load cycles are similar, and the maximum deflection values are around 0.35 mm. This indicates that the number of cycles has a relatively small impact on the trend of deflection variation, while the maximum value of deflection is relatively stable.
(2) After cycling 200 and 500 times, the variation pattern at the joints in the board is different. After 500 cycles, the variation pattern at the midpoint of the plate becomes more and more obvious, while after 200 cycles, the deflection variation curve at the midpoint of the plate is similar to the vibration curve. This indicates that as the number of cycles increases, the deflection change of the pavement panel will be influenced by more factors, and when the number of cycles is small, the deflection change of the pavement panel may be more inclined to exhibit vibration effects.
(3) As mentioned earlier, vibration can have an impact on the road panel. Especially when the number of cycles is small, the deflection change curve at the plate and joint is similar to the vibration curve, which further confirms the impact of vibration on the pavement slab. Therefore, when designing, maintaining, and managing road panels, the impact of vibration factors should be considered.

Acknowledgements. 1. 2023 Chongqing Higher Education Teaching Reform Research Project (Grant Nos:233447) and 2023 Chongqing University of Science and Technology Undergraduate Education Teaching Reform Research Project (Grant Nos. 202342);

2. 2023 Chongqing Construction Science and Technology Plan Project (Funded projects:18) "Research and Demonstration of Key Technologies for Prefabricated Steel Fiber Reinforced Concrete Pavement Panels".

References

1. Tayabji, S., Ye, D., Buch, N.: Precast concrete pavements. Technology overview and technical considerations, pp. 113–125 (2013)
2. SHRP2. Precast Concrete Pavement (2013)
3. Tayabji, S., Tyson, S.: Precast concrete pavement innovations. Concr. Int. **39**(10) (2017)
4. Littleton, P., Mallela, J.: Florida demonstration project: precast concrete pavement system on US 92 (2014)
5. Tayabji, S.: Precast concrete pavement implementation by US highway agencies (2015)
6. Schexnayder, C., Ullman, G.: Pavement reconstruction scenarios using precast concrete pavement panels. Pract. Period. Struct. Des. Constr. **34**(18), 21–32 (2007)
7. Sadeghi, V., Hesami, S.: Finite element investigation of the joints in precast concrete pavement. Comput. Concr. **21**(5), 547–557 (2018)
8. Ioanides, A.M., Korovesis, G.T.: Analysis and design of doweled slab-on-grade pavement systems. J. Transp. Eng. **118**(6), 745–768 (1992)
9. Teller, L.W., Cashell, H.D.: Performance of doweled joints under repetitive loading. Highw. Res. Board (1959)
10. Zhang, X.: Analysis of mechanical properties and engineering application of SFPCP joints under loading. Chongqing University of Science and Technology (2023)

Research on 3D Laser Scanning for Enhancing Production Quality Control of Concrete Prefabricated Beams

Jianwen Chen[1]([✉]), Lei He[1], Zhangming Wang[2], Jianqiang Li[1], Bin Zhang[2], Qian Cheng[2], and Qianglong Qu[2]

[1] Hangzhou Traffic Engineering Group Co. Ltd., Hangzhou 310012, China
1611077661@qq.com
[2] CCCC Highway Bridges National Engineering Research Center CO., Ltd., Beijing 100088, China

Abstract. Based on the Hangzhou-Ningbo Expressway project, with the aim of addressing the drawbacks pertaining to challenging quality control, lengthy production duration, and low cost-effectiveness in the process of manufacturing concrete precast beams, this study employs three-dimensional laser scanning technology as a means of data acquisition in monitoring the quality of concrete precast beam spacing, size, and flatness within precast beam yards. By means of statistical analysis on the collected deviation data, various deviations in key indices of prefabricated components are identified, and the underlying causes for installation discrepancies are summarized. The test results convincingly demonstrate that the utilization of three-dimensional laser scanning technology can significantly enhance both efficiency and accuracy in quality inspections of finalized precast beams. Furthermore, this technique holds tremendous potential for application in quality control measures concerning concrete precast beams, thereby offering valuable technical support towards advancing the development of prefabricated concrete beams.

Keywords: 3D Laser Scanning · Concrete · Prefabricated Beam · Quality Control

1 Introduction

At present, under the support of a series of domestic policies, the practice of assembly construction has emerged as a pivotal driver for propelling the metamorphosis and elevation of the construction sector towards standardization, industrialization, and ecological sustainability [1]. In contrast to conventional cast-in-place concrete structures, assembly bridges possess a distinctive advantage in terms of their heightened industrialization. This notable attribute bears profound significance for the advancement of the bridge industry, as it stems from the prefabrication and fabrication of structural components within dedicated girder yards. Consequently, this approach engenders construction processes that are both remarkably efficient and environmentally sustainable [2]. However,

P. Xiang and L. Zuo (Eds.): PBSFTT 2023, LNCE 382, pp. 305–314, 2024.
https://doi.org/10.1007/978-981-97-5108-2_33

due to factors such as uneven professional ability of construction personnel, immature construction technology, and non-uniform acceptance standards, assembly prefabricated girders often have a series of component production deviation problems in the production process [3].

Three-dimensional laser scanning technology encompasses an automated, all-encompassing, and exquisitely precise method of capturing spatial data [4]. This technology employs a state-of-the-art three-dimensional laser scanner to delicately capture the essence of the target object, entirely devoid of any physical contact. By meticulously acquiring the spatial coordinates of each point gracing the surface of this remarkable entity, it deftly constructs an intricate three-dimensional point cloud model, harnessing the power of its immaculate three-dimensional spatial point coordinate data [5]. This innovative approach transcends conventional surveying and mapping methods, which rely solely on single-point acquisition. Instead, it enables the rapid and high-precision acquisition and presentation of three-dimensional data pertaining to target object surfaces. Ultimately, it culminates in the construction of a comprehensive three-dimensional point cloud model. By utilizing a cutting-edge three-dimensional laser scanner, wide-ranging areas can be scanned with exceptional accuracy and speed, regardless of weather conditions. This direct and high-precision technique effectively circumvents the limitations inherent in data analysis that rely on unilateral perspectives and localized approaches. Distinguished scholars both domestically and overseas have undertaken profound research endeavors grounded in the realm of 3D laser scanning technology, particularly concerning comprehensive deformation monitoring of landslides [6], full-section deformation measurement of tunnels [7], and feature extraction of buildings [8]. However, the practical utility of 3D laser scanning technology as a means for acquiring quality inspection data for assembled bridges is limited.

In accordance with this, we put forth a methodology for quality control in the production of precast concrete girders, utilizing the innovative technology of 3D laser scanning. Drawing upon the context of the precast girder yard within the Hangzhou-Ningbo Expressway Duplex, we present a comprehensive exposition on the application of 3D laser scanning technology in the process of precast girder fabrication. This entails precise evaluation of reinforcement distribution distances, dimensional accuracy of concrete components, and surface uniformity. Furthermore, it enables swift rectification of any quality concerns that may arise during precast component manufacturing, thereby affirming and substantiating the dependability and trustworthiness of 3D laser scanning technology.

2   3D Laser Scanning Technology

2.1   Point Cloud Data Segmentation

To extract intricate geometric details of voluminous prefabricated T-beams, the initial endeavor in unraveling the three-dimensional laser scan data entails point cloud segmentation. By scrutinizing the attribute label information, feature planes and crucial geometric points are ascertained and isolated from the point cloud data. To tackle the diverse spectrum of large-scale prefabricated structures characterized by varying degrees of geometric intricacy and magnitude, scholars both domestically and internationally have devised an array of algorithms for extracting geometric information.

For components with relatively simple geometry, Castillo et al. [9] proposed a point cloud segmentation algorithm based on surface normal estimation and local point connectivity, which can deal with unstructured point cloud data and stably detect corner points and edge data from point cloud data. Pu [10] used a plane growth method to delineate and identify architectural features. In the realm of substantial preassembled constituents, Tan et al. [11] ingeniously amalgamated the region germination algorithm with the Random Sampling Consensus (RANSAC) algorithm to ascertain and delineate the distinctive planar facets of the prefabricated components. Furthermore, they adroitly employed the RANSAC algorithm to deftly expunge any extraneous perturbations that may have occurred in the z-axis. However, this technique proves to be susceptible to noise interference when conducting line searches and lacks the ability to address minute structural elements. Previous investigations have primarily concentrated on sizeable components boasting uncomplicated geometries within pristine indoor settings, thereby leaving a significant research gap regarding the intricate modeling of expansive institutional elements featuring multi-unit amalgamations. Consequently, this study presents an innovative approach for noise reduction and seamless integration of three-dimensional point clouds pertaining to extensive prefabricated T-beam members.

2.2 Point Cloud Data Noise Reduction

The acquisition of point cloud data through 3D laser scanning is, unfortunately, prone to the intrusion of noise. This unwelcome addition stems from a confluence of factors: the precision of the scanner itself, the irregularity in operator technique, the susceptibility of measured surfaces to reflection, and the influence of environmental conditions. Furthermore, variables such as the angle between laser and surface, object movement during scanning, material composition, ambient light levels, temperature fluctuations, and humidity further encroach upon the accuracy of three-dimensional laser scanners. Consequently, discernible noise manifests within the gathered point cloud data. Regrettably, this presence of noise obfuscates key features within said data set; thus compromising not only algorithmic efficiency but also result fidelity.

In this paper, the K-Nearest Neighbors (KNN) algorithm is employed for the purpose of mitigating noise in point clouds. This nonparametric method, encompassing both classification and regression tasks, operates by evaluating the distances between distinct samples within the feature space. The computation of these distances adheres to Eq. (1).

$$d(x, y) = \sqrt{\sum_{k=1}^{n} -(x_k - y_k)^2} \tag{1}$$

In Eq. (1), the distance metric, d(x, y), is computed using the K-Nearest Neighbors (KNN) algorithm to determine the dissimilarity between various samples. The essence of the KNN algorithm lies in its ability to discern patterns within the feature space. If a sample under classification has most of its k-nearest neighbors (i.e., those with the smallest distances) belonging to a particular category, it can be inferred that this sample also belongs to said category and shares similar characteristics with others in that category. In the context of denoising with the KNN algorithm, if the average distance between a point and its k nearest neighbors exceeds a predetermined threshold, it is deemed as noise and subsequently discarded from further analysis.

2.3 Point Cloud Data Stitching

The technique of multi-site cloud splicing has a profound impact on the precision of the point cloud model. The comprehensive point cloud representation of the precast concrete element is encompassed by the diagonal scanning performed at two different stations, resulting in a considerable overlap in point cloud coverage. This allows for automatic splicing, and the usability of the point cloud model is assessed through an analysis of the accuracy of data fusion in a corresponding report. On the other hand, generating a unified point cloud for reinforced members poses greater complexity. To address various working conditions simultaneously, a target sphere is employed as a reference to seamlessly integrate distinct station-based point clouds. The goal is to maintain an average splicing error within 1mm for the target sphere, thus producing an intricate 3D model capturing the section of the prefabricated girder. Figure 1 visually presents representations of both the beam body model and reinforcement skeleton model following their meticulous scanning and seamless integration process.

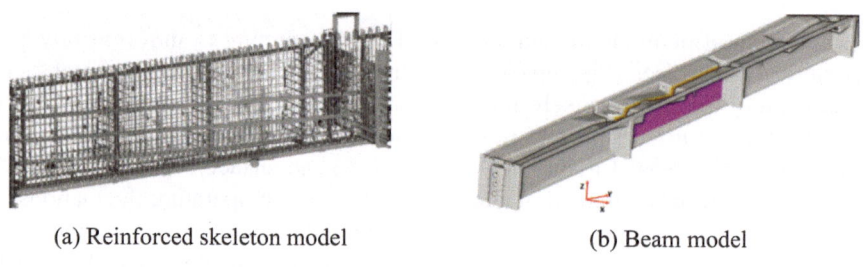

(a) Reinforced skeleton model   (b) Beam model

Fig. 1. Model of reinforcement and beam body under point cloud.

3   Project Overview

The background of this study is the Hangzhou-Ningbo Expressway Duplicate Line, with a total of 6,499 prefabricated T-beams, of which 5,291 are 30 m span T-beams. The parameters for quality control of the final precast T-beams primarily encompass the robustness of the concrete composition, accuracy in the deployment of steel reinforcement, precision in the installation of prestressing pipes, and adherence to other stipulated construction standards. Table 1 detailedly lists installation requirements for concrete. Table 2 detailedly lists installation requirements for rebar.

Furthermore, the examination of the beam's finished product necessitates an assessment of the concrete's surface evenness, ensuring a uniform hue, absence of conspicuous

Table 1. Installation requirements for concrete.

Inspection items			Allowable deviation
Concrete strength (MPa)			Within qualifying standards
Beam length (mm)			±10
Section size (mm)	Width	Dry seam	±10
		Wet joints	±20
	Height	Beams, slabs	±5
	Thickness of top and bottom plates, webs or beam ribs		±5
Flatness (mm)			≦5
Position of cross-tie beams and pre-embedded parts (mm)			≦5
Cross slope (%)			±0.15

Table 2. Installation requirements for rebar.

Inspection items		Allowable deviation (mm)
Spacing of reinforcing bars in the same row of beams and slabs		±10
Spacing of hoop reinforcement and transverse horizontal reinforcement		±10
Reinforcing steel skeleton dimensions	Length	±10
	Width, Height	±5
Bend up rebar location		±20
Protective layer thickness	Beam	±5

construction joints, and the avoidance of honeycomb or pockmarked blemishes. Moreover, the sealing anchor concrete must display impeccable density and a flawlessly level finish. While assessing concrete strength through rebound meter tests remains crucial, most quality control measures pertain to dimensional checks. To expedite the inspection process for concrete T-beams and enhance production line capacity, we have embraced three-dimensional laser scanning technology to monitor precast girder plate quality. Empirical evidence substantiates that the comprehensive 3D model scanning report can supplant traditional quality inspection records for precast girder plates.

4 Data Acquisition and Processing

4.1 3D Laser Scanner

The project utilizes the FARO S350 Plus 350 3D laser scanner, a sophisticated and fully automated device renowned for its outstanding precision. With a weight of 4.2 kg, this portable scanner boasts an omni-directional capability, rendering it suitable for a diverse range of measurement tasks within a distance range extending from 0.6 m to an impressive 350 m. The device's remarkable distance accuracy, which can achieve up to 1 mm precision, enables it to adapt effectively to various lighting conditions while achieving scanning speeds of up to an astonishing 970,000 points per second.

Employing advanced non-contact 3D laser scanning technology, this equipment is capable of rapidly capturing highly precise three-dimensional information across expansive areas. Particularly in the realm of quality control for precast concrete beams, the utilization of this scanner significantly enhances inspection efficiency and accuracy. By configuring the scanning resolution at half its maximum capacity and employing a triple factor increase in sweeping quality, the resulting point spacing reaches an impressive level of precision at 3.1 mm per every ten meters scanned.

(a) Reinforced skeleton   (b) Finished T-beam

Fig. 2. A three-dimensional laser scanner scan

4.2 Measurement Point Arrangement

Given the vast multitude of reinforcement bars and the intricate interlocking phenomenon they exhibit, a solitary measurement often fails to capture the full extent of their three-dimensional characteristics. To mitigate any potential obstruction of the camera's field of view, minimize measurement redundancy, and expedite field data processing speed, a total of twelve meticulously crafted T-beam/reinforcing steel skeleton scanners were strategically positioned, as illustrated in Fig. 3. These scanners were distributed amongst adjacent partitions (two at either end, three in the middle, and four on each side), with an additional two stationed at the extremities of the girder. Furthermore, to ensure utmost precision in point cloud data alignment, numerous reference spheres were thoughtfully installed within the shared visual range of these stations, as depicted in Fig. 2(b).

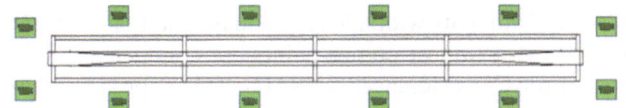

Fig. 3. A three-dimensional laser scanner scans the steel skeleton

5 Data Analysis

5.1 Rebar Size

The 3D point cloud model that was generated underwent slicing at appropriate locations in order to accommodate the rebar sections. This slicing process utilized the Randomized Sampling Consistent (RANSAC) algorithm, which accurately calculated the spacing of the rebars. A total of 120 hoop spacings were randomly selected from various positions within 30 T-beams and compared with the designated value of 150 mm. The final results of this comparison can be observed in Fig. 4. It is evident that the majority of the hoop spacings fall within an acceptable margin of error for rebar installation. Nonetheless, there are a few instances where certain hoop spacings exceed this limit. These discrepancies were identified through a meticulous analysis involving three-dimensional laser scanning and precise calculations regarding the dimensions of the reinforcement cage. Subsequently, prompt adjustments were made to those beams exhibiting significant deviations in their hoop spacing.

Furthermore, a probability density function curve was plotted using the data obtained from these observations. Upon fitting it with a normal distribution, it became apparent that its mean value slightly deviates from the prescribed design requirements. However, it is noteworthy that there is a substantial density of hoop reinforcement, thereby satisfying said requirements.

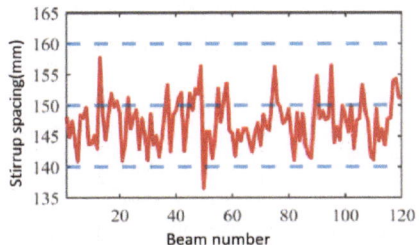

(a) Random sampling point distribution

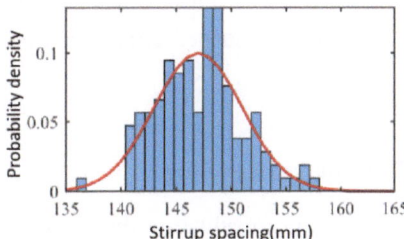

(b) Probability statistics for the normal distribution

Fig. 4. Random sampling of 120 stirrup spacing

5.2 Concrete Size

The spatial configuration of prefabricated girders is assessed employing the slice fitting technique. Initially, the three-dimensional point cloud representation of the completed precast T-beam is sliced along its width, length, and height. For each slice, an automated boundary fitting and extraction of corner points are conducted utilizing the Random Sampling Algorithm for Consistency (RANSAC). This approach allows for the determination of the spatial geometric properties of the precast girders within each slice. Figure 5 presents a plot depicting the girder length, height, and roof width of 30 scanned girders using 3D laser technology. Remarkably, the structural dimensions consistently fall within acceptable construction tolerances.

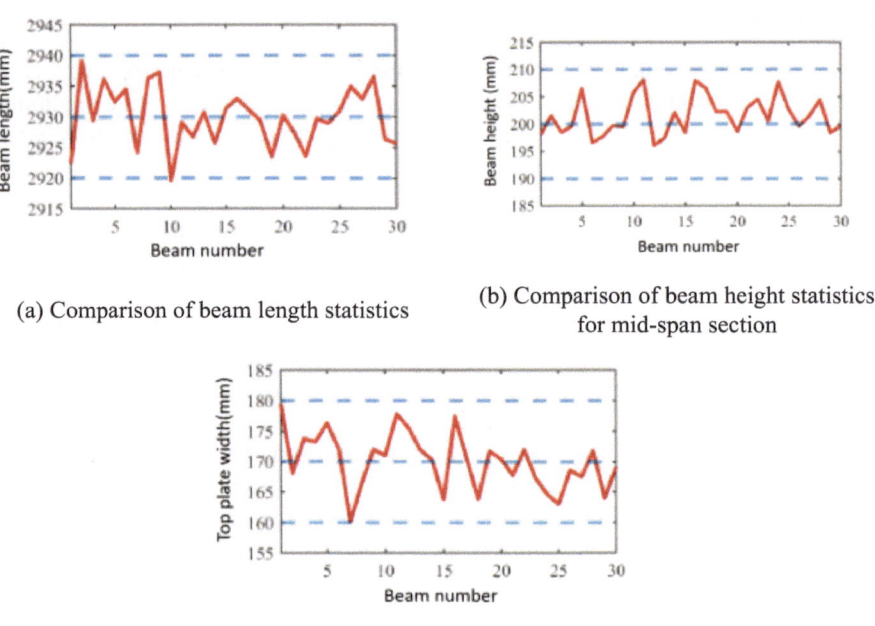

(a) Comparison of beam length statistics

(b) Comparison of beam height statistics for mid-span section

(c) Comparison of top plate width statistics for mid-span section

Fig. 5. 30 pieces of concrete beams and beams long structural dimensions

5.3 Concrete Leveling

In the realm of evaluating the evenness of concrete surfaces, this project employs the method of optimal plane fitting to assess the flatness of precast T-beams. The desired plane is derived from an analysis of various factors including inspection criteria, geometric attributes of the components, normal vectors assigned to each point, and other pertinent characteristics. Subsequently, utilizing the segmented point cloud data, we employ an "adaptive sliding plane method" to determine the optimal reference plane for the surface. Finally, by calculating the distance between each point and this reference plane, we are able to generate a visual representation of flatness distribution.

To exemplify this process, a single prefabricated T-beam has been randomly selected and its web surface flatness is depicted in Fig. 6 using the adaptive sliding plane technique. From this illustration, it is evident that the maximum variation in unevenness on the web surface measures less than 5 mm, thereby satisfying construction requirements. Additionally, in order to address any potential widespread inconsistency in maximum unevenness across all T-beams within a given batch, diligent inspections and prompt cleaning of concrete outer formwork are carried out to ensure uniformity and flatness during pouring procedures.

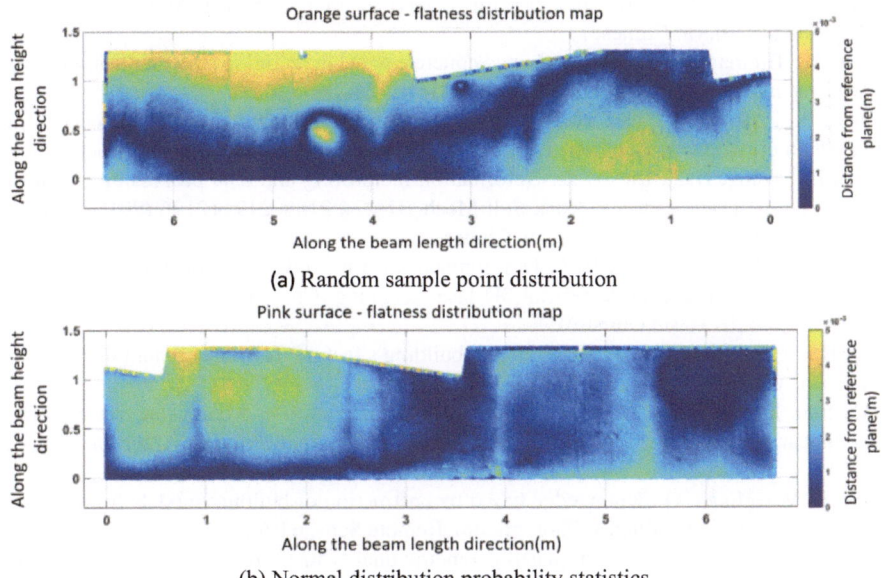

(a) Random sample point distribution

(b) Normal distribution probability statistics

Fig. 6. Surface flatness of concrete

6 Conclusion

This research paper is centered around the implementation of three-dimensional laser scanning technology for the purpose of assessing and managing the production quality of precast concrete girders. The project draws from the Hangzhou-Ningbo Expressway Duplicate Line Ningbo Section Phase II Project as its contextual backdrop and achieves intelligent detection in various aspects, including reinforcement bar dimensions, concrete appearance, and flatness. By employing 3D laser scanning as a means to inspect precast concrete beams, it is possible to entirely supplant conventional quality inspection methods. This innovative approach allows for prompt rectification of construction errors, markedly enhancing efficiency in the quality inspection process while reducing labor costs. As a result, it holds immense potential for widespread application within engineering projects.

References

1. Jiang, Z.L., Zang, G.G.: Research on whole process management of prefabricated intelligent construction of bridge engineering. J. Highw. Eng. **46**(4), 39–45 (2021)
2. Editorial Department of China Highway Journal. Review of academic research on bridge engineering in China 2021. China J. Highw. Transp. **34**(2), 1–97 (2021)
3. Li, W., Kong, D.Y., Tang, D.Q., et al.: Research on construction and installation deviation of prefabricated concrete structure based on 3D laser scanning and BIM. Archit. Struct. **50**(S2), 484–488 (2020). (in Chinese)
4. Seo, H.: Tilt mapping for zigzag-shaped concrete panel in retaining structure using terrestrial laser scanning. J. Civ. Struct. Heal. Monit. **11**(4), 851–865 (2021)
5. Jiao, B., Zhao, J.H., Zhao, H.D.: Virtual assembly technology of prefabricated pavement based on measured point cloud. Transp. Sci. Technol. **2**, 53–57 (2020). (in Chinese)
6. Xu, J.J., Wang, H.C., Luo, Y.Z.: Deformation monitoring and data processing of landslide based on 3D laser scanning. Rock Soil Mech. **31**(07), 2188–2191+2196 (2010).https://doi.org/10.16285/j.rsm.2010.07.011. (in Chinese)
7. Xie, X.Y., Lu, X.Z., Tian, H.Y.: Development of a modeling method for monitoring tunneldeformation based on terrestrial 3D laser scanning. Chin. J. Rock Mech. Eng. **32**(11), 2214–2224 (2013). (in Chinese)
8. Li, B.J., Fang, Z.X., Ren, J.: Extraction of building's feature from laser scanning data. Geom. Inf. Sci. Wuhan Univ. **01**, 65–70 (2003). (in Chinese)
9. Castillo, E., Liang, J., Zhao, H.: Point cloud segmentation and denoising via constrained nonlinear least squares normal estimates. Innov. Shape Anal.: Models Algorithms 283–299 (2013)
10. Pu, S., Vosselman, G.: Knowledge based reconstruction of building models from terrestrial laser scanning data. ISPRS J. Photogramm. Remote Sens. **64**(6), 575–584 (2009)
11. Tan, Y., Li, S., Wang, Q.: Automated geometric quality inspection of prefabricated housing units using BIM and LiDAR. Remote Sens. **12**(15), 2492 (2020)

Study of the Construction Technique of Reinforced Bar Product on Large and Variable Section Pylon

Minghua Ruan[✉], Hang Luo, and Di Wu

CCCC Second Harbor Engineering Co., Ltd., No.5 Branch, Wuhan 430014, Hubei, China
395921117@qq.com

Abstract. With the development of large span bridges, the height of the pylon is getting higher and higher, and the cross-sectional form of the pylon is getting more and more complicated, the construction of the pylon is a high-risk operation. The traditional pylon reinforcement construction uses manual asembling on the pylon, which is a labor-intensive high-altitude operation with high labor intensity, low construction efficiency, and not easy to guarantee quality. The pylon of Yanji Yangtze River Bridge is in the form of an octagonal outer cavity with a circular cross section, and the main reinforcement is arranged in four layers from outside to inside. The maximum diameter of the outer main reinforcement is 50 mm, the outer two layers and the third layer main reinforcement are connected by hoop reinforcement, and staggered arrangement of height directions. Based on the construction of the pylon of Yanji Yangtze River Bridge, the research on the construction technology of reinforcement bar was carried out for the characteristics of the pylon reinforcement, which solved the technical problems such as assembling of reinforcement parts, overall lifting and alignment docking on the pylon, which greatly reduced the labor intensity of operators, reduced the safety risk of overhead work and improved the quality of the pylon construction.

Keyword: pylon · outer cavity with a circular cross section · reinforced bar product

1 Overview

The pylon of Yanji Yangtze River Bridge is a portal frame structure with a height of 184 m. The pylon adopts an octagonal outer section with a circular inner cavity, the outer dimensions of the pylon section are chamfered to 2 m, the inner cavity is circular, the diameter of the upper beam is 3.6 m, the diameter of the middle beam is 5 m, and the diameter of the rest of the section is 6 m. The different diameters adopt a gradual transition in the form of a circular platform. The pylon cross-section is arranged with double layers of vertical reinforcement, of which the outer ring main reinforcement is divided into 50 mm and 40 mm according to different heights. The inner circle is arranged with one layer of vertical reinforcement, and the diameter of the main reinforcement of the inner circle is divided into 40 mm and 32 mm according to different heights. A

P. Xiang and L. Zuo (Eds.): PBSFTT 2023, LNCE 382, pp. 315–324, 2024.
https://doi.org/10.1007/978-981-97-5108-2_34

layer of 32 mm diameter vertical reinforcement is arranged between the inner and outer circles. The hoop reinforcement is divided into 20 mm and 16 mm diameter, and the vertical spacing of hoop reinforcement is 10 cm in the encrypted area of the lower pylon and 15 cm in the rest of the column. The outer and middle layers of the main reinforcing bars are connected by a chain of hoops, and the height direction is arranged according to the plum blossom. The pylon standard section reinforcement plan is shown in Fig. 1.

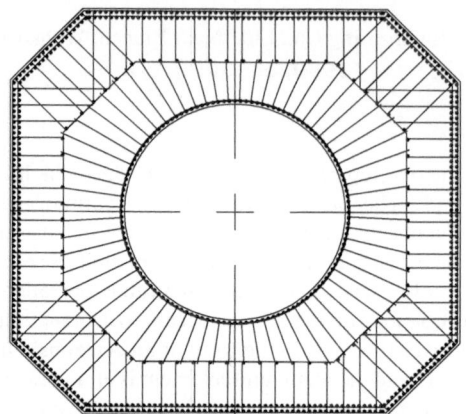

Fig. 1. Standard section steel plan of the Pylon

The most common way of steel reinforcement construction in the pylon is to process the unit parts in the steel processing yard and tie and form all manually on the pylon on site. This project pylon main reinforcement diameter is 50 mm, a 6 m long reinforcement weighs about 93 kg, workers work at height with high risk, high labor intensity, low construction efficiency, quality is not easy to ensure. In the intelligent development trend of modern bridge construction [1–3], we proposed the special action of "mechanization for human, automation for human reduction" and carried out research on the key technology of the pylon reinforcement.

2   Study on the Overall Construction Process of Reinforced bar Product on Pylon

After investigation and research, the pylon reinforced bar product construction process mainly has the following three kinds: 1, based on the steel mesh overall bending forming parts grouping production process; 2, based on the steel piece and block production forming parts grouping production process; 3, The production process of assembing reinforcement by workers under the pylon using moulding frame. As the pylon of this project is a round section with octagonal outer and inner cavity, the outer and middle layers of main reinforcing bars are connected with rings and hoops, and the height direction is arranged according to plum blossom, considering the difficulty of steel parts production, production efficiency and actual production situation on site, there are significant advantages of using the third steel parts construction process [4–6].

The overall construction process of pylon reinforced bar product adopts the intelligent steel processing equipment in the steel processing plant to process the semi-finished products, transport them to the moulding frame in site for assmbling, lift them by special spreader, and use the tapered sleeve to lock the reinforcing steel joint to connect with the pylon reinforcing steel.

3 Key Technology for Construction of Reinforced bar Product on Pylon

3.1 The Design for Moulding Frame

According to the height of the pylon section, the length and width of each section are changing with the slope, after calculation, it can be seen that the length of the 4 large faces of the 6 m standard section of the pylon is reduced by 82 mm per section. The spacing of the reinforcement in each chamfered part of the pylon section is unchanged, and the reinforcement in the middle area of the 4 large faces is closed and divided. The inner cavity is a 6m diameter circle with the same reinforcement spacing. There are 4 layers of main reinforcement in the standard section of the pylon, except for the outer 2 layers where the relative spacing of the main reinforcement position is unchanged, and the relative spacing of the main reinforcement position in the other layers changes with the change of section height [7].

In order to meet the requirements of the above-mentioned changes in the production of reinforcing bars product, a detailed research and design was carried out for the reinforcing bar section moulding frame. The moulding frame mainly includes: outer operating platform, inner ring operating platform, telescopic platform, bottom positioning system, top positioning system, climbing ladder, travel system and strong skeleton. The outer operating platform is moved along the guiding track by the walking motor, and the inner operating platform is fixed to the ground by expansion bolts. Since the spacing of the main reinforcement in the middle layer is 45 cm and the quantity is small, in order to improve the overall stiffness of the reinforcement parts, a strong skeleton is set up between the middle layer and the main reinforcement of the inner ring, with one horizontal connection every 1.9 m, three in total, and each horizontal connection is welded to the main reinforcement of the middle layer and the inner ring. The size of the strong skeleton is adjusted according to the closing of the reinforcement in every other section. The structural layout of the steel bar component binding jig frame is shown in Figs. 2 and 3.This moulding frame can meet the whole pylon reinforcement assembling requirements by adjusting the position of operation platform, bottom positioning system and top positioning system, which has the characteristics of simple operation, strong adaptability and low construction cost [8–10].

The rebar bottom positioning system mainly consists of positioning channel steel, outer ring rebar positioning device, middle ring rebar positioning device and inner ring rebar positioning device. The positioning device is composed of 70×5 mm square steel and round steel pipe, which is made according to the steel design drawing. According to each section of the pylon reinforcement bottom section layout, the bottom reinforcement positioning device is placed by total station, adjusted in place and fixed with the positioning channel steel.

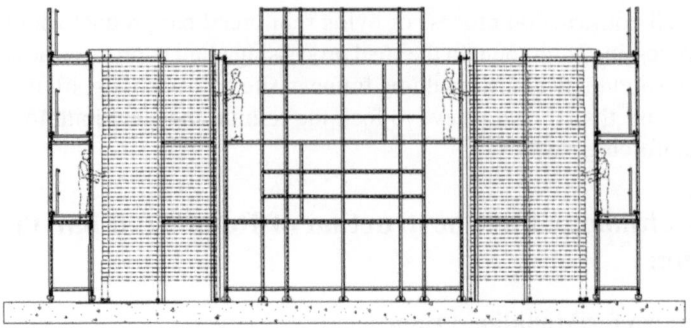

Fig. 2. Elevation layout of moulding frame for reinforced bar product

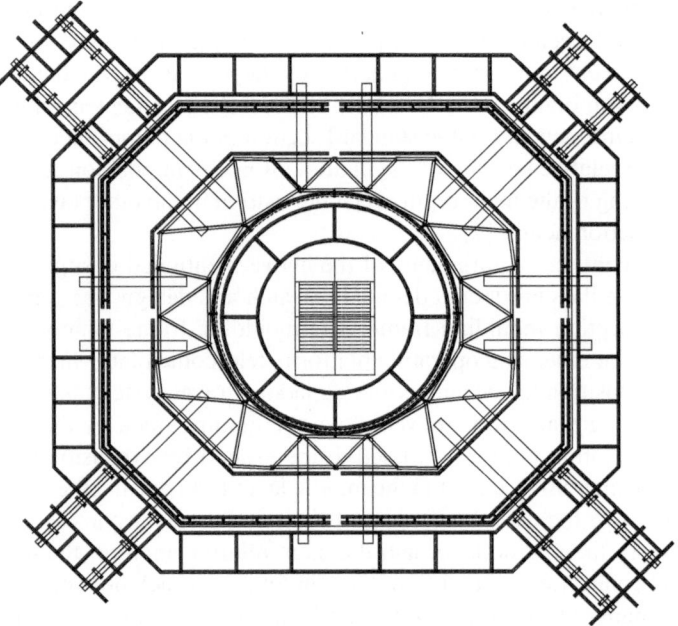

Fig. 3. Plan layout of the top of the moulding frame for assembling reinforced bar product.

The rebar top positioning system is mainly composed of comb plate support beam, outer ring rebar positioning comb plate, middle ring rebar positioning comb plate and inner ring rebar positioning comb plate. The comb plate is composed of 75 × 8 mm angle steel, which is made according to the steel design drawing. According to the top section layout of each section of pylon reinforcement, a total station is used to place the top reinforcement positioning device, which is adjusted in place and fixed with the comb plate support beam.

3.2 Design of Overall Lifting Spreader for Reinforced bar Product

According to the situation of each section of the pylon reinforcement, the spreader was studied and designed in detail to meet the requirements of the overall lifting of reinforcement parts. The spreader is composed of adjustable frame, main load bearing frame, movable distribution beam and movable lifting point, and each part is locked by fine-rolled rebar. The overall lifting structure of the steel reinforcement component in the steel reinforcement department is shown in Fig. 4. The adjustable frame moves along the main load-bearing frame to accommodate the change of center of gravity of the pylon section. The movable distribution beam moves along the main load bearing frame and the movable lifting point moves along the movable distribution beam to accommodate changes in profile and rebar take-up.

Fig. 4. Schematic diagram of the overall lifting spreader structure of the reinforced bar product.

3.3 Fabrication of Reinforced bar Product

According to the elevation position of the top reinforcement of the Nth section on the pylon, the bottom and top section layout of the N + 1th section of pylon reinforcement is drawn, and the initial positioning of moulding frame is done according to the requirement of 50 cm spacing between the outer operation platform and the main reinforcement. Through the N + 1 pylon reinforcement bottom and top cross-sectional layout, the bottom reinforcement positioning device and the top reinforcement positioning comb plate are placed by total station, and after precise positioning, the bottom positioning device is fixed with the bottom positioning channel, and the top comb plate is fixed on the positioning support of the tied tire frame. The reinforcement installation is assisted by pylon crane or truck crane, and the hoop bars are positioned and tied by hoop positioning tooling, and the main bars and hoop bars are welded and reinforced after the assembling is completed. According to the reinforcement assembling process, after the N + 1th section of reinforcement parts are tied, the top surface of the reinforcement parts are remeasured to check the deviation of the reinforcement position from the design position.

3.4 Integral Lifting of Reinforced bar Product

The spreader is connected with the reinforced bar product by 48 lifting points, including 24 lifting points in the outer circle, 16 lifting points in the middle circle and 8 lifting points in the inner circle, and the spacing between lifting points is controlled within 2 m. According to the change of the cross-sectional dimension of the pylon, the FEA software is used to establish the calculation model of the pylon section reinforcement, and the vertical deformation cloud diagram of the whole reinforcement part is shown in Fig. 5. Through the calculation and analysis, it can be seen that the maximum stress of the reinforced parts structure is 127 MPa, the maximum vertical displacement is 9.1 mm, and its structural strength and stiffness meet the requirements of the code. The multiple lifting points arrangement can effectively reduce the lifting deformation and facilitate the adjustment of the parts attitude.

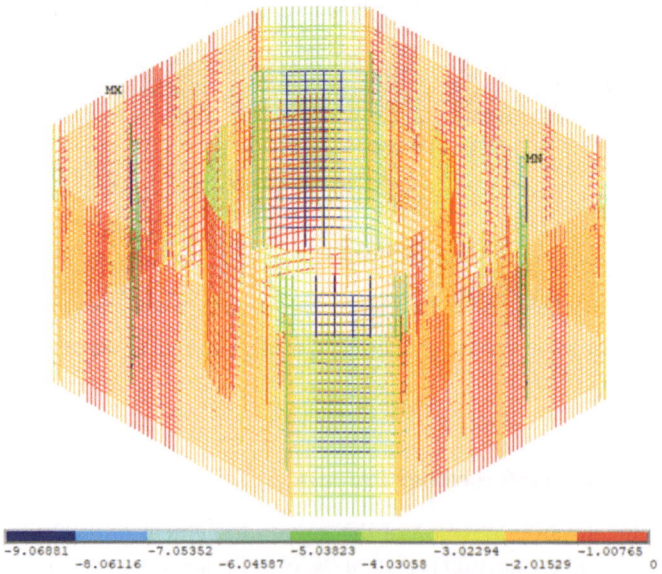

Fig. 5. Cloud diagram of vertical deformation of the whole lifting of reinforced bar product (mm).

The calculation model of lifting spreader is established by finite element analysis software, as shown in Fig. 6. The calculation results of the main components of the spreader are shown in Table 1. From the calculation results, it is known that the maximum stress of the spreader is 145 MPa, the maximum vertical displacement is 7 mm, and its structural strength and stiffness meet the specification requirements.

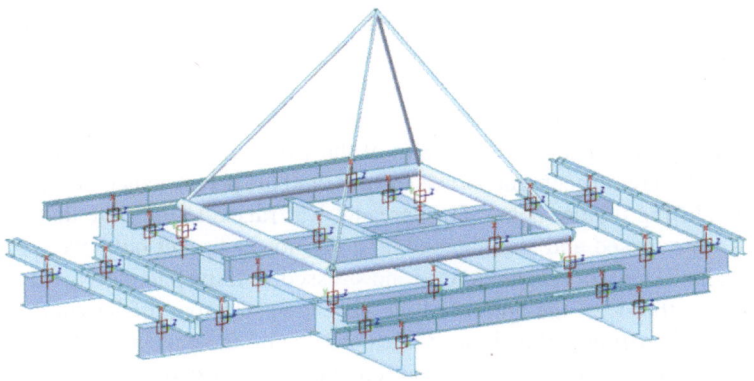

Fig. 6. Calculation model of overall lifting spreader for reinforced bar product.

Table 1. Calculation results of the main components of the spreader

No.	Components	Basic Portfolio		Standard Portfolio
		Shear stress (MPa)	Combined stress (MPa)	Vertical displacement (mm)
1	Load-bearing beam HM588 × 300	37	145	7
2	movable beam 2[28	18	101	4
3	Fixed beam HN400 × 200	9	35	2
4	Connection beam HW200 × 200	8	15	2
5	Flat link and diagonal brace Ø245 × 8	6	34	1

Before lifting, the adjustable frame of the spreader is adjusted to ensure that the center of gravity of the spreader is the same as the center of gravity of the reinforced bar product. The steel reinforcement parts are connected to the main reinforcement by removable steel fixture, which is fixed on the spreader beam. The connection by steel fixture can effectively improve the connection efficiency of lifting points, facilitate disassembly and reusability, and reduce the welding workload of traditional welding lugs as lifting points. The movable lifting point of spreader adopts 5t unloading buckle + wire rope + 5t basket bolt to connect with the lugs of spreader beam, and the horizontal position of the bottom opening of spreader is controlled by adjusting the length of the basket bolt, and the parts are connected after leveling. By controlling the plane relative position relationship between the spreader and the parts, the center of gravity of the parts and the center of gravity of the spreader basically coincide, and the attitude of the parts is adjusted at first, and the attitude of the parts is precisely adjusted by the flower basket bolts.

The reinforced bar product was lifted by XGT2850-120S pylon crane, which has a lifting capacity of 107 tons within 28 m of the working radius, and the actual lifting weight of the steel reinforcement parts is 99 tons, and the performance of the pylon crane can meet the lifting requirements. The spreader was raised to the state that most of the steel ropes were tensed and straightened, and the flower basket bolts were adjusted to ensure that all 48 steel ropes were tensed and stressed. Lift the part slowly off the ground to observe the attitude of the bottom of the part, to ensure that the height difference of the bottom of the part is within 5cm, if it exceeds 5 cm, the part will fall and adjust the flower basket bolt until it meets the requirements. Hang 4 cables at the bottom of the part and lift the part to the pylon.

3.5   Integral Lifting of Reinforced bar Product

In order to improve the efficiency of the alignment on the pylon of the reinforced bar product, 12 positioning flares are set on the pylon reinforcing steel, which are used to guide the precise alignment, including 2 on each large surface of the outermost ring reinforcing steel, 8 in total, and 4 on the inner ring reinforcing steel. One prism is set at each of the 8 corner locations on the top of the reinforced bar product for real-time measurement during the alignment of the reinforced bar product, which is used to guide the precise positioning of the reinforced bar product. By adjusting the flower basket bolts, the radial height difference of all the main bars is controlled within 1cm. Borrowing the cross beam on top of the integrated intelligent pylon building machine, a 5t hand chain hoist was used to match the rebar parts for precise alignment until the design requirements were met. After all measurement points are within the allowable range, taper sleeve and clamp are installed. The taper sleeve connection joint grade is I, the percentage of taper sleeve locking joint area in the same connection section can be 100%, which provides favorable conditions for the construction of parts [11–13].

The four chamfered surfaces are connected simultaneously using the tapered sleeve squeezer to ensure that there is no change in the butt joint attitude of the reinforced bar product. The connection of steel reinforcement components on the tower is shown in Fig. 7. After completion, the main reinforcing bars are then extruded and joined symmetrically along both sides. By pulling white lines to ensure that the sleeve is at the same level after the extrusion is completed. The main reinforcing bars are horizontally corrected by means of auxiliary facilities such as crowbars or hand-held hoists to meet the tapered sleeve connection conditions. Connect all the outermost main reinforcement before loosening the structure, and the remaining tapered sleeves are connected as soon as possible in the subsequent continuous operation of loosening the structure. For the extruded completed sleeve, small calipers are used to test the quality of extrusion.

3.6   Measurement of Reinforced bar Product

According to the location of the tied tire frame of the reinforced bar product, an independent coordinate system and a measurement-specific observation pier were established

Fig. 7. Schematic diagram of the connection on the pylon of the reinforced bar product.

to reduce the errors generated by the turning points as well as the external environment, so as to reduce the accumulation of errors during the release process. In order to summarize the changes in the posture of the middle part of the connection process, five measurements were taken during the unconnected, connected 25%, 50%, 75% and 100% processes of the reinforced bar product. The overall deviation was (0 to 15) mm before connection and (5 to 20) mm after unhooking, and the overall deviation was basically unchanged (considering the measurement error). It was initially determined that the accuracy at the time of starting positioning had a greater impact on the part, and that the part basically stopped deviating during the connection process as the connection stiffness increased.

4   Conclusion

The pylon of Yanji Yangtze River Bridge has an octagonal cross-section and a circular inner cavity, with 4 layers of reinforcement in the standard section and a maximum diameter of 50 mm. Through the pylon steel parts construction technology research, developed a very large variable cross-section pylon steel parts moulding frame and steel parts overall lifting spreader, mastered this kind of cross-section steel parts production method and the pylon buttress precision alignment method, realize the pylon steel parts factory standardized processing, let steel parts assembling does not occupy the pylon construction key line, effectively reduce the labor intensity and overhead construction operation risk, improve the pylon construction quality and quality. Improve the quality and efficiency of pylon construction. Practice shows that the pylon using steel parts construction is conducive to quality improvement and rapid construction, with good social and economic benefits.

References

1. Doi, F., Seto, K., Minami, M., et al.: Vibration control of model structure of large bridge tower: 4th report, intelligent control under construction. J. Antibiot. **63**(609), 1428–1434 (1997). https://doi.org/10.1299/kikaic.63.1428
2. Petroski, H.: Tower bridge. Am. Sci. **83**(2), 121–124 (1995)
3. Ohuchi, H., Scordelis, A.C.: Slender reinforced concrete bridge towers under cyclic lateral load. J. Struct. Eng. **117**(2), 325–342 (1991). https://doi.org/10.1061/(ASCE)0733-9445(1991)117:2(325)
4. Zhang, A., Li, Z.: Key construction techniques for pier reinforcement in long sections and with variable cross-section. World Bridges **47**(4), 37–41 (2019)
5. Wang, X.: Overall design of Zhongshan bridge in Shenzhen-Zhongshan link. Bridge Constr. **49**(1), 83–88 (2019)
6. Chen, S.: Key techniques for construction of substructure of non-navigable span bridge of Hong Kong-Zhuhai-Macao bridge over shallow water area. Bridge Constr. **46**(1), 6–11 (2016)
7. Peng, Y., Ding, S., Ren, M., Li, Z., Zhang, W.: System conception and overall design of the Yanji Yangtze River Bridge in Hubei. Bridge Constr. **52**(03), 1–7 (2022)
8. Yuan, H., Huo, D., Wu, Z.: Key technology of assembling steel reinforcement parts of the cable tower of Lingdingyang bridge for the Shenzhen-China passage. Highway **67**(04), 147–151 (2022)
9. Zhou, M.S., Zeng, W., Feng, S.D., Wang, Y.: Research on forming and component assembly technology of ultra-high cable tower reinforcement mesh. Constr. Technol. **50**(09), 41–44 (2021)
10. Cheng Mao-Lin, W., Zhong-Zheng, Y.-L., Bin, C.: Construction method of assembling super-high steel reinforcement parts based on mesh bending and forming. Highway **67**(01), 145–150 (2022)
11. Wei, L., Cao, M., Cui, B., Xiong, W.: Study of construction method for installing superstructure of Wangdong Changjiang river highway bridge. Bridge Constr. **49**(1), 6 (2019)
12. Wang, D., Han, B.: Key techniques for construction of pylon of channel bridge of Pingtan straits rail-cum-road bridge. Bridge Constr. **49**(3), 5 (2019)
13. Li, J.: Key construction techniques for pylons of main navigational channel bridge of Hutong Changjiang river bridge. Bridge Constr. **49**(6), 6 (2019)

Research on Video-Based Slope Deformation Monitoring Technology

Jinhu Yang[1,2(✉)], Yong Xiao[1,2], Jing Wang[3], and Fuai Jiao[3,4]

[1] CCTEG Chongqing Engineering (Group) Co., Ltd., Chongqing, China
25719918@qq.com
[2] School of Resources and Safety Engineering, Chongqing University, Chongqing, China
[3] Chongqing College of Architecture and Technology, Chongqing, China
[4] Universiti Malaysia Sabah, Kota Kinabalu, Sabah, Malaysia

Abstract. The deformation and instability of slopes pose significant risks to national assets and the safety of people's lives. Reasonable and reliable monitoring techniques are crucial for the prevention and management of slope instability. Conventional slope deformation monitoring techniques have drawbacks such as point-based monitoring, difficulty in maintenance, and high investment costs. The rapid development of video image processing technology has greatly driven the transformation of measurement techniques. Through research on equipment selection for image acquisition, supplementary lighting under low visibility conditions, subpixel edge detection techniques, video interference removal, development of monitoring software, and integration of field power supply systems, a high-precision slope displacement monitoring system based on video images has been developed. The system has been tested and verified for monitoring performance in both nighttime dark environments and daytime conditions. Field experiments were conducted at the Pan Gui Road Station slope of the extension project of Chongqing Rail Transit Line 4, and the deformation values and trends obtained by the system were consistent with the high-precision Leica total station monitoring data. The system enables non-contact, long-distance, and real-time online monitoring of slopes within a measurement range of 100 m, with a monitoring accuracy of about 1 mm, and has significant potential for widespread application.

Keywords: Video Image · Slope Monitoring · Deformation Monitoring · Automated Monitoring · Sub-pixel · Slope Stability

1 Introduction

Slope safety monitoring typically involves various methods such as manual on-site monitoring, GNSS (Global Navigation Satellite System), inclinometers, total station robots, static leveling instruments, 3D laser scanners, synthetic aperture radar, etc. These monitoring methods have their own disadvantages. For example, manual on-site monitoring is labor-intensive and infrequent, GNSS, inclinometers, total station robots, and static leveling instruments provide point measurements and are challenging to maintain, while 3D laser scanners and synthetic aperture radar require substantial financial investment [1–4].

© The Author(s) 2024
P. Xiang and L. Zuo (Eds.): PBSFTT 2023, LNCE 382, pp. 325–334, 2024.
https://doi.org/10.1007/978-981-97-5108-2_35

Since the 20th century, the rapid development of video image processing technology has greatly transformed measurement techniques and is now widely applied in areas such as national defense, aerospace, robot vision, biomedical engineering, industrial product inspection [5]. Image processing monitoring technology uses the information reflected by targets through light, combined with image processing and recognition techniques, to obtain parameters of the target. This process does not require physical contact with the slope and allows for long-distance measurements.

Researchers like Billie F. Spencer Jr. at the University of Illinois have applied video technology to infrastructure inspection and monitoring. Tung Khuc at the University of Central Florida has researched video-based non-point infrastructure displacement monitoring. Maria Q. Feng at Columbia University has studied video-based multi-point displacement measurement for structural health monitoring. In China, researchers have early applications of close-range measurement technology for displacement fields, existing tunnel lining deformations, and new tunnel contour under-excavation measurements [6]. Sun has studied the application of close-range photography technology in slope deformation monitoring in open-pit mining areas [7]. Zhou at Huazhong University of Science and Technology has conducted dynamic displacement monitoring using video images for high-rise buildings and bridges [8]. Gao et al. has used an image monitoring system for engineering monitoring of bridge structures [9]. Researchers have applied photogrammetry technology to terrain measurement, building measurement, foundation pit measurement, and landslide monitoring, achieving an accuracy of 1:10,000 [10]. It is evident that measurement technology based on video images is widely used in various industries.

Video-based slope safety monitoring calculates the displacement information of the object under test by analyzing images before and after changes in the object. Through research on equipment selection for image acquisition, supplementary lighting under low-visibility conditions, sub-pixel edge detection technology, video interference removal, monitoring software development, and integration of field power supply systems, a high-precision slope displacement monitoring system based on real-time video images has been developed. This system enables 24-h real-time online non-contact monitoring of slopes at a long distance.

2   Video Image Monitoring Technology Principle

2.1   Sub-pixel Edge Detection and Positioning Principle

The method of calculating the most suitable position for the target feature from the image is called image target sub-pixel positioning technology. For example, in an ideal situation, a square digital image with a size of 4x4 after digitization has a center coordinate of (1.5, 1.5), as shown in Fig. 1. If the integer pixel value is taken as the center coordinate of the digital image, it will result in a positioning error of 0.5 pixel value. If the average value of the pixel coordinates of the target image is calculated as the center coordinate of the target image, the positioning will be more accurate.

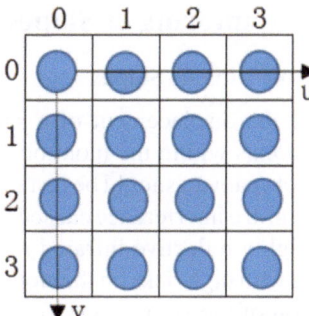

Fig. 1. 4×4 square target

2.2 Principle of Slope Image Displacement Monitoring

To achieve slope image displacement monitoring, the first step is to complete object image measurement, establish an imaging geometric model, and determine the model parameters. This is a crucial step in realizing image detection and positioning, which involves camera modeling and camera calibration. The purpose is to infer three-dimensional information from the image by obtaining the cameras intrinsic and extrinsic parameters, or to calculate the coordinates of the target in the image from the three-dimensional information.

Camera Imaging Model. To establish the geometric relationship between spatial points and their corresponding image points, we use (u, v) to represent the coordinates in the pixel-based image coordinate system, and (x, y) to represent the coordinates in the physical unit-based imaging plane coordinate system. If the coordinates of O1 in the Ouv coordinate system are (u0, v0), and the physical dimensions of each pixel in the x and y directions are dx and dy respectively, then the two coordinate systems have the following relationship.

$$\begin{bmatrix} \mu \\ \nu \\ 1 \end{bmatrix} = \begin{bmatrix} 1/dx & 0 & \mu_0 \\ 0 & 1/dy & v_0 \\ 0 & 0 & 1 \end{bmatrix} \begin{bmatrix} x \\ y \\ 1 \end{bmatrix} \tag{1}$$

Camera Calibration. Camera calibration essentially involves calculating the correspondence between image coordinates and reference coordinates. In the image coordinate system, the imaging plane coordinate system, the camera coordinate system, and the world coordinate system, the first two coordinate systems are two-dimensional, and the image coordinate system is known. The last two coordinate systems are three-dimensional. If the size of the object is known and the world coordinate system is selected with the object as the reference, then the world coordinate system is also known. In most cases, determining the relationship between the world coordinate system and the image coordinate system is referred to as camera calibration.

3 Image Displacement Monitoring of Slopes

3.1 Image Displacement Monitoring of Slopes

In order to accurately measure the displacement of a specific point on a slope, it is necessary to select measuring points within the monitored slope area and stable points outside the area. The measurement camera should be installed in a stable area. To achieve good monitoring results and minimize interference caused by external disturbances, fixed monitoring markers (such as highly reflective infrared reflective markers) are selected and installed at critical positions with high risks on the slope. Then, measurement stations are installed in the stable section of the slope, away from the monitored displacement area.

3.2 Measurement and Calculation of Displacement Images at Slope Monitoring Points

Assuming that the position of the monitoring camera is located in the stable area and remains stationary with the reference point. It is assumed that the position of the monitoring camera and the reference point are located in the stable area and remain stationary.

Based on the principle of light propagation in a straight line, if it is possible to image both the displacement target point and the fixed reference point in the same camera, the actual variation of the displacement point can be measured by observing the relative change ($\delta L = L1 - L2$) of the displacement target point with respect to the reference point in the image. The schematic diagram of the imaging effect is shown in Fig. 2.

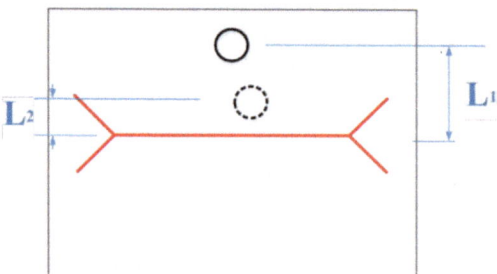

Fig. 2. Image representation of displacement changes at monitoring points

4 High-Precision Real-Time Image Measurement System for Slope Displacement Research and Development

4.1 Overall Design of High-Precision Real-Time Image Monitoring System for Slope Displacement

The overall design of the high-precision real-time image monitoring system for slope displacement is shown in Fig. 3. The entire system consists of an equipment terminal, a data management platform, and accessories.

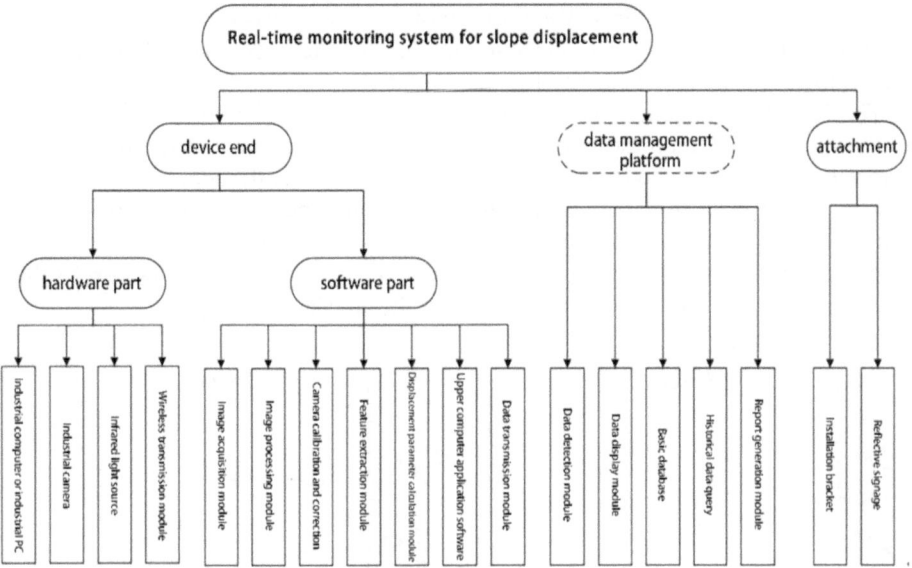

Fig. 3. System Overall Design Proposal

4.2 Research and Development of High-Precision Real-Time Image Monitoring Equipment for Slope Displacement

To meet the requirements of complex monitoring environmental conditions, the project fully considers all-weather real-time monitoring, outdoor power supply, and communication conditions in the system design. The system consists of an image pickup device, low-light compensation device for dark environments, dust and rain protection device, outdoor power supply system, edge computing terminal, and communication transmission device.

The displacement image perception device adopts an integrated design of image acquisition, supplementary lighting, dust-proof, rain-proof, and moisture-proof functions. The displacement image monitoring software is developed using QT software programming to build the system's user interface.

5 Experimental Testing of Slope Displacement Image Monitoring

To verify that the algorithm can achieve micro-displacement monitoring, the project conducted micro-displacement monitoring tests in an underground garage and an outdoor environment. These tests were conducted to evaluate the monitoring performance in both dark and bright conditions during the night and day, respectively.

5.1 Night Testing

Night testing was conducted in an underground parking garage at a distance of 70 m with a field of view of 10 m. Laser infrared illumination was used to provide supplementary lighting to the reference points. The purpose of the testing was to evaluate whether the hardware solution and software algorithms could accurately identify and monitor small displacements of the targets in low-light conditions.

The X-axis movement of the reference points is 1 mm, 2 mm, 3 mm, 4 mm, and 5mm respectively. A monitoring is conducted for every movement. The ellipse fitting is performed on the above image to calculate the coordinates of the center of the reflective markers at different positions during the movement. The experimental results for a monitoring distance of 70 m are shown in Fig. 4. In the figure, the horizontal axis represents the number of experiments, and the vertical axis represents the pixel values.

At a distance of 70 m, the experimental results are shown in Fig. 4.

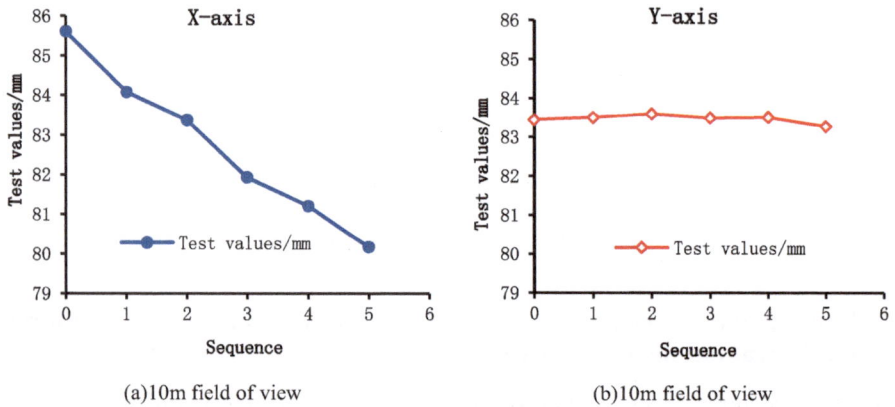

(a)10m field of view   (b)10m field of view

Fig. 4. Monitoring Results of a 70 m Distance from The Center of The Circle Displacement

As the retroreflective marker moves, the corresponding circle center coordinates also shift, and the captured images show evenly distributed spots from the retroreflective marker. By using sub-pixel algorithm detection and recognition, the tiny horizontal movements of the spot's center point can be accurately monitored, with a measured static accuracy of less than 1 mm.

5.2 Outdoor Glare Testing

Under strong sunlight conditions, the field of vision on the highway does not require supplementary lighting. Reflective markers were placed at a distance of 70m, and the camera focal length was set to achieve a field of view size of 10 m. The reflective markers were horizontally moved on the X-axis by 1 mm, 2 mm, 3 mm, 4 mm, and 5 mm, and monitoring was conducted 5 times for each movement. The experimental results of the 70 m monitoring distance are shown in Fig. 5.

From the above image, it can be seen that at a monitoring distance of 70m, significant horizontal displacement changes can be detected, while the vertical displacement remains relatively unchanged. The measured static accuracy is better than 1mm. Therefore, the hardware design and software compatibility of this system can effectively achieve real-time monitoring of slope displacement.

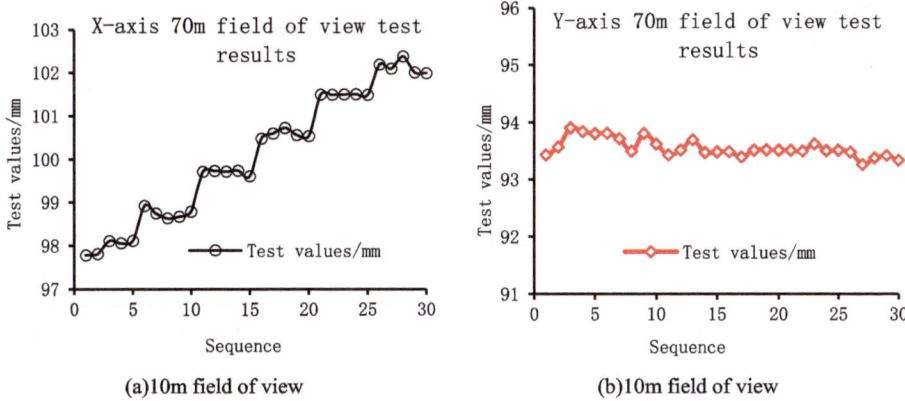

Fig. 5. Monitoring results of a 70m distance from the center of the circle displacement

5.3 Field Experiment

In April 2022, the research team conducted an on-site industrial experiment of a high-precision real-time image monitoring system for slope displacement at the Panguilu Station slope of the west extension section of Chongqing Rail Transit Line 4. The experiment involved one set of high-precision real-time image monitoring system for slope displacement, one set of total station, and several reflective markers.

After one month of video image monitoring and manual monitoring with a total station, the data comparison analysis is as follows.

(1) The lateral displacement range of monitoring point in the video monitoring is between 1.7 mm and 1.1 mm, while the lateral displacement range in the total station monitoring is between 1.4 mm and 1.0 mm. The monitoring comparison curve is shown in Fig. 6(a).
(2) The longitudinal displacement range of monitoring point in the video monitoring is between 9.4 mm and 8.6 mm, while the longitudinal displacement range in the total station monitoring is between 9.0 mm and 8.6 mm. The monitoring comparison curve is shown in Fig. 6(b).

The on-site experimental results show that the high-precision real-time image monitoring system for slope displacement can achieve 24-h real-time online monitoring of slope displacement in engineering slopes. The comparison with high-precision Leica total station monitoring data shows consistent deformation values and trends. The monitoring accuracy is 1 mm within a measurement distance of 100 m, demonstrating its excellent value for promotion.

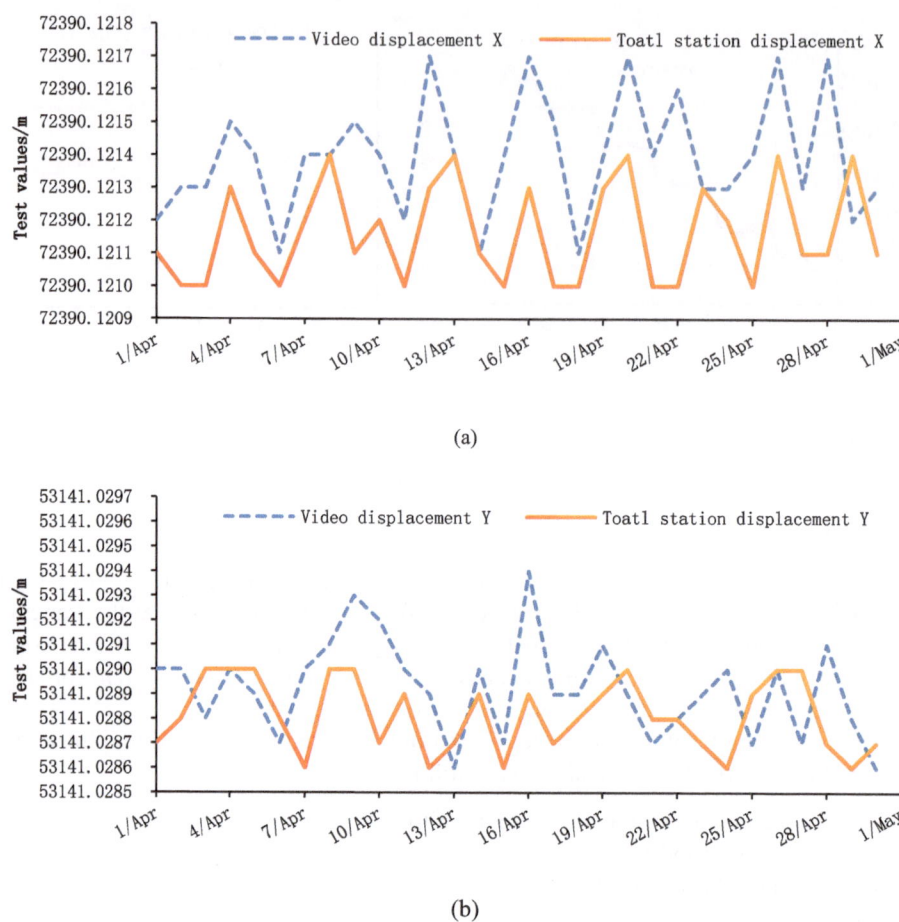

(a)

(b)

Fig. 6. Comparison curve of displacement at monitoring point

6 Summary

Through research on equipment selection for image acquisition, fill light under low vision conditions, subpixel edge detection technology, video interference removal, monitoring software development, integration of field power supply systems, etc., a set of high-precision real-time image monitoring technology equipment based on subpixel edge detection for slope displacement monitoring is formed.

The high-precision real-time image monitoring system for slope displacement can effectively achieve long-distance non-contact 24-h real-time synchronous online monitoring of slopes, with a monitoring accuracy of millimeters. Through the "one machine, multiple uses" function, it can effectively capture and store on-site images triggered by abnormal slope displacement. The system has the characteristics of simple installation layout, high cost-effectiveness, strong applicability, and functional suitability.

The video image displacement deformation monitoring technology is novel, but there are still many aspects worth further research in practical applications. For example, achieving monitoring distances of hundreds or even thousands of meters, expanding the monitoring field of view, improving the accuracy of non-target point monitoring, and identifying radial changes are all key research directions for future development.

Acknowledgement. Supported by independent key project of CCTEG Chongqing Engineering (Group) Co., Ltd., Research on Intelligent Monitoring and Early Warning Technology for Complex Mountainous Environmental Geological Disasters (2023ZDYF15).

References

1. Li, B., Li, X., Rui, H., et al.: Displacement prediction of tunnel entrance slope based on variational modal decomposition and grey wolf optimized extreme learning machine. J. Jilin Univ. (Eng. Technol. Edn.) **53**(06), 1853–1860 (2023)
2. Shi, B., Zhu, X.X., Zhang, C.C., et al.: Rock and soil disaster sensing and application. Sci. Sin. Tech. **53**, 1–13 (2023)
3. Li, H., Lei, Y., Gan, C.: Tunnel deformation monitoring based on vision assistant. J. Mech. Eng. **54**(01), 90–98 (2018)
4. Qin, W., Zhang, Z.H., Li, J.: Research on the design of slope safety early warning system based on video image recognition. China Plant Eng. **09**, 7–9 (2022)
5. Xu, X., Liu, H., Zhao, X.F.: Research on slope instability displacement monitoring technique based on laser spot video identification method. Dalian Univerity of Technology (China); University of Michigan (United States), 9435 (2015)
6. Zhang, Z.M., He, S., He, C.N., et al.: Research on highway slope disaster automated monitoring method based on video image processing. J. Phys.: Conf. Ser. **2095**(1), 012042 (2021)
7. Sun, G.L., Tao, Z.G., Gong, W.L.: Slope disaster monitering and early warning network system and its engineering application. J. China Univ. Min. Technol. **46**(02), 285–291 (2017)
8. Zhou, Y., Wang, X.R., Zhu, Y.P., et al.: Monitoring and numerical simulation of an interbedding high slope composed of soft and hard strong-weathered rock. Rock Soil Mech. **39**(06), 2249–2258 (2016)
9. Gao, J., Shang, Y.Q., Sun, H.Y., et al.: Application of CCD micro-deformation monitoring technology to slope remote monitoring. Rock Soil Mech. **32**(04), 1269–1272 (2011)
10. Meng, Y.D., Cai, Z.L., Xu, W.Y., et al.: A method for three-dimensional nephogram real-time dynamic visualization of safety monitoring data field in slope engineering. Chin. J. Rock Mech. Eng. **31**(S2), 3482–3490 (2012)

334   J. Yang et al.

A Method of Concrete Surface Crack Detection Using an Improved Convolutional Neural Network (CNN) Model

Zhexin He[1](✉) and Huan Zhang[2]

[1] Chongqing College of Architecture and Technology, Mingde Road, Chongqing, China
akulahzx@qq.com
[2] Chongqing Academy of Science and Technology, Yangliu Road, Chongqing, China

Abstract. This essay spotlights concrete crack detection in infrastructure maintenance, highlighting its importance for structural integrity, cost-effectiveness, and eco-consciousness. It delves into various detection methods and introduces an improved VGG-16-based deep learning model with batch normalization, P-ReLU activation, and Adam optimization for better training outcomes. Through experiments on the MendeleyData-CrackDetection dataset, the enhanced model outperforms the original. This study underscores the significance of hyperparameter optimization and algorithm choice in deep learning.

Keywords: concrete crack detection · an improved VGG-16 model · Adam optimziation

1 Introduction

Under the background of intelligent manufacturing in China, Intelligent construction technology is increasingly becoming an important development direction in the field of construction and engineering. Intelligent building technologies combine artificial intelligence, big data, the Internet of Things, and advanced sensing technologies to improve the efficiency, quality, and sustainability of building construction, operation, and maintenance.

Crack detection on concrete surfaces is crucial in maintaining infrastructure and ensuring structural safety used for detecting and repairing cracks in concrete structures early to ensure safety and reliability. Concrete structures such as bridges, buildings and roads are part of the infrastructure of modern society. Cracks that are not detected or not repaired in time may cause serious damage to the structure and even threaten life and property safety [1]. Early detection and treatment of cracks in concrete structures can reduce maintenance and repair costs. Chronic neglect of cracks may lead to more expensive repairs and recovery work [2]. Concrete cracks may cause moisture and harmful substances to penetrate and thus negatively affect the surrounding environment. Effective detection helps to reduce environmental pollution by [3]. Existing concrete surface

P. Xiang and L. Zuo (Eds.): PBSFTT 2023, LNCE 382, pp. 335–345, 2024.
https://doi.org/10.1007/978-981-97-5108-2_36

crack detection methods generally have the following four directions: Visual detection [4], Deep learning methods [5, 6], Image processing technology [7] and Sensor technology [8].

The application of deep learning and image processing technology provides a new hope for crack detection, which can improve the accuracy and efficiency of detection, reduce maintenance costs. Convolutional neural network (CNN) is one of the key factors for deep learning to make breakthrough in the field of image. The LeNet-5 [9], proposed by LeCun et al. in 1998, was an early CNN, which laid the foundation for the digital recognition task. Subsequently, AlexNet [10]'s proposal led to the deep learning research in the field of image recognition, which achieved a significant victory in the ImageNet competition in 2012. Deep learning has been widely used in image classification and object detection tasks. Some important work has included the Faster R-CNN [11], the YOLO (You Only Look Once) [12], and the Mask R-CNN [13]. These models have achieved significant performance improvements in the field of object detection through their end-to-end training methods. In these improvements, the VGG-16 neural network is one of the important milestones in the field of deep learning, which was proposed by Karen Simonyan and Andrew Zisserman [14] in 2014. The network structure includes a 16-layer deep convolutional neural network with a series of convolutional layers, pooling layers, and fully connected layers. VGG-16 demonstrates a broad potential in image processing. It is widely used in image classification tasks, especially achieving excellent performance on large-scale image datasets, such as ImageNet.

This paper aims to propose an improved deep learning model for the automatic detection and identification of concrete surface cracks. The model is based on the convolutional neural network and the attention mechanism, which is able to effectively extract the features of the concrete surface and classify them. The advantage of this model is that it does not require manual annotation of the crack location or complex image preprocessing, thus saving a lot of time and resources. In this paper, we verify the validity and robustness of this model by performing experiments on publicly available concrete surface crack datasets. The contribution of this paper is that it provides a new way to solve the crack problem in infrastructure maintenance and construction engineering, thereby improving safety and reducing maintenance costs.

2   Crack Recognition of the Concrete Surface Based on an Improved Convolutional Neural Network

In the construction research, the accuracy of automatic identification and timely diagnosis of concrete surface cracks has attracted the attention of researchers. To improve the accuracy of crack recognition, reduce the neural network parameters, and improve the training efficiency, the VGG-16 neural network structure was improved. The MendeleyData-CrackDetection [15] concrete crack dataset, including crack and no crack images, covering dry shrinkage, plastic shrinkage, temperature and external load cracks, were used. The dataset includes 40000 RGB images of 227×227 pixels in negative (without cracks) and positive (with cracks), with 20000 per category. The dataset had surface finish and illumination differences, without random rotation, flip, or tilt data enhancement. The data set was divided 7:1 into training set (35000) and validation set (5000).

In the concrete surface crack detection, the original VGG-16 network has problems, such as long training time and general accuracy. Therefore, the network is improved to introduce the batch normalization layer [16], which maps the activation values of each layer to a range of mean 0 and variance 1 to solve the gradient vanishing problem. The batch normalization method improves the network convergence rate and reduces the number of iterations while maintaining the same accuracy [17]. The basic mathematical expression for the method is shown as follows.

For the input m samples $x_i \sim x_m$, the mean value is shown below in (1):

$$\mu = \frac{1}{m} \sum_{i=1}^{m} x_i \tag{1}$$

The variance is shown below in (2):

$$\sigma^2 = \frac{1}{m} \sum_{i=1}^{m} x_i (x_i - \mu)^2 \tag{2}$$

The normalized result is shown below in (3):

$$\hat{x}_i = \frac{x_i - \mu}{\sqrt{\sigma^2 + \varepsilon}} \tag{3}$$

After a batch normalization operation, the mean of the data was adjusted to 0 and the variance to 1. To avoid cases where the inequality does not hold for $\varepsilon = 0$, we introduce the constant ε. However, such an operation may affect the feature distribution of the image, and thus, the need to recover the original feature distribution of the image through scale transformation and offset operation. The specific mathematical expressions are given in the formula (4)–(6).

$$y_i = y_i x_i + \beta_i \tag{4}$$

$$y_i = \sqrt{Var(x_i)} \tag{5}$$

$$\beta_i = E(x_i) \tag{6}$$

These parameters are obtained by learning and training, where E represents the mean and Var represents the variance function.

2.1 LeakyReLU Activation Function and the P-ReLU Activation Function

Most convolutional neural network models will use the ReLU function as the activation function after the convolutional layer. However, due to the characteristics of the ReLU activation function, when the network output is negative, the output value is always 0 after the activation function processing, which triggers the gradient disappearance in subsequent training, that is, the phenomenon of neuron "death". Considering this feature, some researchers will adopt the LeakyReLU activation function to optimize the convolutional neural network model in practice. For example, Chen Mianshu et al. [18] used the LeakyReLU activation function to select the nonlinear activation function in the image classification task based on the convolutional neural network. The mathematical expression of the P-ReLU activation function Formula is shown in formula (7).

$$PReLU(x_i) = \begin{cases} x_i, & x_i > 0 \\ ax_i, & x_i < 0 \end{cases} \tag{7}$$

This function is similar to the ReLU function, that is, when the input value is positive, the original value is directly output, and when the input value is negative or zero, the original value is multiplied by a constant C. The Leaky ReLU function is such a function, where C is a small positive number, determined before training begins. This has the advantage of maintaining the activation of neurons with negative inputs, avoiding the phenomenon of neuronal death, while also increasing the neuronal diversity. According to the above formula, the default value of C is 0.01 [18].

However, there are also some problems with using the Leaky ReLU activation function in the network training. Since the output coefficient of negative input in its mathematical expression is fixed and is not necessarily the most suitable for training effect, if you want to find the best value of coefficient C, you need to do many experiments. To address this issue, this study employed P-ReLU activation function in the network model to replace the original ReLU activation function in the convolutional layer. The P-ReLU activation function [19] is a learnable activation function that automatically adjusts the output coefficient at the negative input based on the training data.

Within the negative interval, the weight of the neurons is controlled by the parameter A. Unlike the LeakyReLU activation function fixed weight, this weight can be learned and dynamically adjusted during training. The variable i represents the different channels. This activation function allows all channels to share one weight, and can set different weights for each channel. The default initial value is 0.25. Because the P-ReLU function still has derivatives at x < 0, no gradient vanishing problem, and the function is non-saturated, it can effectively solve the problem of neurons dying in the negative interval, so as to improve the network performance and accelerate the model convergence to a certain extent. Below is the image of the ReLU activation function versus the P-ReLU activation function [20] (see Fig. 1).

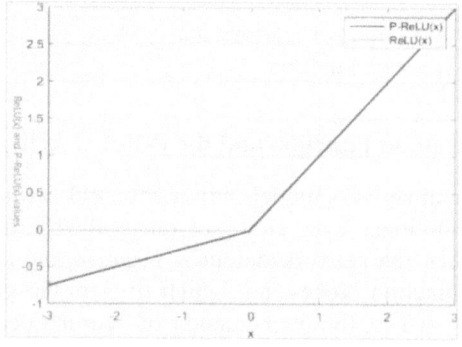

Fig. 1. Convolutional neural network using ReLU(x) versus P-ReLU(x)

This paper is based on the VGG 16 convolutional neural network and preserves the overall structure. A batch normalization (BN) layer was added after each set of convolution operations, replacing the original activation function as P-ReLU to solve the neuronal death problem. The last fully connected layer of the original model has 1000 labels, but there are only four categories in this paper, so we changed the last Softmax classifier to four labels and reduced the number of fully connected layers to two. The dimensionality of the first fully connected layer becomes 4096 and the second is 4, corresponding to four concrete crack detection images. These adjustments simplify the network structure and improve the identification efficiency. The structure of the improved convolutional neural network (CNN) model is shown below (see Fig. 2).

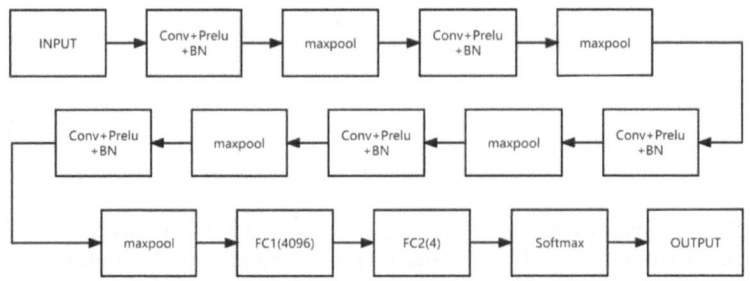

Fig. 2. The structure of the improved convolutional neural network (CNN) model

In the modified VGG-16 convolutional neural network model, the key parameters of each layer are shown in the table below (see Table 1).

Table 1. Parameters of the improved VGG-16 neural network model

Network layer name	Convolution kernel size/step size	Number of convolutional nuclear	Network depth
Conv-1-1+PReLU	3 * 3/1	64	1
Conv-1-2+PReLU	3 * 3/1	64	1
Maxpooling	2 * 2/2	/	/
Conv-2-1+PReLU	3 * 3/1	128	2
Conv-2-2+PReLU	3 * 3/1	128	2
Maxpooling	2 * 2/2	/	/
Conv-3-1+PReLU	3 * 3/1	256	3
Conv-3-2+PReLU	3 * 3/1	256	3
Conv-3-3+PReLU	3 * 3/1	256	3
Maxpooling	2 * 2/2	/	/

(continued)

Table 1. (*continued*)

Network layer name	Convolution kernel size/step size	Number of convolutional nuclear	Network depth
Conv-4-1+PReLU	3 * 3/1	512	4
Conv-4-2+PReLU	3 * 3/1	512	4
Conv-4-3+PReLU	3 * 3/1	512	4
Maxpooling	2 * 2/2	/	/
Conv-5-1+PReLU	3 * 3/1	512	4
Conv-5-2+PReLU	3 * 3/1	512	4
Conv-5-3+PReLU	3 * 3/1	512	4
FC1	/	4096	/
FC2	/	4	/
Softmax	/	4	/

After calculating the loss function, to update the parameters of the network nodes, we adopted the Adam optimizer [21] for training. Compared with the traditional stochastic gradient descent (SGD), the Adam optimizer comprehensively considers the first and second order gradient estimation, thus achieving the model convergence faster and improving the training efficiency. The mathematical expression of the Adam optimization algorithm is shown below in formula (8)–(12).

$$m_t = \mu * m_t - 1 + (1 - \mu) * g_t \tag{8}$$

$$n_t = v * n_t - 1 + (1 - v) * g_t^2 \tag{9}$$

$$\widehat{m}_t = \frac{m_t}{-\mu^t}, \quad \hat{n}_t = \frac{n_t}{1 - v^t} \tag{10}$$

$$\Delta\theta_t = \frac{\widehat{m}_t}{\sqrt{\hat{n}_t + \varepsilon}} + \eta \tag{11}$$

$$\theta_t + 1 = \theta_t + \Delta\theta_t \tag{12}$$

$\theta_t + 1$ represents the weight values of the neural network model in the $t + 1$ round iteration. Meanwhile, m_t and n_t represent the first and second moment estimates of the gradient, respectively. The t and t are correction terms for m_t and n_t, which are the first moment estimate and the second moment estimate of the corrected deviation, respectively. The μ, v and ε are hyper-parameters, usually set as $\mu = 0.9$, $v = 0.999$ and $\varepsilon = 10^{-8}$.

2.2 Setting of the Hyperparameters of the Network Model

In neural network training, the choice of learning rate is crucial. Too much learning rate may lead to oscillations and extended training time, while too little learning rate may lead to slow convergence and local optimal solution problems. Therefore, the rational selection and adjustment of the hyperparameters is the critical task. After multiple training and adjustment, the hyperparameters of the original VGG-16 and the modified model are as follows (see Table 2 and Table 3):

Table 2. The hyper-parameters of the original VGG-16model

optimization algorithm	SGD
learning rate	0.001
batch size	16
Number of learning rounds	50
Loss function	cross-entropy

Table 3. The hyper-parameters of the improved VGG-16model

Optimization algorithm	Adam
learning rate	0.0001
batch size	32
Number of learning rounds	120
Loss function	cross-entropy

Note that the experiments were performed in a GPU environment, using the Adam optimizer to improve the convergence rate. The batch size of the original model was 16, but oscillation problems occurred, so the batch size was increased to 32 to improve stability and efficiency.

3   Experimental Results and Analysis

The evaluation model mainly focuses on the following aspects: training accuracy, decline speed of loss function, model convergence rate, and oscillation existence. We will analyze the effect of the improvement strategy on the experimental results from the perspective of training accuracy. The accuracy reflects the model's ability to identify concrete surface cracks, the loss value reflects the error level of the model in identifying diseases and insect pests, and the oscillation degree reflects the stability and gradient explosion of the model. The following figure shows the training accuracy of the original VGG-16 model (see Fig. 3) and the improved convolutional neural network (see Fig. 4).

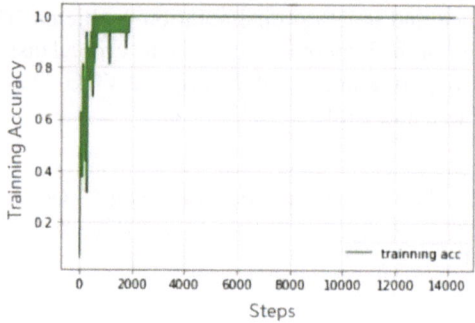

Fig. 3. Prediction accuracy of the VGG-16 original model with the number of training times

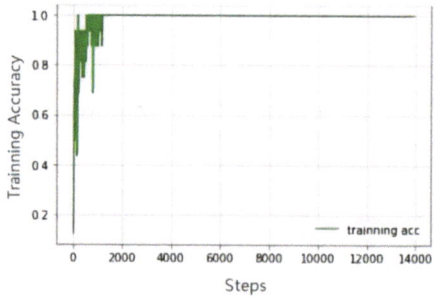

Fig. 4. Prediction accuracy of the improved VGG-16 model with training times

As can be seen from the figure above, after 120 rounds of training, the convergence rate of the improved model and the training accuracy are significantly better than the unimproved model. Because we introduce the batch normalization module and P-ReLU activation function, which effectively improves the convergence rate of the network, avoids large oscillations, reduces the risk of overfitting, and improves the generalization ability of the model. In contrast, the unimproved VGG-16 model failed to perform as well on the accuracy of the model training due to its numerous parameters. In terms of convergence rate, the original model starts to converge at round 15, while the improved model approaches convergence and is faster at round 10. Considering the training accuracy and convergence speed, it can be concluded that in addition to the change of the network training accuracy, it is also necessary to pay attention to the change of the loss function.

For the original, the VGG-16 model and the improved convolutional neural network. From the curve of the loss function, the loss function value of both models decreases rapidly in the beginning stage of training and eventually drops to nearly zero, but the improved neural network model decreases faster. Although the improved model had local oscillations at the beginning of training, which showed slightly worse stability, considering the size of the loss function and the reduction of the training loss function. The loss value changes of the original VGG16 network model (see Fig. 5) and the improved convolutional neural network model during training (see Fig. 6).

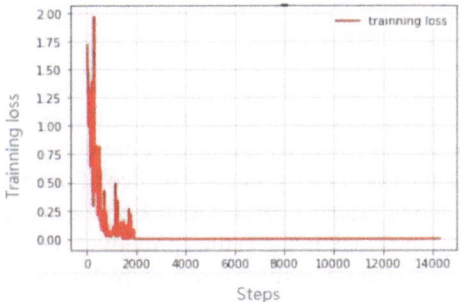

Fig. 5. The original model loss values were varied with the number of training times

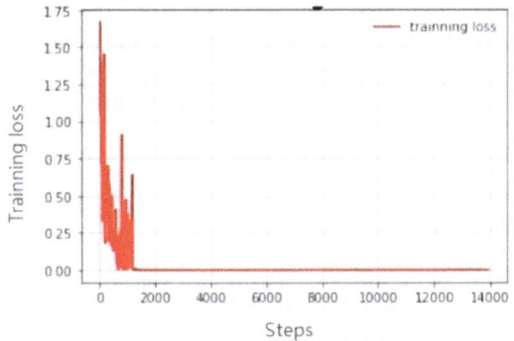

Fig. 6. Improvement model loss values as a function of training times

When training the network model using SGD stochastic gradient descent, the network convergence rate is significantly slower compared to the Adam optimization algorithm. In the experiment, the network starts to converge until the end of training. Finally, the network model does not fully converge and needs to increase the number of training rounds, leading to a significantly longer training time. The final training accuracy was 91.1%, which was lower than the 98.7% for the improved model. The loss function value also fluctuates repeatedly in a certain area until the end of the training, with no level close to zero. Therefore, it is necessary to choose the appropriate optimization algorithm according to the network model and the data set reality to obtain better results.

An improved deep convolutional neural network model based on VGG-16 convolutional neural network to identify four concrete surface cracks. The training accuracy of the improved model reached 98.7%. Improvements include introducing a batch normalization module after each convolutional layer to improve the convergence rate of the model. Meanwhile, the ReLU activation function in the original network was replaced with the P-ReLU activation function to reduce the effect of the ReLU function on the gradient vanishing problem of the network training.

4   Conclusion

As a strong infrastructure country, China is crucial to the timely and effective monitoring of the surface cracks of buildings, which is related to the safety of people's lives and property. However, traditional methods are often time-consuming and laborious, and fail to meet practical needs. Fortunately, the emergence of deep learning techniques has solved this problem. Convolutional neural network model is widely used to identify and classify concrete surface crack images, thus providing reference data for improving process and maintenance. LeNet-5 is the earliest convolutional neural network model, followed by AlexNet, VGG, GoogLeNet, ResNet, etc. These models are deepening, but also bring some problems, such as training accuracy tends to saturation, occupy space and increasing parameters. Therefore, when selecting a network model, we need to focus on network structure, training time, occupancy space and parameters in order to find the most suitable model for this study. In this paper, we improve the original VGG model structure by adding the batch normalization layer and P-ReLU activation function and replacing the SGD optimizer with the Adam algorithm, thus increasing the convergence speed.

References

1. Jin, Y., Ma, W., Li, J., Kovacevic, A.: Electrical resistance–based monitoring of concrete cancer. Constr. Build. Mater. **244**, 119516 (2020)
2. Ma, Y., Zhang, Y., Wang, Z., Wu, J.: Experimental study on the preparation and mechanical properties of laser welded high-strength steel sheet for automobile body. Mater. Sci. Eng. A **764**, 140699 (2021)
3. Matthews, R.D., Fowler, H.: The use of waste tyres in cementitious materials: a review. J. Clean. Prod. **278**, 122756 (2020)
4. Li, J., Li, Z., Wang, X., Xie, Z.: Transfer learning for visual object recognition: a survey. Pattern Recogn. **74**, 59–78 (2019)
5. Park, H., Lee, H., Lee, Y.: Multi-modal factorized bilinear pooling with co-attention learning for visual question answering. Pattern Recogn. 88535–88549 (2019)
6. Tóth, Á., Pálsson, S., Borbála, T., Fülöp, L.: The use of deep learning techniques in the field of image processing: a review. Pattern Recogn. Lett. 109048–109066 (2021)
7. Li, Y., Wang, H., Li, Z.: Bridge crack detection using deep learning: a review. J. Struct. Eng. **147**(1), 04019186 (2021)
8. Wang, H., Li, Y., Li, Z.: Bridge crack detection using deep learning: a comparative study. J. Bridg. Eng. **26**(5), 04020049 (2021)
9. LeCun, Y., Bottou, L., Bengio, Y., Haffner, P.: Gradient-based learning applied to document recognition. Proc. IEEE **86**(11), 2278–2324 (1998)
10. Krizhevsky, A., Sutskever, I., Hinton, G.E.: ImageNet classification with deep convolutional neural networks. In: Advances in Neural Information Processing Systems (NIPS) (2012)
11. Pang, J., Chen, K., Shi, J., et al.: Libra R-CNN: towards balanced learning for object detection. arXiv preprint arXiv:1904.02701v1 (2019)
12. Redmon, J., et al.: YOLOv3: an accurate, fast, and robust real-time object detection system. IEEE Trans. Pattern Anal. Mach. Intell. (2022)
13. Li, Y., Wang, H., Li, Z.: Bridge crack detection using deep learning and image processing: a review. J. Autom. Control Eng. **5**(3), 46–53 (2021)

14. Simonyan, K., Zisserman, A.: Very deep convolutional networks for large-scale image recognition. arXiv preprint arXiv:1409.1556 (2014)
15. Ozgenel, C.F.: Concrete crack image for classification. MendeleyData, v2 (2019). https://doi.org/10.17632/5y9wdsg2zt.2
16. Youjun, Y., Bokai, T., Hongjun, W., et al.: Improved application of VGG model in apple appearance classification. Sci. Technol. Eng. **20**(19), 7787–7792 (2020)
17. Hinton, G., Van Camp, D., Hinton, R.: RMSProp: divide the gradient by a running average of its recent magnitude. arxiv preprint arXiv:1212.5701 (2012)
18. Chen, M., Yu, L., Sang, A., et al.: Multi-label image classification based on a convolutional neural network. J. Jilin Univ. (Eng. Edn.) (3), 10771084 (2020)
19. He, K., Zhang, X., Ren, S., et al.: Delving deep into rectifiers: surpassing human-level performance on ImageNet classification. In: Proceedings of the IEEE International Conference on Computer Vision, p. 10261034 (2015)
20. Hakur, R.S., Yadav, R.N., Gupta, L.: PReLU and edge-aware filter-based image denoiser using convolutional neural network. IET Image Process. **14**(13), 3869–3879 (2020)
21. Kingma, D., Ba, J.: Adam: a method for stochastic optimization. In: Processing of the 3rd International Conference for Learning Representations, pp. 1–12 (2015)

Study on Mechanical and Thermal Properties of Prefabricated Cavity Structure

Lei Sun, Zhenyu Zhang$^{(\boxtimes)}$, Yong Zhan, Qian Wang, and Wu Chen

Wuhan Railway Vocational College of Technology, No.1 Canglong Avenue, Jiangxia District, Wuhan 430200, China
250551197@qq.com

Abstract. In order to improve the contribution of prefabricated buildings in the field of reducing carbon emissions, a hemispherical prefabricated cavity intima structure was designed based on the prefabricated laminated floor technology, which is more in line with the development law of mechanics. From the mechanical properties and thermal properties as the research objects, the full size static loading test and the thermal test of the test house are used as the research methods, and the deflection deformation data of the cavity structure under different loading conditions and the temperature change data of the test house under different external temperatures are obtained. The final result verifies that the fabricated cavity structure has sufficient structural safety from the mechanical properties. Its performance in terms of thermal performance also proves that the prefabricated structure can maintain the thermal insulation capacity similar to that of the concrete structure with thermal insulation layer without any thermal insulation material. It provides theoretical support for further promoting the development of cavity structure.

Keywords: Prefabricated structure · Static loading test · Thermal performance

1 Introduction

In recent years, with the vigorous promotion of the national "dual carbon" policy, the guidance and incentives from local policies, as well as the driving force of the transformation and development needs of the construction and engineering industry, net-zero buildings have gradually become a significant hotspot for the industry [1]. Designing and constructing new materials and structures that balance energy consumption and generation, reducing greenhouse gas emissions and overall energy consumption throughout the construction process and the entire lifespan of the building, and achieving precise monitoring and in-depth data analysis are necessary conditions for achieving net-zero building technology [2, 3].

Building zero carbon is a systematic project, rather than the independent operation of each link and each component [4, 5]. Aiming at the pain points of low energy consumption structure, such as low industrialization degree and limited safety recognition degree, the prefabricated hemispherical cavity plate structure with stable mechanical properties,

P. Xiang and L. Zuo (Eds.): PBSFTT 2023, LNCE 382, pp. 346–353, 2024.
https://doi.org/10.1007/978-981-97-5108-2_37

high safety and reliability and low energy consumption was developed through the combination of prefabricated construction technology and hollow inner membrane structure and optimization and improvement of hollow inner membrane parameters [6, 7]. The main innovation points of the research are: the node construction mode is independent cast-in-place, the semi-spherical hollow inner membrane provides better bearing capacity, the thermal bridge effect of the cavity structure partition plate, the development of efficient green building materials with highly industrialized traditional building materials [8], the application of energy saving and heat insulation technology in the building structure to the prefabricated building system, to achieve the balance and coordination of technical adaptability and economic sustainability [9].

2   Structure Introduction

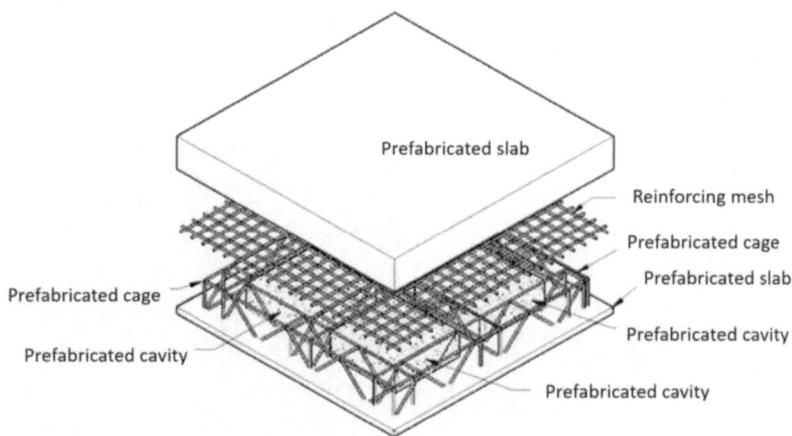

Fig. 1. The structure diagram of fabricated cavity structure

The prefabricated cavity floor structure is made by prefabricated cavity intimal members combined with concrete laminated floor fabrication technology. The specific structure is shown in Fig. 1.

Compared with solid plate, under the premise of similar mechanical properties, the body weight of assembled cavity structure can be reduced by 30%–60%. Save steel 20%–30%, concrete 40%–50%, formwork 60%–70%; Shorten the construction period by 1–1.5 days/flow time [10, 11].

3   Mechanical Property Test

3.1   Experimental Design

In this paper, the full size test of 8700mm×8700mm prefabricated laminated cavity floor is carried out to evaluate the bearing capacity and service performance of cavity floor structure through the observed deflection deformation, stress and strain of each measuring point and the crack development of the floor during load application [12]. The specific static loading test process and measuring point arrangement are shown in Fig. 2.

The specific loading process is as follows:

1. **Preload:** preload before loading, apply the load 0.5 kN/m^2, the duration is 15 min, and then unload to ensure that all the test instruments can work normally.
2. **Formal loading:** formal loading from the plate across the central position gradually spread to all sides, forming uniform loading. The load of the first stage is 1.0 kN/m^2, and then each stage is increased by 1 kN/m^2 until the eighth stage; The ninth stage starts to load in increments of 2 kN/m^2 per stage to the eleventh stage; The twelfth level load is 15 kN/m^2; The final stage load is 16 kN/m^2. After each stage of loading is completed, stand for 20 min to start reading, observe the cracks at the bottom of the board, and make a mark.
3. **Static observation:** Stop loading after loading to the last level of load, and observe and read again after 12 h of load, and then unload to understand the load holding capacity and recovery of the board.

Fig. 2. Test design and measuring point layout

3.2 Experimental Analysis

The test site layout is shown in Figs. 3, 4 and 5. The maximum deflection value at the center of the plate is 13.245 mm, and the standard deflection limit is L/250, that is, 34.8 mm. The deflection of the test plate does not exceed the limit value to meet the bearing capacity requirements, and the crack development also meets the normal use requirements, indicating that the structural safety of the floor can fully meet the use requirements.

The deflection value of the floor gradually decreases from the center point outward and is evenly distributed. The stress characteristics are similar relative to the center symmetric position, and the deformation trend is the same in the vertical and horizontal direction. All these characteristics indicate that the prefabricated composite cavity floor has good integrity and bidirectional stress [13].

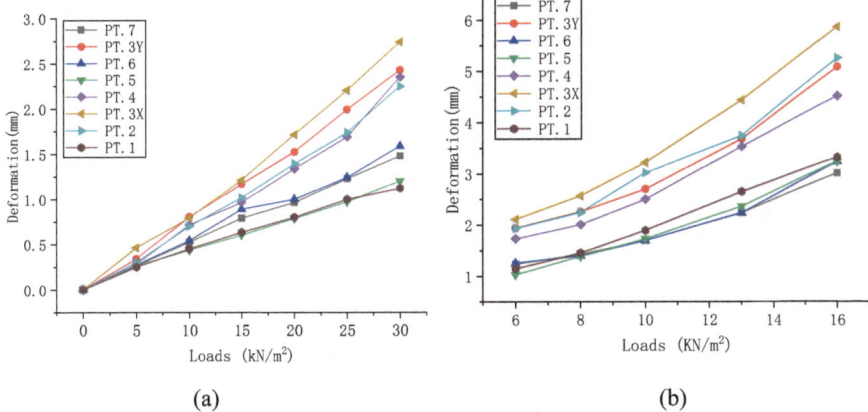

Fig. 3. Deflection deformation curve under different loads

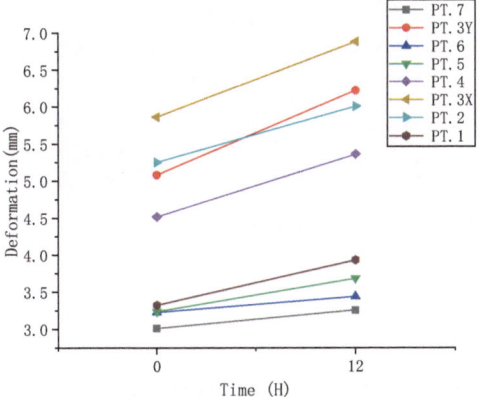

Fig. 4. Deflection deformation after unloading

4   Thermal Performance

4.1   Experimental Design

Two newly built 3 m × 3 m × 3 m experimental prototypes were used for comparative tests, respectively using assembled integral multi-ribbed cavity floor (no insulation layer but waterproof layer in the floor) and traditional cast-in-place concrete floor (with waterproof layer and insulation layer in the floor).

Hygrograph: used to determine the temperature and relative humidity of the environment, where the allowable error of temperature is ±2 °C, and the error of humidity indication is ±5%RH.

Anemometer: An instrument used to measure the velocity of air. Low speed is 0–5 m/s, medium speed is 5–40 m/s, and high speed is 40–100 m/s.

Specific test arrangement:

1. Make two experimental rooms with identical size, dimension and wall thickness.
2. The cavity structure experiment room is not equipped with thermal insulation materials, and the cast-in-place concrete structure experiment room is equipped with roof insulation layer.
3. Set the same orientation, shading and other boundary adjustment,

The test time is different time periods in March, May and August, and the test content is outdoor temperature, indoor temperature of cast-in-place structure and indoor temperature of cavity structure.

Fig. 5. Infrared thermometer

4.2   Test Data Analysis

The temperature data of March and May are listed in Figs. 6, 7 and 8. By comparing the temperature conditions of outdoor, precast cavity experiment and traditional concrete experiment, it can be found that: when the outdoor temperature fluctuates in March, the temperature of prefabricated structure changes more gently, and the overall temperature data are higher than that of traditional concrete structure; In May, when the outdoor temperature is higher, different outdoor temperatures will also bring different performance of the assembled structure. When the outdoor temperature is about 19 °C, the overall temperature of the assembled structure is lower than that of the traditional structure, but when the outdoor temperature exceeds 23 °C, the assembled structure will have a higher room temperature than the traditional structure.

In order to better understand the difference between the performance of the cavity structure and conventional concrete under high temperature conditions, temperature tests were conducted on a single day in August, from 11:40 PM to 15:40 PM, with a time interval of one hour. The obtained internal temperature data of the experimental body are shown in Fig. 9. It can be found that the cavity structure can not exceed the performance advantage of thermal insulation materials for the time being from the temperature comparison in each period, but the cavity structure shows a more gentle performance in the temperature development trend of one day.

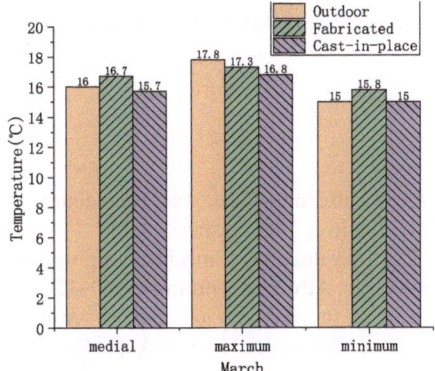

Fig. 6. Temperature data for March

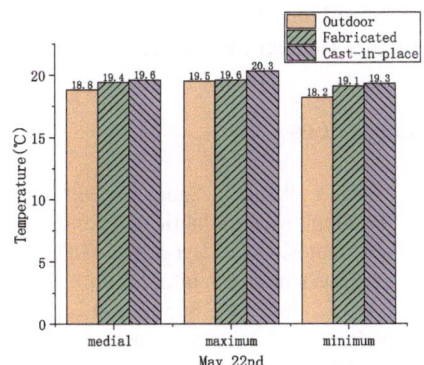

Fig. 7. Temperature data for May 22

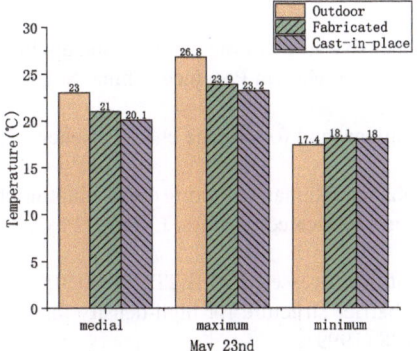

Fig. 8. Temperature data for May 23

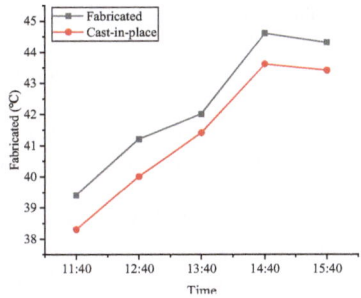

Fig. 9. Temperature data for August

5 Conclusion and Prospect

In this paper, the performance of a prefabricated cavity structure is studied from both mechanical and thermal aspects. The deformation and crack development analysis of the cavity structure is carried out by full-scale static loading test. The test results show that the structure has sufficient bearing capacity and the deformation and crack development meet the requirements of the concrete structure code. Thermodynamic performance is compared between the test house made by prefabricated cavity structure and the house made by traditional concrete structure by means of thermal test. The results show that prefabricated cavity structure can show better ability to cope with temperature changes, and it is more in line with the characteristics of warm winter and cool summer than traditional structure. Prefabricated structures can become poorly equipped to handle high temperatures.

After the preliminary verification that the assembled structure is safe enough, what factors cause the sensitivity of the assembled cavity to high temperature becomes the focus of future research, and how to improve the better temperature control ability of the assembled structure will become the key factor to promote the assembled cavity structure to meet the low-carbon development.

References

1. Wang, Y., Wang, F., Sang, P., et al.: Analysing factors affecting developers' behaviour towards the adoption of prefabricated buildings in China. Environ. Dev. Sustain. (2021)
2. dos Santos, D.F., Tayt-Sohn, M.D.R., Simo, R.A.: Evaluation of mechanical properties in concrete structure reinforced with carbon fiber coating. Adv. Eng. Forum **44**, 119–126 (2022)
3. Bae, A., Jung, E., Seo, B., et al.: Development of injectable water-stop for grouting in underground concrete structure joint. J. Test. Eval.: Multidisc. Forum Appl. Sci. Eng. **1**, 50 (2022)
4. Grinham, J., Yan, B., Malkawi, A., et al.: Corrigendum to zero-carbon balance: the case of housezero. Build. Environ. **207**, 108511 (2022). 212
5. Grinham, J., Fjeldheim, H., Yan, B., et al.: Zero-carbon balance: the case of HouseZero. Build. Environ. (Jan. Pt.B), 207 (2022)
6. Liu, C., Zhang, S., Chen, X., et al.: A comprehensive study of the potential and applicability of photovoltaic systems for zero carbon buildings in Hainan Province, China. Solar Energy (May), 238 (2022)
7. Group, R.E.M.: Zero-carbon housing a key battleground for climate change. Renew. Energy Monit. (2022)
8. Wang, Y., Chen, H., Dong, S., et al.: Surface enhanced Raman scattering of p-aminothiophenol self-assembled monolayers in sandwich structure fabricated on glass. J. Chem. Phys. **124**(7), 163 (2006)
9. Ryu, S.O., Joshi, P.C., Desu, S.B.: Low temperature processed $0.7SrBi2Ta2O9$–$0.3Bi3TaTiO9$ thin films fabricated on multilayer electrode-barrier structure for high-density ferroelectric memories. Appl. Phys. Lett. **75**(14), 2126–2128 (1999)
10. Kumaran, M., Senthilkumar, V., Sathies, T., et al.: Effect of heat treatment on stainless steel 316L alloy sandwich structure fabricated using directed energy deposition and powder bed fusion. Mater. Lett. **313**, 131766 (2022)
11. Cao, Y., Yang, H., Li, H., et al.: Ultralight biomass-based carbon aerogel with hierarchical pore structure fabricated using unidirectional freeze casting and potassium hydroxide activation. Mater. Lett. (Jun.15), 317 (2022)
12. Long, Y., Zhang, G., Lin, S., et al.: Microstructure and mechanical properties of WB2-B4C composites fabricated by boro/carbothermal reduction and SPS. J. Am. Ceram. Soc. (2022)
13. Yuan, Z., Zhao, X.W., Ye, L.: Super-strong polyphenylene oxide/polyphenylene sulfide blend foam fabricated by constructing highly oriented crystalline structure. J. Supercrit. Fluids (2022)

Research on Quality Inspection Method of Reinforcement Sleeve Grouting Joint

Duwen Shen, Qili Gan$^{(\boxtimes)}$, and Langhua Li

Chongqing College of Architecture and Technology, Chongqing, China
277306408@qq.com

Abstract. At present, the connection of vertical members of prefabricated concrete structures in our country is mainly based on reinforcement sleeve grouting technology, which is a hidden project, and the inspection of construction quality determines the evaluation of connection quality. Therefore, this paper introduces the construction quality detection technology of sleeve grouting currently commonly used in our country, and analyzes the applicability and effectiveness of different quality detection methods for reinforcement sleeve grouting joint joints from the aspects of filling degree of grouting material and insert length of steel bar, which determine the quality of sleeve grouting, and combined with the actual detection situation of engineering projects.

Keywords: reinforcement sleeve grouting connection · filling degree · steel bar insertion length

1 Introduction

In recent years, the prefabricated concrete structure with its unique advantages rapidly spread across the country, under the guidance of the government, all places have encouraged and guided its development, the prefabricated building has begun to show results. There are a large number of connection nodes in the prefabricated concrete structure, and the nodes are mostly connected by reinforced grouting sleeve [1]. The connection quality of the grouting sleeve and the construction quality of the installation joint are particularly critical, which directly affects the safety and service life of the whole building. The main factors affecting the connection quality of the grouting sleeve are the filling degree of the grouting, the depth length of the reinforcement and the strength of the grouting material. The main research in this paper is the detection of the grout filling degree and the detection of the insert length of the steel bar in the grout connection of the reinforcement sleeve [2].

Sleeve filling degree refers to the degree to which the concrete injected into the sleeve fills the space of the sleeve. The degree of fullness is critical to the stability and safety of the structure, because if the grouting is not full, it will lead to gaps inside the sleeve, which will reduce the compressive strength of the concrete and the carrying capacity of the structure. In addition, incomplete grouting will also cause quality problems such as water seepage and cracks, which seriously affect the service life of the building. As

P. Xiang and L. Zuo (Eds.): PBSFTT 2023, LNCE 382, pp. 354–361, 2024.
https://doi.org/10.1007/978-981-97-5108-2_38

shown in Fig. 1, the grouting in a practical project is not full, resulting in deformation of the connection part.

Fig. 1. The reinforcement sleeve is not full grouting

At present, the common detection methods of reinforcement sleeve grouting fullness include endoscope method, X-ray method, impact echo method, embedded wire drawing method and embedded sensor method. This article mainly introduces the first two quantitative detection methods, that is, the actual detection of the fullness of the sleeve grouting [3, 4].

The insertion length of steel bar directly affects the anchorage of steel bar and concrete, and determines the load transfer of sleeve grouting. At present, the steel bar insertion length detection adopts the function of showing the distance between the shooting end face of the measurement lens and the selected point, and directly measures the vertical distance between the end end of the connecting steel bar and the shooting end face of the lens, and then considers the vertical gap between the shooting end face of the lens and the center of the grout port, the insertion length of the connecting steel bar can be calculated [5–7].

2   Grout Fullness

2.1   Endoscopic Method

In the case of incomplete filling or leakage of grouting in the grouting connection of the sleeve, due to the action of gravity, the liquid level of the mobile grouting material will be lower than the grouting outlet of the sleeve. According to this feature, endoscopy detection holes are prepared at the height of the grouting outlet of the sleeve. An endoscope with dimension measurement function can be used to quantitatively determine the incomplete range and then calculate the grouting fullness. The endoscopy method is divided into preforming endoscopy method, grouting hole drilling endoscopy method and sleeve cylinder wall drilling endoscopy method, which can be selected according to the shape of the grouting hole. When the pulping hole is straight, the preforming endoscopy method or the pulping hole drilling endoscopy method can be used according to the actual situation, and the sleeve barrel wall drilling endoscopy method can also be used if necessary. When the outlet hole is not linear, the cylindrical wall drilling endoscopy method can be used.

The testing instruments of endoscopy mainly include endoscope, inner snooping head, front looking measuring lens, side looking measuring lens, rigid sleeve and rubber plug. The endoscope is a new type of measuring instrument, which uses a miniature camera and sensor to monitor and measure the internal condition of the building structure in real time, as shown in Fig. 2. The focal length of the inner snoop head is 5 mm–150 mm, the resolution is 1080 mm, and the effective visual distance is 15 cm. The front view measuring lens and side view measuring lens are optical acquisition lenses, respectively, to observe and measure the measurement structure in the front and side. The rigid casing and rubber stopper are used for auxiliary equipment and for the stable line of measurement process.

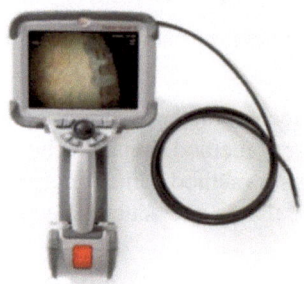

Fig. 2. Endoscope

1. Detection principle of pre-perforated endoscope method: the pre-perforated combination plug is designed. During grout construction, the straw in the film is removed after the slurry is uniformly discharged from the grout outlet, and the film is left to form a detection hole before detection.

    The preparation of the hole must be completed before the detection of the pre-perforated endoscope. The preparation process of the hole: the first step is to connect the straw wrapped with a film to the rubber plug; The second step is to insert the combined plug into the outlet pipe from the surface of the prefabricated member; The third step is to block the rubber plug; The fourth step is to test the hole.

    Detection steps: the first step is to prick the film; The second step use the front view lens to observe the cavity inside the punctured place; The third step is to extend the side-looking three-dimensional measurement lens into the gap between the inner wall of the sleeve and the connecting steel bar and observe it down; The fourth step is to measure the vertical distance between the upper surface of the grout material and the end face of the side-looking 3D measuring lens, and the grout fullness can be obtained according to the conversion of the vertical distance, as shown in Fig. 3.

2. The endoscopic detection procedure of the grouting hole drilling is shown in Fig. 4: the first drilling depth is 20 mm–30 mm, and the grouting material of the whole section of the grouting hole is crushed and cleaned, so that the rubber plug in the probe positioning device can be installed during the detection. Drill holes along the outlet hole, and clean up the grout debris and powder every 20 mm–30 mm forward;

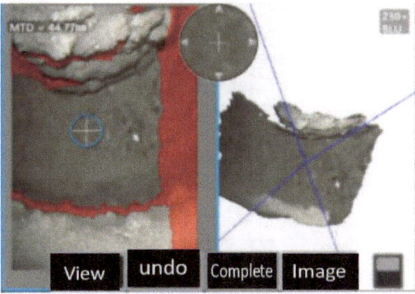

Fig. 3. Test result of preformed hole endoscopic method

When the distance from the sleeve outlet is smaller than 20 mm, every 5 mm forward, clean up the grout debris and powder, observe the drilling condition, until the detection hole penetration; Use the front view lens to observe, if not full and then use the side view measurement lens to measure the length of the grout defect area.

Fig. 4. Endoscopic method of grout hole drilling

As shown in Fig. 5, the grouting of No. 1 and No. 3 sleeves is full, while No. 2 and No. 4 sleeves are not full. The measured data are 23.59 mm and 60.09 mm respectively by direct distance measurement. According to the measured data, the filling degree of sleeve grouting can be directly calculated.

3. Tube wall drilling endoscopic method of detection steps: use the steel bar scanner to determine the position of the sleeve; Partial removal of the concrete protective layer outside the corresponding position of the pulp outlet height of the sleeve, exposing the outer wall of the sleeve; The drilling equipment is first equipped with a metal drill bit to make a hole in the wall of the sleeve, and then replaced with a stone drill bit to continue drilling into the inner cavity of the sleeve 4 mm–6 mm. Use the front view lens for observation. If it is not full, use the side view measurement lens to measure the length of the grout defect area.

2.2 X-ray Method

When an X-ray penetrates the object under test, the intensity of the ray decreases as it passes through the object. The degree of ray attenuation is not only related to the energy of the ray, but also directly related to the property, thickness and density of the measured object, etc. A certain detector (such as: film, IP board, DR Flat panel detector, etc.) is

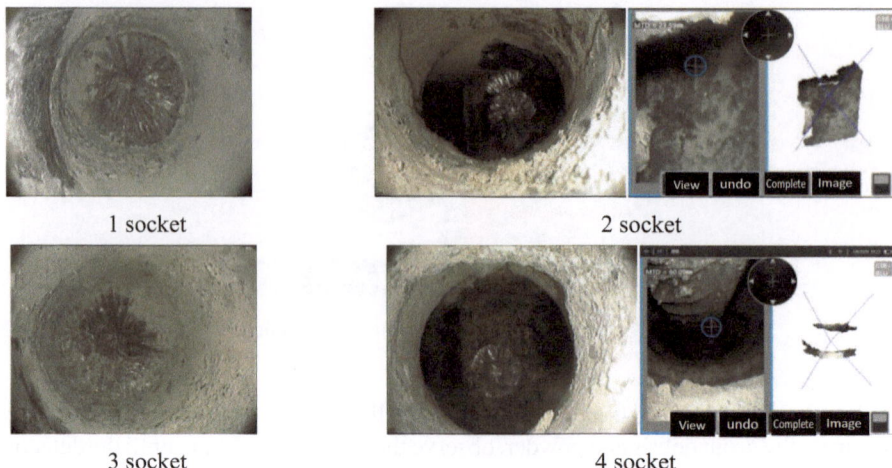

1 socket   2 socket

3 socket   4 socket

Fig. 5. Test result of endoscope method of hole drilling

used to record the intensity of the transmitted ray, and the projected image inside the sleeve can be obtained. Through professional processing software, the full degree of sleeve grouting can be quantitatively detected.

Application conditions of X-ray method: the thickness of the structural layer in the inspected area should not be greater than 200 mm, the sleeve is arranged in a single row or "plum blossom shape", the personnel must withdraw to the radiation safety area, and the X-ray local damage method can be used if necessary.

The equipment of X-ray method is portable X-ray device. The X-ray machine system is mainly composed of a transmitter tube, a console, a laser aligner, a control cable, a power line, etc. The traditional imaging method is industrial film imaging. In recent years, CR (Computed Radiography) and DR (Digital Radiography) techniques are gradually emerging.

The 16# building of a project is a prefabricated shear wall structure with a wall thickness of 200 mm. X-ray method is used to detect the construction quality of the sleeve joints of the 6 storey shear wall components of the 16# building, in order to verify the reliability of the X-ray method in practical engineering applications.

Fig. 6. X-ray detection

The detection procedure is shown in Fig. 6: According to the selected perspective focal length 700 mm, the X-ray machine and imaging device are placed at both ends of the measured wall panel. After taking good radiation protection measures, set the voltage at the console to 280 kV and preheat the machine for 37 min. After the preheating is completed, input the selected perspective parameters in the console: tube voltage 250 kV, tube current 2.0 mA, DR Exposure time (2 × 5) s, film exposure time 10 min. Turn on the ray for perspective lighting and turn off the ray machine after completion.

X-ray method can effectively identify two indexes of grout sleeve - grout fullness and steel bar insertion length, and this method can meet the actual testing requirements.

3   Detection of Steel bar Insertion Length in the Sleeve-Ray Method

After the field splicing of the prefabricated member is completed and before the grouting construction of the sleeve, the insertion length of the connecting steel bar is fixed, and it is convenient for endoscopic detection. The detection principle is to use the characteristics of high dimensional accuracy of the sleeve, the measurement of the insertion length of the connecting steel bar into the measurement of the relative distance between the end of the insertion section of the connecting steel bar and the known reference point in the sleeve, the three-dimensional measurement endoscope accurately measures the relative distance, and the insertion length of the connecting steel bar is calculated.

3.1   Semi-grout Sleeve Detection Method

Taking the center of the grout outlet of the sleeve as the reference point, auxiliary tools are used to pound horizontally from the bottom of the grout hole before testing, quickly determine the relative position of the end of the connecting steel bar and the grout outlet of the sleeve, select the corresponding measurement lens, and combine the later observation results can be divided into the following three working conditions:

The first working condition is that the connecting steel bar is higher than the pulp outlet. It is judged that the end of the steel bar is higher than the bottom of the pulp outlet through the introduction of auxiliary tools; Using the forward looking measuring lens for observation, the end of the steel bar is higher than the top of the slurry outlet of the sleeve; After investigation, the common brand and conventional model of semi-grouting sleeve on the market at present, the end of the connected steel bar under the design anchoring length is located in the range of 5 mm above and below the bottom of the grouting port of the sleeve, and the upper cavity of the grouting port is used as the adjustment margin, so the insertion length of the steel bar can be directly determined to meet the requirements.

The second condition is that the end of the connecting steel bar is at the position of the pulp outlet. Rigid bushings and rubber plugs are used to assist the positioning of the forward-looking measuring lens; The appropriate position is selected in the outlet channel to make the imaging clear and get the 3D image. The point-to-surface measurement function is adopted, in which the surface is formed by the leftmost end point on the outlet, the rightmost end point and any point in the outlet hole on the same horizontal plane. The point is positioned at the end of the connecting steel bar. The relative distance between the end of the connecting steel bar and the horizontal section of the center of the outlet hole is measured, and the insertion length of the connecting steel bar is calculated.

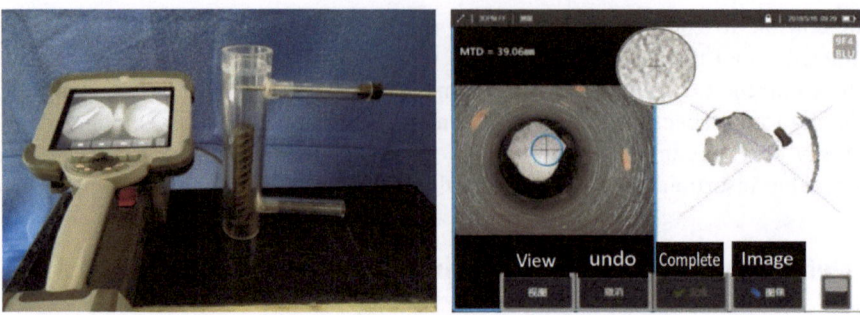

Fig. 7. The end of the connecting steel bar is lower than the outlet

The third condition is that the end of the connected steel bar is lower than the pulp outlet. As shown in Fig. 7, it is judged that the end of the steel bar is lower than the bottom of the pulp outlet through the introduction of auxiliary tools; The side-looking measuring lens is selected, and the rigid sleeve and rubber plug are used to assist the positioning of the side-looking measuring lens; Adjust the insertion depth of the rigid sleeve so that the side view measuring lens is located in the inner cavity of the sleeve to observe and shoot 3D images inside the sleeve; Using the function of displaying the distance between the shooting end face of the measuring lens and the selected point, the vertical distance between the end of the connecting steel bar and the shooting end face of the lens is directly measured, and the vertical deviation between the shooting end face of the lens and the center of the grout port is considered to calculate the insertion length of the connecting steel bar.

3.2   Full Grout Sleeve Test Method

The detection method of full grouting sleeve takes the upper surface of the limit stop card in the middle of the sleeve as the known reference point, extends the inner snooping head equipped with a forward looking measuring lens into the inner cavity of the sleeve, and captures the 3D image of the middle of the sleeve. The relative distance between the end of the connecting steel bar at the installation end and the upper surface of the limit gear card can be measured by the measurement function of the endoscope point to surface, and then the insertion length of the steel bar can be calculated.

The probe should be extended down the gap to the limit stop card in the detection of the full grout sleeve, so that the camera can take a clear picture of the middle of the sleeve and measure the distance. In order to meet the requirements of detection, the lens measuring range should not be less than 80 mm, taking into account factors such as the Angle of view of the front view measuring lens, the thickness of the limit stop card, the adjustment space reserved for the steel bar and the insufficient insertion length.

4   Conclusions

In this paper, combined with the actual engineering sleeve grouting connection method, the detection technology of grouting fullness and steel bar insertion length is explored, the main results are as follows:

(1) The endoscope method can effectively detect the filling and insertion length, and the X-ray method can quantitatively detect the filling. The two test results are in agreement with the actual project.
(2) The endoscope method has the advantages of simple structure, low cost and wide applicability, and can be widely used in practical engineering. X-ray method has higher requirements for equipment and certain technical requirements for operators, which need to be improved and promoted.
(3) The endoscopic method and X-ray method cannot achieve dynamic observation during construction, and the subsequent research direction can extend the post-construction detection to the construction, so as to effectively control the construction quality of the sleeve grouting connection.

References

1. Tao, J., Cao, C., Dong, Z., et al.: Research on grouting quality control and inspection technology of steel bar sleeve in prefabricated building. Chongqing Archit. **21**(10), 55–58 (2022)
2. Yang, J.: Analysis of quality control and on-site inspection technology of prefabricated building Engineering. China Build. Metal Struct. **05**, 144–146 (2022)
3. Kim, Y: A study of pipe splice sleeves for use in precast beam-column connections. The University of Texas at Austin (2000)
4. Tan, S.S.: Development status of prefabricated building node quality inspection in Guangxi. Jiangxi Build. Mater. **07**, 46–47 (2021)
5. Khaled, A., Sami, H., LeBlanc, B.: Horizontal connections for precast concrete shear walls subjected to cyclic deformations part 1: mild steel connections. PCI J. **40**(3), 78–96 (1995)
6. Jiang, M., Liu, J.: Prefabricated construction quality detection method based on BIM explore. J. Intell. City **7**(9), 17–18 (2021). https://doi.org/10.19301/j.carolcarrollnkiZNCS.2021.09.009
7. Ozsoy, U., Koyunlu, G., Ugweje, O.C., et al.: Nondestructive testing of concrete using ultrasonic wave propagation. In: 13th International Conference on Electronics, Computer and Computation (ICECCO), pp. 1–5 (2017)

Application of Ultrasonic Nondestructive Testing Technology to the Horizontal Joint Slurry at the Bottom of Shear Wall

Langhua Li, Duwen Shen[✉], and Qili Gan

Chongqing College of Architecture and Technology, Chongqing, China
547297938@qq.com

Abstract. Ultrasonic nondestructive testing (NDT) technology has a wide range of applications in the connection nodes of prefabricated building components. It is mainly used to test the quality of the connection nodes of prefabricated building components, such as filling degree, anchorage of steel bars, etc. This paper mainly introduces the detection of the filling degree of the shear wall bottom joint slurry. By discriminating acoustic parameters such as "sound velocity, amplitude, waveform" in ultrasonic detection method, the fullness of grouting is detected. According to the examples, the operation steps such as field detection, data processing and analysis and entity excavation verification are summarized, and suggestions are provided for the efficient application of ultrasonic detection technology in the quality detection of fullness.

Keywords: ultrasonic nondestructive testing · prefabricated building · filling degree of grouting

1 Introduction

The integrated concrete building is a development direction and a hot research topic in the current construction field, but its quality control and detection technology is relatively lagging behind. Especially for the grouting quality at the joint of the component, there is no good method that is low cost and can effectively detect the filling degree of the grouting material. In the past, ultrasonic technology has been widely used in concrete defect detection. Using ultrasonic to detect the filling degree of grouting material in prefabricated shear wall is helpful to promote the innovation and development of detection technology, which can improve the quality of assembled integrated concrete building construction and improve the relevant quality detection technology system. The technology is suitable for the ultrasonic detection of filling degree of grouting material in the construction joint of prefabricated concrete prefabricated shear wall. The research results can provide an effective means for the quality control of assembled monolithic concrete buildings and the establishment of testing technology system [1].

P. Xiang and L. Zuo (Eds.): PBSFTT 2023, LNCE 382, pp. 362–369, 2024.
https://doi.org/10.1007/978-981-97-5108-2_39

2 Principle of Ultrasonic Nondestructive Testing Technology

The principle of ultrasonic non-destructive testing technology is to use the characteristics of ultrasonic propagation in the material to detect the internal structure and state of the material [2]. In ultrasonic non-destructive testing, ultrasonic probes are usually used to emit ultrasound to the inside of the material, and then by receiving the reflected ultrasonic signal, judge the defects, cracks, pores and other problems inside the material. When the ultrasonic wave propagates in the material, it will be affected by the internal structure, defects, cracks and other factors of the material, resulting in physical phenomena such as reflection, refraction and scattering. By measuring the characteristics of these phenomena, information such as the location, size and nature of defects, cracks, pores and other problems inside the material can be inferred. Therefore, ultrasonic nondestructive testing technology is an efficient and accurate detection method, which is widely used in the detection of various materials [3].

In this paper, the normal distribution is used as the statistical basis for judging distance. As shown in Fig. 1, it can be seen from the classical normal distribution that the probability of $[\mu - \sigma, \mu + \sigma]$ is 68.26%, the probability of $[\mu - 2\sigma, \mu + 2\sigma]$ is 95.44%, and the probability of $[\mu - 3\sigma, \mu + 3\sigma]$ is 99.74%.

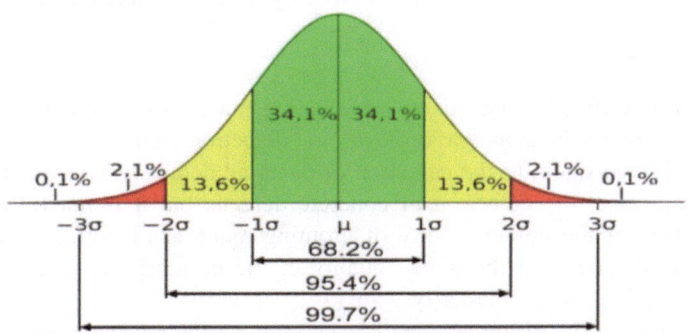

Fig. 1. Classic normal distribution

The probability judgment defect is related to the value of its standard deviation and the number of measurement points. When the standard deviation of the judgment defect is 3 times, the defect will not be misjudged, but it is easy to miss the judgment. When the defect is judged with 2 times the standard deviation, the defect is not easy to miss the judgment, but it will be misjudged. The multiple of standard deviation should be related to the number of measurement points. If judging by sound velocity V, amplitude A, or frequency F, the detection data can be arranged from largest to smallest, and the mean value M_V, M_A, M_F and corresponding standard deviation S_V, S_A, S_F, of the detection data can be calculated respectively, using the formula:

$$X_0 = M_X - \lambda S_X \tag{1}$$

The X in the formula can respectively represent the sound speed V, amplitude A, or frequency F, λ according to the need to query the standard table, calculate the critical value to judge the suspicious value [4, 5].When the detection result data is outside the critical value of the measurement point, judged as abnormal. In addition, the received signals in the ultrasonic detection of the horizontal joint at the bottom of the prefabricated concrete structure and the grouting and anchor joint grouting material defects are comprehensively analyzed and evaluated according to three parameters: sound time, first wave amplitude and waveform [6]. By comparing the normal ultrasonic detection point waveform with the defect ultrasonic detection point waveform, the defect point can be intuitively judged.

In the transmitting and receiving acoustic channels, the difference of time generated in the direct propagation of sound waves or the propagation process of diffraction due to defects leads to the difference of sound velocity in the same ranging. Sound waves in different acoustic interfaces, due to the different acoustic impedance of the medium, resulting in differences in sound energy reflection or transmission, due to the attenuation of sound energy defects lead to a decline in the amplitude of the first wave. Due to the defects of concrete, the sound wave propagation produces reflection, refraction and diffraction, and the superposition results in the phase change of waveform or the distortion of the first wave of the received waveform [7].

3   Detection of Horizontal Joint Slurry at the Bottom of Shear Wall

3.1   Transducer

Although the detection of the quality of grouting material of prefabricated concrete structure can refer to the principle of ultrasonic detection technology of concrete, it is not completely equivalent to the detection of concrete defects. The existing ultrasonic equipment can meet the detection of concrete defects, but it is not fully applicable to the detection of the defect quality of grouting material of prefabricated concrete structure. The detection of the defect quality of the prefabricated concrete structure grouting material has its particularity.

The detection of the filling degree of grouting material has its particularity, such as the equivalent of the defect, the defect in the grouting material is much smaller than the defect in the concrete, but the advantage is that the ultrasonic ranging at the anchor node of the prefabricated shear wall is usually shorter. In view of the short ranging of shear wall and the needs of detecting small and medium-sized defects of grouting material, the suitable transducer is considered to be selected for matching. The transducer with high frequency and small diameter radiating end face is the technical means that must be adopted.

At present, the frequency of the transducer commonly used in engineering testing is 50 kHz, and the wave speed v of the ultrasonic wave is the product of the frequency f and the wavelength λ. When the sound speed of the grout is 4.40 km/s, such as the transducer with a frequency of 50 kHz, its wavelength λ is 88 mm. Because the wavelength is long, the ultrasonic wave is easy to bypass the defect propagation, the sensitivity of the 50 kHz transducer to detect the defect is very low, so the selection of the frequency of the transducer must consider the size of the defect that needs to be detected, usually set the minimum defect that can be measured by ultrasonic detection is approximately the wave length. If the transducer of grout is used in the detection of 500 kHz, its wavelength λ

is less than 9 mm, that is, the use of high-frequency transducer can greatly improve the sensitivity of defect detection. In addition, at present, many domestic ultrasonic instruments factory generally configured 50 kHz transducer diameter of about 40 mm, foreign Switzerland PROCEQ ultrasonic instrument configuration 50 kHz transducer diameter is 50 mm, because the size of the transducer diameter is too large, used for structural grout grouting, ultrasonic detection is more difficult to find less than the transducer diameter defects. Therefore, the domestic first developed for prefabricated concrete structure grouting material grouting quality ultrasonic detection of high-frequency small diameter radial end transducer.

3.2 Simulation Test

Ultrasonic measuring points are arranged at the bottom of the shear wall connecting the horizontal joint pulp. First, a set of symmetric starting points are determined on both sides of the horizontal joint connected at the bottom of the component, and then the positive and negative sides of the component are drawn along the horizontal joint connected at the same direction with 100 mm equal spacing, and the measuring points are numbered according to 1, 2, 3, and ··N in turn. Ultrasonic detection will first transmit and receive transducers are placed at the symmetric starting point number 1 on the bottom horizontal joint on both sides of the positive and negative sides, and then move the transducer in the same direction according to the corresponding number on both sides of the test. After the end of the component detection, you can view the ultrasonic instrument real-time display of the detection data and its statistical summary analysis results, if there is a suspicious test point with a * prompt, you can retest the suspicious test point according to experience or increase the test point around it, judge the scope of the defect, assess the grouting material fullness or whether there is a local cavity.

According to the actual engineering situation of the horizontal joint slurry at the bottom of the shear wall, A simulated shear wall grouting layer defect specimen A was made, and 5 defects were arranged in the specimen: a large defect of 120 mm × 150 mm was set in the center position; Two small rectangular defects with a size of 20 mm × 50 mm are arranged on the left and right sides of this large defect, among which the two defects on the left side are horizontally located at 1/4 and 1/2 of the width of the specimen (perpendicular to the direction of the ultrasonic channel); The two defects on the right are vertically located at 1/4 and 1/2 of the width of the specimen respectively (parallel to the direction of the ultrasonic acoustic path), as shown in Fig. 2.

Ultrasonic measuring points: a pair of measuring points are arranged at each of the 4 small defects on the left and right sides of the specimen, and three pairs of measuring points are arranged at the big defect in the middle of the specimen, that is, a total of 7 pairs of ultrasonic measuring points at the defect. In addition, according to the work experience that the number of measuring points without defects is 1.6 times greater than the number of defect measuring points, 11 pairs of ultrasonic measuring points without defects are arranged, that is, A total of 18 pairs of ultrasonic measuring points on the simulated defect specimen A. The test data according to the ultrasonic detector are shown in Table 1.

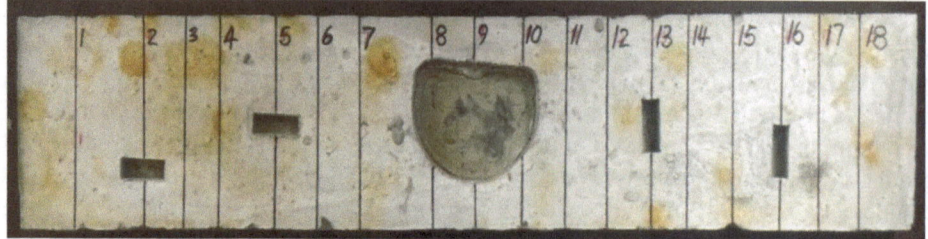

Fig. 2. Photos of the simulated defect position of specimen A and the number of ultrasonic measuring point

Table 1. Test data of 18 ultrasonic test points of specimen A

Number of measuring points	Speed of sound (km/s)	Amplitude (dB)	Station serial number	Speed of sound (km/s)	Amplitude (dB)
A-01	4.284	75.40	A-10	4.083	60.99
A-02	3.902	60.18	A-11	4.167	74.77
A-03	4.209	74.91	A-12	4.167	72.83
A-04	4.212	74.75	A-13	4.038	62.24
A-05	4.092	63.48	A-14	4.153	74.31
A-06	4.228	74.35	A-15	4.196	71.68
A-07	4.141	73.91	A-16	4.02	71.86
A-08	4.104	60.98	A-17	4.02	71.78
A-09	3.576	67.42	A-18	4.02	73.13

3.3 Data Analysis

The 18 ultrasonic measuring points in Table 1 are classified, with an average sound velocity of 4.091 km/s and an average amplitude of 69.94 dB. According to the ultrasonic theory, the defect points are discriminated and the sound velocity or amplitude of the abnormal points is less than the average value, such as A-02, A-05, A-08, A-09, A-10 and A-13 in the table. The result of testing data is consistent with the reality.

In addition, the wave train view of all ultrasonic measuring points corresponding to the serial number of measuring points in Table 1 is shown in Fig. 3. The defect points are judged by observing the ultrasonic wave train view.

In the wave train view of Fig. 3, the ultrasonic first wave starting point of 7 measuring points (2, 5, 8, 9, 10, 13, 16) is obviously backward, indicating that their sound time is too long in the same distance measurement. In Table 1, the (sound velocity, amplitude) data of these measuring points are obviously small. According to the acoustic parameters such as "sound velocity, amplitude, waveform" as the principle of judging the defects of grout, the defects can be clearly judged here. Figures 4, 5 and 6 illustrates the ultrasonic

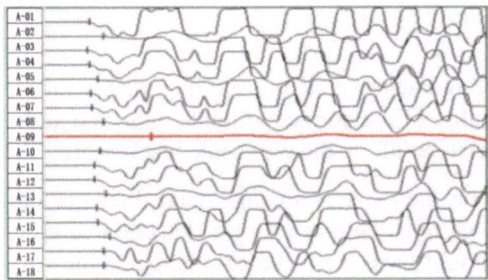

Fig. 3. Wave train view of ultrasonic measuring point

waveforms of A-05, A-09, A-16, and A-04 measuring points (Fig. 7) for comparison, which are typically poor.

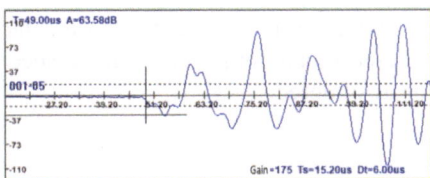

Fig. 4. Defective waveforms at measuring point No. 05

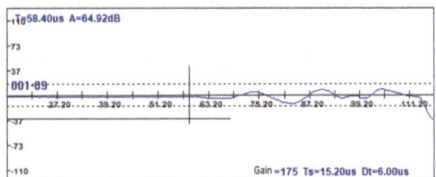

Fig. 5. First wave loss defect waveform at measuring point No. 09

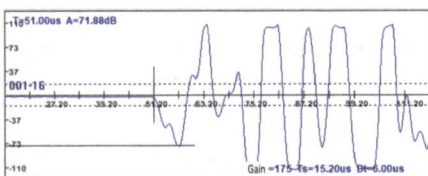

Fig. 6. Defective waveform at measuring point No. 16

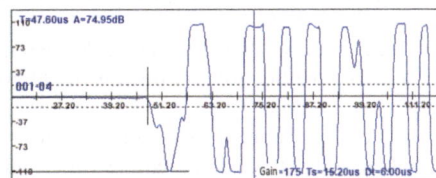

Fig. 7. Comparison waveform without defect at measuring point No. 04

The judgment result printed by the ultrasonic detector is clear and correct and consistent with the actual defect situation. The defect size of the measuring point No. 16 is equivalent to that of the three measuring points No. 03, No. 05 and No. 13; However, the measuring point only shows the sound velocity value is small, and the amplitude dB value is not too small, because the direction (vertical) of the long size defect of the measuring point is in the same direction as the acoustic path of the transmitting transducer and the receiving transducer during the detection, and the ultrasonic wave is easy to diffract. The test results show that when the flat shape defect transmits and receives the acoustic path of the transducer transversely, ultrasonic detection is easy to find, on the contrary, the vertical defect may be missed.

4   Conclusions

In the field of assembling integral concrete building quality inspection technology, aiming at the grouting quality of the horizontal joint connected at the bottom of shear wall of prefabricated concrete building, the ultrasonic test research shows that:

(1) The transducer with high frequency and small diameter radiating end face is the technical means that must be taken for the detection of grouting material fullness. The transducer should have the technical characteristics of short wavelength and high flaw detection sensitivity.
(2) Ultrasonic detection of grout quality according to the "sound velocity, amplitude, waveform" and other acoustic parameters of the principle of discrimination, each ultrasonic measurement point

The wave train view can effectively judge the defects and degree of horizontal joints connected at the bottom of shear walls of prefabricated buildings.

(3) Ultrasonic test results can be more correct to determine whether there are defects, but the lack of quantitative description of defects, can not provide a more specific engineering quantity for the subsequent repair of defects.

References

1. Tong, S., Wang, Z., Shang, T.: Ultrasonic flat test technique of concrete strength. Non-Destr. Test. **1**, 24–27 (2004)
2. Shi, P., Gu, S., Zhang, J., et al.: Research on high frequency and small diameter ultrasonic detection method for grouting quality of bottom joint of prefabricated component. Eng. Qual. **37**(10), 29–35 (2019)
3. Khaled, A., Sami, H., LeBlanc, B.: Horizontal connections for precast concrete shear walls subjected to cyclic deformations part 1: mild steel connections. PCI J. **40**(3), 78–96 (1995)
4. Kim, Y.: A study of pipe splice sleeves for use in precast beam-column connections. The University of Texas at Austin (2000)
5. Zhong, X., Yang, X.: Prefabricated building cladding sealing quality detection technology research. Chinese Build. Waterproof **10**, 11–13 (2017). https://doi.org/10.15901/j.carolcarroll nki.1007-497-x.2017.10.004
6. Gao, R., Li, X., Tong, S., et al.: Research on detection technology of grouting hole at bottom joint of prefabricated shear wall based on improved ultrasonic method. Constr. Technol. **46**(17), 15–19 (2017)
7. Ozsoy, U., Koyunlu, G., Ugweje, O.C., et al.: Nondestructive testing of concrete using ultrasonic wave propagation. In: 13th International Conference on Electronics, Computer and Computation (ICECCO), pp. 1–5 (2017)

Research on the Installation Plan of Stiffening Beam and Bridge Deck for Single Tower and Single Span Steel Truss Suspension Bridge

Xianhui Man, Yingjia Wang$^{(\boxtimes)}$, and Yanan Yi

Chongqing College of Architecture and Technology, Mingde Road, Shapingba District, Chongqing, China
81814810@qq.com

Abstract. Taking a certain super large bridge with a main span of 256 m as the engineering background, the construction sequence of the stiffening beam and bridge deck of a single tower and single span steel truss suspension bridge was studied. The results show that different construction sequences of stiffening beams and bridge decks have a significant impact on the stress of stiffening beam members and suspension rods. Through simulation analysis, it is found that the stiffening beam of the bridge is lifted from the beam end to the B6 closure section, and the construction sequence of the bridge deck is symmetrical from the middle of the stiffening beam span to both ends. The steel truss stiffening beam members and suspension rods are the most reasonable in terms of stress, and the rationality of the installation construction plan is verified through actual measurement and monitoring data.

Keyword: Single tower and single span steel truss suspension bridge · Stiffening beam · Bridge deck slab · Construction sequence

1 Introduction

Suspension bridges have developed rapidly due to their advantages of simple structure, clear stress distribution, and reduced material consumption with larger spans. Suspension bridges fully utilize the advantages of high-strength steel wire tension on the main cable and suspension rods, and their span capacity is the largest among various bridge types, reaching over a kilometer [1].

At present, only some research has been conducted abroad on the construction sequence of stiffening beams and bridge decks for tower single span steel truss suspension bridges. There is little research on the impact of the lifting sequence of stiffened beam segments and the different paving methods of the bridge deck on the stiffened beam [2, 3].

The long span suspension bridge built in China has limited design and construction experience as well as computational research on the unique asymmetric single tower and single span steel truss suspension bridge type. This article mainly studies the construction sequence of this special bridge type stiffening beam and bridge deck [4]. By

© The Author(s) 2024
P. Xiang and L. Zuo (Eds.): PBSFTT 2023, LNCE 382, pp. 370–379, 2024.
https://doi.org/10.1007/978-981-97-5108-2_40

comparing and analyzing the calculation of different construction schemes, the vertical displacement of the main cable, the stress parameters of the steel truss stiffening beam members, suspension rods, and cable towers are controlled, and a reasonable and convenient construction sequence of the stiffening beam and bridge deck is obtained, providing reference for the construction of similar projects.

2 Engineering Background and Research on the Hoisting Sequence of Steel Truss Stiffening Beam Segments

A special bridge adopts a single tower and single span suspension bridge structure with a main span of 256 m, with a bridge deck width of 13.4 m. The main cable saddle body is cast entirely of cast steel, the cable tower is a portal tower with a reinforced concrete structure, and its crossbeam is a prestressed reinforced concrete structure with a rectangular cross-section. The bearing platforms of the two tower columns are designed as a whole. The longitudinal floating system is adopted between the tower beams at the main tower.

The overall layout of the bridge is shown in the following Fig. 1:

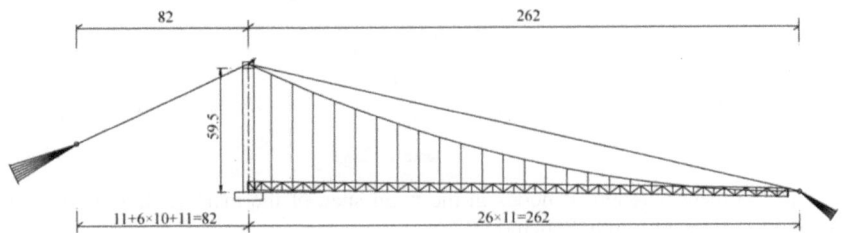

Fig. 1. Overall Layout (Unit: m)

Design speed: 40 km/h; Design load: Highway - Class II.

By using finite element software to establish a finite element model and analyzing the initial equilibrium state of the suspension bridge, the linear shape of the main cable and empty cable states, as well as the stress free length of the main cable and suspension rod, were determined.

These were used as important parameters for construction control. Simulation calculations were conducted on the super large bridge, and the optimal construction sequence for the installation of stiffening beams and bridge decks of a single tower and single span steel truss suspension bridge built in mountainous areas was analyzed [5].Compare and analyze the on-site measured data with the theoretical data calculated by the model, and summarize simple and effective construction methods.

The lifting of steel truss stiffening beams is very important for the construction monitoring of large-span suspension bridges, and the primary problem to be solved is the sequence of lifting steel truss stiffening beam segments [2, 6]. Therefore, when determining the lifting sequence of steel truss stiffened beam segments, the main consideration is its impact on the changes in the main cable shape. This article mainly verifies whether the lifting sequence is reasonable from the following two aspects: ① The absolute vertical deformation of the main cable at each step during the lifting of the stiffening beam segment is relatively minimum. ② During the lifting process, the vertical deformation of the main cable is small.

When determining the lifting sequence of stiffened beam segments, all segments should be freely suspended in a certain order [7]. Option 1: Starting from the support platform of the cable tower, lift the section of the stiffening beam and close it at the beam end. Starting from lifting section B1, to lifting section B26.Option 2: Lift the stiffened beam segment from the beam end until the B6 segment of the stiffened beam is closed.

The displacement changes of the nodes in the main span of the main cable during the lifting stage of each section in Scheme 1 are shown in Fig. 2. (Up is positive, down is negative).

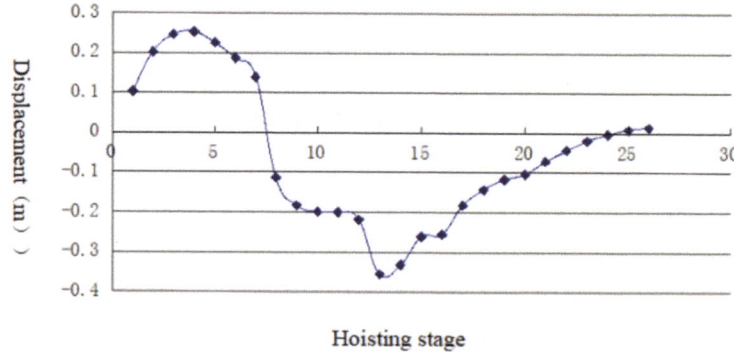

Fig. 2. Displacement diagram of nodes in the main span of the main cable during the lifting process of the stiffened beam in Scheme 1

The displacement changes of the nodes in the main span of the main cable during the lifting stage of each section in Scheme 2 are shown in Fig. 3. (Up is positive, down is negative).

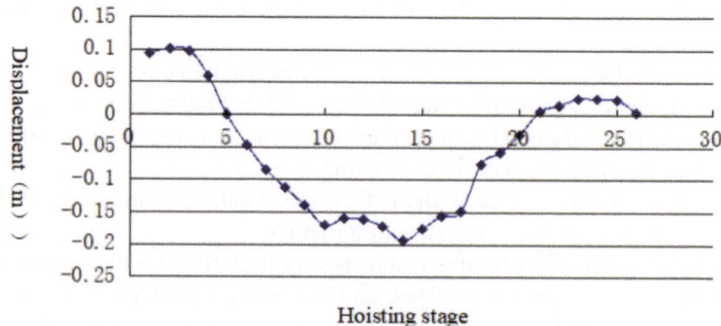

Fig. 3. Displacement diagram of the main cable mid span node during the lifting process of the stiffened beam in Scheme 2

Comparing the displacement change curves of the nodes in the main span of the main cable during the lifting process of the stiffened beam segment between two schemes, Scheme 1 shows that the change in elevation of the nodes in the main cable span during the lifting process of the stiffened beam segment is greater than Scheme 2. According to the principle that the absolute vertical deformation of the main cable in each step of lifting the stiffening beam segment is relatively minimum, and the vertical deformation of the main cable upwards is small during the lifting process, the specific calculation results of the two schemes are compared in Table 1.

Table 1. Comprehensive comparison of displacement changes of main cable nodes in the middle span during the installation of stiffened beam segments

levelScheme	Maximum absolute value of vertical deformation	Maximum vertical upward deformation
Scheme 1	0.3550 m	1.3530 m
Scheme 2	0.1923 m	0.3533 m

From the above table, it can be seen that the maximum absolute vertical displacement and the maximum vertical upward deformation of the main cable at the mid span node of each step of lifting the stiffening beam section in Scheme 1 are greater than those in Scheme 2, namely 0.3550 m > 0.1923 m, 1.3530 m > 0.3533 m. So in Scheme 2, the deformation of the main cable is relatively gentle during the lifting of each section of the stiffening beam. Finally, Option 2 was chosen as the sequence for lifting the steel truss stiffening beam segments.

3 Research on the Sequence of Bridge Deck Pavement

According to calculation and analysis, different bridge deck paving sequences have a greater impact on the stress of stiffening beam members and suspension rods, while the impact on the change of the main cable shape is relatively small. Therefore, this article focuses on the rationality of the stress distribution of the stiffening beam members and suspension rods during the entire construction control process when determining the paving sequence of the bridge deck [8]. The specific division of the bridge deck corresponds to the segment division of the steel truss stiffening beam. Combining the simplicity and convenience of construction, after screening, three paving sequence schemes for the bridge deck were obtained [9].

After selecting three options, calculate the stress changes of the stiffening beam members and suspension rods for each option. The stress changes of the stiffening beam members and suspension rods in the PD1 scheme are shown in Fig. 4, Fig. 5, and Table 2.

According to the calculation, the maximum tensile stress of the stiffened beam member is 288.0 MPa > 200.0 MPa, the maximum compressive stress is 282.2 MPa > 200.0 MPa, and the maximum tensile stress of the suspension rod is 319.5 MPa < 501.0 MPa.

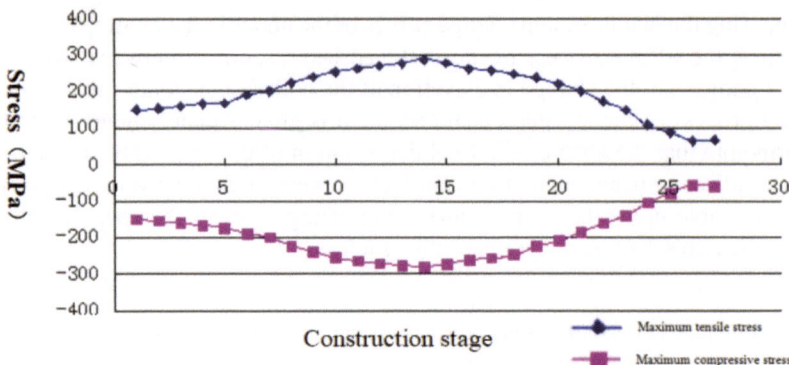

Fig. 4. Maximum Stress Envelope Diagram of Stiffening Beam Member in Scheme 1

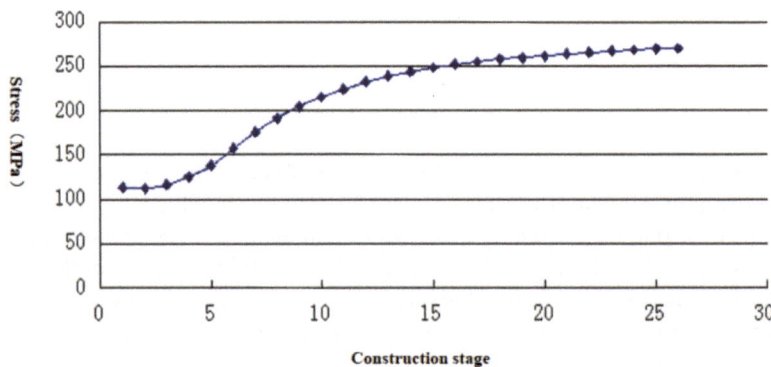

Fig. 5. Maximum stress diagram of suspension rod in Scheme 1

Table 2. Maximum stress values of main components in Scheme 1

Member	Maximum tensile stress (MPa)	Maximum compressive stress (MPa)
Stiffening beam members	288.0	−282.2
Boom	319.5	/

The stress changes of the stiffening beam members and suspension rods in Scheme 2 are shown in Fig. 6, Fig. 7, and Table 3.

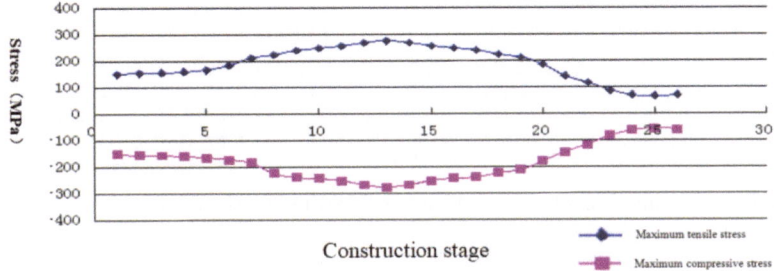

Fig. 6. Maximum Stress Envelope Diagram of Stiffening Beam Member in Scheme 2

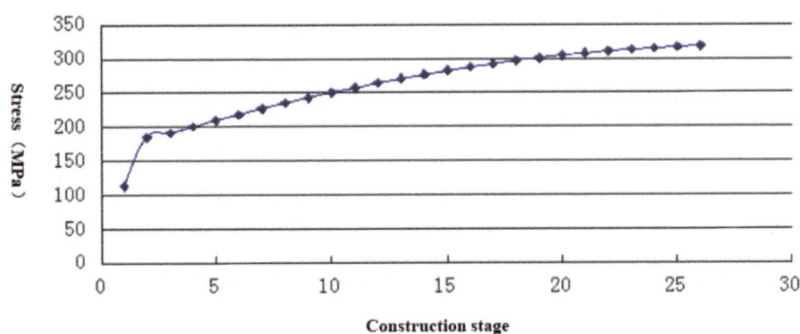

Fig. 7. Maximum stress diagram of suspension rod in Scheme 2

Table 3. Maximum stress values of main components in Scheme 2

Member	Maximum tensile stress (MPa)	Maximum compressive stress (MPa)
Stiffening beam members	278.0	−274.1
Boom	318.0	/

According to the calculation, the maximum tensile stress of the stiffened beam member is 278.0 MPa > 200.0 MPa, the maximum compressive stress is 274.1 MPa > 200.0 MPa, and the maximum tensile stress of the suspension rod is 318.0 MPa < 501.0 MPa.

The stress changes of the stiffening beam members and suspension rods in Scheme 3 are shown in Fig. 8, Fig. 9, and Table 4.

Comparing the stress changes of stiffening beam members and suspension rods in the process of bridge deck pavement with three schemes, it can be seen that different bridge

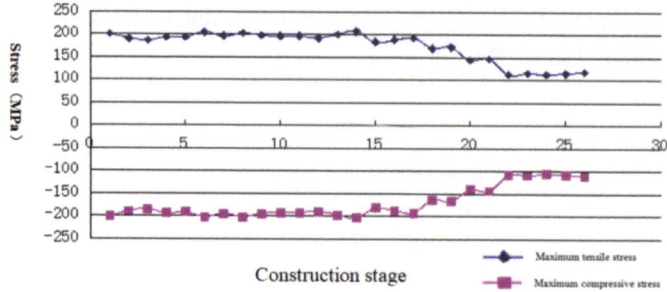

Fig. 8. Maximum Stress Envelope Diagram of Stiffening Beam Member in Scheme 3

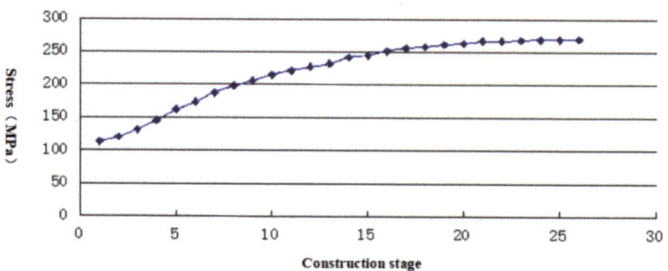

Fig. 9. Maximum stress diagram of suspension rod in Scheme 3

Table 4. Maximum stress values of main components in Scheme 3

Member	Maximum tensile stress (MPa)	Maximum compressive stress (MPa)
Stiffening beam members	207.0	−204.0
Boom	265.4	/

deck pavement sequences have a significant impact on the stress of suspension rods and stiffening beams; In Scheme 1, Scheme 2, and Scheme 3, the stress of the stiffening beam member is relatively high and there is an exceeding limit phenomenon, and the stress of the suspension rod meets the requirements of the specification; However, the relationship between the maximum stress of the stiffened beam members in three different paving schemes is: Scheme 3 < Scheme 2 < Scheme 1, which is 207.0 MPa < 278.1 MPa < 282.2 MPa, and the stress of the stiffened beam members significantly increases. The stress situation of the stiffened beam in Scheme 3 is the most reasonable, so Scheme 3 is determined as the final paving sequence for the bridge deck. During the process of bridge deck pavement, the stress of the stiffening beam members exceeds the limit. Part of the factors is that the paving sequence of the bridge deck is different, and another important reason is that the connection method between the stiffening beam segments adopts rigid connection. Therefore, after determining the sequence of symmetrical paving of the bridge deck from the middle of the stiffened beam span to both ends, the impact of different connection methods between segments during the lifting process of the stiffened beam on the internal force of the structure should be analyzed.

4 Verification of Measured Data

During bridge construction, three test sections are arranged on the steel truss stiffening beam, namely the 1/4, 1/2, and 3/4 cross sections of the steel truss stiffening beam. Four test points are arranged on each test section, and specific sensors are installed on the upper and lower chords of the test section [10]. According to the measured data of the steel truss stiffened beam during construction monitoring, the maximum tensile stress of the chord in the mid span test section is about 118.2 MPa, and the maximum compressive stress is about 124.6 MPa, which is smaller than the theoretical calculation value; During the entire construction monitoring process, the measured stress of the 1/4 and 3/4 span test sections of the steel truss stiffening beam is within 70 MPa, so the construction process is safe and reliable. The comparison between the measured stress and theoretical stress of the 1/2 mid span section of the steel truss stiffening beam during construction is shown in Table 5.

Table 5. 1/2 span mid section steel truss beam chord measured stress (MPa)

Location of measuring points		Measured stress value	Theoretical stress value	Allowable stress value
Mid span section	Top chord tensile stress	118.2	153.0	200.0
	Lower chord compressive stress	124.6	159.6	200.0

From the above table, it can be seen that after the completion of bridge deck pavement, the measured stress value of the chord of the steel truss stiffening beam is less than the theoretical calculation value, which meets the requirements of the specifications and ensures structural safety during the construction process. During the construction process, the measured maximum tensile stress of the suspension rod was 303.21 MPa, which is less than the theoretical calculation value of 310.2 MPa and even less than its allowable stress value of 501.0 MPa. Therefore, the stress of each component in each construction stage meets the requirements of the specifications.

5 Conclusion

This article studies the reasonable construction sequence of the stiffening beam and bridge deck of a single tower and single span steel truss suspension bridge through simulation calculation of a finite element model. When lifting the stiffening beam, the lifting sequence from the beam end to the closure section of the cable tower is determined based on the principle of the relative minimum vertical deformation absolute value of the main cable and the minimum vertical upward deformation of the main cable during each section lifting. When laying the bridge deck, based on the principle of the most reasonable stress situation of the steel truss stiffening beam members and suspension rods, a plan was determined to symmetrically pave from the middle of the steel truss stiffening beam span to both ends. By comparing the measured data and theoretical calculation values during the construction monitoring process, the maximum tensile stress of the upper chord is 153.0 MPa, and the maximum compressive stress of the lower chord is 159.6 MPa, both of which are less than the allowable stress of 200.0 MPa; The measured maximum tensile stress of the suspension rod is 303.21 MPa, which is less than its allowable stress value of 501.0 MPa. Verified that the determined lifting sequence plan for the stiffening beam and bridge deck is reasonable.

References

1. Mengbo, Z.: Suspension Bridge Handbook [M]. People's Communications Press, Beijing (2003)
2. Matsuzaki, M., Uchikawa, C., Mlitamura, T.: Advanced fabrication and erection techniques for long suspension bridge cables. J. Constr. Eng. Manag. **116**(1), 112–129 (2016)
3. Adanur, S., Gunaydin, M., Altunisik, A.C.: Construction stage analysis of Humber suspension bridge. Appl. Math. Model. **36**(11), 5492–5505 (2012)
4. Liu, G., Peng, Y.: Research on erection methods of steel stiffening truss girder for baling river bridge. J. Highway Transp. Res. Dev. (English Edn.) **3**, 50–56 (2010)
5. Wang, Z.: Key Control Techniques for Construction of Long Span Steel Truss Bridge. Chang'an University (2018)
6. JTG-T D65–05–2015. Design Specification for Highway Suspension Bridges [S]. People's Communications Press (2015)
7. Helwig, T.A., Fan, Z.: Behavior of steel box girders with top flange bracing. J. Struct. Eng. **125**(8), 829–837 (2014)
8. Da, W., Wei, W., Lei, W.: Control analysis of excessive jacking construction of suspension bridge main cable saddle. Transp. Sci. Eng. **3**(01), 32–37 (2019)
9. Luo, X.: Analysis of effect of fetching girder from trestle of long-span steel truss bridge under cantilever erection. J. Appl. Mech. Mater. **587**, 1522 (2014)
10. Tianxiang, J.: Analysis of key technologies for construction of long span steel truss suspension bridge. Heilongjiang Transp. Technol. **42**(03), 103–104 (2019)

Research on Design Method of Prefabricated Concrete Structure Based on BIM Technology

Yunlai Zhang[✉] and Xiajuan Shi

Chongqing College of Architecture and Technology, Chongqing 401331, China
349816825@qq.com

Abstract. At present, in the construction of engineering, prefabricated buildings are increasing, this kind of building has a high degree of standardization, fast construction speed and low cost, and has become an important part of the development of modern architecture. In prefabricated buildings, the design of concrete structures has a direct impact on building quality and cost. In this process, through the effective application of BIM technology, it can promote the optimization of concrete structure design, provide reliable guidance for the overall construction design, and promote the continuous improvement of architectural design benefits. This paper analyzes the basic characteristics of prefabricated engineering, analyzes the advantages of using BIM technology in the design of prefabricated concrete structure, and explores the design method of prefabricated concrete structure based on BIM technology. Finally, the application prospect of BIM technology in prefabricated engineering explains the application value of BIM technology in its concrete structure design.

Keywords: BIM technology · prefabricate · assembled type · concrete · structural design

1 Introduction

Prefabricated building is one of the important forms of development in the current construction industry [1]. The pre-fabricated concrete structure design of such buildings determines the efficiency and quality of engineering construction. It is necessary to ensure that the design of each concrete structure is reasonable, can meet the needs of standardized construction, and prevent the assembly problems caused by design errors, which is the key to ensure the construction period is shortened and cost control benefits [2]. At present, the application of BIM technology in the field of construction is also increasing. It is necessary to study the design method of prefabricated concrete structure based on BIM technology for the optimization of construction and the improvement of quality and efficiency of construction projects.

P. Xiang and L. Zuo (Eds.): PBSFTT 2023, LNCE 382, pp. 380–385, 2024.
https://doi.org/10.1007/978-981-97-5108-2_41

2   Basic Characteristics of Prefabricated Engineering

First, design standardization. In order to ensure the "seamless" assembly construction results of prefabricated buildings, each prefabricated component needs to go through a standard design inspection. The standard design harmonizes the production standards and specifications of precast components to ensure that the design errors of concrete structures are within the acceptable range [3]. Such prefabricated parts can be effectively connected during the assembly process to ensure the overall structural stability and quality reliability of the building.

Second, industrialization of prefabricated parts production. In prefabricated buildings, more precast concrete components are needed to meet the needs of a large number of prefabricated engineering buildings, so the industrialization process must be accelerated. In this case, production activities include prefabricated components and the main structure of the building.

Third, professional construction and assembly. The construction of prefabricated buildings does not require a lot of construction personnel, but must ensure its professionalism. The installation personnel must have corresponding qualifications during the installation process to ensure the professionalism of the construction personnel. In addition, high-performance machinery is required during the assembly process to ensure a more precise construction method.

Fourth, building structure integration. The structural integration of prefabricated buildings is also one of its important characteristics, which helps to improve the stability and coordination of the building structure and ensure the integrity and compressive strength of the structure [4].

3   Advantages of Using BIM Technology in Prefabricated Concrete Structure Design

In the construction industry, BIM technology can also play an important role in ensuring project progress, improving project quality, and preventing corruption in project implementation [5]. BIM technology is a method of using information technology and digital models to design, build and manage projects. The industry believes that BIM technology is the second revolution in the construction industry after CAD. BIM is an interdisciplinary discipline, which is the product of the close combination of civil engineering and information technology disciplines [6]. BIM technology can be applied to a variety of civil and road construction projects, bridge construction, waterway construction and other projects. BIM technology is the future development direction of building information technology with its characteristics of large data set, modeling and visualization. The field of civil engineering is a field prone to corruption. The implementation of BIM can simulate the construction process in advance, reduce design defects, reduce rework, effectively improve project quality, and effectively prevent corruption. Through accurate calculation and design of BIM, the cost can be reduced and controlled within a given range [7]. BIM technology has the advantages of "full business coverage, complete process documentation, and traceable results", and it is feasible to apply it to the construction of prefabricated projects.

At present, the application of BIM technology in building construction continues to develop, and its positive role is becoming more and more important. The use of BIM technology in assembly engineering can effectively reduce the time and technical costs of engineers. In prefabricated buildings, using BIM technology for project design can intuitively carry out work and project management, and provide accurate information. With the powerful BIM modeling function, the construction process of prefabricated parts can be simulated, problems that may occur in the construction process can be solved, unnecessary errors and collisions can be avoided, and the construction scheme can be improved, which is a necessary effort to reduce costs. Using BIM technology to design the assembly engineering frame can accurately calculate the project cost and accurately calculate the traditional manual drawing mode, improve the drawing efficiency and save the design time and cost [8]. In addition, the use of BIM technology in prefabricated buildings can calculate and analyze the force, deflection, hanging pull and other information of the building, and can also assist construction planning to avoid unnecessary safety problems. Through effective technical analysis, it can also accurately design the structural design of precast concrete components, and constantly reduce the error of precast parts, which is necessary to improve production efficiency and reduce costs [9].

4   Design Method of Prefabricated Concrete Structure Based on BIM Technology

4.1   Application in Concrete Structure Design

As shown in Fig. 1, it is the concrete deepening design drawing of an assembled engineering project, which aims to use BIM technology to deepen the design of civil drawings, mainly to complete the deepening design of the headroom, beam, plate and column,

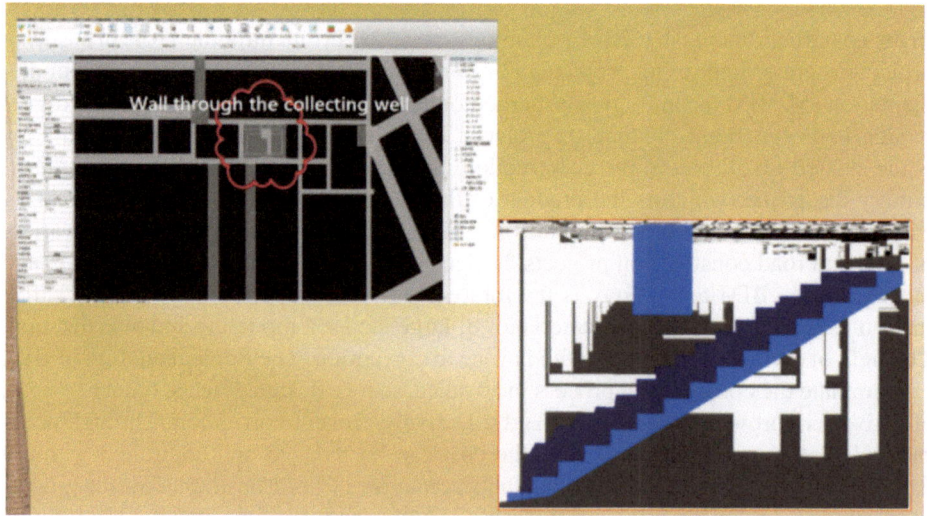

Fig. 1. Design drawing of concrete structure

stairs and water collection Wells in engineering construction, and to give scientific and reasonable guidance to the effective construction of the site construction team.

In addition, BIM technology can also be applied in the steel bar layout design of the project to deepen the design diagram for the steel structure of the project. Building project design drawing based on BIM technology according to the complex and changeable structural characteristics of steel frame column, the use of BIM technology for secondary deepening design, the realization of steel column joints, steel beam joints and other complex parts of the design, processing, installation and optimization problems [10]. The application of BIM technology in other aspects of concrete structure design is shown in Fig. 2 below:

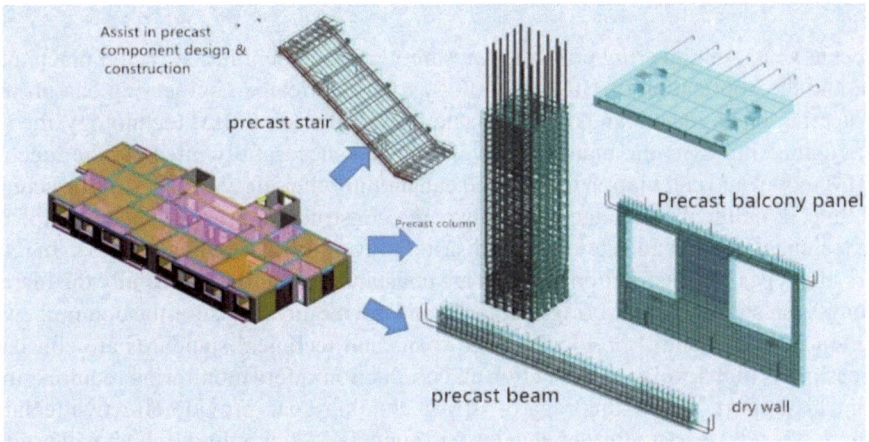

Fig. 2. Application of concrete aided precast design

It can be seen that with the support of BIM technology, many precast concrete parts of prefabricated buildings can be accurately designed and produced. Therefore, BIM technology has important application value for prefabricated engineering.

4.2 Application in Lightweight Model

The use of lightweight BIM model in prefabricated building projects can realize the plane layout and site division of the construction site, and build various temporary facilities by using BIM technology and according to the Construction Plan, so as to display the safe and civilized construction status and CI planning of the site layout more intuitively and clearly [11].

In the application of BIM model, mobile devices are used to visualize the BIM optical model in the field. In addition, the design department adopts 4D construction cycle software for schedule management, which is conducive to the elaboration of construction plans and significantly improves the execution of construction plans. The overall graphic scheme is combined with the BIM model to create a 4D construction cycle, which can fully simulate the entire prefabricated construction process, and accurately control each foundation and each design, so that the construction organization does not have collisions and dead corners [12]. The 3D laser scanner used in the construction of prefabricated projects together with BIM technology can focus on the following tasks: (1) construction quality control; ② Steel structure pre-assembly; (3) Structural deformation monitoring; ④ Professional reverse curtain wall construction, small decoration, etc. With the progress of the project, the completion model of LOD500 will be gradually generated to manage the operation and maintenance of the property, and the intelligent management of the building will be realized by using BIM technology.

5 Application Prospect of BIM Technology in Prefabricated Engineering

In recent years, engineering construction safety management has attracted much attention, and the application of BIM technology to prefabricated engineering can promote engineering management and design to a new level. Based on BIM technology, the relevant engineering dynamic management system can successfully integrate the functions of BIM model and calculation model, and can monitor the safety status of the structure in real time, dynamically manage and analyze the construction progress, and provide reliable technical support and services for disaster prevention and reduction work. In recent years, the types of construction projects are constantly changing, especially the increase of long-span steel structure bridge projects, and the requirements for the comprehensive solution of large control problems, large spans and technical standards are constantly increasing. A high level of construction and production safety monitoring requires strong technical support, and in this regard, BIM technology can provide effective technical support. It is necessary to further develop and apply BIM technology, which will promote the optimization and development of prefabricated building projects.

6 Summary

There are some differences between prefabricated projects and traditional construction projects, which require a large number of prefabricated parts for assembly and construction. In this regard, BIM technology is needed to ensure the structural optimization design of concrete prefabricated parts, ensure the reliable production accuracy of prefabricated parts, and improve the construction quality of prefabricated construction projects.

References

1. Jiaxing, H., Shilong, W.: Teaching practice of prefabricated building construction course based on BIM and CAD technology. Build. Technol. Dev. **50**(8), 15–18 (2019)
2. Fan Wenzhang, X., Jingjing, C.L.: Research on integrated design calculation of prefabricated high pile wharf based on BIM technology. Water Transp. Eng. **5**, 172–177 (2023)
3. Yongmin, L., Han, W., Desxin, Z., et al.: Research on whole life cycle information management platform of prefabricated buildings based on BIM. Build. Econ. **44**(1), 77–83 (2023)

4. Zhang, A., Lan, T., Li, X., et al.: Research on fine installation management of all-dry connected prefabricated floor based on BIM+ 3D laser scanning technology. Steel Struct. **38**(3), 43–50 (2023)
5. Wang Guolin, W., Yunlong, G.L., et al.: Application research of BIM Technology in the design practice of a prefabricated energy-saving building. J. Bonding **50**(2), 180–183 (2019)
6. Wei Bowen, H., Xiaofei, M.Z., et al.: Application practice of BIM technology in the demonstration stage of super high-rise prefabricated buildings: a case study of Nanyobina commercial land project. Build. Econ. **44**(z1), 297–300 (2019)
7. Honge, Y.: Integrated design and research of built-up interior decoration based on BIM technology. China Ceram. **3**, 123–126 (2023)
8. Xingchen, R.: Application of prefabricated BIM technology in building life cycle. J. Railway Eng. **39**(6), 90–94 (2022)
9. Yuan, C., Yang, H., Chang, L., et al.: Research on construction of steel structure prefabricated industrial plant based on BIM technology. Concr. World **1**, 38–45 (2023)
10. Yan, D., Zhang, J., Yang, L., et al.: Application practice of BIM technology in the construction process of prefabricated building engineering. Sichuan Build. Mater. **49**(7), 118–119 (2023). (in Chinese)
11. Guotai, C.A.I.: Application of BIM technology in the construction process of prefabricated steel structure building. Brick World **2**, 70–72 (2023)
12. Min, Q.: Research on the application of BIM technology in the preliminary design stage of prefabricated buildings. China Ceram. **1**, 143–145 (2023)

Contribution of Green Roof in Urban Energy Saving Project

Bingduo Qin, Weina Zou$^{(\boxtimes)}$, Nan Jiang, and Yi Lu

School of Ecological Technology and Engineering, Shanghai Institute of Technology, No. 100, Haiquan Road, Fengxian District, Shanghai, People's Republic of China
Zouwn@sit.edu.cn

Abstract. In this paper, through the simulation of whether the roof greening is arranged on the top of buildings when different weather events occur in summer, it is found that roof greening can produce a local cooling effect of 0.8 °C–1.5 °C and an overall cooling effect of 0.6 °C–1.3 °C in summer, which has a good positive impact on urban energy supply and stability. Through energy consumption simulation, it is found that the arrangement of green roof can produce 3 W/m^2/h cooling effect on the room, and the entire cooling season can reduce the simulated building energy consumption of 21772.8 KWH. It provides relevant data for the restoration of urban ecological environment and the follow-up study of urban energy conservation.

Keywords: Green roof · Energy-saving technology · Building energy efficiency · municipal environment · Ecological benefit

1 Introduction

Today, more than half of humanity lives in urban environments, and about 80 percent of energy consumption occurs within city boundaries, a figure that is rising with the development of the global economy. With the continuous growth of urban energy demand, the environmental problems caused by urban energy supply and energy consumption become more and more serious. With the continuous intensification of climate change, its consequences have brought impacts on human beings, especially the production and life of urban residents, and also put forward higher requirements for urban energy supply and maintenance system. Facing the urban energy dilemma, scholars have carried out corresponding research from different fields. Johari Fatemeh [1] designed and validated a more accurate urban building energy model by improving data quality); At the level of smart power grid, Zhang et al. [2] and Souabi Sonia [3] respectively studied the intelligent scheduling mode of urban EV charging stations and home photovoltaic power generation through LSTM algorithm. Vakili Seyedvahid [4] proposed a hybrid and coding design model applied to smart grid wireless communication to improve the capacity and security of smart grid communication systems. At present, the research on urban energy saving effect focuses more on power, resource allocation model and policy research, and less on improving the capacity and stability of urban energy system

P. Xiang and L. Zuo (Eds.): PBSFTT 2023, LNCE 382, pp. 386–399, 2024.
https://doi.org/10.1007/978-981-97-5108-2_42

through building renovation represented by roof greening. By planting plants on the roof of the building and changing the thermodynamic properties of the underlying surface under the urban microclimate, green roof can significantly improve the urban green area, offset the ecological environment problems in the city caused by the separation of man and nature, and effectively improve the redundancy of energy supply and the stability of the energy system in the city. To a certain extent, the impact of the deterioration of the ecological environment and the change of climate conditions on the city will be weakened [5]. The effect of building energy saving and urban climate improvement brought by green roofs is significant. The research results of Mickovski et al. [6], Bates et al. [7] Kazemi et al. [8] show that the use of urban waste in green roof matrix can reduce the overall carbon emission of the city through the use of construction waste. It can also support green roof plants to improve urban microclimates. In terms of climate regulation, Andrew [9], Wong [10], Elnabawi [11] and Pragati [12] respectively explored the positive significance of green roofs for urban microclimate regulation from field records, model analysis, theoretical research and energy consumption simulation. The degree of adjustment is affected by the form of green roof, green roof plants, urban geographical location and so on.

Under the premise that extreme weather events in the world are becoming more and more serious and frequent, it is an effective means to improve the degree of climate change to improve the stability of urban energy system and reduce the overall urban energy consumption through multidisciplinary joint efforts. However, there are few studies on the effect of green roofs using construction waste substrate on the regulation of urban ecosystem and the contribution level of urban energy conservation under different summer weather events. This experiment simulates the contribution degree of green roofs using construction waste substrate to urban climate and building energy efficiency under the background of extreme meteorological events, in order to provide theoretical basis for the comprehensive development of urban energy conservation technology.

2 Simulation Result Analysis

Envi-met is a commonly used medium and small-scale climate simulation software, which can accurately simulate the influence of architecture, terrain, vegetation and other factors on medium and small-scale climate, and can generate intuitive and accurate simulation maps. Envi-met is one of the preferred software for medium and small-scale climate simulation in the world. In this paper, Envi-met 5.1.1 was used to conduct a simplified modeling of the experimental building and its surrounding environment to analyse the difference in microclimate effects of green roofs under extreme and typical summer weather conditions.

In order to provide more accurate experimental data for the simulation experiment, a long-term green roof temperature monitoring experiment was carried out in Henan Province, China from June to September 2022. The volume ratio of each component of the experimental substrate is peat soil: crushed brick ($\leq$3 mm) = 1:1. The green roof type is mixed type, The vegetation was Osmanthus fragrans (20%), Camellia sasanqua (30%) and Sedum lineare (50%). A small weather station with FT BQX9 was set up in the central open space of the test plot to record meteorological information. The sampling interval was 300 s, and the GIS500 handheld temperature sensor was used to record the temperature of the exposed roof in the experiment site every 15 min as the control group. In order to fully study the energy-saving effect of green roofs in different summer weather events, three typical weather days are set, namely, typical summer day: August 2, 2022, and extreme high temperature day: August 12, 2022. The Person correlation analysis shows that the correlation of meteorological indicators on each meteorological day is $\geq$ 95%. The meteorological data that passed the correlation test were taken as model data to participate in the simulation experiment, and all the meteorological indicators were shown in Table 1. In order to ensure the accuracy of the model simulation results, the Settings of the building and underlying surface in this experiment are the same as the actual situation. All the indicators set are shown in Table 2, and the default values of the system are adopted for the indicators not mentioned.

2.1 Comparative Analysis of Simulated Measurement

Figure 1 shows the measured and simulated temperatures with and without green roofs for each weather event. On the whole, the simulated temperature and the real temperature show the same daily variation law, the measured temperature in the period with solar radiation is slightly higher than the simulated temperature, while the measured temperature in the period without solar radiation is slightly lower than the simulated temperature.

Table 1. Main meteorological data for each meteorological day

	Hot sunny day	Typical summer
Mean air temperature	34.3 °C	31.1 °C
Maximum air temperature	40.6 °C	35 °C
amount of precipitation	-	-
Sunshine duration	14 h	14 h
Total daily solar radiation	19402368 J/m^2	18553344 J/m^2

Figure 2 shows the measured and simulated wind speeds with and without green roofs for each weather event. In the simulation of wind speed, because the model itself has blurred the contingency of some environmental impacts, compared with the complex urban environment, the simulation of wind environment is more difficult. In the process of measurement, wind speed is affected by a variety of external conditions, and its instantaneous change is large and the contingency is large, so ENVI-met cannot simulate the time-by-time dynamic wind speed. Therefore, the simulation of wind speed is not enough to reflect the chance, the wind speed reflects the characteristics of a more gentle, can only reflect the overall change. The difference between the measured wind speed and simulated wind speed on green roof under four simulated conditions is 0.03, 0.07, 0.02 and 0.04 m/s, respectively, which can indicate that the wind speed simulation can reflect the measured situation well to a certain extent.

Table 2. Building setting indicators and values

	parameter			parameter	
	Name	Setting		Name	Setting
Thermal performance parameters W/mk	Roof	–	Green roof parameter	Plant height	0.3 m
	Outer wall	0.9		LAI	3
	Interior door	3		Blade emissivity	0.9
	Entry door	2		Matrix thickness	0.5m
	exterior window	3.2		Matrix dry bulk density	1000 kg/m^3
	Interior wall	0.9		Surface roughness	Rough

2.2 Accuracy Analysis of Simulation Results

In order to evaluate the accuracy of microclimate simulation intuitively and scientifically, Root mean square error (RMSE) and Mean absolute percentage error (Mean absolute percentage error) are introduced here. As a quantitative evaluation index for the accuracy of green roof meteorological simulation, MAPE is calculated as follows.

$$RSME = \sqrt{\frac{1}{n}\sum_{i=1}^{n}\left(y_i' - y_i\right)} \tag{1}$$

$$MAPE = \frac{1}{n}\sum_{i=1}^{n}\frac{\left|y_i' - y_i\right|}{y_i'} \times 100\% \tag{2}$$

In the formula, y_i' is the simulated value, y_i is the corresponding point value measured by experiment, and n is the measured temperature count. According to the studies of Willmott [13] and Salata et al. [14], when RSME value is 0.52 °C–4.30 °C and MAPE

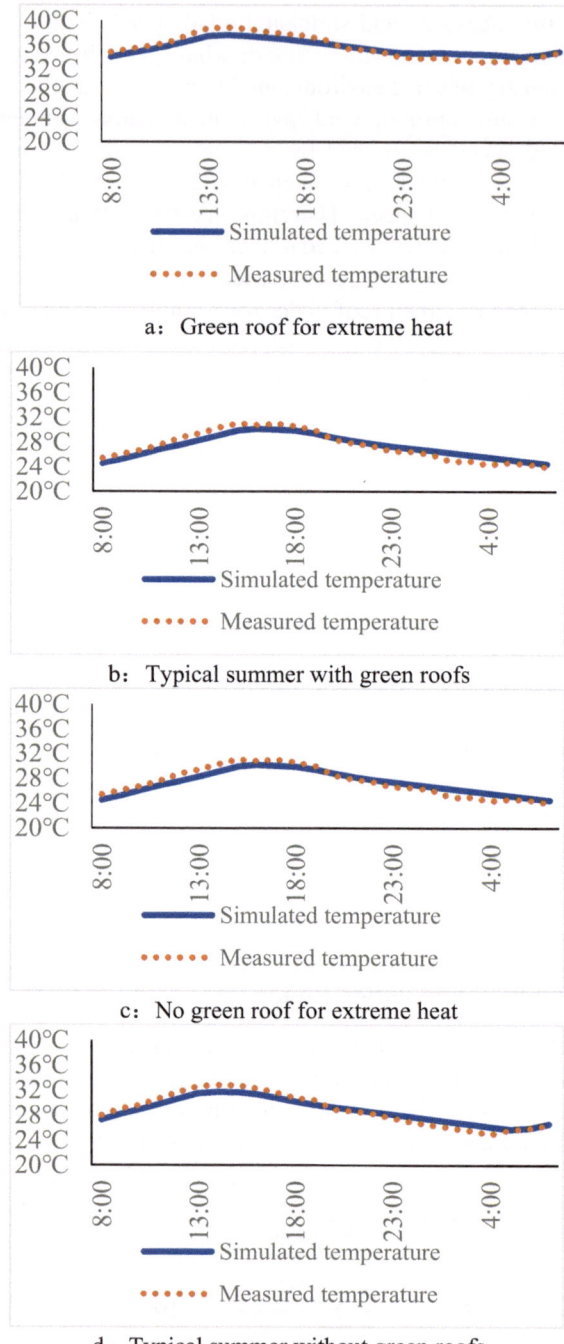

a： Green roof for extreme heat

b： Typical summer with green roofs

c： No green roof for extreme heat

d： Typical summer without green roofs

Fig. 1. Measured and simulated temperature values of different weather events

value is less than 10%, the accuracy of the model is relatively ideal. The MAPE and RSME values of the four simulated weather events in this paper are shown in Table 3. All calculated values meet the accuracy requirements.

Table 3. MAPE and RSME values for simulated weather events

	Index	Value
Extreme heat with green roofs	RSME (Air temperature)	1.63 °C
	MAPE (Wind speed)	3.93%
Typical summer with green roofs	RSME (Air temperature)	1.74 °C
	MAPE (Wind speed)	4.52%
Extreme heat without green roofs	RSME (Air temperature)	2.65 °C
	MAPE (Wind speed)	7.14%
Typical summer without green roofs	RSME (Air temperature)	2.23 °C
	MAPE (Wind speed)	4.32%

3   Simulated Temperature Field Analysis

This section sets four kinds of urban microclimate temperature simulation scenarios for whether there is roof greening when two different weather events occur. The type of roof greening set is composite roof greening, and the surface vegetation is set according to the actual situation. Figure 3–Fig. 4 shows the temperature field distribution of two time nodes at 2:00 and 13:00 respectively under four simulated conditions.

Figure 3 shows the temperature distribution of the green roof at 1.4m altitude during different weather events in the late night. This simulation result can more intuitively and clearly represent the impact of the green roof on the urban microclimate. In the late night, regardless of whether there is a green roof in the building under extreme weather conditions, there are high temperature zones in the building impact zone in the north-south corridor between the buildings. Roof greening is difficult to eliminate the high temperature zone between buildings, but it can significantly reduce the temperature difference between the high temperature zone and the outer area in different weather events. The layout of roof greening in extreme high temperature weather events and typical summer has a cooling range of about 0.8 °C and 1.2 °C, respectively. At the same time, it should be noted that in the case of roof greening, the overall temperature of the environment when extreme high temperature weather events occur and when typical summer meteorological events occur is lower than that in the case of no green roof at the same time, 0.6 °C and 1.0 °C, respectively, and the overall temperature distribution is more average.

Figure 4 shows the temperature distribution at 1.4m altitude of the green roof at noon when different weather events occur. This simulation result shows that the urban microclimate changes completely opposite to that at night when different weather events

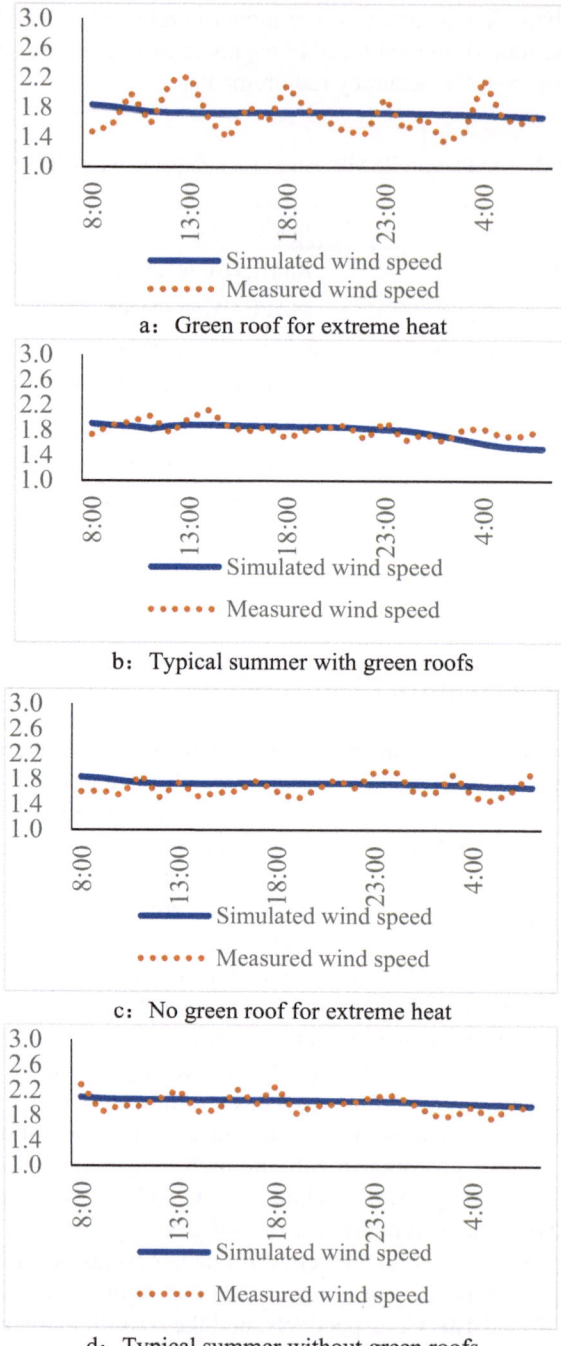

a: Green roof for extreme heat

b: Typical summer with green roofs

c: No green roof for extreme heat

d: Typical summer without green roofs

Fig. 2. Measured and simulated wind speed of different weather events

occur. At noon, no matter whether there is a green roof in the building under extreme weather conditions, there will be a relatively significant low temperature zone on the north, south and east sides, among which the low temperature area on the west side of the building is the largest. Roof greening can significantly improve the range and temperature difference of low temperature zones between buildings, and the arrangement of roof greening in extreme high temperature weather events and typical summer has a cooling range of about 0.9 °C and 1.5 °C, respectively. At the same time, it should be noted that in the case of roof greening, the overall ambient temperature at noon when extreme high temperature weather events occur and when typical summer meteorological events occur is lower than 1.1 °C and 1.3 °C respectively in the same period without green roofs, and the overall temperature distribution is not very average. It should be noted that when typical summer weather events occur, no matter whether roof greening is arranged or not, there will be a significant high temperature range on the south side of the building, and this high temperature area is the highest temperature value in the simulated environment at this time. The temperature in this high temperature area with

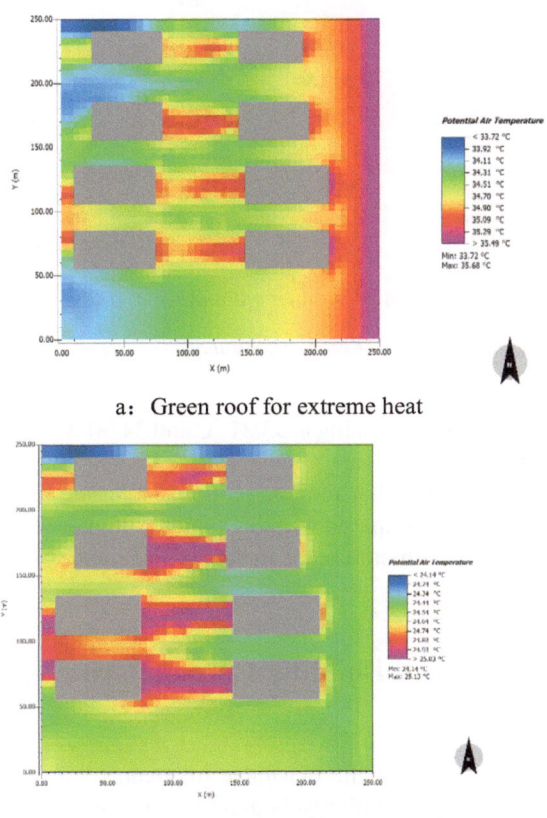

a: Green roof for extreme heat

b: Typical summer with green roofs

Fig. 3. Temperature distribution of 1.4 m at night with four simulated results

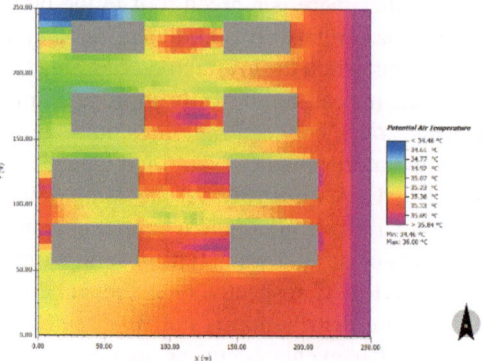

c： No green roof for extreme heat

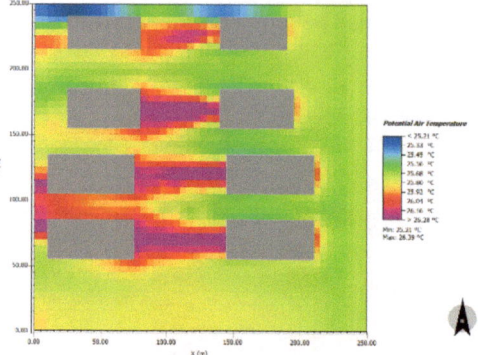

d： Typical summer without green roofs

Fig. 3. (*continued*)

roof greening and without roof greening is 33.21 °C and 34.90 °C, respectively. However, this high temperature region does not appear when extreme hot weather events occur. The reason for this situation is that the temperature increased by 5 °C when extreme hot weather events occurred compared with the typical summer weather events, and the overall increase of ambient temperature led to the disappearance of this high temperature region.

4   Evaluation of Overall Building Energy Saving Effect Caused by Roof Greening

In order to clarify the impact of the existence of green roofs on the overall energy consumption of buildings and quantitatively demonstrate the contribution of green roofs to the energy conservation of urban energy consumption systems, this paper uses energy plus software to conduct a comparative analysis of the energy consumption of building HVAC systems. The model setting and parameter design of EnergyPlus software are

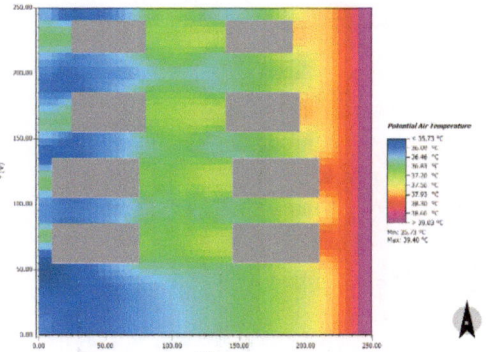

a：Green roof for extreme heat

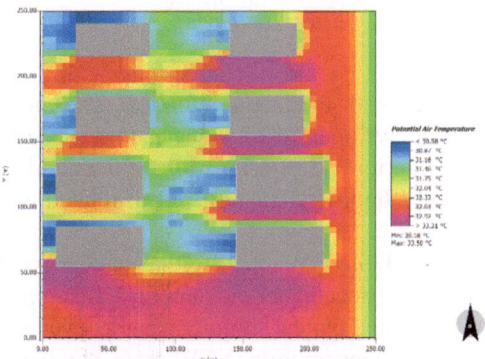

b：Typical summer with green roofs

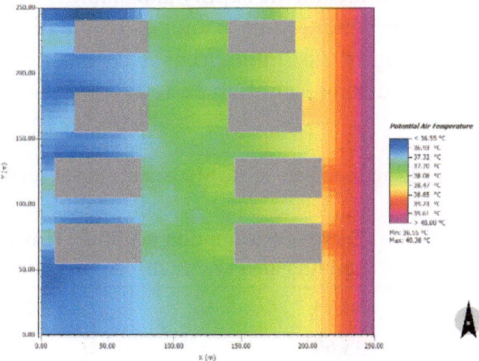

c：No green roof for extreme heat

Fig. 4. Temperature distribution of 1.4 m at noon with four simulated results

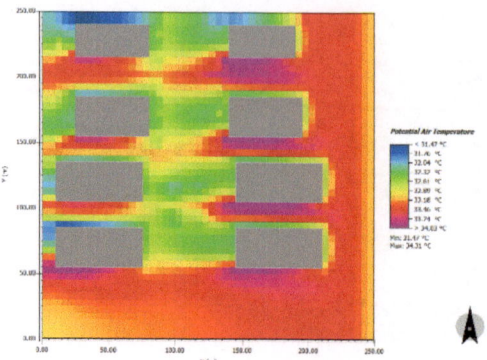

d: Typical summer without green roofs

Fig. 4. (*continued*)

arranged strictly in accordance with the actual situation. The model has seven floors on the ground, with four Windows open on all sides and no shade for the Windows. In the simulation software, the interior and exterior walls of the building are differentiated from the internal and external Windows. The Settings of related parameters are shown in Table 2.

The test results are shown in Table 4. When different weather events occur, green roof has a relatively similar reduction effect on building energy consumption. In both extreme high temperature weather events and typical summer weather events, the reduction effect of green roof on building energy consumption is 3 $W/m^2/h$, and the energy saving effect in extreme high temperature weather events is 0.32 $W/m^2/h$ higher than that in typical summer. This has a very positive effect on the supply and stability of urban power energy system. Based on the typical air conditioning habits of residential users and the 180-day cooling period, each building in the simulation scenario can reduce the electricity consumption of 21,772.8 KWH during a cooling period.

Table 4. Formatting sections, subsections and subsubsections.

	Green roof	Energy consumption per unit building area ($W/m^2/h$)	Energy saving per unit building area ($W/m^2/h$)
Hot and sunny days	Yes	30.45	3.24
	No	33.69	
Typical sunny days	Yes	22.02	2.92
	No	24.94	

5 Conclusion

In this experiment, the method of model simulation was used to select the extreme hot weather and typical summer weather during the experiment for accurate environmental data sampling, and the microclimate characteristics of temperature and wind speed at the 1.4m altitude layer of the community were studied. Based on the premise that the evaluation of the model is in good agreement with the actual measurement, the micro-climate effects of different preset greening schemes in the study area are simulated and analyzed, and the following conclusions are obtained. Combined with the actual mea-surement values, the quantitative evaluation indexes RSME and MAPE for the accuracy of green roof weather simulation were used to evaluate the model fitting and forecasting ability. The results show that the model can better reflect the changes of temperature, wind speed and wind direction in the simulated area, and the evaluation results of the indicators are within the accurate range proposed by previous studies, indicating that it is feasible and scientific to analyze the actual microclimate change through model fitting data. Through the analysis of the measured data and simulated microclimate envi-ronmental characteristics, it can be seen that roof greening and weather events have significant effects on the microclimate environmental temperature in this region. There is an east-west high temperature zone between buildings at night, and the arrangement of roof greening in extreme high temperature weather events and typical summer can reduce the temperature in this high temperature zone by 0.8 °C and 1.2 °C, respectively. Under the influence of roof greening, the temperature in extreme hot weather events and typical summer can be reduced by 0.6 °C and 1.0 °C respectively, and the overall temper-ature distribution is more average. The reason for the high temperature zone is that due to the direction of the building and the installation of surface vegetation, the north-south air circulation is not smooth, and the east and west sides of the building are respectively formed a quiet wind zone. At noon, a low temperature zone appeared on the east and west sides of the building respectively. During extreme high temperature meteorological events and typical summer meteorological events, the temperature decreased by 0.9 °C and 1.5 °C, respectively, compared with the temperature around the building, and the overall ambient temperature decreased by 1.1 °C and 1.3 °C. The main reasons for this are the arrangement of green roofs and the shading effect of buildings and surface plants. When extreme hot weather events occur, due to the comprehensive influence of solar radiation and air temperature, the ambient climate temperature of extreme hot weather in the experimental model is more average, and the cooling effect of roof greening is suppressed to a certain extent. It also causes the high temperature area on the south side of the building formed by the radiation of the building in typical summer days to disappear during the extreme high temperature summer days. The above results show that green roof can change the urban microclimate to some extent by reducing the max-imum daily building temperature, increasing the minimum daily building temperature and smoothing the daily building temperature change process, which has a very positive significance for the supply and stability of urban energy system.

In the simulation experiment of building energy consumption, the significance of roof greening on urban energy consumption is further highlighted. The average energy-saving effect of roof greening in summer can reach $3W/m^2/h$, and this energy-saving effect will be improved in extreme high temperature meteorological events. On the whole, the summer energy saving effect of roof greening on a single building in the test site can reach 21772.8 KWH. For the city, its summer energy saving effect and value are very significant. This indicates that in the future urban renewal process, large-scale roof greening can have a very important positive significance for the stability of urban energy supply system and the promotion of low-carbon society.

References

1. Johari, F., Shadram, F., Widén, J.: Urban building energy modeling from geo-referenced energy performance certificate data: development, calibration, and validation. Sustain. Cities Soc. **96**, 104664 (2023). https://doi.org/10.1016/J.SCS.2023.104664
2. Zhang, Y., et al.: Multistep multiagent reinforcement learning for optimal energy schedule strategy of charging stations in smart grid. IEEE Trans. Cybern. **53**(7), 4292–4305 (2023). https://doi.org/10.1109/TCYB.2022.3165074
3. Souabi, S., Chakir, A., Tabaa, M.: Data-driven prediction models of photovoltaic energy for smart grid applications. Energy Rep. **9**, S9 (2023). https://doi.org/10.1016/J.EGYR.2023.05.237
4. Vakili, S., Ölçer, A.I.: Techno-economic-environmental feasibility of photovoltaic, wind and hybrid electrification systems for stand-alone and grid-connected port electrification in the Philippines. Sustain. Cities Soc. **96**, 104618 (2023). https://doi.org/10.1016/J.SCS.2023.104618
5. Karczmarczyk, A., Baryla, A., Fronczyk, J., et al.: Phosphorus and metals leaching from green roof substrates and aggregates used in their composition. Minerals **10**(2), 112 (2020). https://doi.org/10.3390/min10020112
6. Mickovski, S.B., Buss, K., McKenzie, B.M., et al.: Laboratory study on the potential use of recycled inert construction waste material in the substrate mix for extensive green roofs. Ecol. Eng. J. Ecotechnol. **61**, 706–714 (2013). https://doi.org/10.1016/j.ecoleng.2013.02.015
7. Bates, A.J., Sadler, J.P., Greswell, R.B., et al.: Effects of recycled aggregate growth substrate on green roof vegetation development: a six year experiment. Landsc. Urban Plan. **135**, 22–31 (2015). https://doi.org/10.1016/j.landurbplan.2014.11.010
8. Kazemi, M., Courard, L., Hubert, J.: Coarse recycled materials for the drainage and substrate layers of green roof system in dry condition: parametric study and thermal heat transfer. J. Build. Eng. **45**, 103487 (2021). https://doi.org/10.1016/j.jobe.2021.103487
9. Andrew, M., Edoardo, C.D., Jason, B., Tapper, N.J.: Assessing practical measures to reduce urban heat: green and cool roofs. Build. Environ. **70**, 11–14 (2013). https://doi.org/10.1016/j.buildenv.2013.08.021
10. Wong, N.H., Chen, Y., Ong, C.L., Sia, A.: Investigation of thermal benefits of rooftop garden in the tropical environment. Build. Environ. **38**(2), 261–270 (2003). https://doi.org/10.1016/S0360-1323(02)00066-5
11. Elnabawi, M.H., Saber, E.: A numerical study of cool and green roof strategies on indoor energy saving and outdoor cooling impact at pedestrian level in a hot arid climate. J. Build. Perform. Simul. **16**(1), 18–24 (2023). https://doi.org/10.1080/19401493.2022.2110944
12. Pragati, S., et al.: Simulation of the energy performance of a building with green roofs and green walls in a tropical climate. Sustainability **15**(3), 2006 (2023). https://doi.org/10.3390/SU15032006
13. Willmott, C.J.: Some comments on the evaluation of model performance. Bull. Am. Meteor. Soc. **63**(11), 1309–1313 (1982)
14. Ferdinando, S., et al.: Relating microclimate, human thermal comfort and health during heat waves: an analysis of heat island mitigation strategies through a case study in an urban outdoor environment. Sustain. Cities Soc. **30**, 79–96 (2017). https://doi.org/10.1016/j.scs.2017.01.006

The Modular Protection and Design Method of Traditional Residential Buildings in Bayu Ancient Town

Qiuna Li[1], Jingyuan Shi[2], and Yueyuan Zhao[1(✉)]

[1] Chongqing Vocational College of Architecture and Technology, Chongqing, China
wanerjsmi@163.com
[2] Chongqing Jiaotong University, Chongqing, China

Abstract. This paper analyzes the current problems of the serious damage to the common dwellings and historical key buildings in Bayu ancient town and the insufficient understanding of the complexity of the protection work, established the basic idea of traditional residential building modular protection design, and further proposed the specific modular protection method which is composed of current building element data collation, common residential type renovation design, and special building type key design. The longtan Ancient Town in Chongqing is taken as an example for research and exploration.

Keywords: Bayu Ancient Town · traditional dwellings · modular protection design method

1 Introduction

Bayu ancient town is a general name of the self-organized historical towns scattered in Chongqing and neighboring areas. It is the historical crystallization of the regional settlement and architectural form of Bayu. The settlement pattern of Bayu ancient town is a very special type in Chinese traditional architectural culture, which has been widely concerned by people in recent years [1]. In Bayu area, there is abundant rain, humid climate and large fluctuation of landform. Its traditional folk houses form characteristic Spaces such as sloping roof, inner patio, stilted building and falling and splintering floor, which constitute the traditional building form with rich symbols. The research on the protection of traditional residential buildings in Bayu ancient Town not only inherits the original ecological construction culture, but also effectively maintains the pattern and historical features of the ancient town, and maintains the integrity, authenticity and continuity of traditional residential buildings and settlements [2]. Given that the current research on modular protection design of traditional residential buildings in Bayu is not rich in theory, this article explores targeted modular protection design methods for Longtan Ancient Town in Chongqing, in order to provide experience for the protection of traditional residential buildings in Bayu Ancient Town.

P. Xiang and L. Zuo (Eds.): PBSFTT 2023, LNCE 382, pp. 400–408, 2024.
https://doi.org/10.1007/978-981-97-5108-2_43

2 Practical Problems and Basic Ideas of Traditional Residential Architecture

2.1 Analysis and Research of Practical Problems

Ordinary Dwellings were Badly Damaged

The traditional residential buildings in Bayu ancient Town are mostly brick and wood structures, most of them are aging and damaged. Some of the residential buildings of great conservation significance built in Ming and Qing Dynasties are seriously damaged, while a large number of newly repaired residential buildings are "square boxes pattern" with ceramic tile veneer, which is inharmonious with the overall architectural style of the ancient town. (Fig. 1, Fig. 2).

Fig. 1. Dilapidated traditional dwellings in Longtan Ancient Town of Chongqing

Fig. 2. New tile - tiled dwellings, **Figure** source: Self-shooting

The Historic Buildings were Badly Damaged

In Bayu ancient town, a large number of guild buildings, historical courtyards, ancestral temples, traditional shops and other historic buildings of typical significance have been seriously damaged or lost their functions, and many of them have even ceased to exist. Take Chongqing Longtan Ancient Town as an example, among the historical buildings of high cultural value "seven palaces and ten temples", only the Longevity Palace of Jiangxi Guild hall is well preserved, but the other important architecture are no longer exist [3].

Insufficient Understanding of the Complexity of Conservation Efforts

In the conservation of traditional residential buildings in Bayu ancient Town, the problem of insufficient understanding of complexity still exists. First of all, the complexity analysis of the current situation of the building is not comprehensive enough. In general, there are a large number of traditional residential buildings in Bayu Ancient town, and the current situation is different. Some are well preserved, some are damaged and need to be renovated, and some are so dilapidated that it is difficult to guarantee basic safety [4]. In the protection work, to some extent the comprehensive understanding, combing and analysis of these traditional dwellings with great differences in current conditions are still lacking.

The second is the inadequate protection of the complexity of traditional architectural elements. The traditional dwellings in Bayu ancient town contain a large number of traditional components, the roof of a common residence in Longtan Ancient Town of Chongqing includes four important components: Roof tiles, eaves, overhanging eaves and roof ridge decoration. The rest components, such as door, window, beam, column, balcony, wall base, can be divided into a variety of different types. These traditional architectural elements have gone through historical precipitation and are of high value. If simplify them in the conservation work, the cultural connotation of the ancient town will be destroyed to a certain extent.

The third is lack of pertinence of the complexity of building types, the building types in Bayu ancient town are complex, including not only ordinary folk houses, but also special historical buildings such as all kinds of hall buildings, its use function is different from its historical value, and the protection content that needs to be expanded should not be the same. In the current protection work, some ordinary houses are renovated too complicated, while some special buildings with high cultural value are ignored [5].

2.2 The Basic Idea of Protection Design

The protection design of traditional residential buildings in Bayu ancient town is an important and complicated task. As mentioned above, there are various types of traditional buildings represented by classic dwellings in Bayu area, with different styles of old and new, the differences between ordinary folk houses and special houses such as courtyards and guilds are great, and the traditional architectural symbols are complicated, all of these make the protection and sustainable design of traditional dwellings more difficult to varying degrees. The key to solve practical problems is to find a feasible scientific method and design mode in work. Take the research and practice of protecting Longtan Ancient Town in Chongqing as an example, this paper puts forward the protection idea of **"the data collation of current building elements, the modular renovation design of common residential types, and the key design of special building types"** [6].

3 The Main Content of the Modular Protection Design Method of Traditional Residential Buildings

3.1 Current Building Element Data Collation

Sort Out Status Quo
Through field visits, the main residential houses within the survey scope can be evenly divided into several sections and numbered according to different orientations. For example, the houses on the east side of slate Street in Longtan Ancient Town are divided into E1 to EN, and the houses on the west side are divided into W1 to WN, etc., and the specific Numbers are compiled to facilitate grouping and research. Numbered the traditional residential houses in each block and collected basic data one by one. The basic information of traditional dwellings is sorted out from the aspects of architectural name, architectural form, space and number of floors, main functions, construction age, elevation drawing and preservation status of architectural elements, etc., and detailed data

is recorded in the form of status questionnaire. In particular, the detailed description of the preservation status of multiple architectural elements is the core part of the status questionnaire [7].

Element Classification

It can be divided into several systems according to the actual situation of the main building elements of the present residence. The residential building elements of Longtan Ancient Town can be divided into six systems: roof system, enclosure system, door and window system, beam and column system, balcony system, and other systems. Each system can be subdivided into several subsystems, details as show in Table 1:

Table 1. List of architectural elements of current residential situation in Longtan Ancient Town

System name	The main content
The roof system	The subsystem includes roof tiles, eaves, overhanging eaves and roof ridge decoration
Containment system	It includes the sub-system of wooden wall, masonry wall and special wall. The wooden wall is the main carrier of the traditional characteristics of the ancient town exists in traditional dwellings. Brick and stone walls appeared in late residential buildings, some of the stone walls of mottled texture are very aesthetic. Some traditional dwellings are also built in special ways, such as bamboo walls and adobe walls
Doors and Windows system	Can based on the virtual and real state of doors and Windows, divide into frame door, lattice door pass window subsystem. Frame door made of solid wood, simple appearance, easy to disassemble, most of them used for shop and door of the ground floor. Case door is the wooden door with lattice window; case fan window is the wooden window with hollow out decorative pattern, delicate and beautiful
Post and beam system	It includes wood beam, column body and column base subsystem. The beams and pillars are of full length or half-length logs. Column base is simple stone with little decoration
The balcony system	Can include the subsystem such as wooden balcony, wooden baluster. The wooden balcony of classic folk house and baluster scale are harmonious, simple and elegant. Part of the residential balconies set small eaves for shelter from wind and rain, also classified as this system
Other systems	Including cross - wind window, stone wall - based subsystem. The cross wind window is mostly set above the door, mostly wood grille or texture, with ventilation, decoration, lighting and other multiple functions. The wooden wall of folk houses is mostly carried by stone wall foundation, which is naturally coordinated with stone. A few houses have tiger Windows and other special building components

Chart source: Self-drawing

In the status quo questionnaire, data sorting and recording can be carried out for each building along the street in a typical section according to the sequence of Numbers, so as to serve as the basis for protection and renovation work.

Modular Design Elements

Based on the questionnaire on the current situation of dwellings in Longtan Ancient Town, extracted the architectural elements with traditional features, well preserved status and suitable for construction. Number each element and design the dimension data, and then construct the modular model and describe the specific method (Table 2) and carry out targeted key research to seal the fire wall, Tujia stilted building and other special elements [8].

Table 2. List of Architectural Elements Design and Specific Practices (Part)

Architectural elements	The specific practice of protection design
Roof comb (WJS)	The roof ridge decoration samples were extracted from bayu folk dwellings. According to the different shapes and styles, they were divided into four types and named as WJS-1, WJS-2, WJS-3 and WJS-4. Standard size will be determined after repair. Traditional small blue tiles stack, are axisymmetrical, and not painted paint decoration. The center of the Mosaic or circular pattern, the two sides with tiles or built into the arch, or stacked into a square
Eaves tiles (WD)	Extract circular and cross two types of tiles. The circle is defined as WD-1 with a cloud head pattern. The cross is defined as WD-2 with plant grain on top
Cornice (TY)	Overhanging eaves refers to the part of the roof that overhangs the wall, which is composed of beams and hanging gourds
ChiWen (CW)	ChiWen is a traditional decorative component on the roof ridge of traditional folk dwellings, which is built at both ends of the roof's right ridge, and adopts the traditional stacking of small blue tiles
Fire seal (FHQ)	Extract flat eaves and upturned eaves two types, respectively defined as FHQ-1 and FHQ-2
Wall of wooden (MBQ)	Take local wood and treat it as lumber to retain log texture and color
Masonry walls (ZQ)	Local stone materials and black brick materials are used, which can be divided into ZQ-1 and ZQ-2 types. The former adopts the masonry method of one-ding and one-lining, while the latter adopts the brick-lining method and leaves openings locally
Grille door (GSM)	Local wood processing is adopted, and the components are divided into two sections. The upper part is hollow-out window and the lower part is the baseboard. According to different texture styles of the window, designed as Gsm-1, GSM-2 and GSM-3 elements. Including the center of symmetrical weave, axisymmetric weave and so on
Frame door (KDM)	Adopt local wood processing, no window is set, and the method of adding board and thread is adopted. According to the complexity of thread, it can be divided into TWO types: KDM-1 and KDM-2

(*continued*)

Table 2. (*continued*)

Architectural elements	The specific practice of protection design
The wind window (HFC)	Use local wood for processing. According to different patterns, determine HFC-1, HFC-2 and HFC-3 types, and determine the dimensions of standard components one by one
Grille window (GSC)	Using local wood processing, due to grille texture is different, design six types. The more common type GSC-1 is formed by dividing the window through horizontal and vertical grates, and select a 210MM square light and ventilation hole in the middle
Balcony (YT)	It is made of wood, and the balustrade is carved with hollow pattern. According to different balustrades, it can be divided into YT-1 and YT-2 two types

Chart source: Self-drawing

3.2   Modular Remediation Design of Common Folk Dwellings

The trinity method of "face - line - point" can be adopted in the renovation design of ordinary folk houses in Longtan Ancient Town. (Fig. 3).

The Protection of "Surface"
The protection of "surface" is to renovate ordinary residential houses in the form of list. The element design work mentioned above is applied to the table. The main components of each building are listed in the table with specific renovation measures and corresponding renovation types of elements with corresponding Numbers, so as to achieve the goal of comprehensively covering the renovation work of ordinary residential buildings in the ancient town [9].

The Protection of "Line"
The protection of "line" is to carry out the streetscape renovation design. The key of section design corridor along the street facade, make the effect comparison before and after the transformation. In the renovation of the street style, the element design method is also used to replace or update the building structure part in the street view facade.

The Protection of "Point"
The protection of "point" refers to the transformation of individual dwellings. The typical residential houses are divided into several types, and combining element design to carry out the renovation. Each type can guide the reconstruction design of similar residential buildings. (Fig. 4) The selection principle of residential type includes repair and renovation (Take old repair old), renovation (change new into old) etc., combined with the renovation elements such as the number of floors and rooms (one to three floors/room), special elements (fire sealing wall, etc.), and other elements effects, summarize a variety of different types of residential reconstruction schemes. In addition, in the transformation of "point", should follow the principle of "authenticity", fully respect and retain the

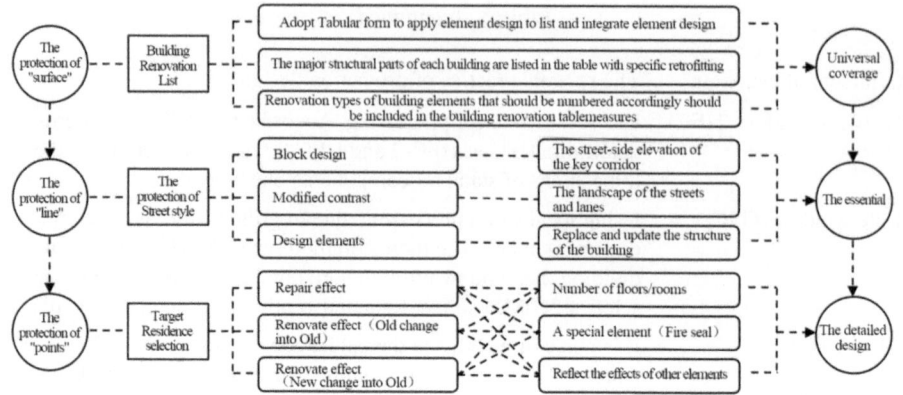

Fig. 3. Protection method diagrams of ordinary folk dwellings in "plane - line - point" **Chart source:** Self-drawing

tangible and intangible cultural elements, and summarize the main points of renovation as follows:

The original wooden components of traditional dwellings should be preserved, and no complex components should be replaced.

The original transverse window, beam and eave structure of traditional folk dwellings should keep the original proportion and form when repairing. Other doors and Windows that need to be replaced should be kept consistent with their original appearance.

The grey brick wall and the traditional stone wall can be preserved, the poor brick wall surface can be transformed with wood, and the uncoordinated hollow brick wall and the white wall can be pasted with wood.

Three or four floors of new buildings within the protected area, the tile facade can be converted to wood. Facade can be added hanging eaves (not outward expansion), for the shutter door. The top floor of a new three-storey building that is too obtrusive can be knocked out.

New buildings with complete forms and coordinated features can be retained within the protected area [10].

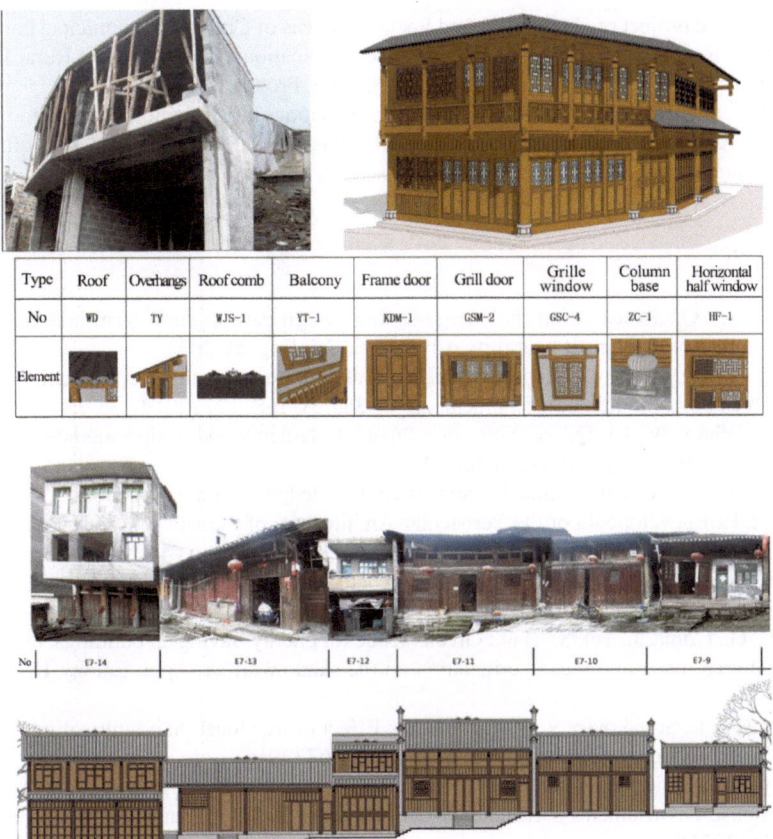

Fig. 4. Design drawing of the renovation of ordinary residential buildings in Chongqing Longtan Ancient Town under the combination of "line-point" **Chart source:** Self-drawing

4 Summary and Outlook

Combined with the case of Longtan ancient town, this paper puts forward the modular protection and design method of traditional residential buildings in Bayu ancient town. The protection of traditional dwellings in ancient town is a complex systematic project. Any problem in any link will greatly reduce the protection effect and even cause irreparable losses. In the specific work, it is necessary to further connect the protection method with the actual operation, such as paying attention to the construction drawing design, paying attention to the degree of completion and refinement of workmanship, and effectively organizing and following up the construction personnel, so as to ensure the real implementation of the design method. In short, the full consideration of the complexity of the overall process of protection is also an important issue that needs to be paid attention to and seriously faced in the future.

Fundings. The project of Humanities and Social Sciences of Chongqing municipal commission of education: Research on the Design Method and Application of Prefabricated Rural Homes in Chongqing from the Perspective of Rural Revitalization, Project No. 23SKGH476.

The project of Humanities and Social Sciences of Chongqing education commission: A Study on the Construction of Genetic Map of Traditional Village Dwelling Culture in Chongqing Driven by Digital Intelligence, Project No. 23SKGH130.

References

1. Shi, J., Li, Q.: Research on the energy-saving design path of rural farm houses under the background of ecological livability. Adv. Mater. Sci. Eng. **10** 2021
2. Shi, J., Li, Q.: Energy saving performance evaluation and planning optimization design of rural residential building environment. Argos **36**(73), 168–177 (2019)
3. Shi, J., Zhao, W., Li, Q.: A study on influential elements and design methods of regional residential areas. Appl. Mech. Mater. **12** 2012
4. Murdoch, J.: The Differentiated Countryside. Routledge, London (2003)
5. Oliver, P.: Encyclopedia of the Vernacular Architecture of the world. Cambridge University Press, New York (1997)
6. Jingyuan, S., Qiuna, L.: Ecological planning path of road traffic in Chongqing villages. WOP in Educ. Soc. Sci. Psychol. **1** (2020)
7. Pérez, G., Coma, J., Sol, S., et al.: Green facade for energy Savingsin buildings: the influence of leaf area index and facade orientation on the shadow effect. Appl. Energy **187**, 424–437 (2017)
8. Bajsanski, I., Stojakovic, V., Jovanovic, M.: Effect of tree location on mitigating parking lot insolation. Comput. Environ. Urban Syst. **56**, 59–67 (2016)
9. Jingyuan, S., Qiuna, L.: Case analysis and system construction of the artistic reuse of waste farmhouse environment: a case study of rural areas in Chongqing. Urban Archit. **18**(19), 90–94 (2021)
10. Nan, L.: On the design of exhibition space in historical buildings through modular thinking: taking the "Danish chair" exhibition at the Danish design museum as an example. Southeast Cult. **S1**, 91–93 (2022)

Analysis and Evaluation Model Study of the Impact of Prefabricated Buildings on Urban Sustainable Development

Jinjing Guo[✉], Yajuan Chen, Jing Hu, and Yaqing Zheng

Chongqing College of Architecture and Technology, Chongqing, China
834131747@qq.com

Abstract. As a new way of building industry that fits the concept of urban sustainable development, prefabricated buildings have a positive impact on urban sustainable development. The influencing factors and influence mechanism of prefabricated buildings on urban sustainable development is theoretically analyzed. In order to elucidate the contribution of prefabricated buildings to urban sustainable development, evaluation model for quantization and empirical study is also established in the paper.

Keywords: Prefabricated Buildings · Urban Sustainable Development · Influence Mechanism · Evaluation Model · Empirical Study

1 Introduction

As an important part of modern industry and urban development, architecture and related industries are crucial to urban sustainable development during planning, design, production, construction, use and maintenance.

Prefabricated buildings have the advantages of modular design, prefabricated materials, assembled construction, and industrialization, which is in line with the development direction of the construction industry, and in line with the concept of urban sustainable development [1]. Promoting prefabricated buildings has become a consensus of urban sustainable development. The impact of prefabricated buildings on urban sustainable development from both qualitative and quantitative aspects is tried to be explored in the paper.

2 Characteristics of Sustainable Development of Prefabricated Buildings

2.1 Characteristics in Planning and Design Stage

Prefabricated buildings that can reduce environmental load and improves industry efficiency have been accepted gradually [2]. From the stage of planning and design, the concept of sustainability is emphasized for prefabricated buildings, such as standardization design, assembly integration and low-carbon strategies, to promote urban sustainable development [3].

© The Author(s) 2024
P. Xiang and L. Zuo (Eds.): PBSFTT 2023, LNCE 382, pp. 409–418, 2024.
https://doi.org/10.1007/978-981-97-5108-2_44

2.2 Characteristics in Construction and Operation Stage

Economic Development

1) High prefabrication rate and assembly ratio of prefabricated buildings can improve the production efficiency and drive industrial upgrading and technological progress, improve the overall level and quality of urban economic development [4].
2) Low labor intensity and short construction period cause improvement of construction quality and benefit.

Environmental Protection

1) Prefabricated buildings mainly have energy saving measures such as exterior wall maintenance structure, efficient energy use and renewable energy.
2) Types of water-saving measures, including water-saving appliances, non-traditional water sources, and water-saving sprinkler irrigation technology, can effectively reduce urban sewage discharge.
3) The functional space of prefabricated buildings is compact and the design is optimized, which can achieve the purpose of improving the utilization rate of land.
4) The prefabricated construction site is mainly operated by dry process, which produces less construction waste and less environmental pollution.

Social Benefit

1) Prefabricated buildings pay attention to the unity of construction and urban surrounding environment, can create activity space, achieve organic integration with urban space and improve the living environment.
2) Prefabricated buildings introduce new development concepts, such as intelligent construction, to solve outstanding problems such as assist in improving urban supporting functions and building sustainable society [5].

3 Influence Mechanism of Prefabricated Buildings on Urban Sustainable Development

Prefabricated buildings involve many aspects such as the progress of buildings and related industries, the increase of resource recycling rate, the upgradation of residents' livability, and economic development [6]. Prefabricated buildings have different degrees of impact on urban sustainable development from three aspects of economy, environment and society, as shown in Fig. 1.

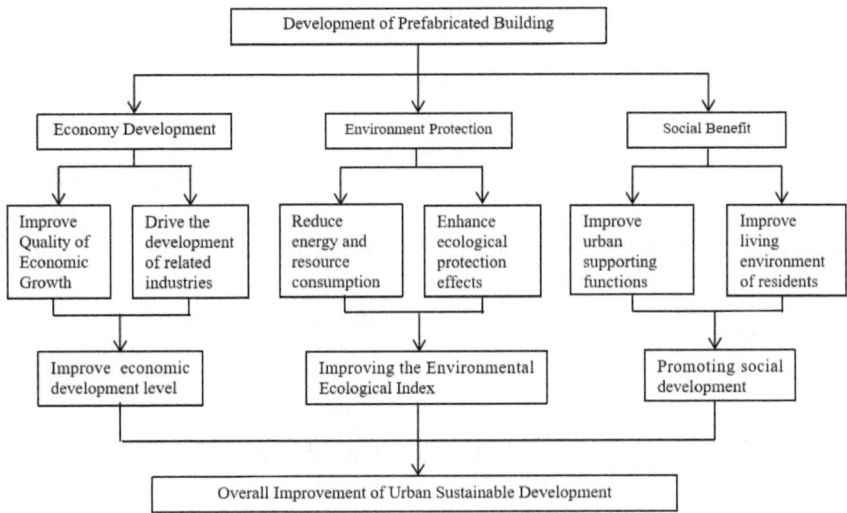

Fig. 1. The Influence mechanism of prefabricated buildings on urban sustainable development

4 Evaluation Model of Prefabricated Buildings Impact on Urban Sustainable Development

4.1 Prefabricated Buildings Development Level Index System

When determining the evaluation index of the development state of prefabricated buildings, the factors such as policy, economy and technology should be mainly considered to achieve wide coverage and high accuracy. Literature research, expert questionnaires and AHP method are used to determine the index weights [7–9], finally form prefabricated buildings development level index system, including 8 one-level indicators and 30 two-level indicators, as shown in Table 1.

$$S_k = \sum_{i=1} A_i \cdot \delta_i = \sum_{i=1, j=1} B_{ij} \cdot \theta_{ij} \tag{1}$$

where, S_k is the development level of prefabricated buildings, and k represents different cities.

As in Table 1, Policies and regulations A_1, production and operation A_8, and market environment A_2 among 8 one-level indicators have the greatest impact on the development level of prefabricated buildings. It is because prefabricated buildings are still in the initial stage of development, which largely depends on the government industrial policy support, and local market, production and operation environment.

Table 1. Prefabricated buildings development level index system

Object S_k	1-level indicator A_i	1-level indicator weight δ_i	2-level indicator B_{ij}	2-level indicator relative weight	2-level indicator combined weight θ_{ij}
Prefabri-cated build-ing develop-ment level index system	Policies and regulations A_1	0.205	Supportive policies	0.506	0.104
			Supervision measures	0.318	0.065
			Completeness of regulations	0.176	0.036
	Market environment A_2	0.152	Industrial layout	0.345	0.052
			Integrity of industrial chain	0.297	0.045
			Enterprises willingness	0.133	0.020
			Market demand	0.225	0.034
	Economic environment A_3	0.114	Regional economic development level	0.273	0.031
			Input-output ratio	0.324	0.037
			Completeness of cost management system	0.231	0.026
			Training of technology talent	0.172	0.020
	Design A_4	0.103	Design technical level	0.222	0.023
			Design standardization	0.275	0.028
			Design systemization	0.272	0.028

(*continued*)

Table 1. (*continued*)

Object S_k	1-level indicator A_i	1-level indicator weight δ_i	2-level indicator B_{ij}	2-level indicator relative weight	2-level indicator combined weight θ_{ij}
			Design integration	0.231	0.024
	Manufacture and transportation A_5	0.095	Production equipment level	0.196	0.019
			Manufacturing and processing technical capabilities	0.192	0.018
			Prefabrication proportion	0.218	0.021
			Prefabrication types	0.212	0.020
			Production standards	0.182	0.017
	Construction A_6	0.089	Technical proficiency of industrial workers	0.252	0.022
			Construction equipment level	0.267	0.024
			Component installation level	0.246	0.022
			Quality control system	0.235	0.021
	Inspection and test A_7	0.088	Technical disclosure level	0.267	0.024
			Component quality control	0.329	0.029
			Building quality inspection	0.404	0.036

(*continued*)

Table 1. (*continued*)

Object S_k	1-level indicator A_i	1-level indicator weight δ_i	2-level indicator B_{ij}	2-level indicator relative weight	2-level indicator combined weight θ_{ij}
	Production and operation A_8	0.154	Perfection of production rules and regulations	0.351	0.054
			Operation management level	0.326	0.050
			Technical level of staff	0.323	0.050

4.2 Evaluation of Urban Sustainable Development Level

The key to scientifically evaluate the level of urban sustainable development is to build comprehensive evaluation index system covering economic, environmental and social aspects [10], as shown in Table 2. The evaluation model includes 3 first-level and 19 s-level indicators, then adopts AHP and entropy method to determine the weight of each indicator. The weight calculation formula is shown below.

$$\omega_i = a\omega_{oi} + (1 - a)\omega_{si} \tag{2}$$

where, $a \in (0,1)$, ω_{oi} and ω_{si} is the weight value assigned to entropy and AHP, ω_i is the weight value obtained by weighted calculation, and i is the index number.

Table 2. Evaluation of urban sustainable development level

Object T_k	First-level indicators C_i	Second-level indicators D_{ij}	Weight ω_{ij}	Index Properties
Urban Sustainable Develop-ment Level	Economy C_1	D_{11} Per capita fixed asset investment(¥/per)	0.0951	+
		D_{12} Per capita actual use of foreign capital ($/per)	0.0342	+

(*continued*)

Table 2. (*continued*)

Object T_k	First-level indicators C_i	Second-level indicators D_{ij}	Weight ω_{ij}	Index Properties
		D_{13} Per capita GDP(¥/per)	0.0859	+
		D_{14} GDP proportion of tertiary industry (%)	0.0522	+
		D_{15} GDP proportion of secondary industry (%)	0.0269	-
		D_{16} Per capita total retail sales of social consumer goods(¥/per)	0.0970	+
	Environment C_2	D_{21} Industrial wastewater discharge (10^4ton)	0.0180	-
		D_{22} Industrial sulfur dioxide emissions(ton)	0.0167	-
		D_{23} Industrial smoke and dust emissions(ton)	0.0156	-
		D_{24} Comprehensive utilization rate of industrial solid waste (%)	0.0416	+
		D_{25} Per capita green space(m^2/per)	0.0625	+
		D_{26} Harmless treatment rate of household garbage (%)	0.0168	+
		D_{27} Green coverage of built-up area (%)	0.0528	+
	Society C_3	D_{31} Population density (persons/km^2)	0.1004	+

(*continued*)

Table 2. (*continued*)

Object T_k	First-level indicators C_i	Second-level indicators D_{ij}	Weight ω_{ij}	Index Properties
		D_{32} Number of doctors per 10^4 persons (per/10^4 persons)	0.0679	+
		D_{33} Per capita road area (m^2/per)	0.0578	+
		D_{34} Ratio of teachers to students in ordinary colleges and universities (%)	0.0414	+
		D_{35} Per capita savings balance at year-end (¥/per)	0.0886	+
		D_{36} Proportion of unemployed people (%)	0.0286	-

4.3　Evaluation Model and Empirical Study

In Table 1, the development of prefabricated buildings mainly can be characterized by 8 first-level indicators. If the correlation between the 8 first-level indicators and the level of urban sustainable development can be derived and analyzed, then the impact of prefabricated buildings on urban sustainable development can be judged.

The multivariate linear function regression equation is used to study the correlation between dependent variables and multiple independent variables, shown as follows:

$$T_k = \alpha_0 + \alpha_1 A_1 + \alpha_2 A_2 + \cdots + \alpha_n A_n \tag{3}$$

where, α_0 is a constant.

SPSS software was used to linear regression analysis of the data, and the independent variables were processed without dimension. The model fitting results were confirmed in Table 3. $R^2 = 0.964$, indicates that the linear relationship between independent variables and dependent variables is close. The Durbin-Watson coefficient is 1.479, which indicates that the fitting effect is good.

Table 3. Model Summary

R	R^2	Adjusted R^2	Standard estimation error	Durbin-Watson coefficient
.993	.986	.964	.30923	1.479

The significance analysis is shown in Table 4, and the significance $P = 0.002 < 0.05$ indicates that the model has significant statistical significance and is effective.

Table 4. Model Significance Analysis

Model	Sum of squares	Degree of freedom	Mean square	F	Significance
Regression	34.676	8	4.41	42.689	.002
Residual error	.399	4	.111		
Total	35.075	12			

Considering the excessive similarity caused by insufficient samples and limited data, the improved least squares estimation method is further used to solve the regression coefficient, and the calculation formula is as follows:

$$\beta = (A^T A + kI)^{-1} A^T T \tag{4}$$

where, k is the regression coefficient, which can be determined when the estimated value of the regression coefficient of each explanatory variable in the model is basically relatively stable. Through SPSS software, k is set in the interval of [0,1] with search step size 0.01 for analysis, all variable coefficients will gradually stabilize with the increase of k value. $k = 0.2$ was selected for regression analysis, and the model $R^2 = 0.975$ was obtained, the independent variables and dependent variables were well fitted, $F = 12.42$, $P = 0.01 < 0.05$, and the significance was good, which once again proved that the model was effective.

The influence formula of independent variable on dependent variable (Formula 5) can be obtained according to regression parameter analysis, and Formula 6 can be obtained by combining Formula 1 and the data in Table 1, therefore, evaluation model can be obtained by combining Formula 5 and Formula 6 listed below.

$$T_k = -2.478 + 0.048A_1 + 0.029A_2 + 0.243A_3 + 0.007A_4 + 0.089A_5 + 0.003A_6 + 0.002A_7 + 0.389A_8 \tag{5}$$

$$S_k = 0.205A_1 + 0.152A_2 + 0.114A_3 + 0.103A_4 + 0.095A_5 + 0.089A_6 + 0.088A_7 + 0.154A_8 \tag{6}$$

Through above evaluation model, it can be found that the development of prefabricated buildings has a positive impact on urban sustainable development mainly by promoting production and operation, economic environment, manufacture and transportation, design and other indicators. The reason may be that China is still in the stage of rapid urbanization development. With the support of the policy, prefabricated buildings are entering a period of rapid development and innovation, and there is significant great progress and visible fast development in the production and operation environment, intuitive economic output, production technology and design means of prefabricated buildings, so as to promote the overall urban sustainable development level.

5 Conclusion

Prefabricated buildings have the advantages of modular design, prefabricated materials and assembled construction, making it easy for the construction industry to achieve industrialization, automation, and high efficiency. The influence mechanism of prefabricated buildings on urban sustainable development is qualitatively analyzed in theory. Prefabricated building development level index system and the evaluation system of urban sustainable development level is both established, then the evaluation model and empirical analysis is performed by linear regression analysis of the correlation between above two systems, which provides a reference for guiding the further development of prefabricated buildings and formulating policies for urban sustainable development.

References

1. Xiang, M.A.: Research on the Development Trend of Greenbuilding and Fabricated Building Under the Background of BIM Technology. Qingdao University of Technology (2018)
2. Wang, Z.: Analysis of sustainable development path for green prefabricated steel structure buildings. China Hous. Facil. 136–138 (2022)
3. Pan W'Gibb, A.G.F., Dainty, A.R.J.: Leading UK housebuilders' utilization of offsite construction methods. Build. Res. Inf. **36**(1), 56–67 (2008)
4. Dongwei, L.I.U., Xue, H.A.O., Ruofan, L.I.U.: Sustainable design method of long-life sustainable housing and integrated construction of Zhejiang Bao-ye prefabricated housing. Archit. Tech. **2**, 58–63 (2021)
5. Lingyan, C.H.E.N.: Research on the sustainable development of prefabricated buildings intelligent construction based on BIM. China Constr. Metal Struct. **44**(1), 184–186 (2023)
6. Jianshu, H.U., Fusheng, L.I., Fushuai, S.U.N.: Research on the construction and application of standard system for prefabricated buildings design. China Stand. **24**, 152–154 (2021)
7. Zheng, Y.: Evaluation of Prefabricated Construction Sustainable Development Research-Hengsheng Building as an Example. Xi'an University of Architecture and Technology (2018)
8. Yao, X.: Research on the Social Impact Evaluation Index System of Prefabricated Buildings Based on Sustainability. Hefei University of Technology (2022)
9. Huixian, H.U.I., Hai, J.I.N., Guangping, L.I.U.: Research on the evaluation of China 's prefabricated buildings based on gray clustering method. J. Eng. Manag. **4**, 26–31 (2019)
10. Su, Y.A.N.G., Hanghang, L.I.: A study of the impact of the renovation of old communities on the sustainable development of cities:empirical test by the DID method. J. Xi'an Univ. Technol. **39**(1), 1–11 (2023)

CFD Numerical Simulation Analysis of Wind Load Based on BIM

Yiping Wang[1] and Bo Wang[2(✉)]

[1] Chongqing College of Architecture and Technology, No. 3, Mingde Road, Chongqing, China
[2] Chongqing Architectural Design Institute Co., Ltd, 31 Renhe Street, Chongqing, China
624234608@qq.com

Abstract. For high-rise circular buildings, based on the values of wind load shape coefficients without similar shapes in the building structural load specifications, this project is based on BIM technology and combines BIM technology with CFD technology. Through CFD numerical simulation method, SST k-w turbulence model is used to simulate the surrounding wind field and 8 wind directions of the simplified single building model, and the overall and local wind loads of the building are studied and analyzed, determine the wind load shape coefficient to provide reference for circular buildings. The conclusion is that the windward walls and lower surfaces of circular buildings are mainly under positive pressure, while the leeward walls, sides, lower surfaces, and roofs are generally under negative pressure; The maximum positive pressure appears on the windward wall surface and lower surface of the outer ring, while the absolute value of the maximum negative pressure of the circular building appears at the edge of the inner and outer rings of the roof.

Keywords: BIM Technology · CFD Technology · Wind Load

1 Introduction

The surface aerodynamic characteristics of high-rise circular buildings under wind load are very complex [1, 2], and wind load has become one of the main loads in structural design and analysis. According to the "Load Code for the Design of Building Structures" (GB50009-2012) [3], there is no similar wind load shape coefficient value. This project is based on BIM technology and combines BIM technology with CFD technology [4, 5]. Through CFD numerical simulation method, SST k-w [6] turbulence model is used to simulate the surrounding wind field and 8 wind directions of the simplified single building model, and the overall and local wind loads of the building are studied and analyzed, To determine the wind load shape factor for design purposes.

2 Project Overview

The project is located in Bishan District, Chongqing, with a building area of 107900 m^2 and a building height of 53 m. It is composed of underground garage, podium and tower. Its main functions are exhibition hall, conference and exhibition, theater, external

P. Xiang and L. Zuo (Eds.): PBSFTT 2023, LNCE 382, pp. 419–425, 2024.
https://doi.org/10.1007/978-981-97-5108-2_45

exhibition service area, etc. The lower part of the tower in this project is a reinforced concrete cylinder, and the upper part is a spatial steel truss. The span of the outer ring main truss is 89 m, and the span of the inner ring main truss is 58 m. The upper plane of the tower is circular, with an inner ring radius of 51.6 m, an outer ring radius of 103.1 m, and a radial width of 51.5 m.

3 Model Establishment and Grid Partitioning

Using the Revit 2016 modeling software, create a building BIM model (see Fig. 1), and simplify the model through GUI operations (see Fig. 2), the model is directly exported to an STL format file through Revit with plugins, and then imported into ICEM CFD for calculation of watershed establishment and grid division [7].

Fig. 1. BIM Model and Renderings.

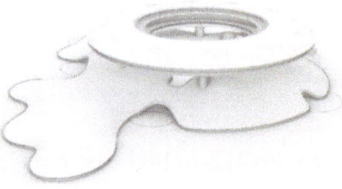

Fig. 2. Target Building CFD Geometric Modeling.

The numerical simulation of the flow around the roof includes two characteristics: on the one hand, the flow around the roof is turbulent, which itself contains vortices of different scales; On the other hand, the flow structure is also a problem of different scales in space. Both have corresponding requirements for mesh generation. Calculate the watershed as 1500 m × 1100 m × 203 m (flow direction x × Direction y × Vertical z) (see Fig. 3). Considering the complex shape of the target building, the grid generation scheme adopts regional blocking technology, dividing roofs or walls with similar sizes and shapes into groups to establish a part, with different parts defining different sizes of grids. Dense unstructured grids are used in the area near the buildings, and hybrid Hexahedron grids are used in other areas, with a total number of about 4 million grids.

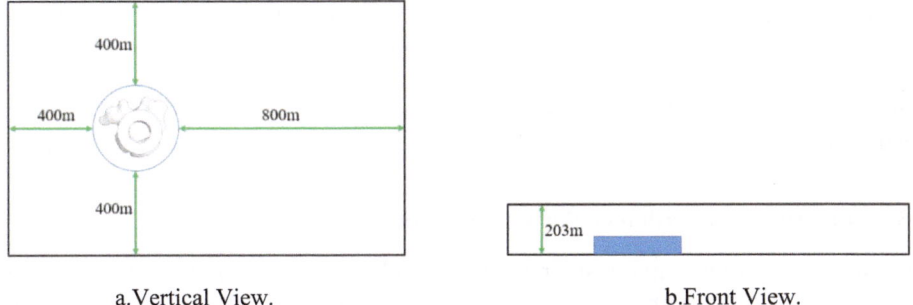

a.Vertical View.   b.Front View.

Fig. 3. Calculate Watershed.

4   CFD Numerical Simulation Analysis of Wind Load

4.1   Selection of Turbulence Models

The wind load numerical simulation of the target building adopts the large-scale general computational fluid dynamics (CFD) software Fluent. The coupling of pressure and velocity is solved using the SIMPLEC algorithm, and the control equation is solved using the Segregated method. The SST k-w model is selected as the turbulence model, which has a wide range of applications and high accuracy. It is a reliable turbulence model for simulating wind loads on building structures. The parameters of the turbulence model are taken from the corresponding values in the UDF file loaded in Fluent. The calculation result is taken as a steady-state result, and the number of iteration steps is taken as 3000. The calculation accuracy is first calculated using first-order accuracy of 1200 steps, and then using second-order accuracy of 1800 steps. The convergence criterion for calculation is taken as a residual value of 5×10^{-4} [8].

4.2   Selection and Handling of Boundary Conditions

Boundary conditions for incoming flow: Velocity inlet is used at the inlet of the flow field, and the average wind profile is represented by an exponential law:

$$U(z) = U_{10} \left(\frac{z}{10} \right)^{\alpha} \tag{1}$$

U_{10} is the average wind speed of the incoming flow at a height of 10 m, calculated based on the local basic wind pressure; α is the ground roughness index [9].

The turbulent kinetic energy and dissipation rate of the inflow surface are expressed as follows:

$$k(z) = \frac{3}{2} \cdot [U(z) \cdot I_u(z)]^2 \tag{2}$$

$$\varepsilon(z) = C_{\mu}^{3/4} \cdot \frac{k^{2/3}(z)}{K \cdot z} \tag{3}$$

In the formula, C_μ is a model constant with a value of 0.09; K is the Carmen constant, with a value of 0.42; $I_\mu(z)$ for the turbulence degree of incoming flow at height.

The outlet surface adopts pressure outlet boundary conditions; Symmetry boundary conditions are used for the upper and lateral surfaces; The use of non slip wall boundary conditions on the surface and ground of buildings [10].

4.3 Numerical Simulation of Working Conditions

This study conducted numerical simulation calculations for the target building in the range of 0°to 360°, with a total of 8 wind directions spaced at 45°intervals. The calculation of different wind directions was achieved through rotating the model.

4.4 Calculation of Average Wind Pressure Coefficient

Calculation of Wind Pressure Coefficient at Measuring Points. The point average wind pressure coefficient C_{pi} on the surface of the structure is obtained from the following equation:

$$C_{pi} = \frac{P_i - P_\infty}{\frac{1}{2}\rho U_G^2} \tag{4}$$

In the formula, P_i is the pressure acting on the measuring point i, and P_∞ is the static pressure at the reference height, ρ is the Density of air, U_G is the average wind speed of incoming flow at the height of gradient wind.

Calculation of Weighted Average Wind Pressure Coefficient for Local Area. The weighted average average wind pressure coefficient C_p of the local area on the structural surface is obtained from the following equation:

$$C_p = \frac{\sum C_{pi} \cdot A_i}{\sum A_i} \tag{5}$$

In the formula, C_{pi} is the average wind pressure coefficient at measurement point i, and A_i is the surface area of the structure represented by measurement point i. Due to the fact that CFD calculation can obtain the wind pressure value at any point on the entire building surface, this project directly provides a contour cloud map of the wind pressure distribution on the building surface, without the need to calculate the weighted average wind pressure coefficient of the area.

4.5 Numerical Simulation Results

Wind Load Indicated by Flow Field and Building. This article uses CFD numerical simulation to calculate the wind speed field around the target building, the average wind pressure coefficient on the building surface, the local shape coefficient, and the distribution of equivalent static wind load under 8 wind direction conditions within the range of 0°to 360°. The wind load is the most unfavorable under the 45° wind direction condition (see Fig. 4).

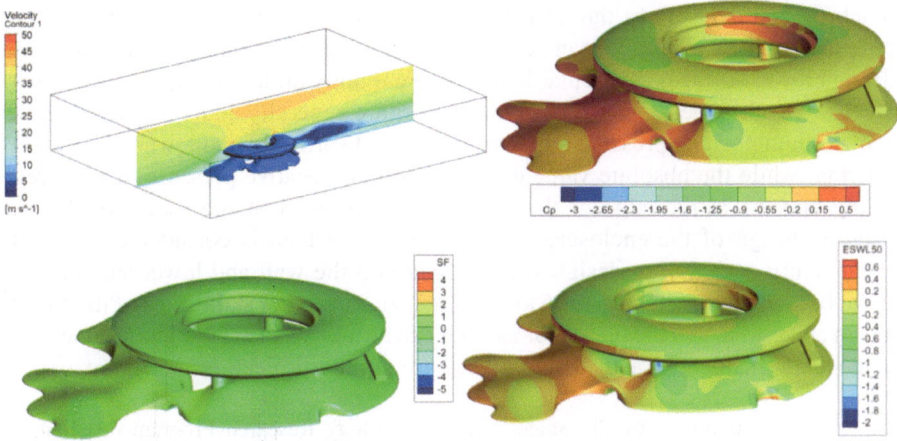

Fig. 4. 45°Wind Direction Working Condition.

Overall Shape Coefficient of the Building.   The overall shape coefficients of the circular building at various wind directions were obtained through CFD numerical simulation calculations (see Fig. 5). It can be seen from the figure that the maximum Drag coefficient of the annular building is 1.12, which occurs under the condition of 45° wind angle. The average value of the overall shape coefficient of circular buildings at various wind directions is about 0.92.

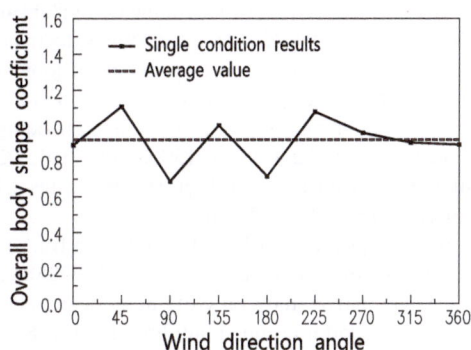

Fig. 5. Numerical Simulation Results of Overall Shape Coefficient of Buildings.

5   Conclusion

In view of the fact that there is no wind load shape coefficient of similar shape in the high-rise ring Building code, based on the BIM technology, this project combines the BIM technology with the CFD technology, analyzes the actual wind load CFD numerical simulation, and determines the design wind load shape coefficient: the overall shape coefficient of the ring building can be 1.2 for partial safety consideration; The windward walls and lower surfaces of circular buildings are mainly under positive pressure, while the leeward walls, sides, lower surfaces, and roofs are generally under negative pressure. The maximum positive pressure appears on the windward wall and lower surface of the outer ring, while the absolute value of the maximum negative pressure of the circular building appears at the edge of the inner and outer rings of the roof; For the wind resistance design of the enclosure structure, the wind load is considered based on a 50 year return period. The design wind pressure on the wall and lower surface of the circular building can be taken as 0.6 kPa, the negative pressure at the edge of the roof can be taken as 1.5 kPa, and the middle of the roof can be taken as 1.0 kPa. These conclusions provide theoretical and practical significance for similar projects in the future.

Fund Projects. Supported by the Science and Technology Research Program of Chongqing Municipal Education Commission (Grant No. KJQN202305203).

References

1. Chen, F.B., Li, Q.S., et al.: Experimental and numerical investigations of the characteristics of wind load acting on complex long-span roof structural. J **30**, 619–627 (2012). http://www.cqvip.com/QK/95593X/20125/43755646.html
2. Dong, X., Ye, J.H.: Flow model of conical vortices on large-span flat roof. J **34**, 13–21 (2013).https://doc.taixueshu.com/journal/20130068jzjgxb.html
3. Ministry of Construction of the China: Load code for the Design of Building Structures. China Architecture & Building Press, Beijing (2012). https://www.docin.com/p-2682232525.html
4. Xu, D.W., Jin, S.C.: Design and construction technology of long span steel structure roof based on BIM model. J. **51**, 90–94 (2021).https://doi.org/10.14144/j.cnki.jzsg.2020.04.053
5. Li, M., Li, Y.C.: Analysis of Indoor Air Flow Organization in Zhaoqing New Area Gymnasium Based on BIM Model (2020). https://doi.org/10.15968/j.cnki.jsjl.2020.11.005
6. Chen, Z.: CFD simulation of wind loads on high-rise buildings with rough walls (2017). https://www.doc88.com/p-9746309443177.html
7. Wu, D.D.: Research on the Application of BIM Based Green Building Analysis and Carbon Emission Calculation (2016). https://www.docin.com/p-2280728863.html
8. Huang, Z., Yan, W.H.: Quantitative study on CFD numerical calculation method for wind load characteristics of finger corridor terminal building. J. **36**, 1–8. https://doi.org/10.19786/j.tzjg.2019.05.001
9. Ministry of Construction of the China: Standard for wind tunnel test methods in construction engineering. China Architecture & Building Press, Beijing (2014). https://www.doc88.com/p-1092124628493.html
10. Nie, S.F., Zhou, X.H.: Numerical simulation of three-dimensional blunt body wind field around high-rise buildings according to CAARC standards. J. **31**, 40–46 (2009). https://www.nstl.gov.cn/paper_detail.html?id=edb96799921acf55031c0f9141b171bc

Application Practice of Locality Prefabricated Residential Design in Rural Revitalization

Yueyuan Zhao[1], Youwu Yuan[1(⊠)], Qiuna Li[1], and Sufa Huang[2]

[1] Chongqing Vocational College of Architecture and Technology, Chongqing, China
en_umu@163.com
[2] Chongqing Bajie Rural Construction Technology Co. Ltd., Chongqing, China

Abstract. Based on the analysis of the advantages and disadvantages of prefabricated residential design in rural revitalization, this paper proposes a strategy for prefabricated residential design in terms of functionality, form, cultural emotions, and other aspects. Analyze the sustainable development of prefabricated residential buildings throughout their entire lifecycle from aspects such as unit type, component parts, materials, and construction, and propose specific methods for standardized and modular design; combined with the traditional architectural characteristics and place spirit of rural areas, showcases specific prefabricated residential design cases, providing a certain reference for the sustainable development direction of prefabricated buildings in rural revitalization.

Keywords: Locality · Prefabricated residential buildings · Rural revitalization · Modular design · Sustainable development

1 Rural Revitalization is Closely Related to the Development of Prefabricated Buildings

1.1 The Country Vigorously Develops Rural Revitalization

In 2021, the State Council issued the "Opinions of the Central Committee of the Communist Party of China and the State Council on Comprehensively Promoting Rural Revitalization and Accelerating Agricultural and Rural Modernization", emphasizing that vigorously developing prefabricated buildings such as steel structures and concrete has the green and environmental characteristics of energy conservation, reducing construction pollution, achieving energy conservation and emission reduction in the construction industry, and sustainable development (Chart 1).

P. Xiang and L. Zuo (Eds.): PBSFTT 2023, LNCE 382, pp. 426–437, 2024.
https://doi.org/10.1007/978-981-97-5108-2_46

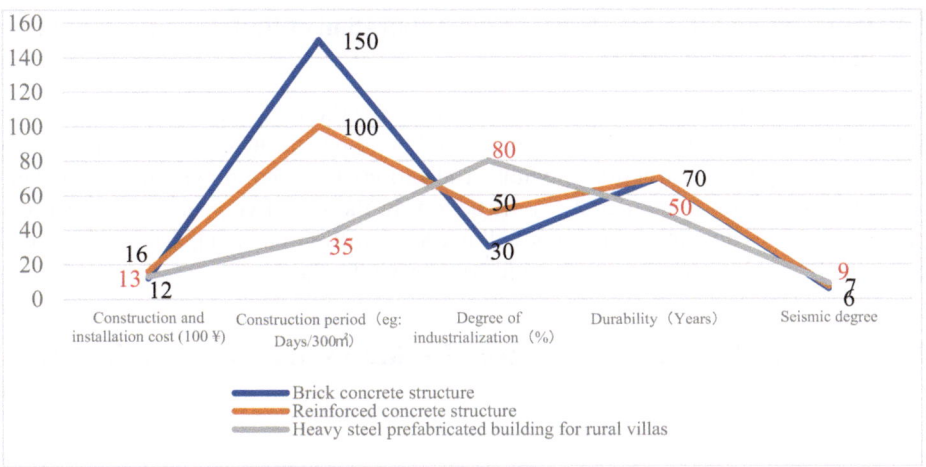

Chart 1. Analysis of Heavy Steel Structure Rural Villa System. **Source:** Summarize and self-drawing based on relevant information

1.2   Pain Points and Difficulties in Rural Revitalization

The overall requirements for rural revitalization are reflected in industrial prosperity, ecological livability, rural civilization, effective governance, and a prosperous life.

However, exploring effective forms of implementation such as the separation of ownership, qualification rights, and use rights of homesteads, the construction of new rural houses is becoming increasingly prominent:

a. Lack of professional participation, unreasonable scheme selection, and lack of overall spatial layout planning;
b. The residential functions are chaotic and mixed, with poor privacy;
c. The energy-saving and seismic performance of residential buildings are poor, and the living comfort is poor;
d. The cost of building construction is high, the construction period is long, the environmental impact is significant, and the waste of resources during the construction process is severe;
e. The technical quality of construction personnel is low, and the quality cannot be guaranteed;
f. Lack of employment opportunities, young and middle-aged people going out to work, and a serious shortage of construction personnel;
g. Non-standard self-built houses, so the construction funds are not supported by financial institutions.

1.3 Prospects and Realistic Issues of the Application of Prefabricated Buildings in Rural Revitalization

The new requirement of the 14th Five Year Plan for the high-quality development of the construction industry is to use prefabricated buildings as the carrier, synergistically promote intelligent construction and industrialization of new buildings, and comprehensively promote rural revitalization construction. In the next five years, the investment scale for rural revitalization in China will exceed 7 trillion yuan [1], and Prefabricated buildings have a huge market demand for rural revitalization.

2 How Prefabricated Residential Buildings are Reflected in Their Locality

2.1 "Locality"

"Locality" mainly refers to the close connection and interaction between buildings and sites, society, and people's lives; All human activities occur in specific environmental and social soils; And it occurs in the specific relationship between "people" and "places" [2].

2.2 Architectural Form

The design of prefabricated residential buildings in the local area fully reflects its local characteristics by extracting recognizable forms of rural buildings, such as the through bucket and dry railing structures in the Bayu area, and the space under the eaves of sloping roofs [3]; Through the design and utilization of courtyard spaces, we aim to enhance the integration of traditional architectural cultural exchange spaces with the functional needs of clothing and grain drying, reflecting the characteristics of adapting to local life; At the same time, by selecting appropriate colors and materials based on local conditions for wood and stone, and matching with the local plant atmosphere, we create a sense of belonging and place spirit in the presence. By combining different forms of residential building types with different building standard modules to meet people's needs [4].

2.3 Functional Utilization

Prefabricated residential buildings in the local area can serve the daily life of villagers (agriculture, daily life) and attract urban people to return to the countryside, creating a hidden urbanization function. Production and life are full of rural sentiments, while infrastructure is modern and more reasonable.

2.4 Humanistic and Emotional Aspects

Paul Oliver scientifically supplemented the concept of vernacular architecture in his "World Encyclopedia of Terrestrial Architecture", proposing to comprehensively study vernacular architecture from multiple perspectives and levels [5]. Through mutual cooperation between craftsmen and designers in rural areas, they jointly participate in the design and construction process, reflecting the geographical, human nature, and attribution of prefabricated residential buildings; There is no need for overly refined, industrialized, and designed construction, breaking away from the overly idealistic and artistic complex, in order to avoid breaking away from the original rural lifestyle and carrying the real life of the vast number of villagers through prefabricated residential buildings.

3   Design Features of Local Prefabricated Residential Buildings

Rural prefabricated residential buildings need to be designed based on the expression of locality, in order to fully express cultural elements, from the following four aspects [6]:

3.1 Standardized Design of Unit Types

The unit area of rural prefabricated residential buildings ranges from 80 to 200 square meters. Guided by the concept of "fewer specifications, more combinations" [7] and referring to the requirements of building materials and structural specifications, the modular induction of the opening and depth dimensions of different functional units of the suite is carried out using "600 mm" as the basic module. According to the "basic room" [8] stereotypes combination method, the standardized units are subdivided into functional spaces such as kitchens, bathrooms, bedrooms, and living rooms, to set reasonable size parameter combination range under unified modulus coordination, simplify the types of components and fittings, and make the house type meet the requirements of standardization and diversification. (Fig. 4).

3.2 Standardized Design of Components

In order to reflect the local characteristics and standardized construction of prefabricated steel structure residential buildings, based on the traditional steel structure system, the standardized design concept is applied to highly integrate the four major systems of prefabricated buildings (building structure system, peripheral protection system, equipment and pipeline system, interior decoration system) by generalizing the components, and adding full decoration to build a new form of prefabricated steel structure building system suitable for the Bayu region.

The main standard components include: the main structure adopts a box shaped steel column, H-shaped steel beam structural system, prefabricated steel stairs, integrated floor panels, roof trusses, integrated roofs, and maintenance structural systems. (Fig. 1).

Based on the standardization of fine decoration style, research and development are carried out to integrate prefabricated technology and standardization, and the unified design of the heaven and earth wall, overall bathroom, and overall kitchen is carried out. (Table 1).

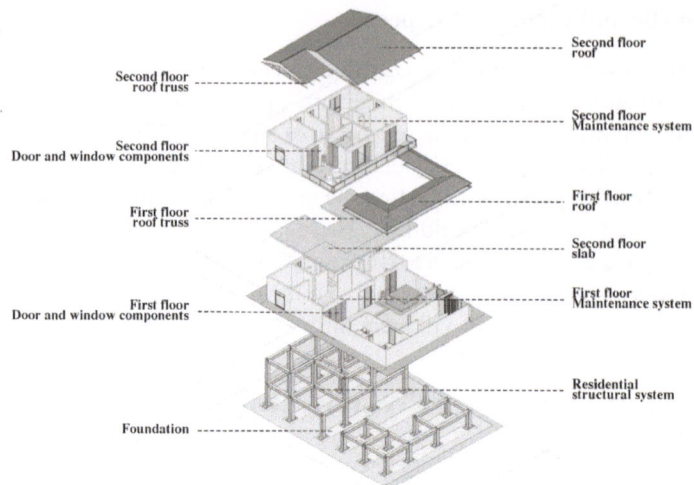

Fig. 1. Standardized Design of Components in Prefabricated Residential Buildings. **Figure source:** Self-drawing

Table 1. Standardized design of interior decoration modules for local prefabricated residential buildings

Bedroom module	Activity room module	Restaurant and Kitchen module	Central room/Living room module	Toilet module
WS-01,WS-02	SF-01,SF-02	CF-01,CF-10	KT-02	WC-01,WC-02

Chart source: Self-drawing

3.3 Standardized Design of Materials and Structures

The prefabricated construction of heavy steel in rural villas embodies assembly and modularization, and the concept of "Combination of old and new" is applied to building materials [9]. Most of them use traditional materials from the local area. Steel structures with H-shaped steel and rectangular steel profiles are used as load-bearing frameworks, and modular insulation and energy-saving composite panels are used as enclosure structures. The heavy steel structural framework is supplemented by insulation and energy-saving composite panels, exterior wall materials, and polymer asphalt tiles, and its architectural performance is derived from the portal steel frame [1]. (Fig. 2).

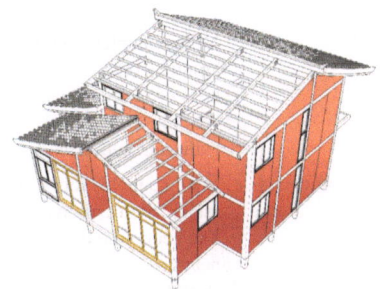

Fig. 2. Material and Structure Display of Prefabricated Residential Buildings in Place. **Figure source:** Self-drawing

3.4 Construction Assembly

The assembly process of prefabricated residential construction includes foundation construction, embedded parts installation, main steel structure installation, wall installation, roof construction, exterior wall decoration, indoor installation foundation construction, embedded parts installation, main steel structure installation, wall installation, roof construction, exterior wall decoration, indoor decoration, and overall delivery, which is very convenient.

4   Design Strategy for Local Prefabricated Residential Buildings

4.1   Sustainable Design Based on the Entire Lifecycle

The traditional residential space is rigid, making it difficult for residents to change the layout of the residential space according to their own needs and preferences. The "variable residence" uses assembly technology to create a design that can be divided and combined, and can achieve multiple combinations. It adapts to different family structures and the different demands of the same family at different stages of life for the residence in a flexible and variable combination.

The two-bedroom layout that adapts to the world of young couples can be transformed into a four-bedroom apartment that fully meets the living needs of three generations living together. The research on household types based on the theory of growth and sustainable development is as follows. (Table 2, Figs. 3 and 4, Chart 2).

Table 2. Functional Design of Sustainable Prefabricated Residential Buildings throughout their Life Cycle

Family stage	Family population and composition	Characteristics and main activities of each stage	Characteristics of each stage and main activity requirements
1. start-up stage (Newly Married Family)	Young couple (2 people)	Young couples centered around work, Less economic burden Working from home, gathering with friends	Independent study, Large living room
2. Upbringing period (Nuclear family)	Young couple, underage children (3–4 people)	Young couples focus on raising their children, The economic burden is heavy, Parent-child activities, receiving family and friends	Young couples focus on raising their children, The economic burden is heavy, Parent-child activities, receiving family and friends
3. Burden period (stem family)	Elderly people, young couples, underage children	Young couples focus on supporting the elderly and raising children, with a heavy financial burden on the school Parent-child activities, family gatherings, and extracurricular activities	There are more bedrooms, larger living rooms, restaurants, and separate courtyards for children's bedrooms and parents' bedrooms to reserve activity spaces. Suitable for aging without consideration
4. Stable period (Nuclear family, backbone family)	Middle aged couple, adult children (3–5 people)	Couples have less work pressure, their children enter society, and their financial burden is relatively light Family gatherings	Independent tea room, study The children's bedroom is arranged separately from the parents' bedroom Aging resistant design
5. Alimony period (Main family, combined family)	Elderly person, middle-aged couple, adult children (4–8 people)	Couples focus on supporting the elderly and their children enter society, Less economic burden, Family gatherings, outdoor activities	There are more bedrooms, larger living rooms, restaurants, and children's bedrooms are arranged separately from their parents' bedrooms, Reserve activity areas in the courtyard. Suitable for aging without consideration
6. Empty nest period (Empty Nest Family)	Middle aged couples or widowed elderly (1–2 people)	Retirement and elderly care for the elderly, There is a certain economic downturn, Family gatherings, outdoor activities	Independent Sunshine Room, Bedroom and Living Room, Emphasize lighting and reserve activity areas in the courtyard Suitable for aging design

Chart source: Self-drawing

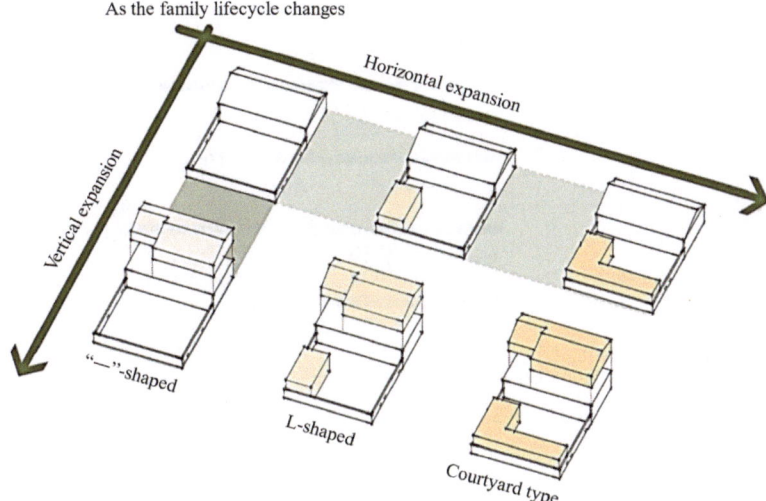

Fig. 3. Evolution of Local Prefabricated Residential Units throughout their Life Cycle. **Figure source:** Self-drawing

Fig. 4. Functional Evolution of Local Prefabricated Residential Houses throughout their Life Cycle. **Figure source:** Self-drawing

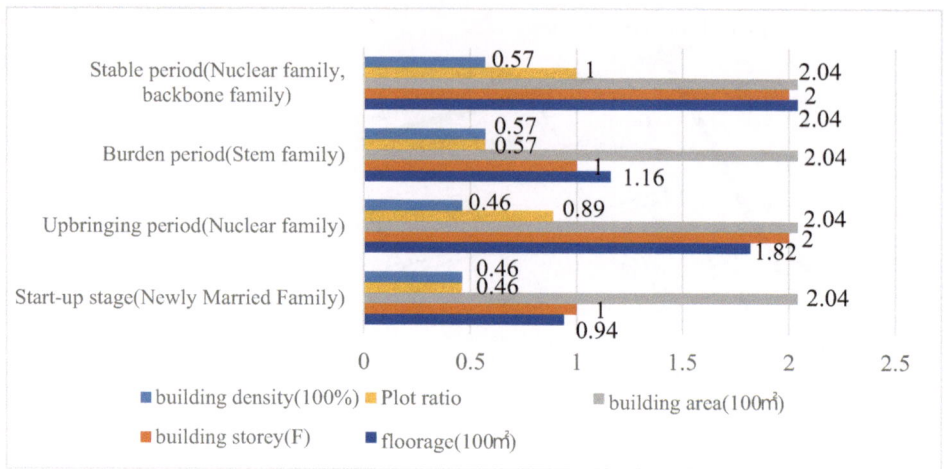

Chart 2. Economic and technical indicators of prefabricated residential buildings with different life cycles. **Source:** Self-drawing

4.2 Prefabricated Traditional Architectural Style Based on Locality

Preserving the unique rural style is the main theme of rural construction. The integration and reconstruction of rural culture and modern life concepts, and the role of rural architecture tends to be a regional and community cultural memory place, innovative architectural construction methods and intangible forms (emotions, culture, and spirit) conveyed in space are indispensable.

Establishing traditional courtyard space forms, sloping roof building facades, and standardized design of local building materials (bamboo, wood, rammed earth, stone, bricks, etc.) in prefabricated residential buildings, in line with local traditional architectural forms; Set up prefabricated cultural squares, gathering places, auditoriums, etc. in the public space of the village. (Figs. 5 and 6).

Fig. 5. Architectural Style of Prefabricated Residential Villages in Local Areas. **Figure source:** Self-drawing

Fig. 6. Architectural Style of Prefabricated Residential Buildings. **Figure source:** Self-drawing

4.3   Respect the Historical Context and Place Spirit of Rural Architecture

In the mountainous areas, buildings usually use the height difference relationship of the site to adopt a "land loving" architectural form, which visualizes the natural structure and expresses the ownership of the land and the "freedom" in space.

Use simple and pure forms to respond to the original pure atmosphere of the site, such as "Stilted building" and "Bench type building", to achieve organic symbiosis with the site.

At the same time, by combining "new" functions with "old" sites, excessive development is avoided. In response to the national homestead policy, each house can only be limited to the existing outline of the homestead, while retaining the original village texture, breakthroughs are made in internal functions to meet the "new" housing needs.

5   Summary and Outlook

The assistance of prefabricated residential buildings in rural construction is an important part of rural revitalization, and it is necessary to cultivate talents in prefabricated building construction, rural renovation, and rural landscape, so that more workers can master their own skills. By building a rural revitalization college, collaborating with universities, and inviting rural revitalization experts, scholars, and frontline practitioners to hold various trainings for the whole province and even the whole country, we aim to provide talent. Introduce cultural and tourism projects to cultivate a group of new era farmers who understand management and management.

At the same time, combined with the integration of BIM technology, green buildings, and ultra-low energy consumption technology, systematic processes such as internet digital display, demand collection, customized design, factory production, visual construction, delivery and use, and post maintenance are utilized to achieve C to F (personal to factory) building transactions.

Integrate online and offline, create a one-stop comprehensive service platform for rural housing construction in China, and optimize prefabricated buildings to assist in rural revitalization.

Ultimately, we will achieve a rural revitalization operation model that integrates rural industries, rural tourism, rural architecture, and rural construction as a whole. Promote the process of rural revitalization in China.

Fundings.

1. The project of Humanities and Social Sciences of Chongqing municipal commission of education: Research on the Design Method and Application of Prefabricated Rural Homes in Chongqing from the Perspective of Rural Revitalization, Project No. 23SKGH476.
2. The project of Humanities and Social Sciences of Chongqing education commission: A Study on the Construction of Genetic Map of Traditional Village Dwelling Culture in Chongqing Driven by Digital Intelligence, Project No. 23SKGH130.
3. The 2021 Higher Vocational Education Teaching Reform Project of Chongqing Education Commission: Research and Practice on the Teaching Reform of Residential Area Planning Curriculum in Higher Vocational Colleges under the Background of Land and Space Planning, Project No. Z213242.
4. Chongqing Education Science Planning Project: Research on the Comprehensive Training Model of Architectural Design Talents from the Perspective of Type Education,Project No.2021-GX-493.

References

1. Weiping, M., Guan Zhixin, X., Shuizhou.: Prefabricated rural residential structural forms based on the construction of beautiful rural areas. Metal struct. Chinese architect. **03**, 107–109 (2023)
2. Xuan, X.: Whose village is the countryside- Local Design of Rural Architecture. Shanghai Cultural and Creative ICC. https://mp.weixin.qq.com/s/A9Ae8L1a6QiM9CbMsZzyIQ
3. Shi, J., Wei, J.: Research on the Construction of Prefabricated Rural Homes in Chongqing Based on SWOT Analysis. In: Proceedings of the 2021 Industrial Architecture Academic Exchange Conference of MCC Construction Research Institute Co. Ltd. (Volume 2), p. 5 (2021)
4. Bribián, I.;Z.: Usón, A.A., Scarpellini, S.: Life cycle assessment in buildings: state-of-the-art and simplified LCA methodology as a complement for building certification. Building and Environment **44**(12) (2009)
5. Oliver, E.: Encyclopeda of vernacuar architecture of the world. Tradit. Dwelli. Settle. Rev. **10**(2), 69–75 (1999)
6. Zhenji, C.: Development path of housing industrialization in China. Construction Technology **45**(07), 582587 (2014)
7. Yue, W.: Research on standardized design and application of prefabricated rural homes - taking the design of fenghuang village in qingdao as an example. Build. Mat. Decorat. **32**, 116–117 (2018)
8. Chen, T.: Research on standardization and diversification of light steel prefabricated rural housing types in Beijing. Beijing University of Technology (2016)
9. Zhu, P.: Research on the design strategy of prefabricated light steel structure farmhouse buildings in minnan region. Chongqing University (2022)

Research on the Cooling Effect
of Environmentally Friendly Semi-flexible
Pavement Material

Dan Hong[1][(✉)] and Dan Liu[2]

[1] Chongqing College of Architecture and Technology, Chongqing 401331, China
155177231@qq.com
[2] Chongqing Yuelai Investment Group Co, Chongqing, China

Abstract. The article studies the cooling mechanism of environmentally friendly semi-flexible pavement by studying the cooling by evaporation of water in the mortar to take away the latent heat of the pavement to form a natural and environmentally friendly water retention and cooling effect. Using a large pore matrix asphalt mixture filled with water retention mortar, simulated indoor sunlight irradiation warming to 60 °C after artificial sprinkling and natural rainfall test method, recorded the asphalt mixture with different void ratio and different grouting amount of specimens before and after the grouting matrix asphalt mixture to do the temperature reduction comparison. The results show that: environmentally friendly semi-flexible pavement materials compared to ordinary asphalt mixture with high reflectivity, less heat absorption, water retention and cooling effect is obvious, to alleviate the urban heat island effect has a certain role in promoting the cooling effect with the size of the void rate or the amount of grouting is positively proportional to the size of the grout.

Keywords: water retaining mortar · Environmentally friendly · Large Pore Asphalt Mixture · cooling effect

1 Introductory

In recent years, with the continuous acceleration of urbanization and the high development of industrial construction, some phenomena have exacerbated the development of the urban heat island phenomenon, such as the reduction of green space; atmospheric pollution; the increase of artificial waste heat from automobiles and air conditioning equipment brought about by the high level of industrial development; the increase in the number of concrete buildings; the change of the wind direction in the city due to the influence of the buildings; and the use of low-reflective paving materials and building materials; The use of paving materials such as asphalt and concrete [1]. At the same time, the external thermalization and the heat generated by indoor machines and equipment lead to an increase in the energy consumption of air conditioning inside the building, which in turn contributes to the increase in the temperature of the city, creating a vicious circle. In addition, asphalt and concrete surface paving contribute to the increase of the

© The Author(s) 2024
P. Xiang and L. Zuo (Eds.): PBSFTT 2023, LNCE 382, pp. 438–445, 2024.
https://doi.org/10.1007/978-981-97-5108-2_47

urban heat island phenomenon. These paving materials have a high heat capacity, and even at night, the heat that is not fully emitted during the day continues to be emitted to the atmosphere, resulting in urban surfaces remaining at high temperatures almost all day long [1]. Therefore, new pavement structures and materials are needed to reduce pavement distress and mitigate the heat island effect, which can improve road performance while suppressing the increase in pavement temperature [2]. Cities are currently facing major and severe threat of environmental stability that many researches focus on finding ways to help save, protect cities, and bring back local biodiversity and human breathtaking places [3].

Water retention and cooling pavement generally refers to a functional pavement that retains water inside the pavement structure, inhibits the rise of surface temperature and reduces air temperature through evaporation of internal water [4]. Sun Gaofeng studied the slag powder, fly ash, slaked lime mixed with water as a water retention emulsion grout infused OGFC matrix formed of water retention pavement materials, indoor simulation of the pavement light and heat environment than dense graded asphalt mixture has a cooling effect of 10 °C [5]. Zhang Liangliang et al. studied the steel bridge deck asphalt concrete pavement layer infused with water retaining mortar as well as the combination of water sprinkling can effectively reduce the pavement creep deformation and rutting damage, and improve the economic benefits [6], Oliveira studied the effect of the use of a cellulose-based water immersion agent on the construction of concrete blocks [7].

Combined with the above research of several pavement cooling ideas, this paper adopts a large pore space asphalt mixture filled with water retention mortar, with reference to the "CJJ/T 206–2013 Urban Road Low Heat Absorption Pavement Technical Specification" [8] of the Appendix C Cooling Effect Measurement Methods, with 220V275W diffused electronic lamps to simulate the sun, to carry out water retention evaporation of the pavement temperature reduction test.

2 Environmentally Friendly Semi-flexible Pavement Material Specimen Production

2.1 Water Retaining Mortar

The water retention mortar used in this paper consists of cement, fly ash, blast furnace slag powder, diatomaceous earth, slow-setting high-efficiency water-reducing agent, microsilica powder and other materials. After the ratio optimization design finally determined to meet the target design requirements of the water retention mortar ratio of cement: fly ash: water: blast furnace slag powder: diatomite: slow-setting high-efficiency water reducing agent: micro-silica powder = 1626.8: 697.2: 1790: 414: 112.8: 27.9: 185.9. Technical requirements to meet the "semi-flexible mix with cementitious grouting materials" (JT/T1238- 2019) [9] specification requirements, water retention mortar performance is shown in Table 1.

Table 1. Comparison of required and measured values of water-retaining mortar performance indexes

Project name	Mobility (S)	Compressive strength (MPa)	Flexural strength (MPa)	Water retention (%)
regulatory requirement	10~14	15~30	>2.0	>20%
measured value	13.6	16.5	3.05	31.6%

2.2 Environmentally Friendly Semi-Flexible Pavement Mixes

In this paper, the environmentally friendly semi-flexible pavement mixture using rubber powder compounded SBS, olefin polymers and petroleum resins to prepare high viscosity modified asphalt to meet the specification requirements; through the ratio design to determine the matrix asphalt mixture of aggregate grading range of SFAC-13, and its specific composition is shown in Table 2; by the Kentenburg fly-away test and Cherenburg precipitation leakage test to determine the grading limit of the optimal oil-rock ratio of 3.4% (the best asphalt dosage is 3.3%) The performance indexes of matrix asphalt mixture meet the technical requirements.

Table 2. SFAC-13 gradation scope of mother asphalt mixture

Sieve size (mm)	19.0	13.2	4.75	2.36	0.6	0.3	0.15	0.075
Percentage of quality passed (%)	100	90–100	10–30	5–22	4–15	3–12	3–8	1–6

2.3 Water Retaining Mortar Production and Forming of Base Asphalt Mixtures

Water retention mortar production according to "highway engineering cement and cement concrete test procedures" (JTG E30–2005) [10] in the cement mortar molding method, the specific steps are first of all will be cement, water retention agent and additives are added to the mixing pot, start the mixer mixing contract for 1~2 min, to be homogeneous, with the mixing to the mixing pot to join the amount of water designed to be added, and add the material used for the time should not be more than 2 min, the water is added to the whole, and then continue to mix the contract for 2 min, turn off the mixer.

Base asphalt mixture according to "highway engineering asphalt and asphalt mixture test procedures" (JTG E20–2011) [11] in the T 0702–2011 asphalt mixture specimen production methods (compacting method) to produce Marshall specimens and T 0703–1993 asphalt mixture specimen production methods (wheel milling method) [12] to produce rutted plate specimens. The base asphalt mixture in 130~160 °C mixing, 120~150 °C under the compacting or rolling molding, Marshall specimen double-sided compacting each 50 times. Mixed rutted specimen plate rolling molding density by compacting Marshall specimen density control, after starting the wheel mill, rutted specimen plate in one direction first 2 round trips (4 times), unloading, and then lift the milling wheel, the specimen will be reversed direction, and then add the same load crushed 12 round trips (24 times), to the specimen to achieve the Marshall standard compactness of $100 \pm 1\%$ until.

Wait until the specimen is cooled to below 60 °C before grouting, when the mortar penetration is complete, use a rubber scraper to scrape off the excess slurry to expose the unevenness of the surface of the asphalt mixture specimen is appropriate. Since the cement mortar will remain on the paving surface, which may affect the appearance of the road surface to a certain extent and reduce the anti-skid performance of the asphalt pavement, the surface can be rinsed with water before the initial setting of the slurry after the cement slurry is grouted. The specimen after grouting will be left at room temperature for more than 24 h, until the slurry hardening, put into the temperature of 20 °C $\pm$ 3 °C, humidity greater than or equal to 90% of the sustenance room sustenance, to the age of the demolding, measured its road performance indicators.

3   Test Methods

3.1   Test Preparation for Water Retention and Cooling of Environmentally Friendly Semi-Flexible Pavements

In order to accurately simulate the sunlight irradiation, so that the water retention and cooling specimens get uniform light, simulating the sunlight of the electric lamp using 220V275W diffusion type electronic lamp, its irradiation height can be automatically adjusted to about 100 cm. The temperature data were recorded by XMTD temperature readout meter with 0.1°C resolution. Adjust the irradiation height, so that the dense standard specimen AC-13 in the electronic lamp directly under the surface temperature of 60 °C after 3 $\pm$ 1 h of irradiation, and fix the height of the electronic lamp.

3.2   Correction Test for Temperature Readout Data

Considering that the thermocouple of the temperature reading instrument will produce test error when connecting the asphalt pavement surface, specially through a cup of high temperature heating of water, with the range of 100 °C thermometer for the comparison of the correction test, and its test data are shown in the following table 3, the data comparison graph is shown in Fig. 1.

Table 3. XMTD temperature correction of the data

	Real-time temperature °C								
XMTD temperature gauge	85.4	59.9	54.2	46.4	43.9	37.4	29.5	25.6	24.8
thermometers	86.4	60.9	55.1	47.2	44.7	38.1	30.3	26.4	25.4
difference (the result of subtraction)	1.0	1.0	0.9	0.8	0.8	0.7	0.8	0.8	0.6

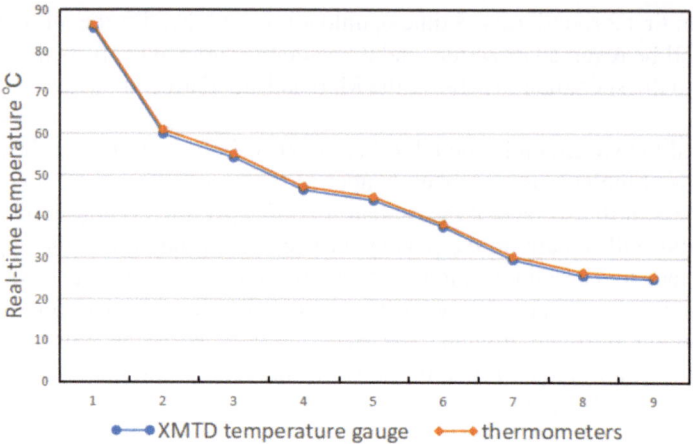

Fig. 1. Correct testing temperature contrast figure

As can be seen in Fig. 1, the difference between the real-time temperature measured by the XMTD temperature reader and the thermometer is not significant, and the XMTD temperature reader is slightly lower than the thermometer, in the range of 0.6 to 1.0 °C.

its average value $\mu = \frac{1}{N} \sum\limits_{i=1}^{N} x_i = 0.822$.

(statistics) standard deviation $\sigma = \sqrt{\frac{1}{N} \sum\limits_{i=1}^{N} (x_i - \mu)^2} = 0.015$.

The standard deviation is less than 0.02, indicating that the difference between the XMTD temperature reader and the thermometer is relatively centralized, and the average value of + 0.8°C can be taken as the correction value of the XMTD temperature reader.

4 Experimental Process and Result Analysis

After the preparatory work, will be immersed in water for more than 12 h to keep the surface of the cooling specimen with a wet towel dry, placed in the diffuse light type electronic lamp directly under the arrangement of the thermocouple position of the temperature readout meter, turn on the power supply of the temperature recorder, 2~3 min, turn on the diffuse light type electronic lamp switch, and at the same time began to record the temperature data.

Comparing the cooling effect under each test condition, the comparison graphs before and after the upper limit grouting are shown in Fig. 2, before and after the median grouting are shown in Fig. 3, before and after the lower limit grouting are shown in Fig. 4.

From Fig. 2, it can be seen that before and after the upper limit grouting has no obvious effect on the cooling effect, and the difference is only 3°C. From Fig. 3, it can be seen that the cooling effect before and after the middle value grouting is relatively obvious, which can reach 7°C. From Fig. 4, it can be seen that the lower limit grouting before and after the cooling effect is more prominent, 5 min after the cooling reached 7

°C, and has been sustained at about 9~10 °C, up to 11 °C, it can be seen that, with the increase of the void ratio, the water retention mortar filling amount increases, its cooling effect is more obvious.

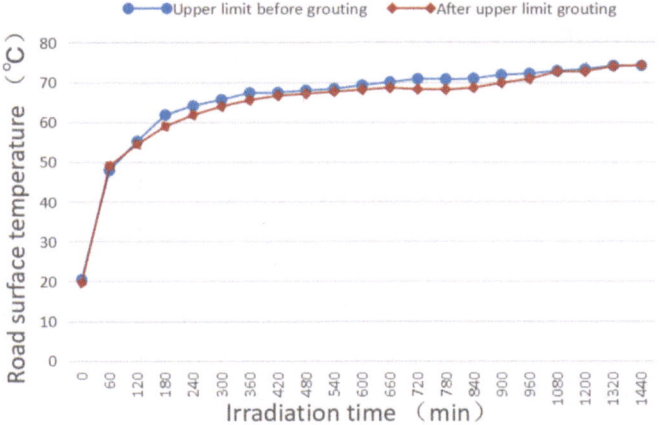

Fig. 2. Upper limit of before and after grouting cooling contrast figure

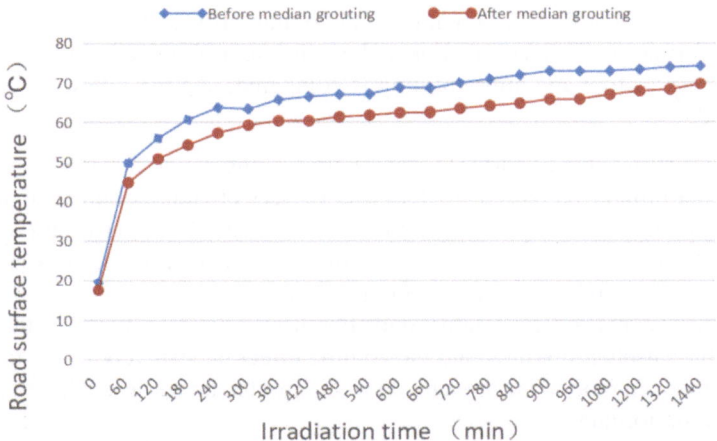

Fig. 3. Median grouting and cooling contrast figure

In the natural phenomenon, the sun exposure time is generally not more than 10 h a day [10], in the morning there is dew, night cooling, rainy weather, water vapor in the atmosphere, rain or artificial sprinkling of water can be stored in the asphalt pavement in the water-preserving mortar, the formation of a pavement is always inside the state of the benign cycle of water, the water-preserving pavement will be a sustained and durable cooling function, to achieve energy-saving and environmental protection of the natural evaporation of the temperature reduction.

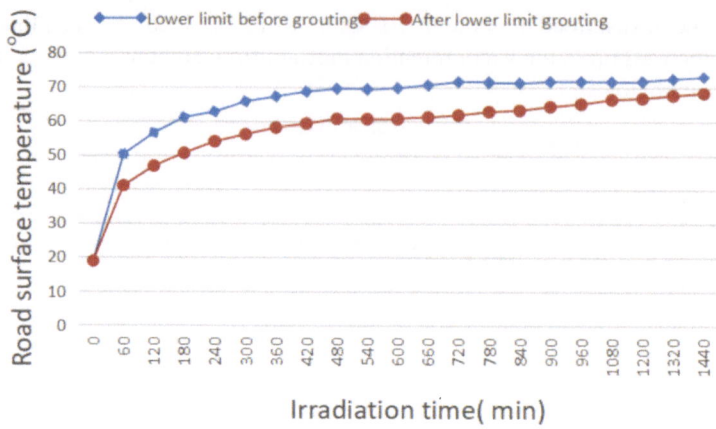

Fig. 4. The lower the temperature before and after grouting contrast figure

5   Concluding Remarks

This paper analyzes the mechanism of water retention and cooling, water storage is achieved by natural rain or artificial sprinkling, cooling is achieved by evaporation of water in the mortar to take away the latent heat of the pavement, forming a natural environmentally friendly water retention and cooling effect. Through the study of the cooling performance of water retention and cooling semi-flexible pavement, the matrix asphalt mixture before and after grouting is compared, and the following conclusions are obtained:

(1) The correction value of XMTD temperature reading meter is $+ 0.8°C$.
(2) The upper limit (void ratio of 16.1%) is not a very obvious cooling due to the small void ratio, and the median (void ratio of 19.9%) and the lower limit (void ratio of 23.4%) have a continuous temperature difference between 10 and 16 °C after 80 min of irradiation, which can last for more than 24 h.
(3) The temperature difference before and after grouting is determined by the void ratio of the base asphalt mixture, and the cooling effect is positively proportional to the size of the void ratio or the amount of mortar grouted under the condition of meeting the road performance.
(4) The water-preserving mortar can circulate and repeatedly absorb and release water, which provides a virtuous cycle of water storage and evaporation for the water-preserving and cooling semi-flexible pavement, and achieves the natural evaporation cooling of energy-saving and environmental protection of the pavement.

References

1. Chen, W.J., Shui, G.Z.F.: Comparison of evaporative cooling effect on the surface of open grain and water retaining asphalt pavement materials. J. Building Sci. **27**, 56–59 (2011). https://doi.org/10.13614/j.cnki.11-1962/tu.2011.04.002

2. Liu, C.X.: Semi-flexible Color Cement Grouted Asphalt Pavement Key Technology Research. D. Chang'an University, Xian (2016)
3. Ayad Eman, H., Rasheed Sarah, E.A.: Sponge cities technology. J. IOP Conference Series: Earth and Environmental Science **1113** (2022). https://doi.org/10.1088/1755-1315/1113/1/012005
4. Li, Y.Y., Xie, L.B., Hu, L.Q.: A review of research on water retaining pavements. J. Centr. Exter. Highw. **37**(1), 53–57 (2017). https://doi.org/10.14048/j.issn.1671-2579.2017.01.013
5. Sun, G.F.: Design of Water Retaining Asphalt Mixture Material Composition and Road Performance Research. D. Chang'an University, Xian (2009)
6. Zhang, L.L., Deng, N.Z., Zeng, Z.R.: Analysis of the cooling effect of water sprinkling on steel bridge deck pavement after filling with water-retaining mortar. J. Railway Construct. **58**(10), 14–18 (2018)
7. Oliveira, A.L., Corrêa, B.P., Ribeiro, I.F.R., Souza, R.A., Calçada, L. M.L.: Influence of water-retaining cellulose ethers on the properties of bedding mortar used in concrete structural masonry. J. Ambiente Construído **15**(3) 57–69 (2015). https://doi.org/10.1590/s1678-862120 15000300026
8. Technical Specification for Low Heat Absorption Pavements for Urban Roads (CJJ/T 206-2013), p. 2. S. China Construction Industry Press, Bejing (2014)
9. Cementitious grouting materials for semi-flexible mixes (JT/T1238-2010), p. 1. S. People's Transportation Press Co., Bejing (2019)
10. Test Specification for Cement and Cement Concrete for Highway Works (JTG 3420-2020). p. 11. S. People's Transportation Press Co., Bejing (2020)
11. Test Specification for Asphalt and Asphalt Mixture in Highway Engineering (JTG E20-2011), p. 12. S. People's Transportation Press Co., Bejing (2011)
12. Song, C.N.: Research on the temperature field of roadbed pavement and working environment temperature index of pavement in desert area. D. Chang'an University, Xian (2006)

Research on the Influencing Factors of Migrant Workers' Transformation into Industrial Workers Based on Intelligent Construction——Taking Chongqing as an Example

YiFan Cao[1], Xu Wang[2(✉)], Yidan Kou[2], and Zixiang Zhi[2]

[1] Chongqing College of Architecture And Technology, University Town, Chongqing, China
[2] Chongqing College of Electronic Engineering, University Town, Chongqing, China
261954380@qq.com

Abstract. A high-quality workforce in the construction industry promotes the development of construction industrialization and intelligent construction. However, there are some difficulties in transitioning from traditional construction workers to industrial workers. This study takes migrant workers in the construction industry in Chongqing as a sample, and conducts empirical research using methods such as literature review, questionnaire survey, SPSS factor analysis, and multiple linear regression analysis. Based on the main influencing factors, strategic suggestions are proposed for your reference.

Keyword: Intelligent construction · Migrant workers · Transformation · Influencing factors

1 Introduction

China has clearly stated in multiple policy documents that it is necessary to "promote green and low-carbon building materials and green construction methods, accelerate the industrialization of new buildings", and vigorously promote industrialized construction methods such as prefabricated buildings [1]. The development of industrialization cannot be separated from the construction migrant workers on the production line. The Ministry of Housing and Urban Rural Development has issued the Guiding Opinions on Accelerating the Cultivation of a New Era Construction Industry Worker Team, proposing that construction industry workers are an important component of China's industrial workers and the foundation for the development of the construction industry. Promoting the gradual transformation of traditional construction workers from migrant workers to industrial workers is not only necessary to adapt to the transformation of the construction industry, but also provides stronger talent support for the sustainable development of the construction industry [2]. Therefore, the research on the transformation of migrant workers in the construction industry has practical significance and research value.

P. Xiang and L. Zuo (Eds.): PBSFTT 2023, LNCE 382, pp. 446–456, 2024.
https://doi.org/10.1007/978-981-97-5108-2_48

2   Analysis of Transition Factors for Migrant Workers

2.1   Identification of Influencing Factors

Literature Research Identifying Factors. For the influencing factors of the transition from migrant workers to industrial workers in the construction industry, scholars have already made a summary, but there are some differences in research perspectives. Starting from 2001, we searched and selected the core journal literature that had been cited more than 5 times, and summarized that the identity contradiction problem of migrant workers who are also workers and farmers is the key (Table 1). The transition of migrant workers to industrial workers is essentially a transformation of their professional and social identities [3]. Focusing on this theoretical foundation, from the perspective of identity transformation, we further selected representative literature in the past ten years, and summarized six types of key factors after sorting: human capital factors, institutional factors, industry factors, enterprise factors, macro-occupational factors and policy factors.

Research Interview Identification Factors. The development of prefabricated buildings in China is in Its early stages, and there is relatively little research on industrial workers [4]. Therefore, team members conducted in-depth visits to relevant management personnel engaged in construction industrialization and industrial workers in representative construction companies, and comprehensively screened the influencing factors from the perspectives of enterprises, workers, and labor subcontracting. Based on the interview results, the identified influencing factors in the literature were ultimately reduced to 30.

Table 1. Literature Identification Impact Factor

No	Influencing factors	No	Influencing factors	No	Influencing factors
1	Payment of wages	11	Work intensity	21	Real-name management
2	Wage Income	12	Occupational stability	22	Labor Contract Signing
3	Gender	13	Living conditions	23	Labor Enterprise Development and Transformation
4	Age	14	Training level	24	Industry Drivers
5	Worker's education level	15	Construction labor system reform	25	Urban Housing
6	Marital status	16	Degree of Migrant Worker Citizenship	26	Logistic Security System
7	Professionalism	17	Level of industrialization of construction	27	Employee Incentive Mechanism
8	Individual's Willingness to Transfer	18	Migrant workers' professionalization level	28	Recruitment Channels
9	Work environment	19	Welfare Benefits	29	Employee Happiness Index
10	Job security	20	Social Security System	30	Management Difficulty

Process of Extracting Data from Questionnaires. Based on the 30 factors identified above, rate the importance of each factor. Based on statistical theory, the designed questionnaire will be in the form of a likert rating scale, set at 1–5 levels (1 = very unimportant, 2 = unimportant, 3 = average, 4 = important, and 5 = very important). During the research process, a total of 900 questionnaires were distributed, and 721 valid questionnaires were initially collected. The respondents in the valid questionnaire not only include migrant workers from traditional construction sites, but also 450 frontline workers engaged in PC component production and prefabricated construction installation workers, 15 project managers in the construction industry, and 51 frontline labor management personnel.

2.2 Selection of Key Factors

Selection Method. Due to the complex and dispersed factors that affect the transformation of migrant workers, it is difficult to extract key factors using traditional regression analysis methods. Therefore, this article uses factor analysis to identify the main influencing factors of the transformation of construction migrant workers. Through SPSS 26.0, a model analysis was conducted on 721 valid questionnaires collected from the survey. The specific analysis elements include reliability testing, applicability testing, factor analysis process, key factor extraction, and importance ranking of key factors.

Reliability Test. Reliability test is an important method to determine whether the data is reliable or not, using this test on the data of 721 valid questionnaires, the value of Clonbach's coefficient (assessment tool) was obtained to be 0.825, which is greater than the basic requirement of 0.7, and meets the basic requirements of the reliability test, which indicates that the data is reliable, and can be analyzed by factor analysis.

Suitability Test. KMO and Bartlett's test of sphericity were used to determine the suitability of the research data for factor analysis. According to the commonly used KMO metrics given by Kaiser: KMO > 0.9 means very suitable; KMO > 0.8 means suitable; KMO > 0.7 means average; KMO > 0.6 means not very suitable; KMO < 0.5 means very unsuitable. As shown in Table 2, KMO takes the value of 0.825, which is greater than 0.8, indicating that it is suitable for factor analysis, and the Bartlett's test of sphericity and the statistical values meet the requirement of significance level.

Table 2. KMO and Bartlett sphericity test tables

Sampling sufficient KMO metrics		0.825
Bartlett's test of sphericity	approximate chi-square (math.)	4990.210
	Degree of freed	528
	Significance	0.00

Key Factor Selection. When extracting key factors, Kunhui Ye [10] proposed that the factor load of key indicators should be greater than 0.5, and the cumulative explanatory variance of key factors should also be higher than 60%. Finally, 16 key factors that meet the above conditions were selected, with a cumulative explanatory variance of 63.794% (>60%) (Table 3). Finally, based on the specific content and internal relationships of factors, they are classified into six key factors, namely human capital factor (F1), institutional factor (F2), industry factor (F3), enterprise factor (F4), macro occupational factor (F5), and policy factor (F6).

Table 3. Extraction of key factors

Key Factors	Number	Loadings	Total Correction Question Correlation	Explained Variance %	Cumulative Explanatory Variance %
F1	1	0.836	0.632	20.457	20.457
	2	0.802	0.712		
	4	0.685	0.734		
	5	0.758	0.736		
	7	0.558	0.702		
F2	19	0.555	0.628	12.124	32.581
	20	0.687	0.598		
	21	0.735	0.549		
	22	0.586	0.589		
F3	23	0.578	0.433	7.257	39.838
	24	0.625	0.692		
F4	30	0.764	0.401	9.975	49.813
F5	18	0.776	0.417	6.852	56.665
F6	16	0.765	0.508	7.129	63.794
	17	0.698	0.578		
	36	0.517	0.568		

Importance of Key Factors. The importance of the above 16 key indicators was calculated and ranked by applying the importance index calculation method. Importance index $P = \sum(kX)*100/5$, where: k is the respondents' rating of an indicator, and the rating weights are from 1 to 5 respectively; $X = n/(N-n)$, N is the number of questionnaires making the same rating of a factor; and N is the total number of questionnaires recovered. At the same time, the arithmetic mean of the coefficients of the indicators within the six key factors was calculated as a way to represent the importance of each key factor, and the final ranking of the importance index was F1 (79.1), F2 (66.85), F5 (59.5), F4 (56.1), F3 (47.95), and F6 (41.6).

3   Empirical Research on the Factors Influencing the Willingness of Migrant Workers to Transition

3.1 Sample Characteristics and Content Analysis

In order to obtain first-hand information and ensure the accuracy of the results, a total of 23 construction sites in Chongqing were visited, 900 construction migrant workers were selected for questionnaire surveys, 623 questionnaires were recovered, and the final number of valid questionnaires was 418 after eliminating invalid questionnaires with low reliability and incomplete data, and the 418 questionnaires were analyzed for measurement. The research object of this paper is mainly the traditional construction site migrant workers, and the survey contains six aspects: age, education level, wage income, wage payment, labor contract signing and willingness to transition [5].

Age. In the 418 sample data involved in this research, the age ranges of 36–54 years old and 55 years old and above accounted for the largest proportion of the number of people, accounting for 61%, close to 2/3 of the total number of people, indicating that the sample of the number of middle-aged and elderly people is larger, and the aging is relatively more serious. Compared with the middle-aged and old-aged groups, the youth group, that is, those aged 25 and below, accounted for a lower proportion, accounting for only 5.5%.

Educational Level. From the survey results of the education level of construction migrant workers, the middle school gradient accounts for the largest proportion, accounting for 52.9%, more than half. in addition the number of people at the level of secondary school, high school as well as technical school is also higher, accounting for 33%, as high as one-third. Elementary school and below accounted for 7.7%, so the overall literacy level of migrant workers in the construction industry remains low.

Wage Income. Through the research, it is found that there is a large gap in the demand for construction workers in Chongqing, and the wages of migrant workers in the construction industry depend firstly on the type of work, with the wages of general handymen ranging from 4,000 to 8,000 yuan, and the wages of reinforcing steel workers, masonry workers and other difficult coefficients and specialties being higher than 8,000 yuan. Secondly, their wages are mainly based on piecework, and the difference in wages for the same type of work mainly lies in the number of pieces completed, with more work getting more pay [6].

Wage Payment. The results of the survey show that because of the national macro-control policies in place, and because of the institutional initiatives for the protection of the rights and interests of migrant workers in recent years, there is basically no wage arrears for construction migrant workers, but there are still 23.7% of the migrant workers who said that they could not get their wages on time [7].

Signing of Labor Contracts. Among the surveyed migrant workers in the construction industry, only 40% of the number of people have signed a contract, less than half. it can be clearly seen that the situation of signing labor contracts by migrant workers in the construction industry is not optimistic.

Willingness to Transform. In order to facilitate quantitative analysis and accurately investigate the degree of willingness of migrant workers, this study set up the questionnaire with five levels, and the results were: very willing (28.2%), willing (21.8%), quite willing (26.1%), not too willing (15.3%), and unwilling (8.6%).

Some of the researched construction migrant workers have the willingness to transition, but in their choice, they are still constrained by factors such as age, education, and wage payment, so it is necessary to analyze the specific significance impact situation of each factor [8].

3.2 Significance Empirical Analysis

Selection and Definition of Explanatory Variables. Based on the results of the factor analysis in Sect. 2, among the main influencing factors (i.e., explanatory variables) extracted for the transformation of traditional construction migrant workers, the influencing factors that ranked in the top five in terms of importance indexes (importance indexes greater than 70) and were easy to analyze quantitatively were selected, which were, in the order of ranking in terms of their importance indexes, age (88.3), wage payment situation (86.1), wage income (80.9), labor contract signing (79.1) and workers' education level (76.3).

Research Assumptions
Assumption 1: Age has a negative effect on the willingness of construction migrant workers to transform, the older the willingness to transform the less strong; Assumption 2:The degree of education has a positive effect on the willingness of construction migrant workers to transition, the higher the degree of education, the stronger the willingness to transition; Assumption 3: Wage income has a negative effect on the willingness to transition of construction migrant workers, the higher the income the less willing to transition; Assumption 4: Wage payment has a negative effect on the willingness to transition of migrant workers in the construction industry, and the willingness to transition is not high if wages can be paid on time; Assumption 5: The contract signing situation has a negative effect on the willingness to transition of construction migrant workers, and the willingness to transition is stronger if the labor contract is not properly signed.

Analysis of Multiple Linear Regression Models. The collected valid data were measured by model analysis, and the multivariate linear regression model was established and econometrically analyzed by SPSS26.0 to analyze the significance of the impact of each factor [9] (Table 4).

Table 4. Means and standard deviations of the variables

Variable	Mean	Standard deviation	Variance	Sample size
Age	2.77	0.863	0.745	418
Wages	4.04	0.879	0.773	418
Wage payment	1.24	0.428	0.183	418
Worker's education level	2.38	0.718	0.516	418
Contract signing status	1.60	0.492	0.242	418

Descriptive statistics of model variables:

Summary of the model: The adjusted R-squared represents the fit of the model, i.e. how well the linear equation responds to the real data. As shown in Table 5, the adjusted R-squared of the determinable coefficient of the model of this study is 0.639, which indicates that this regression equation is able to explain 63.9% of the real data, suggesting that the model has a good explanatory power. Meanwhile, the DW value in the regression model of this paper is 1.992, which is very close to 2, indicating that the residual terms are independent of each other and are not interfered with, which indicates that the data have a good explanatory ability and the regression analysis results are more accurate.

Table 5. Model summary

Model	R	R-square	Adjusted R-square	Error in Standard Estimates	Amount of change in R-squared	Amount of change in F	Degree of freedom 1	Degree of freedom 2	Significance F amount of change	Durbin-Watson (DW)
1	.808a	0.653	0.639	0.770	0.653	46.746	5	124	0.000	1.992

Analysis of variance table: As shown in Table 6, the corresponding F-value in the regression variance is 46.746, which corresponds to a significance of 0.000b, which is less than 0.05, indicating that the regression equation is valid.

Table 6. Analysis of variance

	Model	Sum of Squares	Degrees of Freedom	Mean Square	F	Significance
1	regression	138.715	5	27.743	46.746	0.000[b]
	Residual value	73.592	124	0.593		
	Total	212.308	129			

Regression coefficient table:

Table 7. Regression coefficient

Model	B	Standard error	Beta	t	Significance	Tolerances	VIF
(Constant)	7.965	0.662		12.026	0.000		
Age	−0.808	0.094	−0.543	−8.568	0.000	0.695	1.439
Wage	−0.243	0.081	−0.167	−3.004	0.003	0.908	1.102
Wage Payment	−0.747	0.162	−0.249	−4.621	0.000	0.962	1.039
Payment	0.317	0.114	0.178	2.792	0.006	0.69	1.448
Worker's education	−0.524	0.143	−0.201	−3.674	0.000	0.936	1.069

Multicollinearity test:Multicollinearity generally refers to the strength of linear correlation between explanatory variables. In this paper, the variance inflation factor (VIF) is mainly selected as a way to determine whether there is multicollinearity between variables. The specific determination method to judge the strength of multicollinearity from the perspective of the size of the VIF value is as follows: the larger the VIF is, especially when the VIF $\geq$ 10, it indicates that the multicollinearity between the explanatory variables is more serious; the closer the VIF is to 1, it indicates that the multicollinearity between the explanatory variables is weaker. From Table 7, the tolerance range of the five independent variables is 0.619 ~ 0.932, which is close to 1, and VIF $<$ 10, indicating that the multicollinearity among variables is relatively weak.

Significance analysis: As can be seen from Table 7, at the 10% significance level, the five indicators of age, wage, wage payment, workers' education, and contract signing passed the significance test, and the significance level is high, indicating that the results of factor analysis are more accurate.

Analysis of Results
Age: As shown in Table 7, the regression coefficient value (Beta value in Table 7) of age on the degree of transition intention is −0.543, with a significance level of P = 0.000 $<$ 0.05, meeting the requirements for significance level. It indicates that age affects the willingness to transition of migrant workers in the construction industry to a greater extent under the same conditions. The regression coefficient is negative and the absolute value is large, indicating that there is a significant negative correlation between age and willingness to transition.

Wages: The value of regression coefficient of wage on the degree of willingness to transition is −0.167 with significance level P = 0.003 $<$ 0.05, which satisfies the requirement of significance level. It indicates that under the same conditions, the degree of influence of wages on the willingness to transition of migrant workers in the construction industry is greater. The negative regression coefficient indicates that wages are negatively correlated with the willingness to transform.

Wage payment situation: The value of regression coefficient of wage payment situation on the degree of willingness to transform is −0.249, and the significance level P =

0.000 < 0.05, which satisfies the requirement of significance level. It indicates that under the same conditions, the wage payment situation has a greater degree of influence on the willingness of transformation of migrant workers in the construction industry. The negative value of the regression coefficient indicates that the willingness to transform is lower for the migrant workers who can get their wages on time.

Educational level of workers: The value of regression coefficient of age on the degree of willingness to transition is 0.178, and the significance level $P = 0.006 < 0.05$, which satisfies the requirement of significance level. It indicates that under the same conditions, the degree of education of workers has a greater degree of influence on the willingness of transformation of migrant workers in the construction industry. The regression coefficient is positive, that is, the higher the education level of the construction migrant workers' willingness to transition is higher, and there is a significant positive correlation between the education level and the willingness to transition, which supports the original hypothesis.

Contract signing situation: The value of regression coefficient of contract signing situation on the degree of willingness to transform is -0.201, with significance level $P = 0.000 < 0.05$, which satisfies the requirement of significance level. It indicates that under the same conditions, the contract signing situation has a greater degree of influence on the willingness of transformation of migrant workers in the construction industry. The negative value of the regression coefficient indicates that the transformation willingness of construction industry migrant workers who do not sign labor contracts is stronger.

4 Conclusions and Strategies

Further analysis through empirical evidence found that the younger the age, the higher the education level and the lower the wage of construction migrant workers, the greater the willingness to transition; And the older, the less educated, and the higher the wages of the construction industry migrant workers, the smaller the willingness to transition; at the same time, the wage payment situation and contract signing situation have a negative impact on the willingness to transition of the construction migrant workers, i.e., the willingness to transition is stronger for those migrant workers whose wages can not be paid on time and at the same time do not have a labor contract signed, and the results of the analysis provide ideas for the path of the transition. In view of the difficulties in transition, this paper provides the following theories and suggestions for the transition of future migrant workers in the construction industry:

Firstly, establish policy pathways to attract young migrant workers to transform. Increase the implementation of policy publicity, popularize and update the knowledge of industry development for migrant workers, and promote the active transformation of young individuals. At the same time to promote the realization of industrial workers own process, improve the construction industry migrant workers vocational training system, the establishment of intelligent construction industrial workers training base, increase the attractiveness of the industry to form a benign transformation closed loop.

Secondly, lowering the cost of transition and increasing the transition of migrant workers. Establish a system of subsidies and salary guarantees for high-paying skilled trades to allay their transition concerns and address the skills gap for industrial workers. And set up an incentive mechanism to boost the transformation of lower-wage construction migrant workers.

Thirdly, improving the remuneration system to reduce the loss of migrant workers. By improving the pay and benefits system and implementing a system of payment of wages by banks, the long-standing problem of wage arrears will be eliminated, and their livelihood needs will be met while providing them with appropriate social security, thus reducing the outflow of rural migrant workers and increasing the base for transition [10].

Fourthly, standardize the entry process and increase the contract signing rate for migrant workers. Specialized departments have been set up to regulate and institutionalize the management of construction industry workers, and simple labour contracts have been introduced to increase the supervision of the labour market in the construction industry and to ensure that the basic interests of migrant workers are protected.

Fifthly, a long-term training system should be established. Vocational education and continuing education for migrant workers in the construction industry should be implemented to help migrant workers in the construction industry to become quality construction industrial workers in the new era and to efficiently complete the transition from migrant construction workers to industrial workers.

References

1. Po, L.: Research on incentives to stabilize migrant workers in the construction industry. Hebei University of Technology (2014)
2. Jian, S.: Transformation of Migrant Workers: Searching for Spring in Transformation. China Construction Daily 2016-06-13
3. Zhu, M.: Research on the Mechanism and Countermeasures of the Industrialization of Migrant Workers in China's Construction Industry. Chongqing University (2019). https://doi.org/10.27670/dcnki.gcqdu.2019.000562
4. Chen, D.: Research on the Transformation of Migrant Workers in the Construction Industry under the Prefabricated Building System. Hubei University of Technology (2020). https://doi.org/10.27131/d.cnki.ghugc.2020.000306
5. Xu, Z. A study on the obstructive factors of migrant workers in the construction industry transforming into industrial workers based on rooted theory. Beijing Univ. Architect. Architect. (2022). https://doi.org/10.26943/d.cnki.gbjzc.2022.000517
6. Jianshe, Z., Hou Fang, X., You, Z.Y.: Analysis of the wage level of migrant workers in the construction industry. Construction Economics **35**(11), 30–33 (2014)
7. Peng, H.: A study on the restrictive factors and countermeasures for the industrialization of migrant workers in the construction industry. Dalian University of Technology (2022). https://doi.org/10.26991/d.cnki.gdllu.2022.002911
8. Nunnally, J.C.: Psychometric theory. 2nd ed., McGraw Hill, USA. one thousand nine hundred and seventy-eight
9. Field, A.P.: Discovering Statistics using SPSS [M], 2nd edn. Page, London (2005)
10. Kunhui, Y., Lili, W., Bingheng, L.: Key factors in decision-making of bidding prices for private engineering projects in China. Construction Economics **09**, 54–57 (2012)

Research on the Maturity of Intelligent Construction Management in the Whole Process of Prefabricated Buildings

Huili Zhang[✉]

Chongqing College of Architecture and Technology, Chongqing, China
17115677@qq.com

Abstract. Intelligent construction is an important means of industrialization in new construction, which can realize the whole process of prefabricated buildings, all-round information, integration, digitalization and intelligent management, and reshape the value chain and supply chain system of the construction industry. At present, the relevant research on prefabricated intelligent construction is relatively fragmented and independent, and there are few studies on the intelligent construction management of the whole process of prefabricated buildings. So it is difficult for us to accurately estimate the level of intelligent manufacturing capabilities of contractors in prefabricated buildings. Starting from prefabricated buildings and intelligent construction under the background of new-type building industrialization, this paper constructs the maturity model, maturity evaluation index and evaluation standard of the intelligent construction management of the whole process of prefabricated buildings. It is expected to be helpful for the evaluation and improvement of the level of intelligent construction management in the entire process of prefabricated buildings.

Keywords: Prefabricated · Intelligent Manufacturing · Whole process · Maturity

1 Introduction

Since the implementation of the Guiding Opinions of the General Office of the State Council on Vigorously Developing Prefabricated Buildings, the industrialization of new buildings represented by prefabricated buildings has been promoted rapidly, playing an increasingly important role in promoting energy conservation and emission reduction in the construction industry, improving resource utilization efficiency, and achieving sustainable social development [1]. It is an inevitable trend and top priority for the future development of the construction industry. Intelligent construction is an important means of industrialization in new construction, which can realize the whole process, of prefabricated buildings, comprehensive informatization, integration, digitization, and intelligent management, and set the structure, system, application, management and optimization combination as one, with the comprehensive wisdom of perception, transmission, memory, reasoning, judgment and decision-making, reshaping the value chain and supply

© The Author(s) 2024
P. Xiang and L. Zuo (Eds.): PBSFTT 2023, LNCE 382, pp. 457–467, 2024.
https://doi.org/10.1007/978-981-97-5108-2_49

chain system of the construction industry [2]. The current research on prefabricated buildings in China mainly focuses on the cost-effectiveness, quality, and industrial chain of prefabricated buildings, and rarely uses information means to comprehensively consider the intelligent construction of the whole process of prefabricated building design, processing, transportation, lifting and operation and maintenance [3]. Due to the lack of standards for the classification of informatization, digitization, and intelligence levels in prefabricated buildings, it is impossible to accurately estimate the level of contractors' intelligent manufacturing capabilities in prefabricated buildings [4]. Overly pursuing high-tech intelligence, digitization, and information technology may result in unsatisfactory manufacturing and construction results [5]. How to scientifically and accurately estimate the intelligent construction level of contractors throughout the entire process of intelligent buildings, so as to promote the intelligent construction ability of prefabricated buildings, has become one of the important research directions.

2 Basic Framework of Maturity Model

At present, enterprise organization/project maturity models mainly include two types: two-dimensional structure and three-dimensional structure. The two-dimensional maturity model represented by CMM includes two aspects: maturity level and maturity elements. The maturity model represented by OPM3 not only includes maturity level and project management domain process, but also adds project organization management level. In the intelligent construction management of prefabricated buildings throughout the entire process, in addition to maturity levels and project management process areas, it is also necessary to add dimensions for the entire process management of prefabricated buildings. Therefore, this study adopts the OPM three-dimensional maturity model structure. The maturity model for intelligent construction management throughout the entire process of prefabricated buildings includes three dimensions, as shown in Fig. 1.

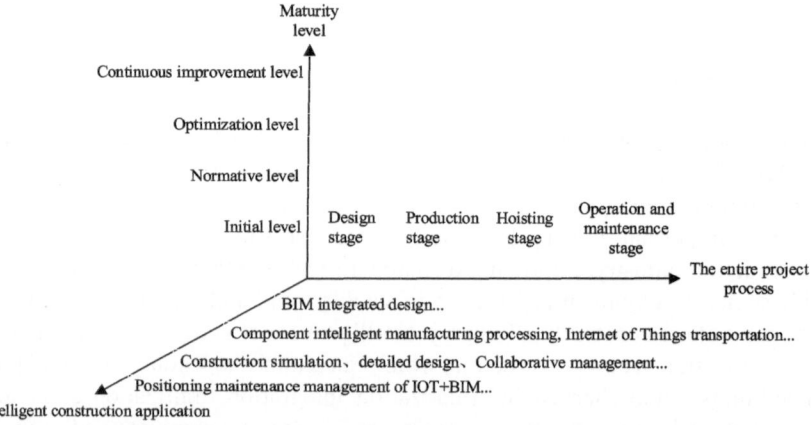

Fig. 1. The **Maturity** Model Structure of Intelligent Construction Management for the Whole Process of Prefabricated Buildings

2.1 Dimension of the Entire Project Process

The maturity model of intelligent construction management throughout the entire process of prefabricated buildings runs through the four stages of design, production, hoisting, and operation and maintenance [6]. Through intelligent construction management throughout the entire process, relevant data is obtained, collected, analyzed, and integrated for each stage of prefabricated buildings, in order to promote the reasonable connection of work in each stage of prefabricated buildings and improve the efficiency of project management in prefabricated buildings.

2.2 Maturity Level Dimension

The whole process of intelligent construction management maturity of prefabricated buildings is divided into four levels: initial level, normative level, optimization level and continuous improvement level, which are continuously upgraded and optimized in order from low to high, and different levels of maturity have their own characteristics. The initial level has basic intelligent construction application conditions, including intelligent construction application environment, management system, and technical support. At the standard level, it has been standardized, and has been significantly improved in the quantitative application of prefabricated whole process intelligent construction. In the optimization level, the whole process of intelligent construction management of prefabricated buildings is quantitatively analyzed, and the standard quantitative evaluation system is used to describe it. Continuous improvement level can improve the intelligent construction management capability of prefabricated buildings throughout the entire process through continuous optimization and improvement, and bring significant long-term economic benefits.

2.3 Dimension of Intelligent Construction Application

With the whole process of prefabricated buildings as the main line, the intelligent construction management of the whole process of prefabricated buildings is realized through the comprehensive application of information technology such as BIM and big data. During the design phase, BIM technology is used for integrated design, such as collaborative design, intelligent splitting, automatic drawing, etc. In the production stage, intelligent manufacturing technology is used for component processing, and Internet of Things technology is used for transportation management. During the lifting phase, construction simulation, detailed design, collaborative management, progress, cost, quality, safety management and so on will be carried out. In the operation and maintenance stage, the Internet of Things technology and BIM technology are used for spatial positioning management and maintenance management.

3 Maturity Evaluation System

3.1 Maturity Evaluation Index

Based on the aforementioned maturity model structure of the whole process intelligent construction management of prefabricated buildings, evaluation indexes should be analyzed and selected from the aspects of the whole process management of prefabricated buildings and the application of intelligent construction. In this study, the application of intelligent construction technology is matched with the requirements of prefabricated buildings, and evaluation indexes are selected from four aspects: integrated design, intelligent construction of components, hoisting construction and operation and maintenance management. According to this, the previous research literature was retrieved and sorted out, the relevant research results were summarized, and the evaluation index system as shown in Fig. 2 was constructed.

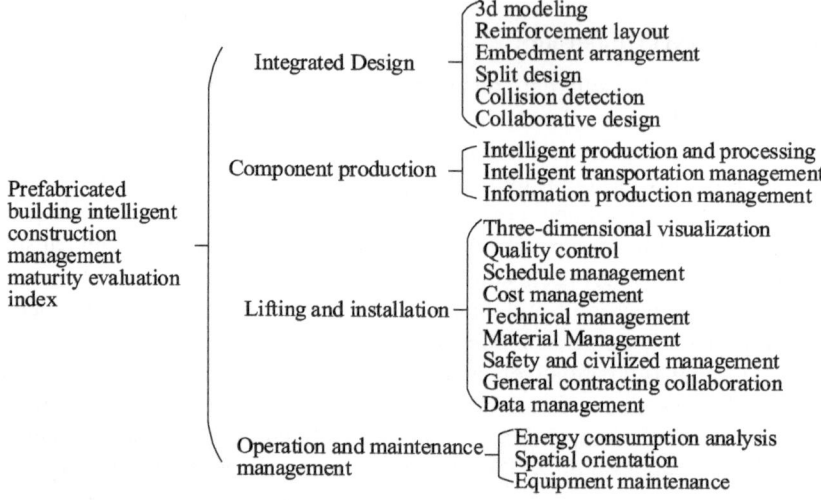

Fig. 2. Maturity evaluation indicators for intelligent construction management throughout the entire process of prefabricated buildings

(1) **Integrated design.** Utilizing the integration characteristics of multi-specialty systems in prefabricated buildings and the advantages of BIM technology in information integration, integrated design is carried out to achieve the goal of information exchange and sharing among various specialties, entire processes, and participating departments, and to complete scheme design, construction drawing design, collaborative design, and deepening design [7]. The focus of intelligent manufacturing in this stage mainly includes using 3D modeling to improve component performance and design accuracy, realizing the practical display of steel reinforcement arrangement, using BIM technology to scientifically set the associated parameters and global parameters for embedded component layout, splitting each component using a 3D model, using collision detection to avoid conflicts between different components, collaborative design management between different disciplines and within disciplines.

(2) **Production and transportation.** In the process of production, processing and transportation of prefabricated building components, intelligent technologies such as machine learning, big data, Internet of Things, and cloud computing are integrated to

achieve full automation and intelligent flow of production and transportation of pre-fabricated building components. The intelligent production and processing of prefab-ricated components can be realized by automatically identifying and extracting component information with computer-assisted manufacturing and BIM technical information. Utilizing the Internet of Things and the Beidou positioning system to form pre-fabricated component QR codes and RFID cards, achieving intelligent transportation management [8]. Connect the information of the buyer, the producer and the carrier, and establish an integrated information management system of procurement, production and transportation.

(3) **Lifting management.** Hoisting management is a link in the construction of prefabricated buildings. In this stage, BIM technology is mainly used to optimize the management of the construction cost, schedule, quality and safety of prefabricated build-ings, and improve the efficiency of construction management. According to the "China Construction Industry Information Development Report", BIM application scope and content of construction enterprises are described, which are summarized in nine aspects: 3D visualization, quality, schedule, cost, safety and civilization, technology, materials, general contracting collaboration and data management.

(4) **Operation and maintenance management.** In order to improve the operational efficiency of prefabricated buildings, the operation and maintenance stage after the com-pletion of the prefabricated building mainly utilizes BIM technology and Internet of Things technology to conduct spatial positioning and maintenance management of the construction equipment. Under the background of "dual carbon", higher requirements have been put forward for building operation. BIM technology is used to collect, analyze, and monitor energy consumption data of buildings in order to better grasp the energy consumption of equipment in different parts.

3.2 Maturity Level Assessment

In this paper, a score interval is adopted to evaluate the maturity level, and a certain score interval is given for each maturity level. By referring to relevant literature, research results and consulting expert opinions, it is shown in Table 1.

Table 1. Maturity level evaluation interval.

Maturity level	Initial level	Normative level	Optimization level	Continuous improvement level
Fractional interval	[60, 70)	[70, 80)	[80, 90)	[90, 100]

3.3 Maturity Evaluation Weight

At present, the weight calculation methods of indicators mainly include Analytic Hier-archy Process (AHP), Sequential Relationship Method, Entropy Value Method, Direct Weighting Method, etc. The objective weighting method represented by Entropy Value

Method requires a large number of original data variables, while the maturity evaluation indicators summarized in this paper are mostly qualitative evaluation indicators, and the direct weighting method needs to be used to calculate weights. Both Analytic Hierarchy Process (AHP) and Sequential Relationship Method (SRM) use evaluators to judge the relative importance relationships between different indicators, and calculate weights through the isogeneity of the relative importance of indicators. However, the judgment of the relative importance of indicators is difficult to grasp accurately, and in engineering practice, direct weighting methods are more commonly used.

Optimization is carried out as follows: (1) According to the index system shown in Fig. 2, experts are invited to score weights, including four first-level indicators of integrated design, production and transportation, lifting management and operation and maintenance management, and their respective second-level indicators are at a total of 5 levels. (2) Weights x_i are given according to the traditional expert scoring method for the above 5 levels. (3) The mean weight of the i-th index is calculated as $\bar{x}_i$, and obtain x_i corresponding to min $|x_i - \bar{x}|$ as the final weight. This method eliminates the extreme subjective opinions of experts, and selects the weight that most people agree with as the final result.

When using the above method to prepare and distribute a survey questionnaire, considering the strong professionalism of prefabricated buildings, in order to ensure the scientific and reasonable results of the survey, it is recommended to select personnel who are very familiar with the management of prefabricated building projects as the survey subjects. This article focuses on personnel engaged in the design, production, construction, operation and maintenance of prefabricated buildings, as well as related scientific research consulting. A total of 142 survey questionnaires were distributed, and 119 questionnaires were collected, of which 115 were valid. The final weight results of the maturity evaluation indicators for intelligent construction management throughout the entire process of prefabricated buildings are shown in Table 2.

3.4 Maturity Evaluation Criteria

By consulting relevant literature and consulting expert opinions, and referring to policy documents and normative standards related to prefabricated and BIM technology, maturity level evaluation standards corresponding to each maturity evaluation index have been developed. As shown in Table 3. After obtaining the corresponding score based on the maturity evaluation index standard, combined with the weight, the corresponding level of maturity for intelligent construction management in the entire process of prefabricated buildings can be calculated, and key improvements can be made for weak links.

The whole process of intelligent construction management maturity of prefabricated buildings constructed in this paper is divided into four levels: initial level, standard level, optimization level and continuous improvement level. By consulting relevant literature and expert opinions, and referring to the policy documents and specifications related to assembly and BIM technology, the maturity level evaluation standard corresponding to each maturity evaluation index is developed. See Table 3. After the index score is obtained according to the maturity evaluation index standard, the corresponding level of intelligent construction management maturity of the whole process of prefabricated

Table 2. Weights of maturity evaluation indicators for intelligent construction management throughout the entire process of prefabricated buildings.

Primary indicators	Weight	Secondary indicators	weight
Integrated Design	0.3	3d modeling	0.25
		Reinforcement layout	0.1
		Embedment arrangement	0.1
		Split design	0.1
		Collision detection	0.2
		Collaborative design	0.25
Component production	0.2	Intelligent production and processing	0.4
		Intelligent transportation management	0.3
		Information production management	0.3
Lifting and installation	0.35	Three-dimensional visualization	0.2
		Quality control	0.1
		Schedule management	0.1
		Cost management	0.1
		Technical management	0.1
		Material Management	0.1
		Safety and civilized management	0.1
		General contracting collaboration	0.1
		Data management	0.1
Operation and maintenance management	0.15	Energy consumption analysis	0.4
		Spatial orientation	0.3
		Equipment maintenance	0.3

buildings can be calculated by combining the weight, and the key improvement is made for the weak links.

Table 3. Maturity evaluation criteria for intelligent construction management throughout the entire process of prefabricated buildings

Whole process	Evaluation indicator	Evaluation criteria
Integrated Design	3d modeling	Model accuracy(0–40),Model information richness(0–20),Component scope rationality(0–20),Data interoperability(0–20)
	Reinforcement layout	Reinforcement design compliance(0–40),Detailed design(0–30),Visualization(0–30)
	Embedment arrangement	Compliance(0–40),Association Parameter and Global parameter scientificity(0–30),Scheme effect display(0–30)
	Split design	Reasonably and accurately split design parameters(0–50),Component independence(0–50)
	Collision detection	Single professional collision detection(0–15),Inter-professional collision detection(0–15),Functional collision detection(0–15),Procedural collision detection(0–15),Extended collision detection(0–15),Pipeline optimization and structural layout(0–25)
	Collaborative design	Process output automatic publishing rate(0–20),Automatic statistics rate of process execution status(0–20),Real time KPI statistics rate for key business(0–20),Business object database processing rate(0–20),Coverage rate of process anomaly monitoring(0–20)
Component production	Intelligent production and processing	Automatic recognition and extraction of BIM model information(0–30),Smart factory management ability(0–30),Digitization of production lines(0–20),Intelligent production line(0–20)

(*continued*)

Table 3. (*continued*)

Whole process	Evaluation indicator	Evaluation criteria
	Intelligent transportation management	Digital information identification of components(0–40),Transportation positioning system(0–30),Automatic recognition and warehousing capability(0–30)
	Information production management	Procurement Planning System(0–30),Warehouse management system(0–40),Production Execution Management System(0–30)
Lifting and installation	Three-dimensional visualization	BIM site layout(0–20),Collaborative modeling(0–20),Visualization technology capability(0–20),Hoisting simulation(0–20),Collaborative construction application(0–20)
	Quality control	Site survey and review(0–30),Planning support and hanger(0–30),BIM quality management system(0–40)
	Schedule management	Virtual simulation(0–30),Progress playback(0–30),Dynamic schedule control(0–40)
	Cost management	Comparison of engineering multiple calculations(0–20),Cost management and material management(0–20),Dynamic analysis and management(0–20),Cost management(0–15),BIM based settlement(0–15),Quantity Price Query(0–10)
	Technical management	Collision Checking(0–15),Construction scheme simulation(0–15),Scheme Optimization(0–15),Simulation of bidding scheme(0–10),3D visual disclosure(0–10), Professional disclosure(0–15),Management disclosure(0–10), Construction animation(0–10)
	Material Management	Logistics management informatization level(0–100)

(*continued*)

Table 3. (*continued*)

Whole process	Evaluation indicator	Evaluation criteria
	Safety and civilized management	Construction site safety management(0–50),On-site monitoring and feedback(0–50)
	General contracting collaboration	Integrated management(0–30),Professional construction management(0–30),Owner communication management(0–20),Subcontractor communication management(0–20)
	Data management	Digitalization of document management(0–50),Document sharing(0–50)
Operation and maintenance management	Energy consumption analysis	Automatic collection of energy consumption data(0–30), Energy consumption analysis(0–30),Capable of warning and identification(0–40)
	Spatial orientation	Spatial orientation(0–30),Information query and sharing(0–30),Monitoring and automatic warning system(0–40)
	Equipment maintenance	Device Information Platform(0–100)

4 Conclusion

In recent years, the country has vigorously promoted prefabricated building projects in order to achieve significant results in energy conservation and emission reduction in the construction industry. With the support of relevant policies, prefabricated buildings have good market prospects. In the context of new building industrialization, this paper integrates intelligent construction into prefabricated buildings, studies and constructs a maturity model for the entire process intelligent construction management of prefabricated buildings, and explores and analyzes the structure, maturity level, and maturity evaluation of the maturity model for the entire process intelligent construction management of prefabricated buildings. A maturity model for intelligent construction management throughout the entire process of prefabricated buildings was designed from three dimensions: process management, maturity level, and intelligent construction application. The entire process includes four stages of design, production, hoisting, and operation and maintenance of prefabricated buildings. Maturity levels from low to high are initial level, specification level, optimization level and continuous improvement level. The application of intelligent construction includes four aspects: integrated design, component production, lifting construction and operation and maintenance management. Based on this, 21 maturity evaluation indicators have been refined, and evaluation standards for each maturity evaluation indicator have been developed.

References

1. Hong, J.K., Shen, G.Q., Mao, C., et al.: Life-cycle energy analysis of prefabricated building components: an inputeoutput-based hybrid model. Journal of Cleaner Production, 2198–2207 (2016)
2. Zhenqiang, Q., Wei, C.: Research on maturity evaluation of whole process engineering consulting based on AHP extension measurement model. J. Eng. Manage. **36**(02), 35–40 (2022)
3. Jiang, J.: A Preliminary Study on the management of university infrastructure projects based on BIM technology under the background of intelligent construction. Modern Urban Research 38(2), 60–63 (2023)
4. Jeong, J., Hong, T., Ji, C., et al.: An integrated evaluation of productivity, cost and CO2 emission between prefabricated and conventional columns. J. Clean. Prod. 2393–2406 (2017)
5. Johari, S., Jha, K.N.: Exploring the relationship between construction workers' communication skills and their productivity. J. Manage. Eng. 37 (2021)
6. Chen, Z., Xiaodong, W., Chen, C.: Development and application of prefabricated buildings in the context of intelligent construction. Building Structure **51**(22), 168 (2021)
7. Liu, Y.: Whole Process Management of Prefabricated Buildings and Application of BIM Technology. China Kitchen and Bathroom 0061-0063 (2022)
8. Liu, K., Zhu, K., An, Y., et al.: Discussion on the Necessity of Whole Process Management of Prefabricated Buildings. Brick and Tile 49–50, 52 (2022)

Operational Energy Saving and Carbon Reduction Benefits of Concrete MiC Building's Envelope

Miaomiao Hou[1,2], Yaoyu Lin[1], Qiong Wang[2], Xiaolu He[2], Yiqian Zheng[1], and Pengyuan Shen[1(✉)]

[1] Harbin Institute of Technology, Shenzhen 51800, Guangdong, China
shenpengyuan@hit.edu.cn

[2] China State Construction Hailong Technology Company Limited, Shenzhen, Guangdong, China

Abstract. Modular integrated Construction (MiC) is an off-site construction method of the highest level and has been fast developed to meet the demand for fast, low-cost and eco-friendly construction. As the construction industry is the main contributor to global energy consumption and carbon emissions, MiC has great potential to improve its thermal and energy performance through better envelope design. However, fewer existing papers study the environmental performance of MiC during the operational phase, especially for concrete MiC. Therefore, this research aims to investigate the environmental performance of MiC during the operation phase through a case study of a real project and a comparative LCA analysis. The simulation modeling was calibrated and validated with real-time measurement readings. The temporal and spatial variations of the indoor thermal environment were analyzed. Finally, the energy consumption was simulated and a whole-life-cycle assessment was conducted, to provide a holistic quantitative energy saving and carbon reduction analysis for MiC compared to normal prefabricated construction. This research can help to improve the envelope design guidance of MiC and the resulting environmental performance.

Keywords: Modular integrated Construction (MiC) · Building Envelope · Operational Phase · Energy Saving and Carbon Reduction

1 Introduction

1.1 Background

The construction industry significantly impacts the environment, consuming 40% of final global energy consumption, and contributing to 39% of total CO_2 emissions [1, 2]. In China, it accounts for an even higher percentage of energy consumption (46%) and CO_2 emissions (51%) [3, 4]. The energy consumption and CO_2 emissions from the operation phase constitute around 80% of a building's life-cycle total amount [5].

P. Xiang and L. Zuo (Eds.): PBSFTT 2023, LNCE 382, pp. 468–478, 2024.
https://doi.org/10.1007/978-981-97-5108-2_50

The conventional "on-site" construction method, which refers to the process that a building is constructed on the construction site, does not have an advantage in meeting the demand of fast, low-cost and eco-friendly construction [6]. Among off-site construction methods, Modular integrated Construction (MiC) is categorized into the highest level, which means modules are fabricated in an off-site factory, where all the structural, mechanical, plumbing, electric and decorative work is done (about 85–90%), then delivered to the construction site and assembled to form the building [7].

1.2 Literature Review on MiC's Environmental Performance

As China has introduced policies to promote the development of MiC, it is necessary to investigate its environmental benefits, so there is some research on the environmental performance of MiC [8]. However, fewer existing papers focus on the operation phase of MiC [9, 10]. Studies present the full life cycle assessment for residential and commercial buildings and found the pre-fabricated steel-framed modular used more operational energy than the conventional concrete building [11, 12]. The thermal performance of a building's envelope significantly impacts the cooling or heating load that takes up most of the energy use [13]. MiC has great potential to improve its thermal and energy performance because of the higher performance envelope, higher manufacturing quality and precision, easier dismantling and reuse [14].

The primary focus of this research is on investigating the environmental performance of MiC during the operation phase through a case study of a real project and implementing comparative LCA analysis of common prefabricated construction methods and MiC. This paper provides: (1) a systemic method of real-time tests and simulation to evaluate the operational environmental performance of MiC; (2) envelope design guidance of MiC and the resulting quantitative environmental benefits.

2 Experiment

2.1 Case Study

The Longhua Zhangkengjing project is a social housing project in Shenzhen, China. It is the first high-rise concrete and the tallest MiC in China. It consists of five 28-story residential buildings, each with a height of 99.7 m. Also, it is the first MiC built in the whole process of green construction in China and was completed in only one year. A 35 m^2 housing unit facing Southwest is selected for the investigation (Fig. 1).

Shenzhen has a warm and humid subtropical climate (Köppen classification *Cwa*) and belongs to hot summer and warm winter areas based on the thermal design code of civil buildings. The average highest and lowest dry-bulb temperature is around 32 °C and 13 °C, necessitating cooling from May to November while no heating is needed. Overall, it is necessary to allow enough daylight indoors while reducing solar radiation which may increase the cooling load. Therefore, the envelope is designed to reduce the heat gain indoors (Fig. 2).

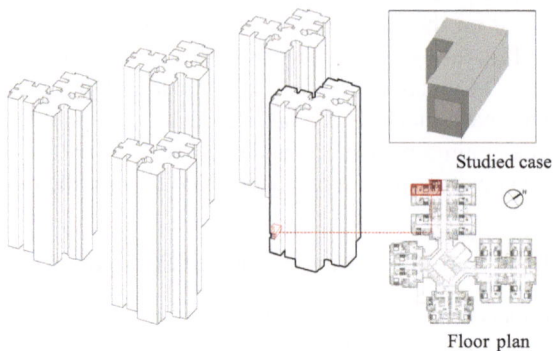

Fig. 1. Axonometric drawing of the studied 35 m^2 housing unit

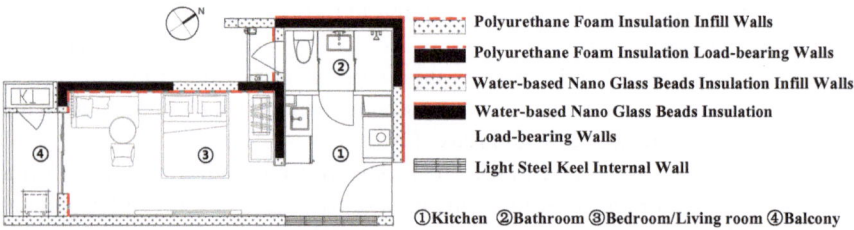

Fig. 2. The construction of the walls of the experimental room

2.2 Experiment Setting of the Real-Time Test

The real-time test was conducted from 17 to 22 June 2023, when it was cloudy and rainy. Ten sensors (Hobo MX1104, 20°–70 °C, ±0.20 °C for temperature sensors; 0–167731

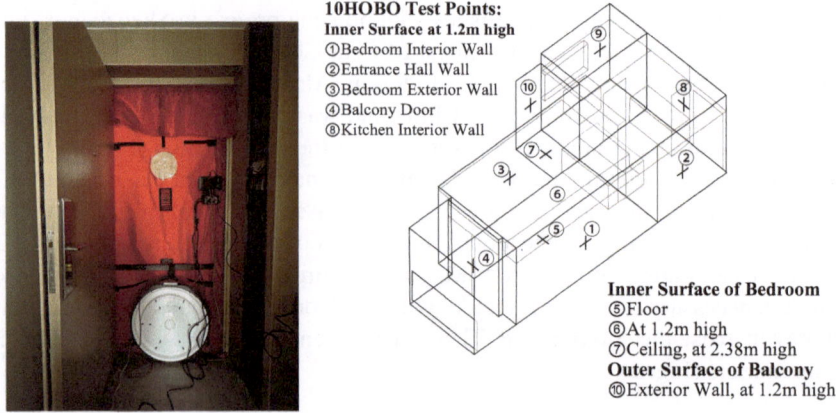

Fig. 3. (left) Indoor view when conducting blower door testing; (right) the layout of sensors

lx, ±10% for light sensors) were set on to measure the temperature and the illuminance of each surface. Sensor 6th and 10th were set to measure the indoor and outdoor air temperature separately. The blower door testing was conducted to test the air tightness of the unit (Fig. 3).

2.3 Simulation Method

Ladybug Tools, plug-ins on the Rhino/Grasshopper platform with engines such as Energyplus for energy simulation and Daysim for daylighting analysis, were used in this study. Firstly the sky model was built for ray-tracing simulation. Solar radiation, dry bulb temperature, and relative humidity are from the 10th Hobo (114.04°N, 22.70°E) and are validated with the meteorological data of Longhua District from the Meteorological Bureau of Shenzhen Municipality (within ±1 °C), while other parameters are from the SWEAR epw file. The intensity of infiltration is from the blower door test. The simulation interval is 2 h during June 17th 00:00–22nd 00:00. Table 1 shows the parameter settings.

Table 1. Thermal and optical parameters of the materials

		Energy Model	Daylight Model
Analysis Result		Outdoor and indoor surface temperature, Air temperature	Point-in-time illuminance (0.05 m * 0.05 m)
Wall/Other materials	Interior Wall	1. Light Steel Keel Wall (U:0.34, R:2.79) 2. 120 mm Concrete Infill Wall (U:0.78, R:1.12)	1. Bedroom Wooden Wall (ρ:0.58) 2. Bedroom White Painting Wall (ρ:0.72) 3. Entrance Hall Wooden Wall (ρ:0.58) 4. Kitchen Wall (ρ:0.80) 5. Bathroom Wall (ρ:0.6) 6. White Painting Wall (ρ:0.75)
	External Wall	1. 230 mm Concrete Infill Wall (U:0.18, R:5.53) 2. Polyurethane Foam Insulation Infill Walls (U:0.24, R:3.99) 3. Polyurethane Foam Insulation Load-bearing Walls (U:0.75, R:1.17) 4. Water-based Nano Glass Beads Insulation Infill Walls (U:0.25, R:3.78) 5. Water-based Nano Glass Beads Insulation Load-bearing Walls (U:0.89, R:0.97)	**External Wall:** 1. Grey Stone Paint Façade(ρ:0.23) 2. White Painting Wall(ρ:0.75) **Other materials:** 1. Stainless Steel Decorative Stripe (ρ:0.72) 2. Frosted Silver Mirror (ρ:0.89) 3. Bedroom Wood Cupboard (ρ:0.58) 4. Mirror Cabinetry (ρ:0.72) 5. Artificial Stone Countertop (ρ:0.8)

(continued)

Table 1. (*continued*)

		Energy Model	Daylight Model
Door	Wooden Door	(U:3.00, R:0.17)	Wooden Door of the Entrance (ρ:0.58)
	Glass Door	(K:2.76, SHGC:0.34, VLT:0.54)	1. Frosted Glass Door (T:0.77) 2. External Glass Door (T:0.54)
Floor/Ceiling (Boundary Condition: Adiabatic)		190 mm Prefabricated Building Block (U:3.22, R:0.18)	1. Bedroom Floor (ρ: 0.58) 2. Kitchen Floor (ρ:0.6) 3. Balcony Floor (ρ:0.58) 4. Bathroom Ceiling (ρ:0.87) 5. White Painting Ceiling (ρ:0.75)
Intensity of Infiltration		0.0001 (m^3/s per m^2 facade)	–

3   Calibration and Validation

The measured readings were used to compare with the simulations for the same point, and the average difference was calculated in Table 2, which was within the acceptable range (± 1.6 °C for temperature and ± 16 lx for illuminance except for the 4th point).

Table 2. The differences between the simulated and the measured readings for all sensors.

Position	1	2	3	4	5	6	7	8	9	Average
Temperature Difference (°C)	0.8	0.9	0.8	0.7	0.7	1.6	1.0	0.8	1.2	1.2 ± 0.5
Illuminance Difference(lux)	11	8	11	101	16	14	9	9	16	12 ± 4 (Except④)

The simulated and measured illuminance fit well while most deviations occur at noon, which may be a result of the difference between the actual global illuminance and the data from the 10th Hobo. The 4th Hobo's data may be impacted by the aluminum window frame that blocks sunlight coming indoors.

The simulated and measured temperature of the No. 1, 2, 3, 5, 7, and 8 test points are basically identical, and those of the 4th and 9th points are within the acceptable range (±1.2 °C). The simulated temperature is the average of the entire surface, different from the measured value at a point, contributing to the deviation. The 6th point exhibits the largest deviation, because the simulated value is the average air temperature of the whole room, which shows a larger fluctuation than that of the center point (Fig. 4).

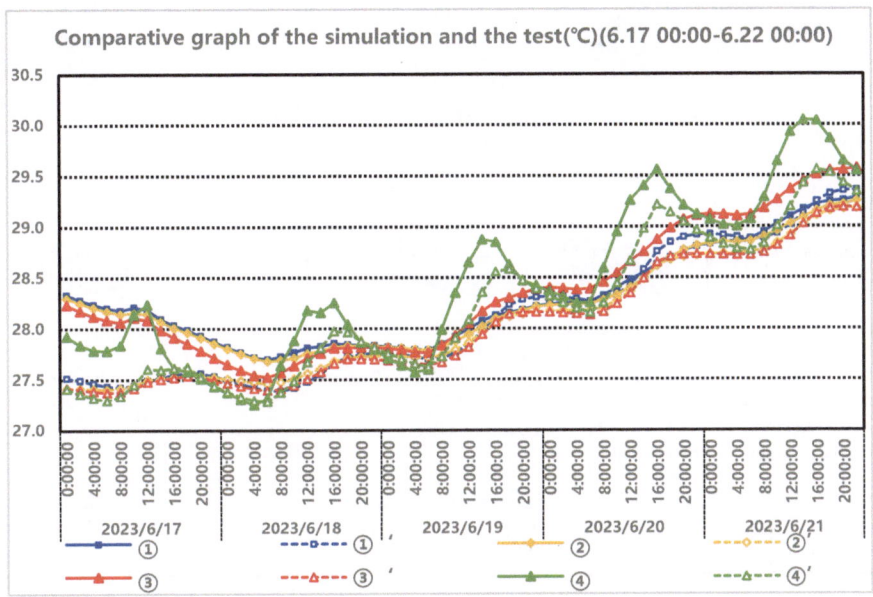

Fig. 4. Analysis via real-time temperature measurement and numerical simulation.

4 Result

4.1 Temporal Variations

The illuminance distribution of each surface of the room on 21 June is selected for visualization and analysis in Table 3. As it was cloudy, the interior of the studied room was hardly exposed to direct sunlight, and only a little bit of reflected light came indoors at 14:00 and 18:00 with the peak illuminance occurring at 14:00.

Table 3. Ray-tracing visualization and illuminance gradient map at 06:00, 10:00, 14:00, 18:00

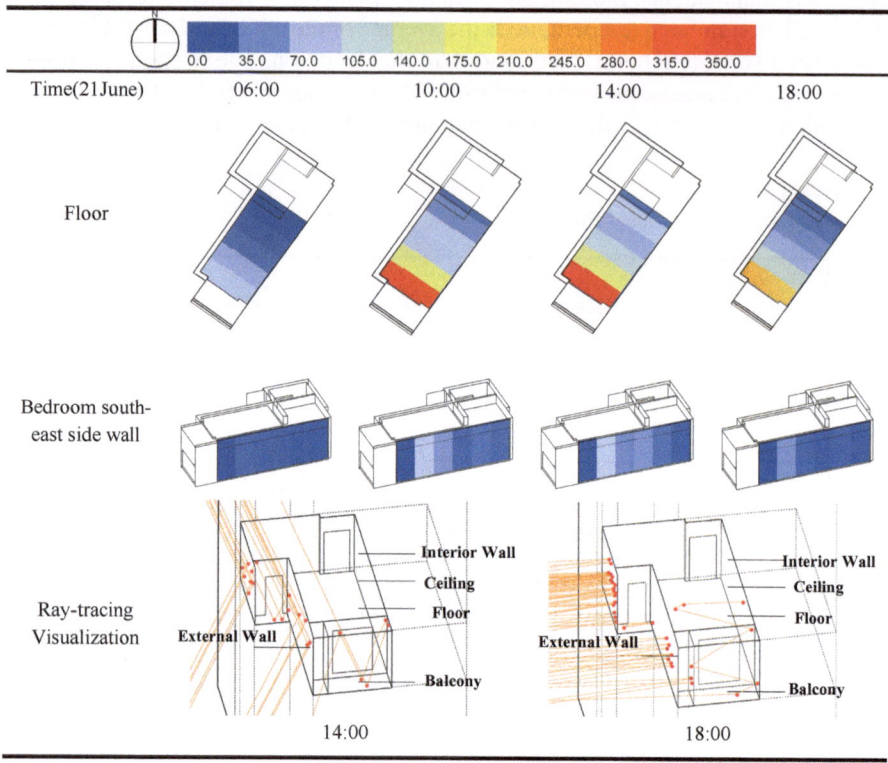

During the measurement period, the outdoor ambient temperature was mainly in the range of 25–34 °C. As shown in Fig. 5a, the indoor temperature at all points shows less fluctuation than outdoor (10th), and only the 4th and 9th points fluctuated more obviously than others. The external wall (3rd Hobo) receives the greatest influence from the outdoor environment.

Figure 5b reflects three exterior wall surface temperatures in the bedroom. The bedroom exterior wall (3rd Hobo) and bathroom exterior wall (9th Hobo) were exposed to direct solar radiation according to the ray-tracing in Table 3, but 3rd Hobo's temperature fluctuation amplitude is more moderate, and its peak appearance time is lagged behind outdoor temperature and earlier than the bathroom wall (9th Hobo). On June 16th - 18th, the bathroom wall (with the 9th point on the inner side) had the lowest indoor temperature and hardly changed with the outdoor temperature, while on the 19th-21st, it began to fluctuate with the outdoor temperature with lag. The reason is the better thermal insulation performance of the insulation infill walls (3rd Hobo) while the better heat storage performance of the heavy shear wall (9th Hobo).

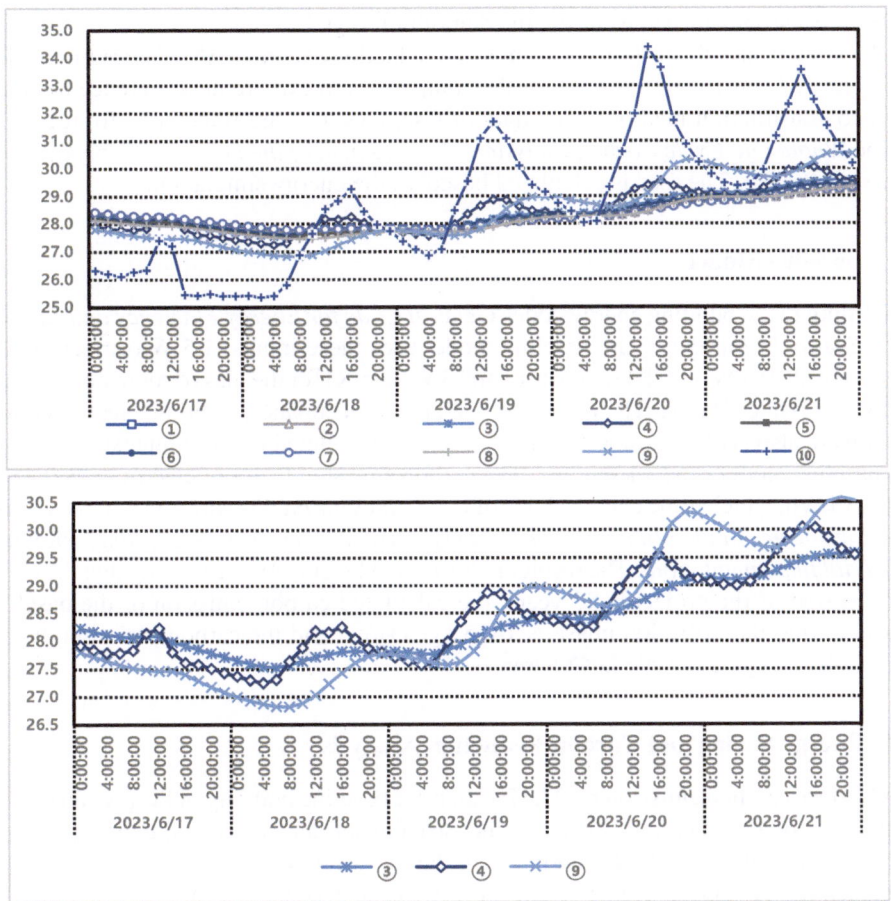

Fig. 5. Variations of temperature from Hobo, (a) above, all sensors, (b) bottom, No. 3, 4, 9

5   Comparative LCA Analysis

The carbon emission calculation of the Longhua Zhangkengjing Project in Shenzhen considers five phases: raw material extraction and manufacturing, transportation, construction, building operation and demolition. The formula for calculating the total carbon emissions is as follows.

$$Ctotal = Csc + Cys + Cjz + Cyy + Ccc \qquad (1)$$

where Ctotal = total carbon emissions ($kgCO_2$);

Csc = carbon emissions from the raw material extraction and manufacturing ($kgCO_2$);

Cys = carbon emissions from the transportation of materials ($kgCO_2$);

Cjz = carbon emissions during the construction ($kgCO_2$);

Cyy = carbon emissions during the operation ($kgCO_2$);

Ccc = carbon emissions during the demolition ($kgCO_2$).

In order to investigate the carbon emission reduction of concrete MiC, a traditional PC prefabricated construction is used for the comparison to explore the difference between the two construction methods. The data of Csc, Cys, Cjz are referenced from *Report on the Whole Life Cycle Carbon Emission Measurement of the Longhua Zhangkengjing Housing Project*. The Cyy is calculated based on the energy simulation.

5.1 Baseline Model

According to the building energy codes *General Code for Energy Efficiency and Renewable Energy in Buildings,* the baseline model uses these inputs: 1.5 $W/(m^2 \cdot K)$ for the outdoor wall, 0.4 $W/(m^2 \cdot K)$ for the roof and 3 $W/(m^2 \cdot K)$ for the glass door in hot summer and warm winter B area. The parameters related to occupancy, lighting, electric equipment were also set based on the code above. The calculated end-use intensity (EUI) of the baseline model is 62.7 kWh/ m^2. The carbon emission factor of electricity is 0.4512 $kgCO_2$/kWh, which is the average level of *Guidelines for the Preparation of Greenhouse Gas Inventories at the Municipal and County (District) Levels in Guangdong Province (for Trial Implementation)*. The result of Cyy is 28.3 $kgCO_2/m^2$. Assuming that the operation period of the project is 50 years, the calculated carbon emission of the baseline model during the operation phase is 1415 $kgCO_2/m^2$ and the carbon emission during the whole life cycle is 1996 $kgCO_2/m^2$.

5.2 Energy Saving and Carbon Reduction Analysis

MiC reduces carbon emission by 23.47% than the baseline building of PC prefabrication in the construction phase. It is because industrial means are considered to be adopted as far as possible in the early design phase in the top-level design, intelligent and refined management are implemented in the processing and construction, and reducing the generation of on-site waste from the source is prior. As a result, the waste of materials produced was reduced by 80% compared to conventional on-site construction.

The carbon emission intensity of the building is 26.2 $kgCO_2/m^2$, and it is 1310 $kgCO_2/m^2$ over a 50-year operation period. Because of the improvement of the MiC envelope's thermal performance, the carbon emission intensity can be reduced by 2.1 $kgCO_2/m^2$ compared to the baseline model, resulting in a reduction ratio of 8%. Overall, the whole life cycle carbon emissions of concrete MiC with a fifty-year building operation period is 1883 $kgCO_2/m^2$, which is reduced by 6% compared to the baseline model (PC prefabricated construction).

6 Conclusion

This paper provides: (1) a systemic method of real-time tests and simulation to evaluate the operational environmental performance of MiC; (2) envelope design guidance of MiC and the resulting quantitative environmental benefits through a real case study. The studied concrete MiC has the envelope of higher thermal performance, resulting in 8% lower operational energy consumption and carbon emissions compared to the baseline building of the normal prefabrication construction method. The whole life cycle carbon emissions reduction ratio of the concrete MiC case is 6% per unit area. Even so, there is still room for improvement in further investigation of the environmental performance of MiC. For instance, more cases of different programs, floor areas, and locations can be studied through this method.

References

1. Bilal, M., et al.: Big data architecture for construction waste analytics (CWA): a conceptual framework. J. Build. Eng. **6**, 144–156 (2016)
2. Edwards, B.: Rough Guide to Sustainability: A Design Primer, 4th edn. RIBA Publishing (2014)
3. CABEE. Research year book of China building energy consumption and carbon emission (2022)
4. Hu, S., Jiang, Y., Yan, D.: China Building Energy Use and Carbon Emission Yearbook 2021: A Roadmap to Carbon Neutrality by 2060. Springer, Cham (2022). https://doi.org/10.1007/978-981-16-7578-2
5. Gustavsson, L., Joelsson, A.: Life cycle primary energy analysis of residential buildings. Energy Build. **42**(2), 210–220 (2010)
6. O'Brien, M., Wakefield, R., Belivean, Y.: Industrializing the Residential Construction Site Report, Department of Housing and Urban Development, Office of Policy Development and Research, Washington, DC (2000)
7. Kawecki, L.R.: Environmental performance of modular fabrication: calculating the carbon footprint of energy used in the construction of a modular home. Arizona State University (2010)
8. Abdelmageed, S., Zayed, T.: A study of literature in modular integrated construction—critical review and future directions. J. Clean. Prod. **277**, 124044 (2020)
9. Kamali, M., Hewage, K.: Life cycle performance of modular buildings: a critical review. Renew. Sustain. Energy Rev. **62**, 1171–1183 (2016)
10. Kamali, M., Hewage, K., Sadiq, R.: Conventional versus modular construction methods: a comparative cradle-to-gate LCA for residential buildings. Energy Build. **204**, 109479 (2019)
11. Aye, L., Ngo, T., Crawford, R.H., Gammampila, R., Mendis, P.: Life cycle greenhouse gas emissions and energy analysis of prefabricated reusable building modules. Energy Build. **47**, 159–168 (2012)
12. Faludi, J., Lepech, M.D., Loisos, G.: Using life cycle assessment methods to guide architectural decision-making for sustainable prefabricated modular buildings. J. Green Build. **7**(3), 151–170 (2012)
13. Li, J., Lu, S., Wang, W., Huang, J., Chen, X., Wang, J.: Design and climate-responsiveness performance evaluation of an integrated envelope for modular prefabricated buildings. Adv. Mater. Sci. Eng. **2018**, 1–14 (2018)
14. Nguyen, T.D.H.N., Moon, H., Ahn, Y.: Critical review of trends in modular integrated construction research with a focus on sustainability. Sustainability **14**(19), 12282 (2022)

Research on Intelligent Construction of Building Information Modelling Based Air-Supported Membrane Structure for Urban Sports Arenas

Zhongyuan Tian, Lili Zhu, Ming Shang, Xing Li$^{(\boxtimes)}$, Sensen Wang, Wenhui Lai, and Xinglun Feng

SpaceZC(Shenzhen) Intelligent Technology Co., Ltd., Shengzhen, China
147095671@qq.com

Abstract. The continuous development of Building Information Modeling (BIM) technology has propelled the traditional construction approach towards high-quality and high-efficiency intelligent construction. Air-supported membrane structures, due to their environmental friendliness, convenience, low cost, and suitability for large spans, offer an ideal solution for constructing lightweight venues in the densely populated city centers. This paper integrates BIM technology to conduct research on intelligent construction for air-supported membrane structures in urban sports arenas. By combining project cases, various aspects of intelligent construction analysis are performed, including BIM collaborative design, collision check, and construction simulation. The feasibility and applicability of BIM technology in the construction application of air-supported membrane structures are explored, providing insights for the integration of BIM technology and intelligent construction of air-supported membrane structures.

Keywords: BIM Technology · Air-supported membrane structure · Intelligent construction · Intelligent control system

1 Introduction

With the continuous development of technology in various industries and the advancement of advanced production processes in China, the level of technological innovation in the field of construction industry has also been improving. Air-supported membrane structures, as a new environmentally friendly architectural product, have been widely applied in various scenarios, particularly in urban sports stadiums located in city centers [1]. They provide a new architectural model for low-carbon and smart city construction. As a type of non-linear form, air membrane architecture exhibits streamlined characteristics on its membrane surface, serving as one of the architectural expressions of curved surfaces. The application of BIM technology in architectural design plays a significant role in the scheme and construction drawing design of non-linear architectural forms. It offers new solutions and construction techniques for unconventional architectural design, driving the transition from traditional two-dimensional design to three-dimensional design [2]. By utilizing the concept of parametric design, BIM technology

© The Author(s) 2024
P. Xiang and L. Zuo (Eds.): PBSFTT 2023, LNCE 382, pp. 479–486, 2024.
https://doi.org/10.1007/978-981-97-5108-2_51

provides intelligent construction solutions for air-supported membrane structure, significantly improving design and construction quality and efficiency. This paper focuses on urban air-supported membrane structure sports arenas as the research object, combining BIM technology to explore the ideas and application conditions for intelligent construction, providing insights into expanding the application scope of air-supported membrane structures in various scenarios and enhancing the level of intelligent design and construction.

2 Literature Review

2.1 Research on Air-Supported Membrane Structures

An air-supported membrane structure in architecture refers to a building form that utilizes flexible membrane materials to create a certain shape, structural stability, and usable interior space under the influence of internal air pressure [3]. The structural principle involves continuously supplying air into the relatively enclosed interior space of the structure through an inflation system [4]. By adjusting the airflow, a dynamic equilibrium is achieved between the air entering and exiting the membrane, while maintaining a certain internal and external pressure differential (generally around 500 Pa). This pressure differential ensures that the interior space of the membrane meets the requirements, maintains the stability of the overall form, and withstands various static and dynamic loads during the building's usage. The specific structural principle is illustrated in Fig. 1.

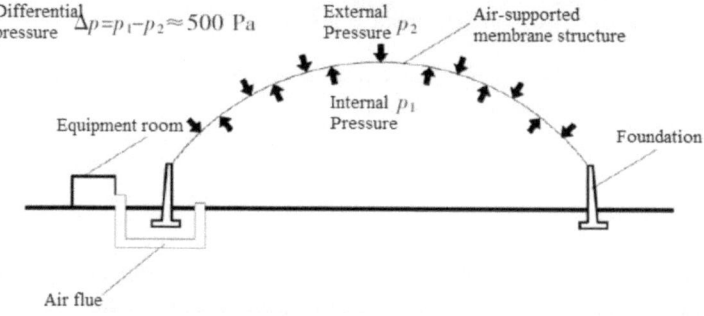

Fig. 1. Diagram of air-supported membrane structure

Air-supported membrane structures maintain their shape through pressure differentials, which gives them significant advantages in terms of interior space layout and span. They are particularly suitable for efficient traffic arrangements within the building. Additionally, unlike traditional load-bearing structures, the membrane material generates uplift forces on the building foundation. This feature can reduce the cost of foundation and ground treatment in areas with poor soil conditions or weak bearing capacity. Furthermore, in larger air membrane structures, components such as steel cables are often combined with the upper membrane. These components are prefabricated, promoting energy and material savings. They also allow parallel construction with other civil engineering components, minimizing construction time. In terms of sustainability and material reuse, air-supported membrane structures exhibit low-carbon, green, and environmentally friendly characteristics.

In summary, this structural form offers advantages such as low cost, short construction periods, reusability, and easy installation and maintenance. It is inherently advantageous for large-scale storage buildings in coal mines. However, it is important to note that this structure requires continuous pressure differential support, meaning that air needs to be constantly supplied to maintain internal pressure, resulting in increased operational costs. Additionally, the current lifespan of membrane materials (generally around 20–25 years) does not match the lifespan of the underlying civil structures. Therefore, there is a need to replace the membrane material once it exceeds its service life, which can impact the building's maintenance. Nevertheless, considering the entire life cycle of a building, air-supported membrane structures still offer unique advantages compared to traditional structural forms, particularly for large-span and large-area requirements. Their lightweight, easy installation, and low-cost characteristics make them an ideal solution for constructing and renovating lightweight sports arenas in densely populated urban areas.

2.2 Research on BIM Intelligent Construction

Intelligent construction is a new concept, which from the extension of smart city and intelligent building, which is to extend the "wisdom" and "intelligence" to the construction process of engineering projects, resulting in the concept of intelligent construction [5]. It is generally believed that intelligent construction is to realize the intelligent management of the construction process, improve the intelligent level of management in the construction process, and reduce the influence of human factors on the construction in order to achieve better construction goals through the full utilization of intelligent technologies and related products in the construction process. Intelligent construction includes at least three aspects of the elements: First, the intelligent construction process is the process of realizing the intelligent construction of engineering projects, through the introduction of new equipment and technology to reduce the dependence on human. It can achieve visual, accurate process management and control. Second, intelligent construction is the technical realization process of making full use of building information model(BIM), information technology and related intelligent products. Third, the form or result of wisdom construction in the application process is visual and data transmission [6, 7].

3 Methods and Data

3.1 BIM Intelligent Construction Methodology

With the continuous development of BIM technology in the field of construction engineering, a comprehensive application system has been established, spanning the entire construction process. Based on the models provided by BIM, various software platforms encompassing architectural design, structural design, facility management, construction, cost management, and more have been developed [8].

In Revit, all elements can be parameterized, meaning that the relationships between all elements in the model can be coordinated and managed using Revit's provided features for change management [9]. The relationships between elements can be automatically created by Revit or manually created by designers during the project development phase. Revit shares some design concepts with traditional CAD but also has some differences. Unlike traditional CAD methods that rely on sketches to create models, Revit directly incorporates real-life elements such as beams, columns, and slabs into the 3D model, making the design process more straightforward and efficient. Once the 3D model is created, various drawings, 2D views, 3D views, and engineering material schedules can be generated from it. In Revit, all drawings, 2D views, 3D views, and engineering material schedules are different representations of the same entity, the model. Information related to the engineering project is automatically collected in detail and drawing views and is synchronized across all representations of the project.

3.2   BIM Intelligent Construction Modeling

The form of pneumatic membrane structures is usually curved and highly irregular. Therefore, manual modeling methods are difficult, and due to the complex nature of node details and the diverse connection methods between different main materials, data synchronization between disciplines is often challenging during the drawing change process [10]. Using structural models to simulate the construction process only allows observation at the unit level, making it difficult to observe the detailed construction of components and nodes. However, a three-dimensional solid model based on Revit can detect the detailed construction of components and nodes, as well as the cross-overlap between components and nodes, while also containing information on internal forces within the structure. Therefore, it is a more realistic method for simulating the construction process. Traditional modeling typically only considers the three-dimensional model and cannot generate information that supports tensioning construction processes. Integrating the time dimension into BIM enables better coordination between design and construction disciplines and provides a more accurate simulation of the construction process.

4   Case Study

4.1   Project Overview

The project is located at a university in Nantong City, Jiangsu Province, China. The pneumatic membrane structure sports arena is situated within the university campus. The designed building area is approximately 5000 square meters. The surrounding area already has teaching facilities and living spaces, which have been in operation for several years. Due to constraints related to the foundation and relevant policies, the traditional architectural engineering approach was not suitable for this project. Therefore, the pneumatic membrane structure method was chosen for construction. The sports arena features a double-layer membrane structure. The outer membrane is made of PVF-coated material with a light transmission rate of 8–10% and decorative colored stripes made of PVDF-coated membrane material. The inner membrane is also made of PVDF-coated material with a light transmission rate of 40%. The bottom five meters of the inner membrane utilize blue PVDF membrane material (as shown in Fig. 2). The project utilizes high-strength architectural membrane material with a surface coating that provides excellent weather resistance, corrosion resistance, and self-cleaning functionality, ensuring a clean and aesthetic appearance throughout its service life.The project incorporates five major systems: an embedded JRT mesh cable system, a snow melting system, an air cushion insulation system, a PM2.5 air filtration system, and a fire protection system:

Fig. 2. Interior and exterior views of a University Gymnasium Project (Photo source: Zhongcheng Space (Shenzhen) Intelligent Technology Co., LTD)

This study utilizes Revit software for modeling and analysis, primarily focusing on the civil engineering discipline and the pneumatic membrane material aspect. Based on the site and interior layout requirements of the gymnasium, the Revit software is employed to establish the Building Information Model (BIM), as shown in Fig. 3:

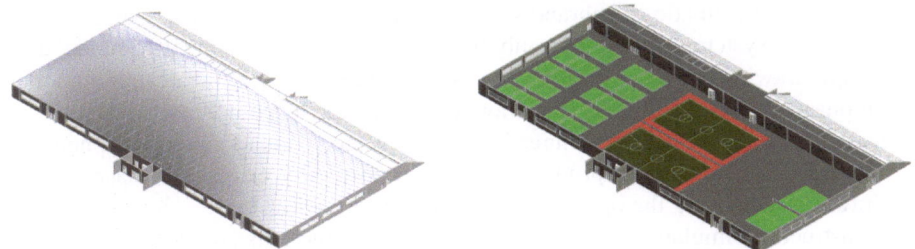

Fig. 3. BIM Model of project

4.2 Project BIM Intelligent Construction Analysis

(1) BIM Collaborative Design: BIM provides a model-based foundation for collaborative design and construction. The BIM working framework for the entire construction project, which includes various stakeholders such as the owner, design team, and construction team. The full-process BIM collaborative design framework of this project is shown in Fig. 4:

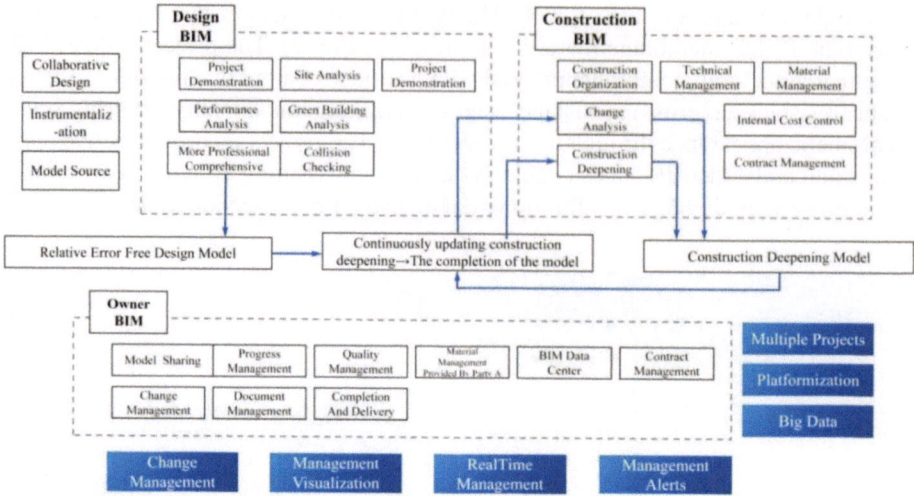

Fig. 4. Multi-stakeholder collaborative based on BIM established in this Project

(2) Construction Drawing Design: With the involvement of various primary materials in the pneumatic membrane structure, a high-precision BIM model allows for direct extraction of 2D drawings and automatic quantity calculations, thereby enhancing the efficiency of quantity takeoff. Simultaneously, during the forward design of the pneumatic membrane structure using BIM methods, the drawings and quantities as design deliverables can be directly linked to the BIM model, allowing for automatic adjustment of corresponding data when changes occur.

(3) Collision Check: The design of this sports stadium mainly includes architectural sub-items, pneumatic membrane structure, and MEP (Mechanical, Electrical, and Plumbing) systems. Since the membrane structure is created in the BIM model using design drawings as input information to accurately represent the design outcomes, emphasis is placed on conducting drawing verification for complex spatial areas such as the foundation and the connection between the foundation and the pneumatic membrane. This aims to promptly identify and rectify errors and clashes in the drawings, achieving the optimization of construction drawings.

(4) Construction simulation and technical disclosure. In the actual implementation of the project, the three-dimensional coordinates of the positioning points are automatically extracted from the design model. The components or installation status are measured using a total station to verify the three-dimensional dimensions (as shown in Fig. 5). The actual three-dimensional coordinate deviations are recorded to provide a basis for pre-adjustment. The data is then imported into the BIM system to ensure that the controlled dimensions are within the error range. Key construction areas such as the connection between the membrane and the foundation, and the fixation of steel cables are fitted, achieving visualized technical disclosure.

(5) Intelligent production and intelligent operation and maintenance based on BIM. Based on BIM technology, all model data from design to construction to operation and maintenance are integrated to carry out intelligent monitoring of the operation

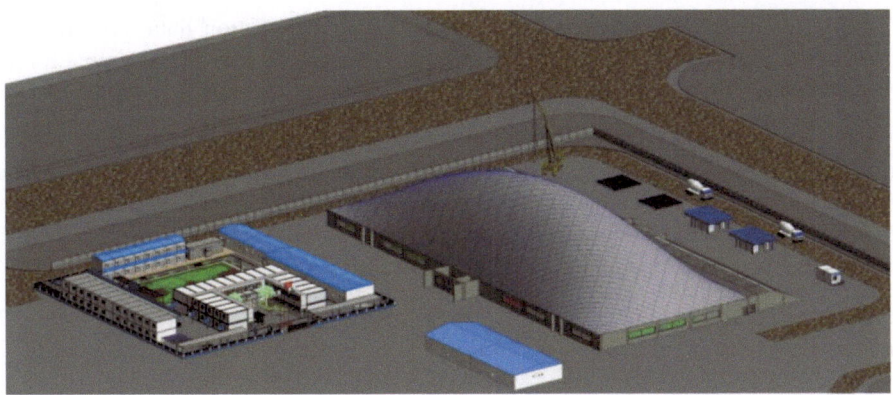

Fig. 5. Construction simulation based on BIM technology

and maintenance of the project. In combination with the intelligent control system developed by ourselves(as shown in Fig. 6), BIM model is imported for visual monitoring and various data analysis, so as to provide an important guarantee for the safe operation of the project.

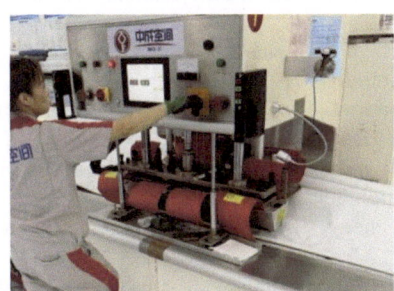 

Fig. 6. Intelligent production and intelligent operation and maintenance of project (Photo source: Zhongcheng Space (Shenzhen) Intelligent Technology Co., LTD)

5   Conclusion

The following conclusions were drawn from the research and analysis of BIM in intelligent construction of pneumatic membrane structure sports arenas:

(1) Using BIM software, engineers can establish the project's BIM model during the design phase and perform parameterized design by creating families through secondary development in software such as Revit.

(2) The collaborative mode of BIM provides a communication platform for multiple participants, and integrates economic and technical indicators into the building model through parameters. This allows for the visualization of indicator data when adjusting and modifying the 3D model, and provides data support for intelligent construction.

(3) BIM technology provides engineers with collision detection data. And the construction simulation provides a visualization of intelligent construction for the operation.

With the continuous development of membrane materials and improvement in structural design, the application scenarios of pneumatic membrane structures will continue to expand, placing higher demands on construction quality and efficiency. BIM technology provides a sustainable data base for the high-quality development of air-supported membrane structure design and intelligent construction.

References

1. Jingke, H., Chenyu, W., Chang-Richards, A., et al.: A spatiotemporal analysis of energy use pathways in the construction industry: a study of China. Energy **239**, 122084 (2022)
2. Xu, J.: Lastest development of integrating building information modelling (BIM) for green building. J. Hum. Set West China **35**(06), 17–23 (2020)
3. Fang-Hui, L., Dong, W., Shi-Zhao, S.: Form finding of air-supported membrane structures. J. Harbin Inst. Technol. (2003)
4. Jinju, Z., Guoqing, Z., Yunji, G.: Numerical study on the effects of opening form on the deflation for an air-supported membrane structure. Case Stud. Therm. Eng. (2019)
5. Lei, W.: Research on design optimization of green building based on BIM technology. Housing **08**, 104 (2020)
6. Changsaar, C., et al.: Optimising energy performance of an eco-home using building information modelling (BIM). Innov. Infrastruct. Solutions **7**, 2 (2022)
7. Niu, Y., Lu, W., Chen, K., et al.: Smart construction objects. J. Comput. Civ. Eng. **30**(4), 04015070 (2016)
8. Sun, C., Liu, Q., Han, Y.: Many-objective optimization design of a public building for energy, daylighting and cost performance improvement. Appl. Sci. **10**, 2435 (2020)
9. Vítor, P., et al.: Using BIM to improve building energy efficiency – a scientometric and systematic review. Energy Build. **250** (2021)
10. Wang, L.J., Huang, X., Zheng, R.Y.: The application of BIM in intelligent construction. Appl. Mech. Mater. **188** (2012)

Simulation Study on the Application Effect of Passive Green Building Strategy in Sinking Soil-Covered Space

Chang Liu[1], Quan Jing[1], and Xianfeng Wu[2(✉)]

[1] China Construction Technology Group Co., Ltd., China Architectural Design and Research Institute Co., Ltd., Beijing 100044, China
[2] Faculty of Architecture, Civil and Transportation Engineering, Beijing University of Technology, Beijing 100124, China
xwu@bjut.edu.cn

Abstract. Yangshao Culture Museum of Yangshao Village national archaeological park (hereinafter referred to as the Museum) fully displays its characteristics, stimulated the regeneration of urban negative space by creatively adopting a sinking layout to hide modern building volume and introducing roof cover to reconstruct the regional landscape and public activity places. The museum meticulously balances the structural heat transfer coefficient with the surrounding sunshade considerations, utilizing software simulations to discern building morphology and intricate components. This approach guarantees the efficacy of the passive green building strategy.

Keywords: passive green building · indoor and outdoor environment study · simulation of application effect

1 Introduction of the Project Site

Zhengzhou, located in the cold area, has a continental monsoon climate of the north temperate zone [1]. The dominant wind direction in spring and summer is southerly wind with the average wind speed being 3.4 m/s and 2.8 m/s respectively. The dominant wind direction in autumn is northerly wind with the average wind speed being 1.8 m/s, while the average wind speed is 3.5 m/s (see Fig. 1) in winter (westerly wind or west-northwest).

Zhengzhou exhibits distinctive traits, notably featuring its peak monthly total radiation during the months of May to July, with a radiation magnitude approximately reaching 550 MJ/m². November sees the minimum radiation with the radiation amount being about 180 MJ/m². The annual total radiation is about 4730 MJ/m² (see Fig. 2). The area is hot in summer and cold in winter. The annual average temperature is about 14.5 °C, and the average humidity is about 65%. The highest monthly average temperature is 27 °C, while the lowest is 1.5 °C. The highest monthly average humidity is 82.1% while the lowest is 51.1% (see Fig. 3 and Fig. 4). The average annual precipitation is about 610 mm. And the precipitation decreases in spring and increases in summer with July and August accounting for about 50% of the annual precipitation (see Fig. 5).

P. Xiang and L. Zuo (Eds.): PBSFTT 2023, LNCE 382, pp. 487–502, 2024.
https://doi.org/10.1007/978-981-97-5108-2_52

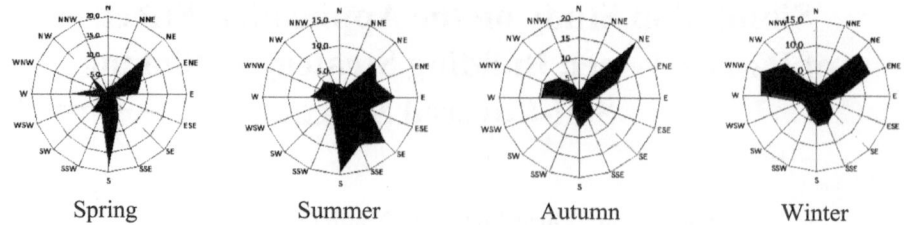

Spring   Summer   Autumn   Winter

Fig. 1. A figure caption is always placed below the illustration. Short captions are centered, while long ones are justified. The macro button chooses the correct format automatically.

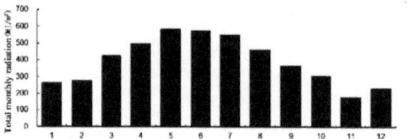

Fig. 2. Monthly total radiation in Zhengzhou

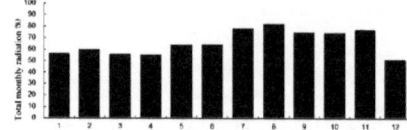

Fig. 3. Monthly total radiation in Zhengzhou

Fig. 4. Monthly total radiation in Zhengzhou

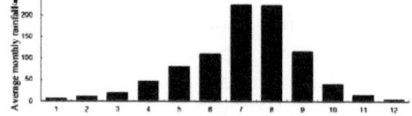

Fig. 5. Monthly total radiation in Zhengzhou

2 Passive Green Building Strategy

Through the analysis of the climate characteristics of Zhengzhou, the passive green building strategies such as sinking type, covered soil type and regular shape are suitable for the Central Plains area [2, 3], which is also the main feature of museum design.

2.1 Sinking Space

The site is a landfill formed many years ago with a depth of 6.4 m. There is a lot of construction waste and domestic waste with uneven nature, so they can not be used as the basic bearing layer. After concession, the excavation and clearance of waste can be carried out in an orderly manner to avoid disturbing the underground cultural relics. The height of the foundation is the same as the depth of garbage removal, which avoids the waste caused by the excessive amount of backfill earthwork and integrates with the atmosphere of the ruins park as a whole, dispelling the volume of modern buildings and avoiding the sense of outburst and pressure of ground buildings.

2.2 Soil-Covered Architecture

In order to further integrate into the landscape of the park and resolve the building volume, especially the sense of conflict between conventional building roof materials and landscape green space, planting roofing is adopted, which puts forward higher requirements for structural load, landscape maintenance, and roof waterproofing. To ensure the basic requirements of plant growth, the minimum thickness of soil cover should be 700 mm to ensure thermal insulation and save building energy consumption.

2.3 Regular Shape and Thermal Insulation Materials

The museum is a regular and square architecture to avoid excessive body changes and reduce the figure coefficient. The spatial scale meets the needs of site value interpretation and communication.

3 Outdoor Environment Study

3.1 Wind Condition Study

The number of tall buildings in the city is increasing with each passing day, and the completion of these buildings will greatly change the wind environment of the city. On the one hand, the tall and dense buildings reduce the ventilation and self-purification ability of the city, and aggravate the air pollution and heat island effect under the condition of low wind speed; on the other hand, when the wind speed is high, the local strong wind will be produced around the tall buildings, making outdoor activities uncomfortable and unsafe, and even causing pedestrian wind environment problems [4, 5]. The periphery of the park is a new urban area under construction, and the construction of high-rise buildings will gradually increase in the future. The sinking space of the museum will become an important wind environment adjustment node in this area. In this paper, the three-dimensional velocity field and temperature field inside and outside the building are accurately simulated by hydrodynamics (Computational Fluid Dynamics: CFD). The flow and heat transfer simulation software based on the CFD principle is used as a simulation tool to study the evaluation of the outdoor wind environment and the prediction of indoor natural ventilation potential [6, 7]. According to Fig. 1, the wind environment in Zhengzhou is conducive to the optimization of the wind environment control of the outdoor site of the project and does not affect the comfort of outdoor activities and building ventilation (see Fig. 6).

First of all, Phoenics is selected as the simulation tool, and the site-building model is established by Autocad [8, 9]. In the process of establishing the site model, the building model and the selection of the calculation area are reasonably simplified. Second, according to the Special Meteorological data set for Building Thermal Environment Analysis in China, the evaluation scene of wind speed and wind pressure is determined, and according to the meteorological characteristics of the project site, three typical working conditions in summer, winter and transition season are analyzed respectively. Finally, the evaluation of the outdoor wind environment of the site mainly takes the Beaufort wind grade as the standard, and the potential of indoor natural ventilation takes the wind pressure difference of the building facade as the standard. The greater the pressure difference is, the more beneficial it is to indoor ventilation [10], but if the wind pressure difference is too large, it will cause damage and shedding of doors, windows and exterior decoration, and the project location is close to cold areas. Excessive wind pressure difference will also increase cold air infiltration, increase heating energy consumption and indoor discomfort. Therefore, the potential of natural ventilation should be moderate.

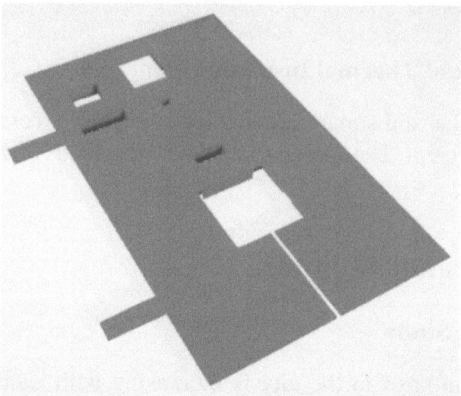

Fig. 6. Monthly total radiation in Zhengzhou

Table 1. Season condition simulation in outdoor wind environment.

No.	Season	Wind speed (m/s)	Wind direction
1	Summer	3.1	S
2	Winter	4.0	WNW
3	Transition season	3.6	NE

A large area of vortex area appeared in the inner courtyard in summer. In most other areas, the wind speed is low, which is about 0.3–1.8 m/s. The maximum wind speed is 2 m/s, and the ventilation condition is good (see Table 1/Fig. 7 and Fig. 8).

The distribution of the flow field around the area is balanced in winter. The wind speed in most areas is 0.3–2.5 m/s, with the maximum wind speed being about 3.7 m/s, the average wind speed about 1.4 m/s, and the maximum wind speed in the central rest area 1.5 m/s, which will make people feel comfortable. The winter wind speed magnification factor of the project pedestrian area is less than 2. The overall situation is good with only one large part of the wind speed magnification factor appearing at the ascending ramp. The area is very small, and the wind speed amplification factor is not more than 2, which does not affect the outdoor comfort of the pedestrian area (see Fig. 9, Fig. 10 and Fig. 11).

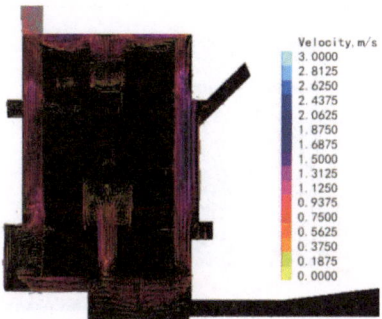

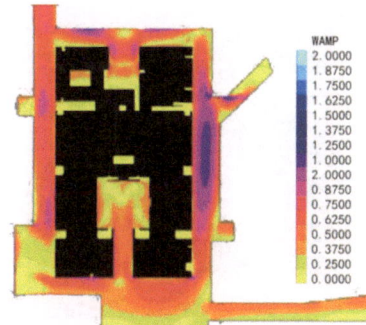

Fig. 7. Vector diagram of wind speed at the altitude of 1.5 m in summer

Fig. 8. Cloud map of wind speed at an altitude of 1.5 m in summer

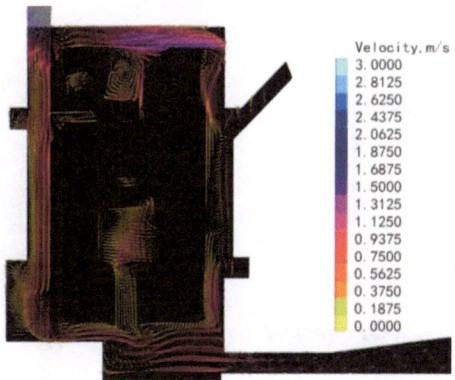

Fig. 9. Vector diagram of wind speed at the altitude of 1.5 m in winter

The distribution of the flow field around the transition season is relatively uniform, with the local wind zone and eddy current zone. The wind speed in most areas is 0.3–1.8 m/s with the maximum wind speed being 3.7 m/s, which is conducive to outdoor heat dissipation and pollutant dissipation (see Fig. 12 and Fig. 13).

In summer, the wind pressure on the facing surface of the building is between 2.5–4.5 Pa, and the windward pressure of more than 50% of the facades exceeds 0.5 Pa, which is a prerequisite for the formation of good indoor natural ventilation (see Fig. 14 and Fig. 15).

The maximum pressure difference between the front and leeward sides of the building in winter is about 1.6–4.9 Pa. The building is a single-row building with moderate ventilation capacity and no excessive cold air penetration (see Fig. 16 and Fig. 17).

During the transition season, the wind side of the building surface is about 0.6–3.1 Pa, and the windward pressure of more than 50% of the facades exceeds 0.5 Pa, which is a prerequisite for the formation of good indoor natural ventilation (see Fig. 18 and Fig. 19).

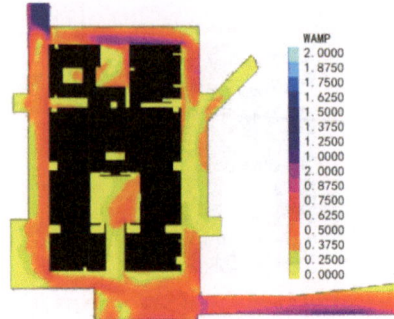

Fig. 10. Cloud map of wind speed at an altitude of 1.5 m in winter

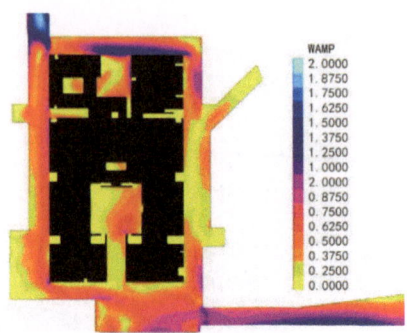

Fig. 11. Cloud map of wind speed magnification factor at the altitude of 1.5 m in winter

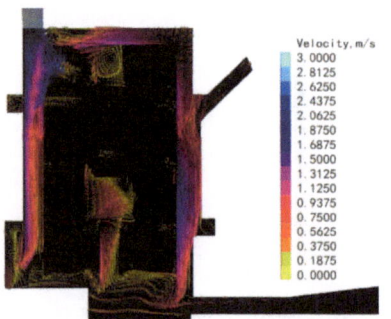

Fig. 12. Vector diagram of wind speed at the altitude of 1.5 m in transition season

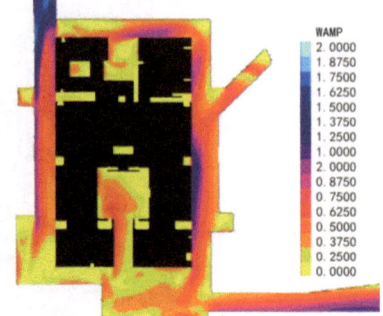

Fig. 13. Cloud map of wind speed at an altitude of 1.5 m in transition season

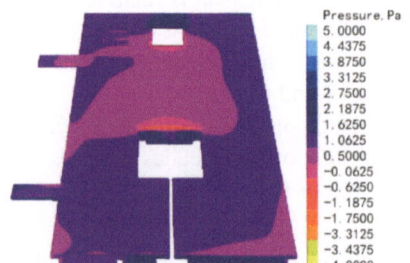

Fig. 14. Pressure distribution on the windward side of buildings in summer

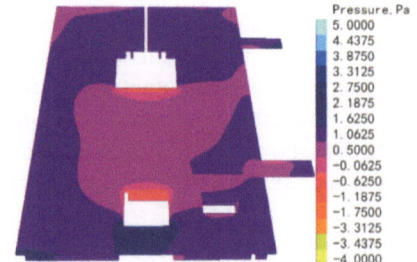

Fig. 15. Distribution map of leeward pressure of buildings in summer

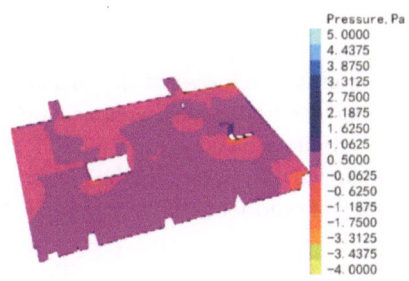

Fig. 16. Pressure distribution on the windward side of buildings in winter

Fig. 17. Distribution map of leeward pressure of buildings in winter

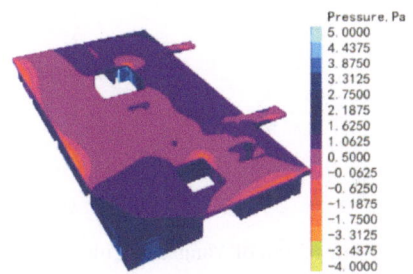

Fig. 18. Pressure distribution on the windward side of buildings in transition season

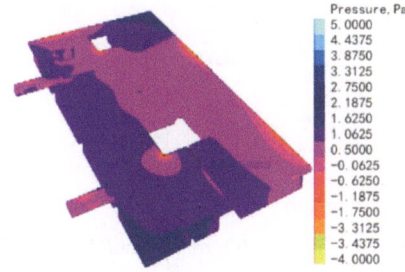

Fig. 19. Distribution map of leeward pressure of buildings in transition season

To sum up, the outdoor wind environment of the site is relatively comfortable. In summer and transition season, the indoor and outdoor pressure difference in 50% of the building opening area exceeds 0.5 Pa, so it is suitable to adopt natural ventilation measures.

3.2 Thermal Environment Study

According to software simulation and analysis, the gray-black part of the shaded area of the building is shown in the following picture, and the non-shaded area is shown by the red line (see Fig. 20 and Fig. 21). The outdoor activity area is 14,231 square feet, of which the shaded area is 7,220 square feet, and the shaded area of trees in the non-shaded area is 67.51 L. The coverage rate of the shaded area of outdoor activity area is 51.21% (see Fig. 22 and Fig. 23).

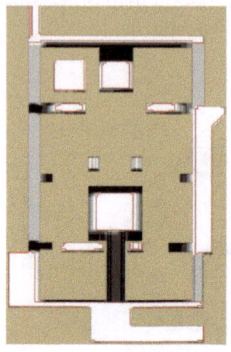

Fig. 20. Distribution map of building shadow area

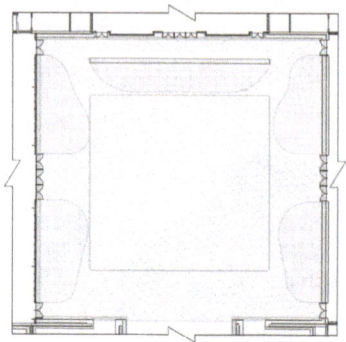

Fig. 21. Planting Map of Yangshao Culture-themed Courtyard

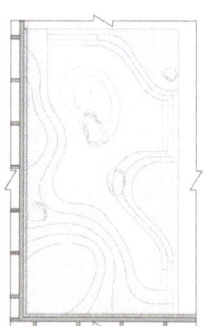

Fig. 22. VIP courtyard planting map

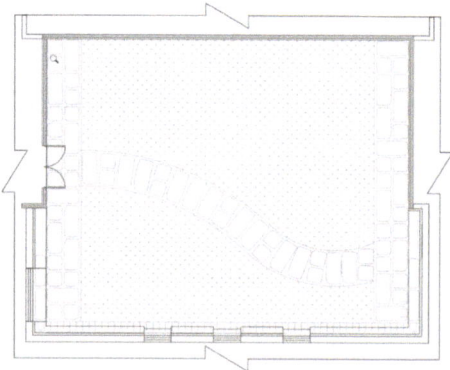

Fig. 23. Office planting map

The total roof area is 13,339.7 m^2, the green space area is 12,254.7 m^2, the roof greening rate is 91.87%, and the total roof area with horizontal projection area and solar radiation reflection coefficient is more than 75% (see Table 2).

Table 2. Season condition simulation in outdoor wind environment.

Plant	Adult crown width (m)	Adult projection area (m²)	Total amount	Plant amount outside the shaded area	The shaded area (m²)
Pinus tabulaeformis A	6	28.26	1	1	28.26
Pinus tabulaeformis B	5	19.625	2	2	39.25
The total shaded area outside the shaded area					67.51

4 Indoor Environment Study

4.1 Sunlight Analysis

The amount of radiation in summer is relatively large, and the use of shading measures in summer is beneficial to reduce the energy consumption of air conditioning [5]. In order to maintain the integrity of the appearance of the museum, the appearance of the offices uses the same decorative concrete louver as the main curtain wall material of the museum, which is analyzed by the project team to ensure the office requirements.

According to the statistics of the above results, the space area of the first floor and the second floor of the project to participate in daylighting calculation is 1,863 square meters, with 1,159 square meters, or 62.2% of the total, meeting the standard. The statistical results are shown in the table below. Office and conference rooms can avoid direct sunlight, and the lighting is mainly through the diffuse reflection through the external decoration of the concrete grille. In addition, the louver of the office area avoids window direct light, indoor lighting is mainly diffuse light, and the staff background can face the window (see Fig. 24).

4.2 Anti-condensation and Thermal Bridge Analysis

In order to maintain a comfortable indoor environment in summer and winter, air conditioning and heating are needed most of the time. At the same time, the transition season has the temperature and humidity conditions of making good use of natural ventilation. The internal space of this scheme is relatively complete, and there is a skylight in the hall of the exhibition hall, so anti-condensation measures are adopted in the construction of the outer wall and roof to ensure that when the inner surface temperature 'twn' of the outer structure is higher than the indoor dew point temperature tl in winter, which will not lead to condensation on the inner surface of the outer structure (see Table 3, 4, 5, 6, 7, 8, 9 and Fig. 25, 26, 27, 28, 29, 30, 31, 32).

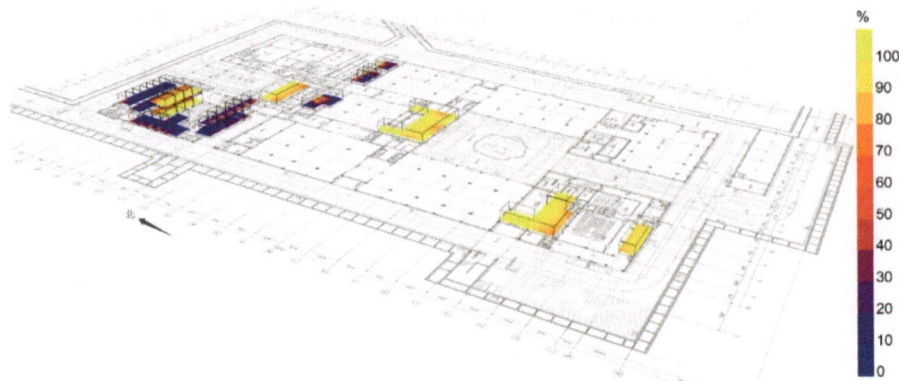

Fig. 24. Simulation results of annual dynamic natural lighting in main functional space

Table 3. Calculation table of the inner surface temperature of a roof

No.	Room	Indoor calculated temperature ti °C	Roof structure type	Total thermal resistance Ro	Internal surface thermal resistance Ri	The inner surface temperature of the enclosure structure θc	Condensation status
1	Foyer, exhibition hall, public toilet	18.00	R01	3.33	0.11	17.20	N
2	Offices, conference rooms, restaurants	20.00	R01	3.33	0.11	19.22	N
3	General warehouse, cinema, multi-function hall	20.00	R01	3.33	0.11	19.22	N

4.3 Change of Inner Surface Temperature

According to Fig. 5, there is abundant precipitation in Zhengzhou, which provides better conditions for planting roofs. The thickness of the roof of the museum is more than 700 mm, and the thermal insulation of the external wall is also thickened, so the main rooms and corridors of the office area adopt natural ventilation. After simulation, it is confirmed that the indoor temperature changes with the outdoor temperature in a reasonable range to reduce the energy consumption of air conditioning. The heat transfer process of the exterior wall and roof of the enclosure structure is usually regarded as a one-dimensional unsteady heat conduction problem without an internal heat source. The numerical analysis method is used to calculate the boundary conditions and calculation

Table 4. Inner surface temperature of exterior wall

No.	Room	Indoor calculated temperature ti °C	Roof structure type	Total thermal resistance Ro	Internal surface thermal resistance Ri	The inner surface temperature of the enclosure structure θc	Condensation status
1	Foyer, exhibition hall, public toilet	18.00	W01	2.17	0.11	17.20	N
2	Offices, conference rooms, restaurants	20.00	W01	2.17	0.11	18.81	N
3	General warehouse, cinema, multi-function hall	20.00	W01	2.17	0.11	18.81	N

Table 5. Inner surface temperature of exterior window

No.	Room	Indoor calculated temperature ti °C	Roof structure type	Total thermal resistance Ro	Internal surface thermal resistance Ri	The inner surface temperature of the enclosure structure θc	Condensation status
1	Foyer, exhibition hall, public toilet	18.00	C01	0.42	0.11	16.40	N
2	Offices, conference rooms, restaurants	20.00	C01	0.42	0.11	13.80	N
3	General warehouse, cinema, multi-function hall	20.00	C01	0.42	0.11	13.80	N

parameters according to the provisions and appendices of the Code for Thermal Design of Civil Buildings GB50176-2016 (see Fig. 33 and Fig. 34).

Table 6. Inner surface temperature of thermal bridge column

No.	Room	Indoor calculated temperature ti °C	Structure type of non-thermal bridge parts	The total thermal resistance of non-thermal bridge parts Ro	The total thermal resistance of thermal bridges R´o	Correction coefficient	Internal surface thermal resistance Ri	The inner surface temperature of the enclosure structure θc	Condensation status
1	Foyer, exhibition hall, public toilet	18.00	H01	1.92	1.47	0.87	0.11	16.10	N
2	Offices, conference rooms, restaurants	20.00	H01	1.92	1.47	0.87	0.11	18.30	N
3	General warehouse, cinema, multi-function hall	20.00	H01	1.92	1.47	0.87	0.11	18.30	N

Table 7. Inner surface temperature of thermal bridge beam

No.	Room	Indoor calculated temperature ti °C	Structure type of non-thermal bridge parts	The total thermal resistance of non-thermal bridge parts Ro	The total thermal resistance of thermal bridges R´o	Correction coefficient	Internal surface thermal resistance Ri	The inner surface temperature of the enclosure structure θc	Condensation status
1	Foyer, exhibition hall, public toilet	18.00	H02	1.82	1.47	0.87	0.11	16.60	N
2	Offices, conference rooms, restaurants	20.00	H02	1.82	1.47	0.87	0.11	18.29	N
3	General warehouse, cinema, multi-function hall	20.00	H02	1.82	1.47	0.87	0.11	18.29	N

Table 8. Inner surface temperature of the thermal bridge floor

No.	Room	Indoor calculated temperature ti °C	Structure type of non-thermal bridge parts	The total thermal resistance of non-thermal bridge parts Ro	The total thermal resistance of thermal bridges R´o	Correction coefficient	Internal surface thermal resistance Ri	The inner surface temperature of the enclosure structure θc	Condensation status
1	Foyer, exhibition hall, public toilet	18.00	H03	1.75	1.47	0.87	0.11	18.50	N
2	Offices, conference rooms, restaurants	20.00	H03	1.75	1.47	0.87	0.11	18.28	N
3	General warehouse, cinema, multi-function hall	20.00	H03	1.75	1.47	0.87	0.11	18.28	N

Table 9. Inner surface temperature of thermal bridge joint of external floor slab

No.	Room	Indoor calculated temperature ti °C	Structure type of non-thermal bridge parts	The total thermal resistance of non-thermal bridge parts Ro	The total thermal resistance of thermal bridges R´o	Correction coefficient	Internal surface thermal resistance Ri	The inner surface temperature of the enclosure structure θc	Condensation status
1	Foyer, exhibition hall, public toilet	18.00	H04	2.27	1.47	0.87	0.11	16.80	N
2	Offices, conference rooms, restaurants	20.00	H04	2.27	1.47	0.87	0.11	18.32	N
3	General warehouse, cinema, multi-function hall	20.00	H04	2.27	1.47	0.87	0.11	18.32	N

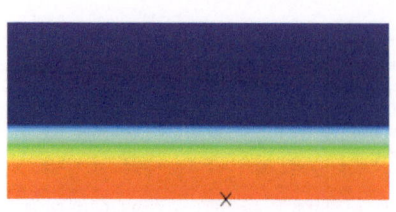

Fig. 25. Temperature field distribution map of RO1 roof (the most disadvantageous) XTimn = 17.2 °C

Fig. 26. Distribution diagram of temperature field of W01 exterior wall XTimn = 17.1 °C

Fig. 27. Temperature distribution on the inner surface of C01 outer window XTimn = 16.4 °C

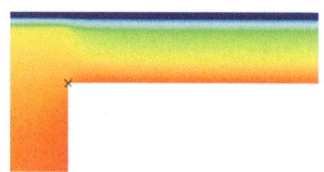

Fig. 28. Temperature distribution of the most unfavorable inner surface of the thermal bridge column XTimn = 16.1 °C

Fig. 29. Temperature distribution of the most unfavorable inner surface of the thermal bridge beam XTimn = 16.6 °C

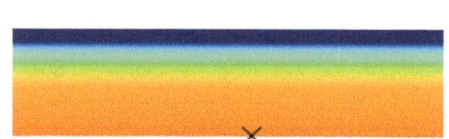

Fig. 30. Temperature distribution of the most unfavorable inner surface of the thermal bridge floor XTimn = 18.5 °C

Fig. 31. Temperature distribution of the most unfavorable inner surface of the thermal bridge joint of the external floor slab XTimn = 16.8 °C

20.0°C 18.4°C 16.9°C 15.3°C 13.7°C 12.2°C 10.6°C 9.0°C 7.5°C 5.9°C 4.3°C 2.8°C 1.2°C -0.4°C -1.9°C -3.5°C

Fig. 32. Temperature distribution of the most unfavorable surface

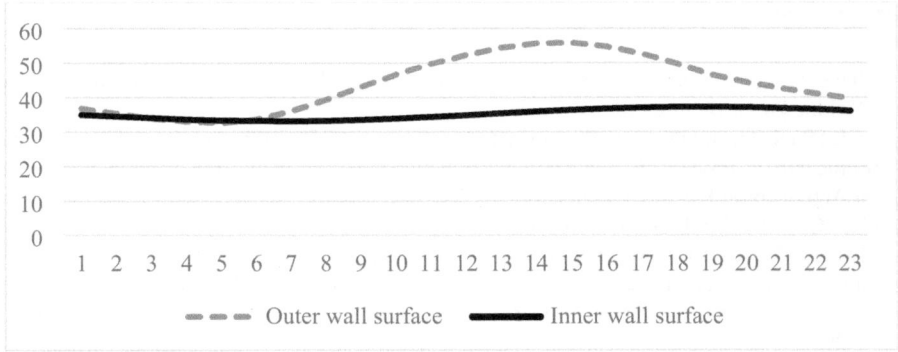

Fig. 33. Temperature distribution of the most unfavorable surface

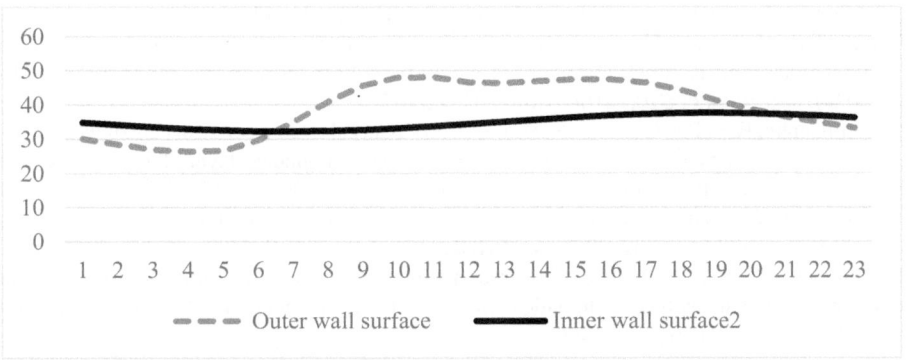

Fig. 34. Indoor and outdoor wall temperature distribution of the wall

5   Conclusion

This project combines the interpretation and communication of site value, urban negative space regeneration, and passive green building strategy. During the scheme design phase, CFD and other software were used to simulate and evaluate key indicators such as wind environment, light environment, and thermal environment, providing clear and accurate guidance for scheme deepening. This paper provides an example and reference for the passive green ecological building design in the Central Plains.

Acknowledgment. The green building simulation of this project is completed by China Architectural Design and Research Institute Co., Ltd.

Fund.   Major Science and Technology Research Project of China Construction Technology Group "Research on Urban Renewal Design Methods and Implementation Technologies in the Context of High Quality Development" (Z2022J15).

References

1. Zhang, Q.Y., et al.: China Standard Meteorological Database for Architecture, p. 27. China Machine Press, Beijing (2004)
2. Li, E.: Study on architectural climate in Zhengzhou. In: Architecture and Urban Physical Environment in the Process of Urbanization: Collection of Essays in the 10th National Conference on Architectural Physics, Guangzhou, pp. 332–325 (2008)
3. Gao, Q.Y.: Research on green design strategy of building exterior structure in Zhengzhou. Master's Degree Thesis of Zhengzhou University, pp. 15–22 (2014)
4. Tang, H.: Study on spatial optimization design of high-rise residential area in Zhengzhou based on microclimate effect. Master's Degree Thesis of Zhengzhou University, pp. 10–13 (2020)
5. Wen, H., Fang, L.: Construction and ventilation evaluation effects of urban wind corridors. Master's Degree Thesis of Chongqing University, pp. 15–23 (2022)
6. Hamed, H., Behrad, E., Reza, Z., Rahmat, S.-G., Navid, M.: Comparative CFD-DEM study of flow regimes in spout-fluid beds. Particuology **85**, 323–334 (2023)
7. Reineking, L., et al.: Convective drying of wood chips: accelerating coupled DEM-CFD simulations with parametrized reduced single particle models. Particuology **84**, 158–167 (2024). https://doi.org/10.1016/j.partic.2023.03.012
8. Soprunenko, E.E., Perminov, V., Reyno, V.V., Loboda, E.L.: Convective drying of wood chips: simulation of impact assessment of crown forest fires on boundary layer of atmosphere using software PHOENICS. Atmos. Ocean Opt. **9680**, 96805Y-96805Y-6 (2015)
9. Kumar, M., Borghain, A., Maheshwari, N.K., Vijayan, P.K.: Simulation of natural circulation in a rectangular loop using CFD code PHOENICS. Kerntechnik **76**(2), 93–97 (2011)
10. Taniguchi, K., et al.: Influence of external natural environment including sunshine exposure on public mental health: a systematic review. Psychiatry Int. **3**(1), 91–113 (2022). https://doi.org/10.3390/psychiatryint3010008

Study on the Three-Dimensional Place Making in the Core Area of Urban Central Railway Station in the Space Narrative Context

Yuan He[1], Duwen Shen[1(✉)], Lili Peng[1], and Chongshi Xie[2]

[1] Chongqing College of Architecture and Technology, Mingde Road, Shapingba District, Chongqing, China
277306408@qq.com
[2] Chongqing Architectural Design Institute Co., Ltd., Renhe Road, Yuzhong District, Chongqing, China

Abstract. The core area of the urban central rail transit station is the area with the highest concentration of urban population and the most prominent contradiction. Aiming at the problems of fragmentation of above-ground and underground space, low quality of underground space environment, and lack of sense of place, this article puts forward the concept of three-dimensional place making. By introducing the analysis method of spatial narrative, and taking its advantages in shaping the sense of place, integrating the information of place, and constructing the semantics of place, this paper studies the characteristics of the place making of Shenzhen Huaqiangbei Station and Tokyo's Shibuya Station, explores the spatial narrative logic that can effectively connect the above ground and underground, proposes the design strategy of three-dimensional place making, and explores the possibility of quantitative study of urban three-dimensional space in station area. The study found that the place making of station area should be based on the characteristics of its own space environment. In the area with high three-dimensional degree, complex space environment and large scale of development, it is appropriate to take the walking network as the framework and use deconstructive spatial narrative technique to create places with "multiple themes, multiple scenes and multiple elements" by segmenting and partitioning, so as to realize the organic integration of various spatial elements. This paper, for the first time, applies the spatial narrative theory and method to the study of place making in the core area of rail transit station, and explores a new way for the study of three-dimensional urban design in the urban railway station area.

Keywords: The Core Area of Urban Central Railway Station · Place Making · The Space Narrative · Digitization of Three-dimensional Space

1 Introduction

Until January 2023, China's rail transit mileage has exceeded 9,584 km and set up 5609 stations [1]. With the help of compact cities and TOD theory, China has ushered in a golden period of rapid rail transit construction. As the core of the urban central area,

© The Author(s) 2024
P. Xiang and L. Zuo (Eds.): PBSFTT 2023, LNCE 382, pp. 503–513, 2024.
https://doi.org/10.1007/978-981-97-5108-2_53

the core area of the urban rail center station has a high degree of population aggregation and a complex urban space environment. It is an important public space node in the city and a place to carry people's social interactions and various activities [2]. Place making is an important research content of urban design (Wang Jianguo, 2017) [3]. In the core area of the rail station, relying on the three-dimensional space on the above-ground and underground, by creating an attractive, distinctive, and rich connotation three-dimensional space environment and social cultural atmosphere, it can effectively improve the urban spatial environmental quality of the core area and enhance the sense of belonging in their place.

Relevant domestic research: In June 2023, searching CNKI database with the subject heading "rail transit station /station area/TOD+place/place making" found 8 journal papers (2 core papers), 7 doctoral theses and 12 master theses. Only Xu Leiqing and Liu Nian et al. (2015) studied the relationship between the public space of the station domain and urban vitality from a micro-scale [11], and Wang Chengfang (2013) proposed the introduction of the "urban catalyst" theory into the design strategy of the site area site creation [9]. Based on the comprehensive retrieval of Chinese databases and literature analysis, it is found that the current three-dimensional research on the rail station area is mainly based on the qualitative research of land use [5], spatial development mode [6], and walking system [7]. Related research focuses on urban design methods, strategies [4] and the compilation of guidelines [8]. Only Gui Wangyang proposed the environmental design strategy of "core construction to enhance the vitality of the station". There is a lack of systematic research on the three-dimensional space environment of the rail transit station area, especially the three-dimensional place construction.

Relevant foreign research: Based on English database, searched the core of Web of Science database with the subject heading "rail transit station /station area/TOD+place/place making" found 16 related papers. The above research focuses on the study of the road network structure of the station area [10], the scale of the plot [12], the degree of spatial openness [13], the interaction mechanism of traffic and urban space [14]. Only Yuan proposed to build an intensive, efficient, ecological, and technological sense of place in urban underground space, and build a "double-layer" urban space system above and below the ground [19]; Sun uses virtual reality technology to study the interface form and space experience of underground squares [20].

Comprehensive domestic and foreign relevant research can be known: there is no special research on the construction of three-dimensional places in the station area, and a few papers on the urban space environment/landscape design of the station area are mainly characterized by description. There is also a lack of new research paradigms and methods in the three-dimensional space organization and place construction of railway station area.

At present, the method of spatial narration has been widely confirmed because it helps to promote the logical integration of urban spatial elements, strengthen the place recognition and environmental characteristics of urban space (Shi Fei et al., 2014) [16]. However, it has not been applied to urban space research in the rail transit station area.

Aiming at the problems of the above and underground space separation, low quality of underground space environment and lack of sense of place in the railway station area, this paper takes Shenzhen Huaqiangbei Station and Tokyo Shibuya Station as examples, creatively introduces the spatial narrative method into the research on the place making of three-dimensional space in the station area, and explores a new way for the realization of the above-ground and underground integration in the core area of railway stations.

2 Brief Analysis of the Concept of Spatial Narration

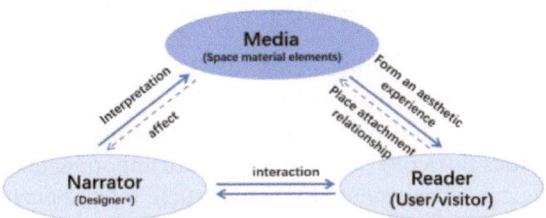

*After the period of modernism, users also acted as spatial narrators widely, and the influence of the government, developers, and builders as narrators cannot be ignored.

Fig. 1. The interrelationship of the three elements of spatial narration. Source: self-drawing according to references [16]

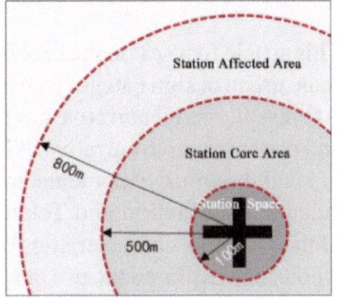

Fig. 2. Core area of urban central railway station. Source: self-drawing

With the spatial transformation of the humanities, the study of the spatial dimension of narration has gradually received attention and has been introduced into contemporary urban studies. Through the integration with behaviorism and postmodernism ideas and methods, a spatial narrative method that guides urban design, architecture and landscape design is formed [16]. Spatial narrative means that the narrator uses spatial material elements as a medium and uses various relevant expression methods to tell stories, events or experiences to the recipient. At the same time, the recipient can self-interpret the space based on the narrative media and the individual's different knowledge systems.

The narrator, medium and reader are the basic elements of spatial narration. Among them, the media, as the spatial material element chosen by the narrator to tell the story [17], is the core of spatial narration and the focus of architectural and urban space research. Through users' interpretation of narrative media in space, a series of memories and associations of relevant events and plots are generated, to arouse the formation, reproduction and sublimation of creators/users' aesthetic experience, and to construct the place attachment relationship between subject and object [17] (Fig. 1).

3 The Core Area of Urban Central Railway Stations

Central stations are railway stations that undertake the functions of city-level centers and sub-centers. In principle, it is an interchange station of multiple rail transit lines, positioned as a regional public service center, and the core area of the rail station encourages the development and construction of urban complexes (also the area with the highest demand for underground space utilization). The "Guidelines" divide the urban central railway station area into four spatial scope levels: (1) 0–100 m above ground and underground space; (2) 100–500 m core influence area; (3) 500–800 m influence area; (4) 800–1500 m secondary influence area [14]. This paper takes the 500 m range of the rail station as its core area, and uses the space narrative theory to study the integration of ground and underground places in this area (Fig. 2).

4 Research on the Three-Dimensional Place Making in the Urban Central Railway Station Area Based on Spatial Narration

This article focuses on the urban space of the rail station area, which belongs to the meso-scale urban design category, so it focuses on the "structure" aspect of the spatial narrative. At present, spatial narrative has gradually developed into two types: structuralist narrative and deconstructive narrative. "Semantics (lexicon)" and "structure" are important means of spatial narrative elements and logical arrangement [15]. This article took Shenzhen Huaqiangbei Station and Tokyo Shibuya Station as examples, focused on the analysis of the narrative characteristics of the place and the relationship between the part and the whole, and explored the use of space narration in the research of place making possibility.

4.1 Structuralist Spatial Narrative—Study on the Characteristics of Shenzhen Huaqiangbei Railway Station

Basic Situation of Huaqiangbei Railway Station

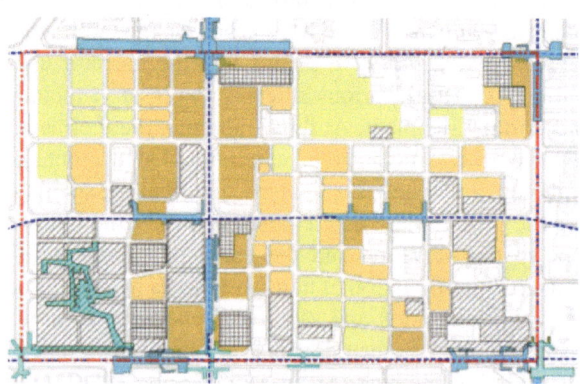

Fig. 3. Commercial space and railway stations of Huaqiangbei. Source: reference [18]

The Huaqiangbei railway station area is a concentrated commercial and densely populated area in Shenzhen, with a very vibrant urban space. This area gathers the related stations of Shenzhen Metro Line 1, 2, 3, 7. The ground is a commercial pedestrian street integrating urban landscape and leisure, with a total length of about 1 km. The underground space is located below the commercial pedestrian street on the ground, starting from Shennan Middle Road in the south and to the end of Huaxin Station on Line 7 in the north. It has a total length of 830 m, a width of 28.1 m, and a total construction

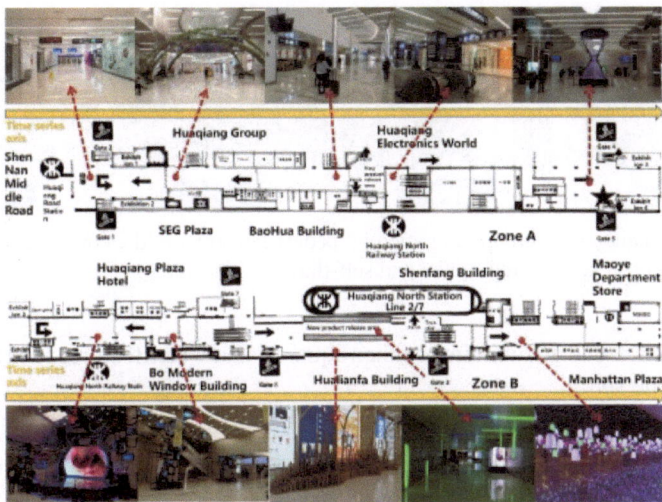

Fig. 4. The spatial narrative clues of the underground commercial street. Source: self-drawn by the author

area of 70530 m². This article focuses on the research object at the station core area: a north-south ground-underground integrated "belt" commercial street (Fig. 3).

The Narrative Characteristics of the Station Core Area

Huaqiangbei Commercial Street spreads out in a north-south belt on the ground and underground. It is affected by the narrow east-west direction, and its place making conforms to the linear narrative method. In structuralist narratives, the internal synchronic relationship between spatial elements is emphasized, and the inner logical relationship between the surface structure (the combination of spatial text carriers) and the deep structure (the inner logic of the meaning of spatial text carriers) is explored.

Construction of Underground Space
Take the section from Huaqiangbei Station Line 2/7 to Shennan Middle Road as an example: After exiting the subway, passengers pass through a hall and walk into an open underground plaza. The bright and flowing roof and vegetated arcades weaken the closure of the underground space. Then the space changes into one side or both sides of the underground street for shops. Montage is used in small local spaces to create a dream space through light and shadow changes, enriching the space environment at the interface between the street and the subway (Fig. 4).

Place Creation of Ground Space
The processing of above-ground space nodes is mainly concentrated at the entrances and exits of underground commercial streets. The above-ground space presents a spatial sequence of "beginning segment -- guide segment -- climax segment -- end segment" through repeated underground entrance/exit structures and set urban sculptures at important nodes and the end of pedestrian streets, which are highly similar to traditional Chinese chapter novels (Fig. 5). Under the principle of unification, the ground space is

Fig. 5. Space sequence of ground commercial pedestrian street in the core area of Huaqiangbei station area. Source: self-photographed and self-drawn by the author

partially dotted with dry fountains, sunken squares, and water features, which play a role in enriching the space.

4.2   The Space Narrative of Deconstructivism Spatial—Study on Site Characteristics of Shibuya Station in Tokyo

Basic Situation of Shibuya Station
Shibuya Station is the second largest transportation hub in Tokyo area after Shinjuku Station. Nine lines intersect here, and the number of transfers exceeds 3 million per day. The Shibuya mega-level station city integrated development project, which is expected to be completed before 2027, includes an area where many functions such as rail transit stations, commerce, office, and culture are gathered, and is one of the most important transportation hubs in central Tokyo.

The Narrative Characteristics of the Station Domain Space
The space narrative dominated by deconstructive narrative focuses on the interaction mechanism between the informal structure of the place and the specific reader, and the relationship between the meaning of the place and the social context [17]. The place making in the core area of Shibuya Station uses a deconstructive narrative method. Instead of setting a distinct spatial sequence, it presents the characteristics of rich and diverse spatial elements, complex and diverse structural relationships, and dynamic and random spatial activities. Compared with the grand structuralist narrative with clear logical levels, Shibuya Station's place making pays more attention to diversified and "small" narratives, and emphasizes the experience of participants through the uniqueness of each venue and space.

Place Making of Three-Dimensional Pedestrian Space
As Shibuya's valleys vary greatly in topography, a four-story spatial walk system from underground to above ground was established, connecting subway stations, ground, corridors, and sky corridors, and relying on a pedestrian network to place making. Outdoor and semi-outdoor escalators and stairs guide the flow of people, and combine the wave-shaped canopy and the large cantilever structure to construct the open space of the interchange square, thereby making the pedestrian space rich in hierarchy and mobility.

By liberating the roof space, creating a multi-level roof plaza realizes the interaction between the sky and the ground. After completion, it will become Japan's largest rooftop observation plaza and become a new type of urban base, forming the urban underground Three-dimensional base surface system on the ground and in the air.

Place Making of Vertical Space

In the vertical direction, through the changes of color and spatial form, a rich sense of place in the vertical direction is created. In addition, the "information ring" integrated with escalators in the urban core realizes the temporal and spatial interweaving and variability of the vertical height space and the horizontal space through the constantly changing time, orientation, and seasonal images on the circular motif.

4.3   Place Making Strategy for the Core Area of Urban Central Railway Station

Choose an Appropriate the Space Narrative Method According to the Characteristics of the Site Area

It is found that: in the station core area, for linear underground streets, ground commercial streets, air traffic corridors, and underground passages, the use of structuralist narrative methods for place making has advantages. For the station space with a high degree of three-dimensionality, a very high degree of networking, and an extremely complex and rich spatial form, the pedestrian network is used as the skeleton, and the deconstructive the space narrative method is used to divide and partition to make a "multi-theme, multi-scene, multi-element" place more reasonable.

Pay Attention to User Experience and Participation

As the most densely populated area, the core area of the urban central rail station needs to create an affinity for the station space environment to provide people with more resting places in the "hurried" urban space. Relying on the three-dimensional space on the ground, ground and air, through appropriate the space narrative techniques, story layout and scene creation are used to create attractive, distinctive, and rich connotation three-dimensional space environment and social cultural atmosphere, and improve the site space.

Realize "Multi-voice Harmony" Through the Diversification of Spatial Elements

In the past, the construction of place making emphasized the meaning of places, but in the face of the complex urban environment, three-dimensional spatial form and larger development scales in the site area, overemphasizing the meaning of design and the structure of the story would be counterproductive. It is possible to realize the coordination between the various parts by dividing the urban space into parts, using the horizontal and vertical traffic lines to connect in series, and learn from the diversified concept of deconstructive narrative to create rich urban spaces and activity venues.

5   Outlook: Quantitative Research and Exploration of Urban Three-Dimensional Space in the Railway Station Core Area

In order to further explore the quantitative research method of spatial narrative in the future, it is necessary to first carry out the digitization of urban space in the core area of the railway station. This paper probes into the digital research method, which is carried out according to the following steps:

5.1 Construct the Three-Dimensional Spatial Database and Collection Spatial Information

By collecting the material space data of the core area of urban central railway station, such as land, building, road, open space, underground passage, underground street, etc. Non-material spatial data, such as population, economy, city-level statistical data, were processed by CAD+3D GIS technology to establish a three-dimensional spatial database of the core area of the central railway station. Taking the core areas of important central rail stations in Tokyo, Osaka, Chicago, Guangzhou, Shenzhen and Hong Kong as typical cases, the method of "digital map + literature collection + field research" is used to collect their three-dimensional spatial information (Table 1).

Table 1. Construction of the spatial database in the core area of railway station

3D GIS Spatial Data		
Spatial site data (S-site)	Spatial attribute data (P-people, L-land, TR-transit)	
S1 Location	P1 Number of inhabitants	L6 Floor area of building
S2 Underground passage	P2 GDP	L7 Total land area
S3 overpass	P3 City level	L9 Number of plots in the core area
S4 bus stops	L1 Land use nature	L10 Various functional areas
S5 Railway Station	L2 Area of each building (above and underground)	TR5 Road network density
S6 Plot size	L3 Number of floors per building	
S7 boundary dimensions of underground space	L4 Open space area	
Urban big data		
M1 Open Street Map	M4 Google MAP	M3 Google Street View

Source: self-drawn by the author

5.2 Establishment of the Urban Space Vector Information Database (Above Ground-Underground Space) in the Core Area of Central Railway Station

Data processing:

(1) **Collection and storage of vector data (SHP surface file):** The DWG file is based on the relevant map base map and the comprehensive field investigation. Take a station as the center and a 500 m radius to ensure that the figure size is consistent with the actual size. Then, the DWG file is imported into ArcGIS, and the vector image data of the target area is saved in the target folder in the form of SHP surface files.

(2) **Cleaning of vector data:** In order to make the data meet the experimental require-
    ments, it is necessary to clean the data in ArcGIS through the "domain buffer" tool,
    the "spatial connection" tool, the property table, etc., and check whether there is any
    error by comparing with the existing map. Finally, the generated target SHP surface
    file is saved in the target folder.

(3) **Establish building height-depth information base:** Data source and processing:
    As a basic input to the model to obtain the building volume. Measure the height and
    depth of relevant buildings within the core area through field research or web search
    and add the height value of corresponding buildings in the attribute table of the SHP
    plane file. The SHP surface file obtained by the relevant data platform comes with the
    building height-depth attribute value, and checks whether it is wrong by comparing
    with the existing map (Fig. 6).

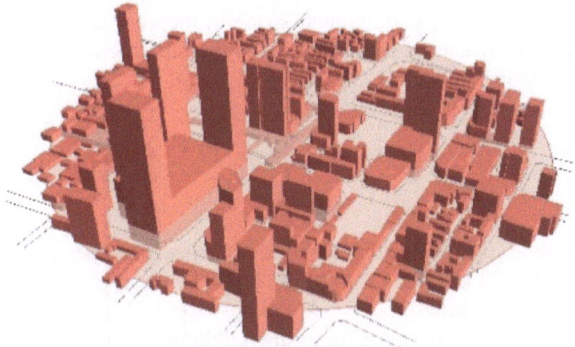

Fig. 6. The urban spatial model generated by ArcGIS. Source: self-drawn by the author (taking
Chunxilu Station, Chengdu as an example)

(4) **Derive the urban space vector information table:** The related spatial attribute
    values in the generated SHP surface file are exported and processed into a csv file
    that can be edited with Excel.

5.3 Three-Dimensional Model Generation of Urban Spatial Vector Information in the Core Area of the Central Railway Station

After obtaining the urban three-dimensional space vector information of the core area of
the central railway station, ArcGIS is used to visualize the urban space vector information
database of the core area of the central railway station, realizing the digitization of the
urban space of the station, and laying the foundation for the next quantitative study of
spatial narrative.

6 Conclusion

Spatial narrative helps to interpret and present hidden information about time, events,
experience, and memory in the core area of the station; effective narrative clues can
integrate the fragmented information of urban space in the site area, and strengthen the

sense of place in the space environment. This paper uses spatial narrative techniques to systematically study the place construction of Shenzhen Huaqiangbei Station and Tokyo Shibuya Station. The place making of the rail transit station core area should select the appropriate spatial narrative method according to its own spatial environmental characteristics. When the spatial environment of the site area is complex, the degree of three-dimensionality is high, and the development scale is large, the organic integration of various spatial elements can be achieved by dividing the urban space into zero and using horizontal and vertical traffic flow lines to connect in series. Furthermore, this paper also explores the quantitative research of urban three-dimensional space in the core area of Urban Central Railway Station, and lays a foundation for the quantitative research of future spatial narrative by constructing three-dimensional spatial information database and three-dimensional model of urban spatial vector information.

Acknowledgement. This article is supported by project "Research on three-dimensional design Mechanism of above-ground and underground space in the core area of Chengdu-Chongqing central rail transit station" of science and technology research program of Chongqing Education Commission of China. (No. KJQN202105203).

References

1. Ministry of Transport of the People's Republic of China. The operating mileage of national rail transit reached 9584 km [EB/OL] (2023). https://www.mot.gov.cn/jiaotongyaowen/202301/t20230112_3737524.html
2. Bertolini, L.: Spatial development patterns and public transport: the application of an analytical model in the Netherlands. Plan. Pract. Res. **14**(2), 199–210 (1999). https://doi.org/10.1080/02697459915724
3. Wang, J.: Digital Urban Design Based on Human-Computer Interaction. High-End Forum on the Frontier of Urban Design Development of the Chinese Academy of Engineering (2017)
4. Tang, Y., Yuan, H.: Development mechanism and urban design method of underground space in commercial center. J. Underground Space Eng. **15**(05), 1306–1315 (2019). CNKI:SUN:BASE.0.2019-05-005
5. He, D.: The spatial development and evolution of site areas and the reorganization of land use patterns under the influence of TOD. Planner **33**(04), 126–131 (2017). https://doi.org/10.3969/j.issn.1006-0022.2017.04.021
6. Gui, W.: Research on the Integral Development Path of Large-scale Railway Passenger Station Space. Southeast University (2018)
7. The concept of "Super Pedestrian District" [DB/OL] (2019). https://mp.weixin.qq.com/s/vi8CclTjMDn2r7sPS2Fg8Q. Accessed 12 Dec 2019
8. Su, B.: Urban design based on public transportation-oriented development: taking Changsha Xiangjiang waterfront as an example. Urban Dev. Res. **20**(11), 91–99 (2013). https://doi.org/10.3969/j.issn.1006-3862.2013.11.015
9. Wang, C.: Research on Land Use Optimization Strategy of Guangzhou Rail Transit Station. South China University of Technology (2013)
10. Zhou, W., Yang, J., Ge, T., Xu, C.: Study on the "3D" development model of urban land: a planning concept based on reducing the demand for motorization. City Plan. **36**(10), 51–57 (2012)

11. Xu, L., Liu, N., Lu, J.: The influence of public space density, coefficient and micro-quality on urban vitality—microscopic observation of Shanghai rail transit station area. New Build. (04), 21–26 (2015). https://doi.org/10.3969/j.issn.1000-3959.2015.04.005
12. Wang, C., Sun, Y., Zhang, C., Huang, Y., Li, M.: Planning and design of rail transit station area based on "node-location" characteristics. Planner **30**(10), 30–34 (2014). https://doi.org/10.3969/j.issn.1006-0022.2014.10.005
13. Wang, X.: Analysis of "spatial differentiation" of influence domain of rail transit stations and related urban design applications. Chongqing University (2016)
14. Vale, D.S., Viana, C.M., Pereira, M.: The extended node-place model at the local scale: evaluating the integration of land use and transport for Lisbon's subway network. J. Transp. Geogr. **69**, 282–293 (2018). https://doi.org/10.1016/j.jtrangeo.2018.05.004
15. Herman, D. (ed.): Narratologies: New Perspectives on Narrative Analysis. Ohio State University Press, Columbus (1999)
16. Shi, F., Gao, C., Meng, L., Jiang, Z.: The origin of spatial narrative method and its application in urban research. Int. Urban Plan. **29**(06), 99–103+125 (2014)
17. Zhang, N., Liu, N., Shi, G.: Interpretation of narrative space design. Urban Dev. Res. **16**(09), 136–137 (2009). https://doi.org/10.3969/j.issn.1006-3862.2009.09.025
18. He, S., Yang, S., Tang, Z.: Space narrative of traditional belief places—taking the ancient city of Chaozhou as an example. Mod. Urban Stud. (08), 17–23 (2016). https://doi.org/10.3969/j.issn.1009-6000.2016.08.003
19. Yuan, H., He, Y., Wu, Y.: A comparative study on urban underground space planning system between China and Japan. Sustain. Cities Soc. **48**, 101541 (2019). https://doi.org/10.1016/j.scs.2019.101541
20. Sun, L., Feng, L., Zhang, Y., et al.: Research on correlation between underground squares' interface morphology and spatial experience based on virtual reality. Int. J. Pattern Recogn. Artif. Intell. **34**(3), 2050004 (2019). https://doi.org/10.1142/S0218001420500044

The Integration of Prefabricated Construction and Vertical Greening
Realizing the Future of Green Cities

Jingjing Sun[(✉)] and Fang Zhou

Chongqing Energy College, No. 2, Fuxing Avenue, Shuangfu New District, Jiangjin District, Chongqing 402260, China
646103998@qq.com

Abstract. To harness the synergy of prefabricated construction and vertical greening, this study focuses on two key aspects: component fusion and technological refinement, and the evaluation of ecological benefits and carbon sequestration potential. This involves optimizing materials, construction methods, and digitalization, along with analyzing carbon sequestration and ecological gains. Emphasizing cohesive design and systematic integration is crucial, with system integration and intelligentization serving as cornerstones for harmonized collaboration between these domains.

Keywords: Prefabricated Construction · Vertical Greening · Ecological Benefits · Carbon Sequestration Assessment · Integrated Design

1 Introduction

In recent years, the global community has been addressing climate change and growing environmental challenges. The traditional construction sector significantly contributes to these problems through carbon emissions and resource consumption. Prefabricated construction and vertical greening have emerged as technologies to reduce carbon footprints and improve urban environments [1]. Prefabricated construction, with modularization and factory production, efficiently uses resources and reduces carbon emissions. Vertical greening enhances urban climates and ecological benefits by greening vertical spaces [2].

Domestic and foreign scholars have extensively researched prefabricated construction and vertical greening. Germany promoted these technologies in the early 20th century, and the American "PCI Design Handbook" advocated them in 1971. China has adopted these techniques, especially in regions like the Yangtze River Delta, Pearl River Delta, and Beijing-Tianjin-Hebei, driven by the "Twelfth Five-Year Plan." With advancements in construction and green tech, prefabricated construction and vertical greening continue to excel.

P. Xiang and L. Zuo (Eds.): PBSFTT 2023, LNCE 382, pp. 514–521, 2024.
https://doi.org/10.1007/978-981-97-5108-2_54

This study explores integrated design and optimization for prefabricated construction and vertical greening, focusing on their roles in ecological benefits and carbon sequestration. Through systematic analysis of existing literature, we highlight their importance in urban sustainability and climate mitigation. The paper covers material selection, construction innovation, system integration, and intelligentization in this context.

2 Integrated Design and Technological Optimization of Components

2.1 Optimization of Materials and Components

Prefabricated construction, as an eco-friendly and efficient building method, has gained widespread attention and research. Integrated design of prefabricated structures and greening modules is a key focus area. This approach enhances building aesthetics, promotes energy efficiency, and supports sustainable environmental development. It optimizes urban space use, intensifies greening efforts, and mitigates urban heat island effects, enhancing the urban living environment. Integrated designs often include elements like green roofs and vertical green walls, reducing stormwater runoff, enhancing rainwater utilization, and acting as biofilters to purify the air, improving urban air quality [3]. Integrated greening modules also enhance sound insulation in prefabricated buildings, providing a quiet and comfortable living environment. Implementing such designs requires comprehensive consideration of factors like plant selection, green structure design, and post-construction maintenance, ensuring sustainability and effectiveness. Figure 1 is a real-life photograph of the integrated assembly of prefabricated building panels and green modules.

Fig. 1. Real-life photograph of the integrated assembly of prefabricated building panels and green modules

Fig. 2. Real-life photograph of an ecological wall

Advances in materials science enable extensive use of new materials in prefabricated construction and vertical greening. New materials enhance structural strength and durability while maintaining lightweight efficiency. In vertical greening, they offer diverse, efficient solutions. For example, lightweight, water-efficient planting substrates improve water use and extend plant life [4]. They also boost energy efficiency and promote resource recycling, reducing environmental impact. Green concrete and recycled plastics, for instance, cut carbon emissions and energy consumption. Vertical greening benefits from materials like ecological wall systems and green roofing, fostering a greener, sustainable urban construction model. Figure 2 is a real-life photograph of an ecological wall.

2.2 Innovation in Structure and Construction Techniques

Leveraging Building Information Modeling (BIM) technology, designers optimize design, reducing construction time and costs for efficient green building implementation [5]. BIM supports personalized vertical greening designs, including plant selection, green layer design, and irrigation planning. This enhances ecological value and integrates information across phases (design, construction, maintenance) for full lifecycle project management. Combined with technologies like big data and AI, BIM enables real-time monitoring and intelligent building performance optimization, fostering deeper integration and innovation in green building and vertical greening. Figure 3 is a design illustration of prefabricated construction and vertical greening based on BIM technology.

Fig. 3. Design illustration of prefabricated construction and vertical greening based on BIM technology

Modular design significantly improves component production and installation efficiency, reducing construction time and costs. It ensures component quality, enhancing overall structure quality and durability. Modular methods minimize waste and environmental pollution on construction sites, offering an eco-friendly and sustainable alternative to traditional methods. Modular design enhances energy efficiency and supports green building standards. As the construction industry increasingly focuses on sustainability and environmental protection, modular design and rapid assembly will see more research and application. Advancements in technology will further innovate and benefit the construction sector, promoting a greener and more efficient direction.

2.3 System Integration and Intelligentization

Integrated building energy management systems are transforming green strategies in prefabricated buildings, emphasizing energy efficiency and sustainability. The construction sector is a significant energy consumer, making it vital for environmental protection and energy conservation. This integrated system enables advanced energy monitoring and management, accurately tracking building energy consumption patterns. Real-time data collection and analysis help stakeholders understand energy consumption trends and formulate rational strategies [6]. By monitoring and analyzing energy parameters, building energy usage can be optimized for efficiency. Integrated energy management systems enhance energy efficiency and eco-friendliness in prefabricated buildings, reducing energy wastage and incorporating green materials and technologies for greener construction. This signifies a shift toward a more intelligent and eco-friendly construction industry. Figure 4 represents a schematic diagram of the basic components of the electrical management system within the building energy management system.

Integrating smart systems into greening optimizes maintenance. Advanced sensors and data analysis cut costs, ensure plant health, and boost eco benefits through real-time monitoring. More data aids greening project decisions, boosting efficiency.

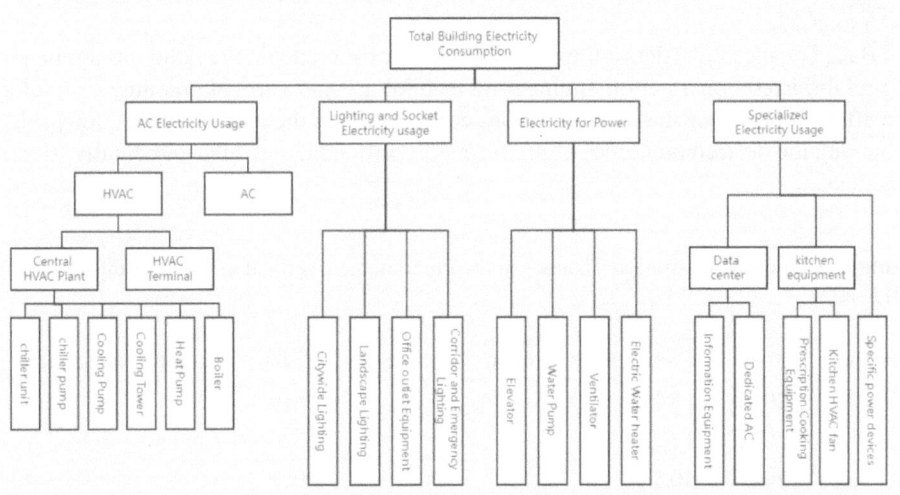

Fig. 4. Schematic diagram of the basic components of the electrical management system within the building energy management system

3 Ecological Benefits and Carbon Sequestration Potential Assessment

3.1 Carbon Sequestration Potential Assessment

In the implementation of modern construction projects, it is of paramount importance to accurately evaluate the carbon footprint of prefabricated building materials and construction processes. Utilizing Life Cycle Assessment (LCA) and other pertinent tools, an in-depth analysis of carbon emissions during various materials and construction phases can be conducted, thereby identifying possible strategies and methods to reduce carbon emissions. This not only facilitates achieving higher environmental benefits in construction projects but also provides robust support and basis for policy formulation and carbon market transactions.

Vertical greening has been demonstrated to be an effective method for carbon sequestration and emission reduction. Research indicates that through rational vegetation selection and management, vertical greening can achieve notable carbon emission reduction efficacy, including sequestering CO_2 from the atmosphere through photosynthesis and reducing building energy consumption [7]. Furthermore, it can ameliorate urban microclimates, mitigate urban heat island effects, consequently decreasing the usage of air conditioning and energy consumption, thereby further reducing carbon emissions.

In constructing a specific prefabricated building, vertical greening was applied to the exterior facade walls, covering 31.9% of the total wall area. This project has been running for 7 years, with ongoing upgrades and energy-saving measures, leading to a gradual decrease in the building's carbon sequestration capacity. Table 1 presents the changes in carbon sequestration for the prefabricated vertical greening building from 2015 to 2022.

Based on the above data comparison, it can be observed that the judicious application of prefabricated construction engineering technology and vertical greening technology can effectively reduce the overall carbon sequestration of the entire construction project, achieving the desired outcomes of green, energy-efficient, and environmentally friendly practices.

Table 1. Carbon Sequestration Changes in the Prefabricated Vertical Greening Building During 2015–2022

Number	Year	Carbon Sequestration	Number	Year	Carbon Sequestration
1	2015	13.24 t	5	2019	8.21 t
2	2016	11.28 t	6	2020	7.76 t
3	2017	10.86 t	7	2021	6.64 t
4	2018	9.91 t	8	2022	5.39 t

In 2015, the carbon sequestration represents the carbon sequestration level before the construction of the prefabricated vertical greening building.

In evaluating carbon sequestration in prefab buildings and vertical greening, long-term effects mustn't be neglected. Prolonged monitoring and analysis of these systems yield precise insights into their carbon impacts and environmental effects. This knowledge informs future research and practice, promoting better utilization of carbon sequestration for sustainable urban development. Ecological Benefit Analysis.

3.2 Ecological Benefit Analysis

Rapid urbanization exacerbates the urban heat island effect, posing threats to urban environments and human well-being. Prefabricated construction and vertical greening offer an effective solution for greener, more sustainable urban development. Green roofs and wall greening mitigate solar radiation's heating impact, lower air temperatures, reduce energy use, and expand green spaces, fostering ecological balance and a healthier urban environment.

Prefabricated buildings have distinct design and structural features that influence the local microclimate. Vertical greening enhances a building's microclimatic effects. Plant transpiration lowers temperatures and raises humidity, improving the urban microclimate. Vertical greening also acts as natural insulation, reducing cooling and heating needs, cutting energy use and carbon emissions. Research confirms its effectiveness in filtering air pollutants like PM2.5, enhancing air quality. Combining prefabricated buildings and vertical greening benefits creates a healthier, greener urban environment for residents.

In a prefabricated construction project covering 3000 m^2 of land and comprising a 57.9-m-tall building with 12 above-ground and 2 underground floors, vertical greening using honeysuckle and ivy covers 41.7% of the exterior wall area. After 2 years since completion, an ecological analysis indicates that the project has optimized indoor temperature and humidity compared to similar local buildings. It has also notably reduced carbon emissions, energy consumption, noise pollution, and indoor PM2.5 levels when compared to buildings of the same type and specifications.

Based on the above data comparison, it can be concluded that the judicious application of prefabricated construction engineering technology and vertical greening technology can effectively enhance the overall ecological benefits of the construction project, simultaneously achieving green, energy-efficient, and environmentally friendly effects, while further improving the living experience for people (Table 2).

Table 2. Provides a comparison of the ecological benefits between this prefabricated vertical greening building and local buildings of the same type and specifications:

Number	Project	This building	similar local buildings
1	Average Natural Indoor Temperature in Summer	25.2 °C	26.7 °C
2	Average Natural Indoor Humidity in Summer	63%	60%
3	Average Natural Indoor Temperature in Winter	16.7 °C	14.5 °C
4	Average Natural Indoor Humidity in Winter	53%	47%
5	annual average carbon emissions	3.41 t	5.22 t
6	annual average electricity consumption	12000 W	15400 W
7	annual average indoor environmental noise	34.4 dB	43.9 dB
8	annual average indoor PM2.5 concentration	4.7 $\mu g/m^3$	7.9 $\mu g/m^3$

4   Conclusion

The growth of prefabricated construction and vertical greening necessitates improved design and construction standards. Standardized and modular designs are proven to enhance project efficiency and quality. Thus, extensive research to enhance these standards is crucial for sustainable development. A multi-tiered standard system ensures safety, environmental protection, and economic benefits, promoting the adoption of prefabricated construction and vertical greening [8].

To boost prefabricated construction and vertical greening, governments must collaborate with industry and research institutions. Together, they should create regulations and policies that foster industry growth, legal protection, and technological innovation. These measures will promote the adoption of prefabricated construction and vertical greening, advancing green and sustainable urban development.

Acknowledgement. Supported by: Chongqing Municipal Education Commission, Science and Technology Research Programme, 2023, Research on Zero-Emission Campus Construction Based on Plant Community Optimisation (Project Title: KJQN202305605, Chair: Jing Sun). No.: KJQN202305605, host: Jingjing Sun).

References

1. Pérez, G., Rincón, L., Vila, A., González, J.M., Cabeza, L.F.: Green vertical systems for buildings as passive systems for energy savings. Appl. Energy **114**, 881–889 (2014)
2. Minunno, R., O'Grady, T., Morrison, G.M., Gruner, R.L., Colling, M.: Strategies for applying the circular economy to prefabricated buildings. Buildings **8**, 125 (2018)
3. Wang, P., Wong, Y.H., Tan, C.Y., Li, S., Chong, W.T.: Vertical greening systems: technological benefits, progresses prospects. Sustainability **14**, 12997 (2022)
4. Prodanovic, V., Zhang, K., Hatt, B., McCarthy, D., Deletic, A.: Optimisation of lightweight green wall media for greywater treatment and reuse. Build. Environ. **131**, 99–107 (2018)
5. Li, L., Shen, F.: Green building electrical energy saving design and feasibility study and solution of energy management system. Mod. Archit. Electr. **1**(133), 8–12 (2021)
6. Maier, D.: Perspective of using green walls to achieve better energy efficiency levels. A bibliometric review of the literature. Energy Build. 112070 (2022)
7. Ling, T.Y.: Rethinking greening the building façade under extreme climate: attributes consideration for typo-morphological green envelope retrofit. Clean. Circ. Bioecon. **3**, 100024 (2022)
8. Adams, K.T., Osmani, M., Thorpe, T., Thornback, J.: Circular economy in construction: current awareness, challenges and enablers. Proc. Inst. Civ. Eng. Waste Resour. Manag. **170**, 15–24 (2017)

Application of BIM + AR Smart Assembly Decoration Technology Research on Rural Residential Environment Construction

Yao Lu$^{(\boxtimes)}$, Sixuan He, and Yang Li

Chongqing College of Architecture and Technology, Shapingba District, Chongqing, China
517370173@qq.com

Abstract. With the popularization and development of prefabricated buildings in rural areas, the application of prefabricated decoration technology in rural residential environment is becoming more and more important, which has also attracted great attention from academia and industry. However, due to many factors, such as low integration, high cost and low market share, the application of this technology in rural residential environment has obvious obstacles. In this paper, based on the modern digital era, firstly, the relevant theories are extensively studied and sorted out, and then the application of BIM + AR intelligent assembly and decoration technology in rural process is studied by means of practical investigation, empirical research and functional analysis. The research shows that the optimization of application mode can better promote the marketization process of assembled decoration technology and help to optimize the construction of rural human settlements. However, to complete the comprehensive optimization of rural human settlements in rural revitalization, further technology market popularization and market adaptation time are required.

Keywords: BIM + AR · Assembled decoration technology · Rural living environment

1 Introduction

Popularization and development of modern fabricated buildings in rural areas. With the promotion of rural revitalization, assembled decoration technology has gradually begun to be applied in rural human settlements. Factory production and prefabricated buildings have become the new goals of the industry development, which is the fourth stage of the development of the architectural decoration industry, that is, the industrialization stage [1]. According to the data released by the Ministry of Housing and Urban-Rural Development, in 2019, the construction area of prefabricated decoration in China was 45.29 million square meters. In 2021, China's assembly decoration area was 83.08 million square meters. Assembled decoration technology has been relatively mature, but users still choose less. There are various obstacles in the adoption of prefabricated buildings, such as low integration, high capital cost, low market share and so on, and the adoption of prefabricated building technology lags behind expectations [2]. This is

© The Author(s) 2024
P. Xiang and L. Zuo (Eds.): PBSFTT 2023, LNCE 382, pp. 522–528, 2024.
https://doi.org/10.1007/978-981-97-5108-2_55

also the main problem of assembly and decoration technology in rural areas. Therefore, we should learn from the mature mode of BIM + AR in the application of prefabricated buildings, and optimize the problems of low integration, high cost and low market share in the application of prefabricated decoration technology. Research on the application of BIM + AR smart assembled decoration technology for rural living environment; Because of the lack of the latest literature. The research results can be used as relevant research materials, providing new ideas and methods for rural revitalization and digital rural construction.

2 Technical Characteristics

BIM technology is a combination of building information models that digitally complete all kinds of design and management work from design to construction, and is widely used in the field of assembly building construction. Augmented Reality (AR) technology visualizes the process of construction inspection, upgrading and maintenance, which is more convenient for construction management and collaborative design. In the future, the home improvement industry will move towards category differentiation, and standardization and customization are two major trends. Although the government of China has formulated various policies to promote the adoption of prefabricated building technology, theoretical research has fully proved that the adoption of innovative technologies will help to achieve sustainable development [3]. The integration of BIM + AR has been gradually transformed in the field of tooling, and more CAD + Sketchup mode are adopted in home improvement construction. Due to industrialization, the home improvement industry has gradually transformed into assembly decoration, and there is an excessive phenomenon in the construction party after the transformation. CAD + Sketchup mode is used for assembly decoration, CAD drawing and calculation of materials, and sketchup visualization effect. The relationship between the two is that Sketchup's model depends on CAD drawings for modeling, which will reduce integration and the information management degree of assembled decoration. Table 1 below shows the degree of matching between technical mode and project management. Subsequently, the degree of construction connection and construction management decreased, resulting in cost being out of control, and many such phenomena are also one of the factors that slow the growth of market share. Therefore, BIM + AR mode is a better combination way to connect with prefabricated decoration and give full play to its advantages.

Table 1. The matching degree between technical mode and project management.

Project management / Technical mode		Design control								Construction supervision				Maintenance management			
		Measurement data	Modeling space	Hydroelectric facility	Material technology	Collision detection	Review and modify	cost control	Drawing output	Project scheduling	Construction inspection	Error recheck	Error rework	Data archiving	After sale maintenance	Upgrade	Reutilization
Cooperation	BIM	T	T	T	T	T	T	T	T	T	T	T	T	T	T	T	T
	AR	T	T	T	T	T					T			T	T	T	
Cooperation	CAD	T		T	M		L	L	T		L	M	M	M	M	T	L
	Sketchup		T														
Matching degree T:tall M:middle L:low																	

3   Rural Technical Standards

With the influx of young residents from rural areas into cities, the rural labor force is insufficient [4]. The population living in rural areas has also decreased dramatically. With the promotion of the rural revitalization strategy, the relationship between workers, farmers and cities will be further adjusted, the gap between urban and rural areas will be narrowed, and all people who go out will gradually return to their hometowns. Long-term living in the city, adaptation to the urban living environment, and the difference between returning to the rural living environment will make many people feel uncomfortable. The improvement of the living environment in rural areas is the main considerations for most people who are preparing to return to their hometowns for development.

With the promotion of rural revitalization, the communication of information and the improvement of infrastructure construction, many residents in modern rural areas choose prefabricated decoration. However, due to low integration, high cost and low market share, most people's understanding of assembled decoration is still a technical way, which is different from the traditional material construction technology. The assembly and decoration process of CAD + Sketchup mode can not reflect the value of digital information management and construction management of BIM + AR assembly and decoration. Through a questionnaire survey of residents who have used prefabricated decoration in rural areas, Table 2 shows users' cognition of the advantages of prefabricated decoration. Residents only know the construction achievements, but the management effect of BIM + AR is not reflected, and the price advantage is not reflected, which also proves that the phenomenon of CAD + sketch as the basis of construction management is still common.

Table 2. User' cognition of advantages of assemble decoration.

Project Kind	Construction period	cost comparison	Environmental protection degree	Construction standard	Data accuracy	integration	Information management
Rural assembled decoration	Short cycle	Low	Dry construction, less pollution.	Brand standard craft	High demand	Tall	Tall
Traditional rural decoration	Long cycle	Tall	Dry and wet operation, serious pollution	The construction level depends on the master	Low requirements	Low	Low
Degree of knowledge		☐ Know		☐ Not know		☐ Uncertainty	

4 Technical Integrity

Compared with the traditional decoration, the assembled decoration guided by BIM+AR technology can further accurately construct consumables, thereby reducing unnecessary waste of building materials and reducing engineering costs [5]. However, the funds brought back by modern rural residents when they go out to work in their hometowns are not enough. Many residents are willing to assemble low-level materials and even construct a part of space first. The objective factors make the design and construction process of BIM+AR impossible. Therefore, at the present stage of technology application, it is very important to follow the integrity of design and construction management to ensure the core value of technology Most of the construction problems and out-of-control costs are caused by unreasonable supply chain planning, poor communication among stakeholders and insufficient workflow control [6]. The Chongqing Municipal Commission of Housing and Urban-Rural Development and the Municipal Finance Bureau issued the Notice on Launching the Pilot Work of Prefabricated Construction of Rural Houses, which pointed out that the standardized, integrated and modular decoration mode should be promoted, and the application of technologies such as the integration of materials, spare parts, equipment and pipelines should be promoted to improve the level of prefabricated decoration. What the policy shows is the standard level that the assembled decoration should reach, indicating that the technology has not been changed.

5   Service Object Selection

The Chongqing Municipal Commission of Housing and Urban-Rural Development and the Municipal Finance Bureau issued the "Notice on Launching the Pilot Work of Assembled Farmhouse Construction", which pointed out that the overall decoration of assembled farmhouses and the integrated design and collaborative construction of the main structure should be implemented. According to CCPA statistics, in the first half of 2022, the newly-started prefabricated buildings in China accounted for more than 25% of the newly-started construction area, with a total area of 2.4 billion square meters. According to the 14th Five-Year Plan issued by the Ministry of Housing and Urban-Rural Development, by 2025, prefabricated buildings will account for more than 30% of new buildings. Half of China's 700 million farmers' houses are uninhabitable, and there are at least 200,000 buildings in each farmhouse, with a total market value of more than 20 trillion. Combined with the policy of integrated design and collaborative construction, BIM + AR intelligent assembly decoration can cooperate with assembly building construction and give priority to technology implementation among assembly building users.

6   Local Cooperation

According to incomplete statistics, there are more than 100 enterprises involved in the field of prefabricated interior decoration in China. According to the relevant calculation model, it is estimated that by 2025, the market scale of assembled decoration in China will reach 632.7 billion yuan, with a compound annual growth rate of 38.26%. However, the demand of assembly decoration market is still greater than the industrial scale, and the transportation cost of a large number of engineering projects has become a large proportion. When BIM + AR assembled decoration technology is applied in rural areas, due to the complex rural road conditions, large transport vehicles can not be used in some areas, which increases the engineering cost. Table 3 points out that the price difference between urban prefabricated decoration and traditional decoration at this stage is not very obvious. If the cost increases too much because of rural road conditions, it will affect the business volume. Therefore, in the process of project design, the cooperation of the local assembly materials industry should be strengthened to reduce transportation costs.

Table 3. Comparison of the cost of urban assembled decoration and traditional decoration.

Project Kind	High-grade decoration	Mid-range decoration	Low-grade decoration	Cheap rental house
Assembled decoration	1600-1700 yuan/m² or more	1100-1200 yuan/m².	800-900 yuan/m²	500-600 yuan/m²
Traditional decoration	More than 1,800 yuan/m²	1100-1500 yuan/m².	800-1000 yuan/m²	500-600 yuan/m²

7 Sustainable Development

On June 30,2021, in the Technical Standard for Assembled Interior Decoration issued by the Ministry of Housing and Urban-Rural Development, the official definition of assembled interior decoration was: following the principle of separation of pipes and structures, adopting an integrated design methods, coordinating partitions with wall systems, ceiling systems, floor systems, kitchen systems, bathroom systems, storage systems, indoor door and window systems, equipment and pipeline systems, etc., and adopting dry construction method to decorate the parts produced by the factory. The state clearly defines the assembled interior decoration, with the purpose of making it clear that the service life of the assembled interior space will be significantly shorter than that of the main building, and it is necessary to establish the assembled interior decoration to meet the needs of convenient adjustment, disassembly, maintenance and renewal. The use of BIM + AR technology in the assembly decoration construction is the premise to ensure the construction and subsequent sustainable development, and it is also the guide to comprehensively connect with the rural assembly decoration market and cover the high, middle and low-end markets in the future.

8 Conclusion

Under the traditional construction mode, there are often problems of high rework, high cost and waste of resources in interior design, but in order to achieve the final interior design effect, we have to make a compromise [7]. The integration of prefabricated buildings and prefabricated decoration, from design mode to construction management, requires decision makers to make dialectical and systematic trade-offs before making appropriate decisions [8]. Technology is more logical, and the construction and application of rural residential environment should follow the law of technology application. Don't change the progress of the project at will because of interests. With the popularization of technology market and the warming of rural market in the future, it is necessary for enterprises to gain sustainable competitive advantage by adopting innovative technologies [9]. At this time, we can improve the rural low-end assembly decoration market, reflect the application value of BIM + AR technology, and help the rural revitalization strategy and provide support for rural revitalization and digital rural construction through the rural development experience of technology.

References

1. Zhang, X.Y.: Research on Industrial Production of Architectural Decoration. Nanjing University of Science and Technology, Jiangsu (2018)
2. Wuni, I.Y., Shen, G.Q.: Barriers to the adoption of modular integrated construction: systematic review and meta-analysis, integrated conceptual framework, and strategies. J. Clean. Prod. **249**, 119347 (2020)
3. Xie Y.J.: New thinking of rural prefabricated buildings: rural renaissance forum Qinyuan Summit, Changzhi (2019)
4. Luo, L., Jin, X., Shen, G.Q., et al.: Supply chain management for prefabricated building projects in Hong Kong. J. Manag. Eng. **36**(2), 05020001 (2020)
5. Lin, Y.H.: The influence of prefabricated buildings on modern architectural design from the perspective of project management. Arch. Eng. Technol. Des. **04**, 108 (2018)
6. Mei, Z.S.: The influence of the rise of prefabricated buildings on interior design. Guide Dev. Build. Mater. **16**, 227 (2021)
7. Abdul Nabi, M., El-adaway, I.H.: Modular construction: determining decision-making factors and future research needs. J. Manag. Eng. **36**(6), 04020085 (2020)
8. Luo, T., Xue, X., Wang, Y., et al.: A systematic overview of prefabricated construction policies in China. J. Clean. Prod. **280**, 124371 (2021)
9. Wong, L., Leong, L., Hew, J., et al.: Time to seize the digital evolution: adoption of blockchain in operations and supply chain management among Malaysian SMEs. Int. J. Inf. Manag.Manag. **52**, 101997 (2020)

Periodic Mist Spray's Dynamic Effect on Outdoor Micro-environment and Thermal Perception

Pin Wang[✉], Sumei Lu, Xiaowei Wu, Jun Tian, and Ning Li

Dongguan University of Technology, Dongguan, Guangdong, China
wangpin@dgut.edu.cn

Abstract. Mist spraying is an active cooling technology to alleviate heat stress during hot summers. There is no clear conclusion as to when mist spray should be used and its cooling potential in hot-humid regions yet. A periodic mist spraying system was set up, and environmental measurements coupled with questionnaire surveys were conducted, investigating the dynamic effect of spraying on the micro-environment and thermal perception. The results showed that elevated ambient temperatures could lead to a more substantial cooling impact, with a maximum cooling value of 5.68 °C. The increase in thermal comfort due to the mist outweighed the decrease in thermal sensation. The study indicated that the mist spray system should be activated if the ambient temperature reached 32.5 °C. Spraying could help local residents maintain a physiological state close to slightly hot and neutral comfort when the ambient temperature exceeded 34°C. The findings provide valuable guidance for the application of mist spray system in practical engineering scenarios in hot-humid areas.

Keywords: Periodic Mist Spray · Cooling Potential · Thermal Perception

1 Introduction

Urban heat island and extreme heat have been occurring more frequently in cities across the world [1], increasing the discomfort of people and leading to a rise in the rate of heat-related mortality [2]. Passive and active cooling are the two main methods for adjusting the human thermal environment [3]. Passive cooling technologies, such as retro-reflective materials and green vegetation, are less effective for rapid cooling of specific areas. In contrast, the active cooling technology adopting artificial equipment allows for fast and stable cooling of the environment [4].

The mist spraying is an active cooling technology based on the principle of evaporative cooling [5], which can effectively affect micro-climate [6]. Ulpiani [7, 8] and Giuseppe [9] conducted experiments in Ancona and Roma cities in Italy and found that the spraying could reduce air temperature by up to 7.9°C in hot-dry climates. Sureshkumar [10] carried out experiments in India. Yamada [11] and Oh [12] executed experiments in Japan. Zhang [13] and Huang [14] completed their experimental studies in Qingdao

© The Author(s) 2024
P. Xiang and L. Zuo (Eds.): PBSFTT 2023, LNCE 382, pp. 529–538, 2024.
https://doi.org/10.1007/978-981-97-5108-2_56

and Shanghai cities in China. Many studies showed that spray was more effective in hot-dry regions, while some papers had also demonstrated the positive effect of spray in hot-humid areas. Mist spray also affects the thermal perception of the human body. Ulpiani [7] and Vanos [15] believed that people showed high satisfaction with the wetness of the mist area in hot-dry climates. While in hot-humid climates, Desert [16] and Wong [17] found that severe discomfort was increased due to increased moisture levels of the skin. The effect of increased humidity caused by mist spraying on thermal perception has not been consistently observed in related studies. The benefits of mist spray for enhancing physiological comfort could vary for people in different regions and climates [18]. As Ulpiani [5] and Meng [19] concluded, the cooling potential was affected by climatic factors and weather conditions as well as the adaptation of local population. On the question of determining the suitable conditions for the application of spray, Huang [20], Yamada [11] and Zheng [21] claimed different conclusion. It is crucial to conduct targeted experimental research.

There is no clear conclusion as to when spray should be used and its cooling potential in hot-humid regions yet. Consequently, this study aimed to investigate the relationship among the weather indicator, spray cooling effects, and improvements in human thermal perception. Considering the potential damage to the spray equipment's lifespan and the increase in water consumption, an intermittent spraying aligned with practical application scenarios was adopted to avoid excessive accumulation of water vapor in the air. This study established a periodic mist spray system to analyze the dynamic effect on the microclimate when the spray was on, running, and off within all the experimental periods. The research explored how the ambient thermal environment impacts the efficacy of misting. Additionally, participants entered the mist at different periods and assessed their thermal perception levels every minute. The dynamic influences of the spraying environment on the thermal sensation and thermal comfort of local residents were evaluated in different ambient thermal environments. The findings are expected to provide guidance for the application, control, and adjustment of the mist spray system in outdoor engineering scenes.

2   Methods

2.1   The Mist Spray System

The mist spray system was set up before the experiment. After the water was filtered, it entered the system through the inlet pipe. Following the start-up and debugging of the host machine, the water was pressurized and passed through the high-pressure outlet pipe. Finally, the mist was ejected from high-pressure nozzles (see Fig. 1).

The experiment period was set to 30 min. The mist system ran non-stop for the first 25 min of each period and stopped running in the last 5 min. A lightweight square shed was constructed with a side length of about 2.20 m. Four nozzles were installed near the center of the shed, 2.30 m above the ground. The height adopted based on the reference to the work of Su [22]. Experiments were conducted in and around this shed (in Dongguan City, China, during June 27–29, 2022), including environmental measurement and questionnaire survey.

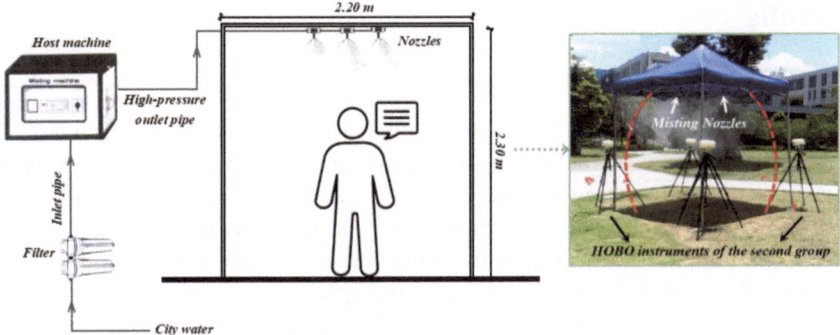

Fig. 1. Schematic illustration of the mist spray system.

2.2 Environmental Measurement

The instrument used in the environmental experiment to measure air temperature and relative humidity was a HOBO logger, which recorded data at intervals of 30 s. The probe of the instrument was positioned 1.10 m above the ground. A radiation box with three blades providing shelter was placed outside the HOBO instrument. Two groups of HOBO instruments were utilized in the experiment. The first group was positioned in the center of the spray area to measure the temperature & humidity changes within the mist. The second group was placed at the four corners on the outer edge of the spray area to measure ambient temperature & humidity changes surrounding the mist (see Fig. 1). By comparing the measurement results of these two groups, the cooling & humidification effects of the mist could be analyzed. A weather station (Watch Dog, WD for short) was also included in the study, recording solar radiation, ambient temperature, wind speed, and other background environmental data at intervals of 10 min. It was located in an open area 25 m away from the mist spray area, with no obstruction from buildings within a 45-m radius.

2.3 Questionnaire Survey

Dozens of participants were invited to the experimental site. Each participant was given one questionnaire form with three parts: 1. Personal information including gender, age, and height; 2. Thermal perception questions before participants entered the mist area; 3. Thermal perception questions after participants entered the mist area. Thermal perception aspects included thermal sensation, thermal comfort, and preferences for temperature & humidity. Thermal sensation evaluation adopted 7-point scales from Very hot (3) to Cold (−3). Thermal comfort also employed 7-point scales from Very comfortable (3) to Very uncomfortable (−3). In addition, the 5-point scale from Substantially increase (2) to Substantially decrease (−2) was used to rate the preferences for temperature and humidity.

After filling in their personal information, participants completed their first assessment outside the spray area, then entered the mist and performed the second assessment, and at one-minute intervals thereafter. The mist spraying kept running continuously for each participant during the questionnaire assessment.

3   Results

3.1   Experimental Results and Analysis

The experiment lasted from 13:00 to 16:00 on June 27, 8:30 to 16:00 on June 28, and 8:30 to 13:00 on June 29, including 30 experimental periods. During the experiment, the maximum solar radiation was 967 kW/m^2, with the highest ambient temperature of 36.9°C, while the average wind speed was 0.18 m/s. This result suggests a typical summer heat climate in Dongguan City. Since there was little wind on the days of the experiment, the mist did not travel beyond the spray area. The measurement results from the weather stations were compared with the mean results from the second group of HOBO instruments located at the four corners of the outer edge of the spray area. The maximum differences in temperature and humidity were found to be 0.53°C and 6.46%, respectively, while the average differences were 0.53% and 3.21%, respectively. Measurement results from HOBO instruments showed consistency to the weather stations, indicating that they were minimally affected by the mist. Therefore, they can be considered as representative of the ambient temperature and humidity outside the mist spray area for comparison purposes.

Figure 2 compares the temperature and humidity inside and outside the spraying area. The variations are essentially the same, with a decrease in air temperature and an increase in relative humidity. Additionally, the periodic changes in temperature and humidity align with the time periods of the mist spray system. At the start of each period, the air temperature drops, and the relative humidity rises. When the system stops running after 25 min of each period, the air temperature increases, and the relative humidity decreases, approaching the values observed outside the spray area at the end of the period. Throughout the experiment, the spray reduces the air temperature by approximately 0.12°C to 5.68°C, with an average decrease of 2.42°C. Moreover, the relative humidity in the mist area increases by around 0.04% to 22.21%, with an average humidification of 9.02%.

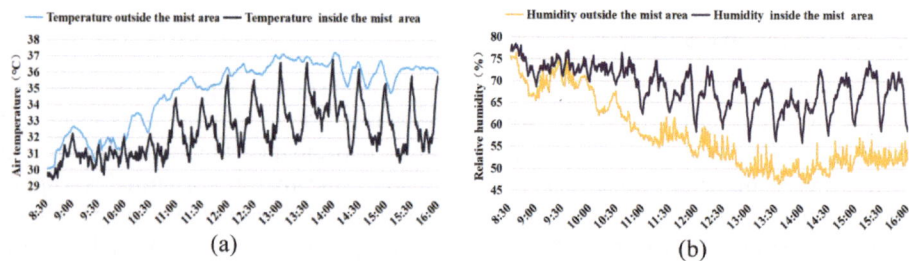

Fig. 2. Comparison of the temperature & humidity inside and outside the mist area during the experiment: (a) Air temperature on June 28th; (b) Relative humidity on June 28th.

Figure 3 analyzes the relationship between ambient temperature and the magnitude of cooling & humidification. The significance (p-value) is reported using a T-test, as are the graphs below in this article. The data indicate a strong correlation, which aligns with the findings of Desert [16]. The analysis demonstrates that as the ambient temperature increases, the cooling and humidification effect of the mist spray becomes more pronounced. For every 1 °C increase in ambient temperature beyond 31 °C, the temperature reduction caused by spraying increases by approximately 0.5 °C. At an ambient temperature of around 31.28 °C, the mist spray can achieve a temperature reduction of about 1.0 °C. While at an ambient temperature of about 35.48°C, the mist spray can lower the temperature by approximately 3.0 °C.

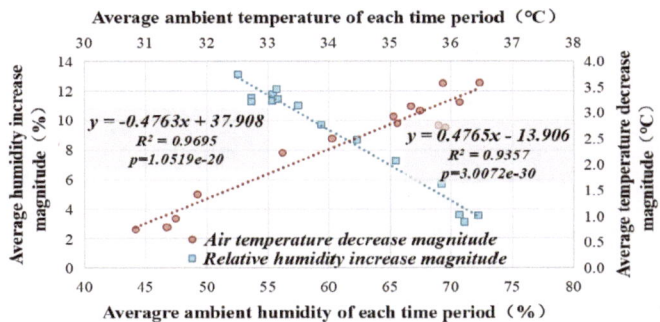

Fig. 3. Relationship between ambient temperature and magnitude of cooling & humidification.

3.2   Results and Analysis of the Questionnaire Survey

A total of 57 participants (29 males and 28 females) were invited to participate in the experiment, with 86% aged between 19 and 29 (8 persons aged under 20, 41 persons aged 20–29, 5 persons aged 30–39, and 3 persons aged over 40), entering the mist area randomly under different ambient environmental conditions.

Figure 4 shows thermal perception rating results at different times throughout the questionnaire survey. Before participants entered the mist, 89% of the participants felt their thermal sensation above 2 (Hot), and 77% reported 3 (Very hot), with an average rating result of 2.65. In terms of thermal comfort, 79% of the subjects selected -2 (Uncomfortable), with 53% selecting -3 (Very uncomfortable). The average rating result was -2.23. A total of 98% of the participants believed that the air temperature should be reduced. 86% of the participants were adaptive to the ambient humidity, while 14% were looking forward to a slight increase in humidity.

Among the participants who had just entered the misting area, only 2% reported their thermal sensation 3 (Very hot), while 39% selected 1 (Slightly hot). The average rating was 1.19. During the subsequent 5 min, the percentage of participants rating 0 and 1 increased to 35% and 54%, respectively. The average thermal sensation rating gradually decreased to 0.65. No participants reported 3 after 1 min in the mist. Regarding thermal comfort, after entering the misting area, 30% of the participants reported below -1 (Slightly uncomfortable), 35% selected 0 (Neutral), and another 35% selected higher than 1 (Slightly comfortable). The average rating was 0.05. Over the next 5 min, the percentage of participants rating 0 and -1 decreased to 5% and 7%, respectively. No participants reported -2 (Uncomfortable) after 2 min in the mist. As the participants continued to stay in the mist, the percentage of participants reporting 2 (Comfortable) gradually increased, reaching 37% towards the end. The average rating for thermal comfort eventually reached 1.42. It seemed that there was a larger range of change in thermal comfort (−2.23 to 1.42) compared to thermal sensation (2.65 to 0.65), indicating that the mist spraying provided a greater improvement in thermal comfort than in thermal sensation.

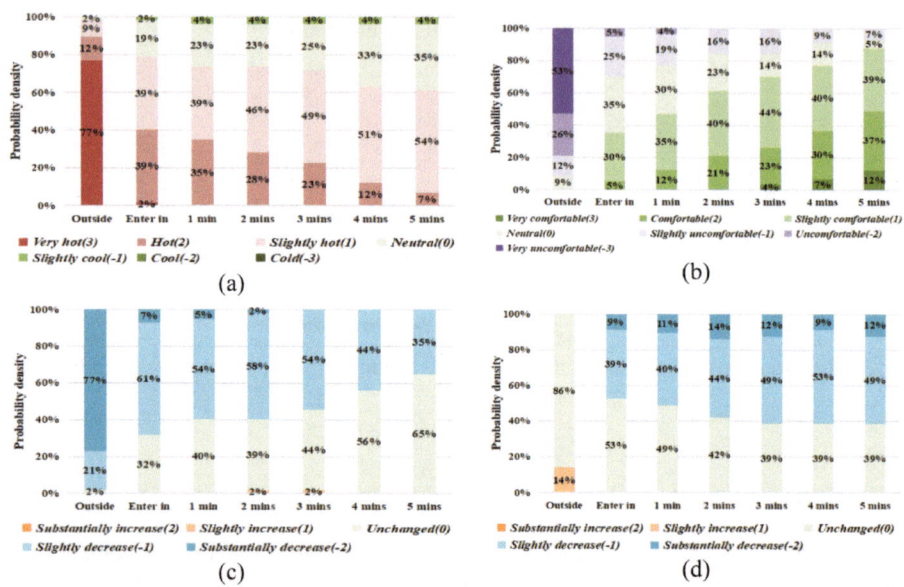

Fig. 4. Questionnaire results at different moments: (a) Thermal sensation; (b) Thermal comfort; (c) Temperature preference; (d) Humidity preference.

When entering the mist area, 68% of the participants believed that the air temperature should be reduced, while 32% voted to remain constant. However, the desire to lower the air temperature steadily decreased over the next 5 min. Despite 61% of participants feeling "Slightly hot" or "Hot", only 35% believed that the air temperature should be "Slightly decreased". On the contrary, the percentage of participants who believed that the temperature could remain constant increased to 65% after staying in the mist for 5 min, indicating that they were gradually adapting to the temperature in the mist environment. Regarding humidity preference, most participants were comfortable with the ambient humidity outside the misting area. When entering the spraying, the percentage of participants who felt the humidity should remain unchanged dropped to 53%. Over the next 5 min, the percentage of participants who desired a slight decrease in humidity increased to 49%. Participants were able to feel the humid environment. As they spent more time in the mist, their demand for reduced humidity increased. However, participants tend to have a high tolerance for high humidity. By the 5-min mark, with 61% of participants were requesting lower humidity, and only 7% feeling "Slightly uncomfortable".

3.3 Comparative Analysis of Experimental and Questionnaire Results

The relationship between thermal perception evaluation results and ambient temperatures was analyzed. Figure 5 illustrates that there is a correlation between them prior to participants entering the mist area. When the ambient temperature exceeds 31.3 °C, most participants feel "Slightly hot". When it exceeds 32.6 °C, most participants feel "Hot". Additionally, when the ambient temperature surpasses 32.4 °C and 34.4 °C, most participants feel "Slightly uncomfortable" and "Uncomfortable", respectively. Once the subjects entered the mist, the correlation weakened. When the temperature is above 31.7 °C, most participants feel "Slightly hot". People are more likely to identify "Hot" outside the mist area at the same temperature, with a sudden increase in the tolerance for hot environment after entering the mist area.

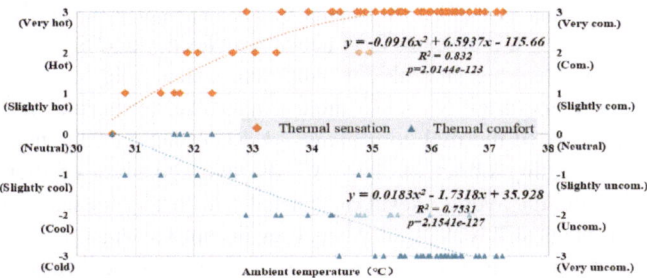

Fig. 5. The relationship between evaluation result and ambient temperature outside the mist.

The change in participants' thermal perception upon entering the spray area is related to the immediate temperature difference between inside and outside the area, as shown in Fig. 6. The increase in thermal comfort outweighs the decrease in thermal sensation. For every 1 °C increase in temperature difference, the thermal sensation decreases by 0.53, while thermal comfort increases by 0.64. The thermal sensation rating decreases for 95% of participants, leading to a more comfortable feeling for all individuals. Even though the air temperature inside the mist remains relatively high after cooling, participants still feel more comfortable.

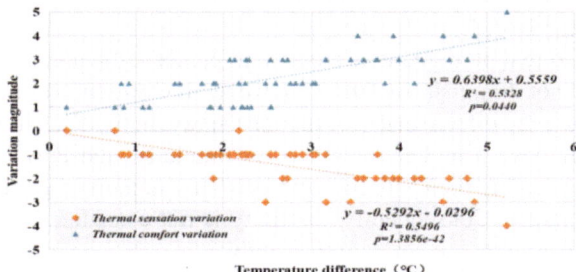

Fig. 6. The relationship between the magnitude of variation in thermal perception and the instantaneous temperature difference.

By combining the results of Figs. 3, 5 and 6, it could infer suitable ambient thermal environmental conditions for the mist spray, as well as the potential for cooling and improvement in thermal perception for the local population.

Previous research [11, 20] has suggested that temperature above 30 °C is favorable for spray application. However, this study shows that when the ambient temperature is 30 °C, the average level of thermal sensation is −0.29, and the average level of thermal comfort is 0.44. This indicates that local residents, who are accustomed to hot weather, can tolerate such outdoor thermal conditions. Although the mist can lower the temperature by approaching 0.40 °C, its necessity appears to be relatively low.

When the ambient temperature reaches 32.5 °C, the average level of thermal sensation is 1.88 (approaching Hot), with the average level of thermal comfort −1.03 (around Slightly uncomfortable). At this point, it is advisable to consider utilizing the mist spray as it can lower the temperature by 1.58 °C. After participants enter the mist area, the average level of thermal sensation decreases to 1.02 (Slightly hot), while the average level of thermal comfort increases to 0.54 (between Neutral and Slightly comfortable). 32.5 °C may be considered as an ambient temperature threshold.

At an ambient temperature of 34 °C, the average level of thermal sensation is 2.64, while the average level of thermal comfort is −1.80. When participants enter the mist, which provides a cooling effect of 2.30 °C, the average level of thermal sensation decreases to 1.39, and the average level of thermal comfort increases to 0.23. Although most of the participants still feel slightly hot, they do not feel uncomfortable. This indicates that the mist system is effective. Spraying could help local residents maintain a physiological state close to slightly hot and neutral comfort.

4 Discussions

Since the amount of spray is assumed to be constant and completely evaporates, the amount of cooling and humidification should not change with the ambient temperature changes. However, in the experiment, some of the mist might be attached to the subjects' hair, skin or surrounding objects, such as the surface of the radiation box and the ground surface, without evaporating. In other words, the amount of spray is not equal to the amount of evaporation. Due to the changing of the participants and environment, the actual amount of mist evaporation was changed, that may be the reason that the amount of cooling and humidification varied with the ambient air temperature, which is consistent with Zhang's research [13].

It was observed that participants were able to perceive the humid environment, and this sensation did not diminish over time. However, the presence of moisture in the air did not compromise the effectiveness of the mist spray. Individuals living in hot-humid climates generally have a higher tolerance for humidity, and they provided positive feedback when evaluating their thermal comfort. This supports the findings from other studies in the literature. Consequently, humidity is not considered a key indicator for triggering the activation of the spray in this study.

The experiment showed that the temperature variation and changing rate caused by the on-off operation of the system were influenced by the ambient temperature. The running and stopping time of the spray may also lead to different experimental results.

In order to fully explore the potential of spraying, the number of subjects should be increased to enhance the credibility of the research conclusions, and the participants' age, height, and weight can be more diverse in future study.

5 Conclusions

This paper studied the effects of the periodic mist spray on the outdoor thermal environment and the thermal perception of local residents. Main conclusions are enumerated as follows: The spray could reduce the temperature by an average of 2.42 °C, with a maximum cooling value of 5.68 °C. In hot weather, the participants felt a greater improvement in thermal comfort than in thermal sensation due to the mist spray, and the pleasure brought by the cooling continued to increase as they stayed longer in the mist. Higher ambient temperatures led to a more substantial cooling and humidification effect. The mist spray system should be activated if the ambient temperature reached 32.5 °C, and the mist could help local residents maintain a physiological state close to slightly hot and neutral comfort.

In conclusion, the mist spray is expected to improve the outdoor thermal environment more efficiently and provide people more comfortable experience. Further research will be conducted to facilitate the application, control, and adjustment of the mist spray system in actual outdoor scenes.

Acknowledgments. This work was supported by the Basic and Applied Basic Research Foundation of Guangdong Province [grant number 2020A1515111115]; the National Natural Science Foundation of China [grant number 52178275]; and the Humanities and Social Science Research Program of Ministry of Education [grant number 22YJAZH112].

References

1. Deilami, K., Kamruzzaman, M., Liu, Y.: Urban heat island effect: a systematic review of spatio-temporal factors, data, methods, and mitigation measures. Int. J. Appl. Earth Obs. Geoinf. **67**, 30–42 (2018)
2. An der Heiden, M., Muthers, S., Niemann, H., et al.: Heat-related mortality an analysis of the impact of heatwaves in Germany Between 1992 and 2017. Deutsches Arzteblatt International **117**, 603 (2020)
3. Li, Z., Chow, D.H.C., Yao, J., et al.: The effectiveness of adding horizontal greening and vertical greening to courtyard areas of existing buildings in the hot summer cold winter region of China: a case study for Ningbo. Energy Build. **196**, 227–239 (2019)
4. Oropeza-Perez, I., Ostergaard, P.A.: Active and passive cooling methods for dwellings: a review. Renew. Sustain. Energy Rev. **82**, 531–544 (2018)
5. Ulpiani, G.: Water mist spray for outdoor cooling: a systematic review of technologies, methods and impacts. Appl. Energy **254**, 113647 (2019)
6. Wang, J.S., Meng, Q.L., Yang, C., et al.: Spray optimization to enhance the cooling performance of transparent roofs in hot-humid areas. Energy Build, **286**, 112929 (2023)
7. Ulpiani, G., Di Giuseppe, E., Di Perna, C., et al.: Thermal comfort improvement in urban spaces with water spray systems: field measurements and survey. Build. Environ. **156**, 46–61 (2019)

8. Ulpiani, G., Di Perna, C., Zinzi, M.: Water nebulization to counteract urban overheating: development and experimental test of a smart logic to maximize energy efficiency and outdoor environmental quality. Appl. Energy **239**, 1091–1113 (2019)
9. Di Giuseppe, E., Ulpiani, G., Cancellieri, C., et al.: Numerical modelling and experimental validation of the microclimatic impacts of water mist cooling in urban areas. Energy Build. **231**, 110638 (2021)
10. Sureshkumar, R., Kale, S.R., Dhar, P.L.: Heat and mass transfer processes between a water spray and ambient air. Appl. Therm. Eng. **28**, 361–371 (2008)
11. Yamada, H., Okumiya, M., Tsujimoto, M., et al.: Study on cooling effect with water mist sprayer: measurement on global loop at the 2005 world exposition. In: Architectural Institute of Japan Conference Summaries D-1, pp. 677–678 (2006)
12. Oh, W., Ooka, R., Nakano, J., et al.: Environmental index for evaluating thermal sensations in a mist spraying environment. Build. Environ. **161**, 106219 (2019)
13. Zhang, M., Xu, C., Meng, L., et al.: Outdoor comfort level improvement in the traffic waiting areas by using a mist spray system: an experiment and questionnaire study. Sustain. Cities Soc. **71**, 102973 (2021)
14. Huang, C., Ye, D., Zhao, H., et al.: The research and application of spray cooling technology in Shanghai Expo. Appl. Therm. Eng. **31**, 3726–3735 (2011)
15. Vanos, J.K., Wright, M.K., Kaiser, A., et al.: Evaporative misters for urban cooling and comfort: effectiveness and motivations for use. Int. J. Biometeorol. **66**, 357–369 (2022)
16. Desert, A., Naboni, E., Garcia, D.: The spatial comfort and thermal delight of outdoor misting installations in hot and humid extreme environments. Energy Build. **224** (2020)
17. Wong, N.H., Chong, A.Z.M.: Performance evaluation of misting fans in hot and humid climate. Build. Environ. **45**, 2666–2678 (2010)
18. Dhariwal, J., Manandhar, P., Bande, L., et al.: Evaluating the effectiveness of outdoor evaporative cooling in a hot, arid climate. Build. Environ. **150**, 281–288 (2019)
19. Meng, X., Meng, L., Gao, Y., et al.: A comprehensive review on the spray cooling system employed to improve the summer thermal environment: application efficiency, impact factors, and performance improvement. Build. Environ. **217** (2022)
20. Huang, C., Cai, J., Lin, Z., et al.: Solving model of temperature and humidity profiles in spray cooling zone. Build. Environ. **123**, 189–199 (2017)
21. Zheng, K., Ichinose, M., Wong, N.H.: Parametric study on the cooling effects from dry mists in a controlled environment. Build. Environ. **141**, 61–70 (2018)
22. Su, M.F., Hong, B., Su, X.J., et al.: How the nozzle density and height of mist spraying affect pedestrian outdoor thermal comfort: a field study. Build. Environ. **215** (2022)

Design of the Prefabricated Building PC Component Visualisation Platform

Guomin Liao$^{(\boxtimes)}$ and Jing Hu

Chongqing College of Architecture and Technology,
No.3 Mingde Road 401331, Chongqing, China
1643493318@qq.com

Abstract. This paper adopts the Django framework and B/S architecture, uses the Web Browser as the client, and applies Pycharm compiler, Python programming language, and MySQL database to design a prefabricated building PC component visualization platform. According to the requirements of prefabricated building PC components, the overall structure of the platform is designed, which is divided into the data layer, the business layer, and the interface layer, mainly achieving 3D visualization of PC component, the data management, the alarm management, the information management, and the equipment maintenance functions. The prefabricated building PC component visualization platform combines data visualization and other technologies to govern visually the storage and management process of PC components, the lifting and installation management process of PC components, and the operation and maintenance management process of PC components.

Keywords: Prefabricated building · PC · Visualization platform · Visualization management

1 Introduction

The development of prefabricated buildings in China began in the 1950s, and by 1983, a total of 924 volumes of general standard atlas of buildings had been compiled, and a large number of prefabricated buildings had been built [1]. According to the requirements of the national medium - and long-term development plan for the construction industry, the proportion of prefabricated buildings in the new construction area must reach 30% by 2025 [2]. Prefabricated building is a system engineering that assembles prefabricated components on the construction site through system integration, achieving the prefabrication of the main structural components of the building. Non load-bearing retaining walls and internal partitions are not built and fully decorated [3, 4]. Compared to traditional building methods, as the factory has already produced prefabricated concrete structures in advance according to building requirements, it only needs to transport PC components to the site for assembly, and then use concrete for pouring. At present, China's construction industry has extensively produced and used prefabricated concrete structural components on a large scale. By applying PC components, the quality and

© The Author(s) 2024
P. Xiang and L. Zuo (Eds.): PBSFTT 2023, LNCE 382, pp. 539–547, 2024.
https://doi.org/10.1007/978-981-97-5108-2_57

efficiency of construction in China can be greatly improved, which is in line with the development concept of green and environmentally friendly architecture in China [5]. Prefabricated building construction has been fully promoted by China at the moment. The modernization of the construction industry is based on the concept of green and sustainable development. Prefabricated buildings have significant characteristics of energy conservation and environmental protection, which can vastly reduce construction waste [6].

With the continuous development of prefabricated concrete buildings, the prefabrication and assembly rates of prefabricated concrete buildings are increasing, the construction scale is becoming larger, and there are more and more types of prefabricated components. The entire process involves the exchange and update of multiple types of information on PC components from production to storage, transportation, and arrival at the construction site. Moreover, various types of information are complex and chaotic, making it difficult to collect, analyze, and utilize this information comprehensively and timely, If there is information bias in PC components at a certain stage, it will have an impact on the construction phase and is not conducive to the development of automated assembly [7, 8].

This paper utilizes the open-source web application framework Django and MySQL database to design a visualization platform for prefabricated building PC components, achieving three-dimensional visualization and data management of prefabricated building PC components in the storage management process, the prefabricated component lifting and installation management process, and the prefabricated component operation and maintenance management process.

2 Platform Development Environment

Computer Hardware: 11th Gen Intel (R) Core (TM) i5-1135G7 @ 2.40 GHz 2.42 GHz, with 8.00 GB of memory.

Operating System: Win10 Home Chinese version.
JDK Environment: Java SE Development Kit (JDK) Version 11.
Python Version: Python 3.7.9.
Web Server: Tomcat 9.0.
Developing IDE: pycharm professional 2020.1.1.
Database: MySQL 5.7.
Browser: Google Chrome browser recommended

3 Design of the Overall Structure of the Visualization Platform

3.1 Platform Objectives

This paper mainly designs a multi-perspective visualization platform for the prefabricated building PC components, which includes data query, update, maintenance, and fault warning. The specific objectives are: ① to intuitively understand the shape, size, and position of PC components; ② facilitate the addition, deletion, modification, and maintenance management of data; ③ convenient for early warning and monitoring, fault diagnosis, and maintenance of PC components.

3.2   Overall Structure Design of the Platform

The platform of this paper is built on the Django framework and B/S architecture (Browser/Server mode), utilizing the Python language, the PyCharm compiler, and the MySQL database to design.

Django Framework

Django is an open-source web application framework based on the MTV pattern, written in Python language [9]. It adopts the framework mode of MTV, namely Model, View, and Template. Django with comprehensive functions provides various components for web application development. These components are open-source as well as inheritable, and can be modified for use, effectively improving the reusability of the code; Django's Model layer comes with a built-in database ORM component, providing an easy database interface for developers, and allowing them to directly define models as well as objects without relying on other database access technologies during the development process. In addition, it can realize basic addition, deletion, and modification operations while creating the data tables rapidly [10, 11].

Due to the open source and strong scalability characteristics of the Django framework, this paper utilizes Django as a web application framework for designing a prefabricated building PC component visualization platform.

Python Programming Language

Python is a high-level, interpretive programming language that supports object-oriented programming paradigms and can use object-oriented features such as categories, objects, and inheritance for programming. It has a rich and powerful standard library, providing a large number of modules and functions, which can facilitate the development of various tasks, such as the file operations, the network programming, the graphical interfaces, etc.

Due to the characteristics of Python language such as easy to learn, read as well as maintain, extensive standard libraries, portability, and scalability, and its ability to be combined with the Django framework, this paper utilizes Python language for program development.

PyCharm Integrated Development Environment

PyCharm is an Integrated Development Environment (IDE) specifically designed for the development of the Python language. It is developed by JetBrains and provides rich functions and tools to facilitate developers in writing, debugging, and managing Python projects.

During the platform development process, Python is used as an object-oriented high-level programming language, and PyCharm is a Python IDE created by JetBrains. Therefore, this paper chooses PyCharm as the integrated development environment for the platform.

MySQL Database

MySQL is an open source Relational Database Management System (RDBMS) developed by MySQL AB in Sweden, acquired by Sun Microsystems in 2008, and later acquired by Oracle. MySQL is widely used in various web and enterprise applications due to its high performance, reliability, and scalability.

During the operation of the platform, the addition, deletion, modification, and querying of PC component data in prefabricated buildings is a very important part. Based on the open source and low cost characteristics of MySQL, MySQL is used as the database designed for the platform, and Navicat graphical management tool is used to manage MySQL.

Overall Platform Structure

The overall structure of the platform is mainly divided into three layers: the data layer, the business layer, and the interface layer. The data layer chiefly includes the system's access and processing of MySQL databases, as well as some operations such as the data cleaning, the data upload, and the data visualization. It can display real-time information related to prefabricated building PC components on the visualization platform. The business layer is divided into functional modules and Django structure. The functional modules mainly include the 3D visualization of the PC components, the data management, the alarm management, the information management, and the equipment maintenance. The interactivity is reflected in the charts, allowing users to have a more intuitive understanding of various information and operating status of PC components. The Django structure is designed from front-end to back-end based on its MTV model (i.e. Model + Template + View design pattern) during the platform design process. The interface layer largely refers to the pages and content presented after users log in and enter the platform, including those modules such as browser, HTML, and CSS. The overall structure of the platform is shown in Fig. 1.

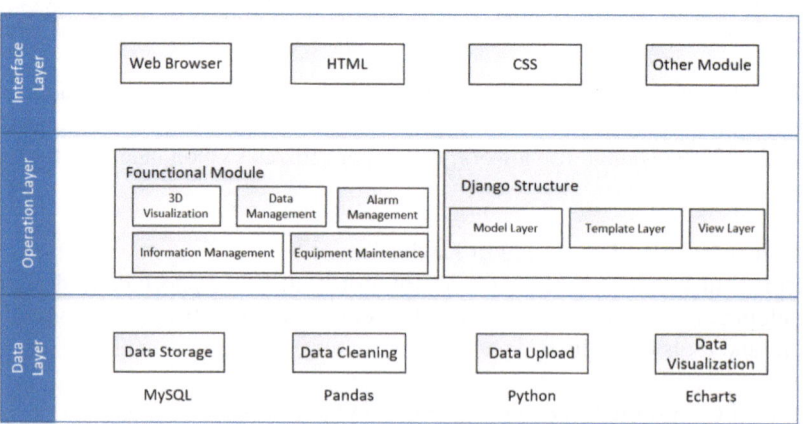

Fig. 1. Overall Structure of the Platform

4   Visualisation Platform Function Design

4.1   3D Visualization

Introducing 3D visualization technology into the prefabricated buildings PC component visualization platform, its main functions include the visual display, the space analysis, the process management, and the fault diagnosis.

The visual display is the process of showing the model of the prefabricated building PC components in three-dimensional form, allowing users to intuitively understand the shape, size, and position of the components. Through visual analysis in three-dimensional space, problems can be identified and resolved in a timely manner to avoid errors in the actual construction process. The process management is to associate construction processes and progress with three-dimensional models of components, displaying construction progress and process completion in real-time. It helps engineering management personnel comprehensively grasp project progress, adjust construction plans timely, and ensure construction progress and quality. The fault diagnosis is to help users diagnose and analyze the faults of components. By observing the 3D model of the component, damage, wear, or other problems of the components can be detected, and the appropriate measures can be taken to repair or replace it in time. The functional module of 3D visualization is shown in Fig. 2.

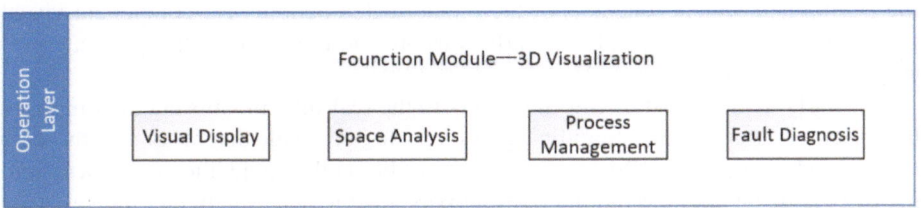

Fig. 2. Structure of 3D visualization function

4.2   Data Management

Data management is an extremely important function in the prefabricated building PC component visualization platform, mainly including the data storage and sharing, the data integration, the data analysis and mining, the data visualization, and the data security as well as permission management.

The data storage and sharing is aimed at avoiding duplicate data entry and transmission, improving data reliability and consistency. The data integration is the integration of data from different sources, to achieve integrated management and comprehensive analysis of data, and provide comprehensive data support for decision-making. The data analysis and mining is the process of discovering patterns and trends in a large amount of data through statistics, analysis, and modeling, extracting valuable information, helping to optimize the design and production process of components, and improving construction efficiency and quality. The data visualization is to display the data by the visual method in the form of charts, graphs, or animations,which is possible to better discover the relationships and rules of the data, thus making more accurate decisions. The data security and permission management is a function that provides data security and permission management, ensuring the confidentiality and integrity of the data. By encrypting, backing up, and controlling permission on data, it can prevent data leakage and tampering, and protect data security of users. The data management function structure is shown in Fig. 3.

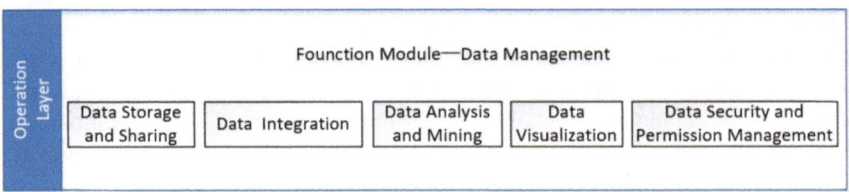

Fig. 3. Data Management Function Structure

4.3 Alarm Management

The alarm management functions of the prefabricated building PC component visualization platform mainly include the early warning and monitoring, the fault diagnosis, and the safety management.

The early warning and monitoring refers to the real-time monitoring of component data, such as temperature, humidity, pressure, etc., for fault diagnosis and the early warning and monitoring. When the data exceeds the set threshold, the system will automatically issue an alarm to remind relevant personnel to take timely measures to avoid potential problems and safety risks. The fault diagnosis is the process of identifying potential faults and problems by analyzing component data. The safety management is achieved by monitoring the data of components, such as tilting and deformation. When abnormal situations occur, the system will automatically issue an alarm to remind relevant personnel to take safety measures to ensure the safety of construction personnel. The alarm management function structure is shown in Fig. 4.

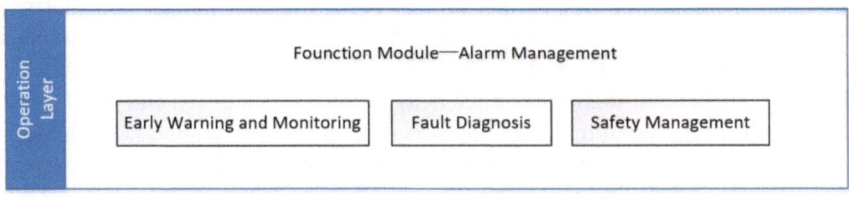

Fig. 4. Alarm Management Function Structure

4.4 Information Management

The information management functions of the prefabricated building PC component visualization platform mainly include the component information management, the construction schedule management, the quality control, and the maintenance and upkeep management.

By managing component information, it is convenient to query and trace the source and usage of components, improving the efficiency of construction and maintenance. Through the management of construction progress, problems during construction can be identified and solved in time, ensuring the smooth progress of construction. Through the management of quality data, quality problems can be promptly identified and resolved, thus optimizing the quality and reliability of components. By managing maintenance and upkeep information, maintenance and upkeep work can be carried out timely, thus extending the service life of components, and reducing maintenance costs. The information management functional structure is shown in Fig. 5.

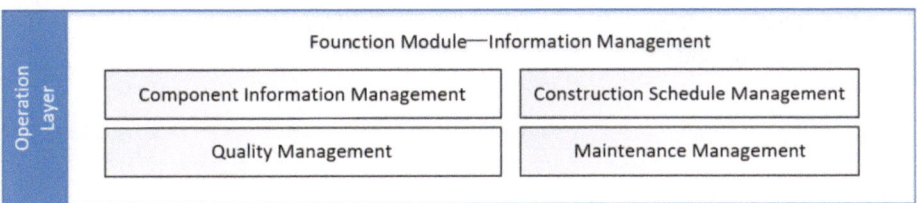

Fig. 5. Information Management Functional Structure

4.5 Equipment Maintenance

The equipment maintenance functions of the prefabricated building PC component visualization platform include the preventive maintenance, the maintenance schedule, the maintenance record, and the equipment optimization.

The preventive maintenance is the process of predicting the lifespan and maintenance cycle of components by analyzing their data. The maintenance schedule and the maintenance records are formulated based on the usage and maintenance needs of the equipment. The equipment optimization is the process of analyzing the usage and maintenance records of equipment, understanding its operational status and lifespan, optimizing its use and configuration, extending its lifespan, and reducing maintenance costs. The functional structure of equipment maintenance is shown in Fig. 6.

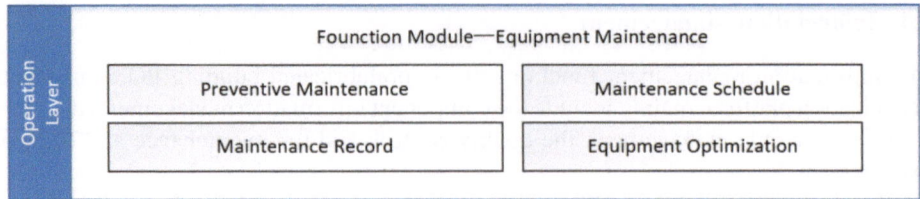

Fig. 6. Equipment Maintenance Functional Structure

5   PC Component Visualization Platform

After completing the design of the prefabricated building PC component visualization platform, the platform is developed by using HTML, CSS, JavaScript, etc. The home-page page of the platform's web client is mainly used to welcome and guide users to view the functional classification of prefabricated building PC components after logging in. As shown in Fig. 7, after entering the homepage, users can click to enter the functional classification sub-page, user center sub-page, and query sub-page of prefabricated building PC components.

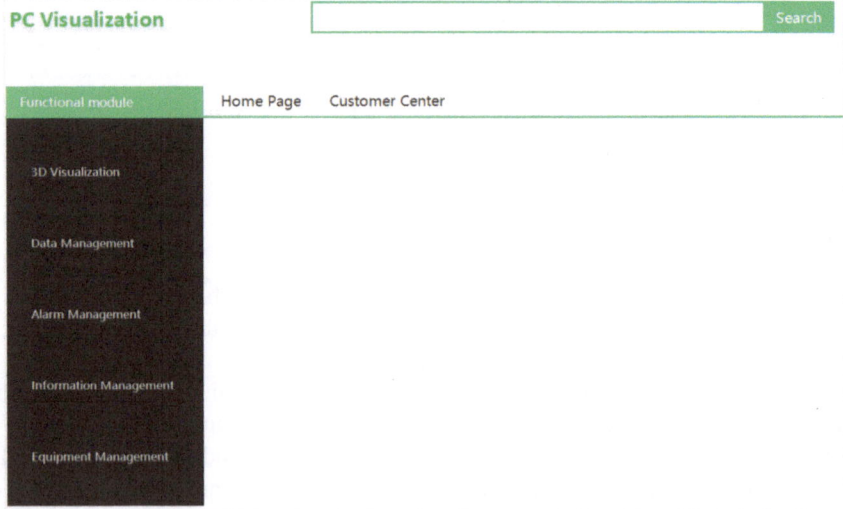

Fig. 7. The Prefabricated Buildings PC Components Visualization Platform Interface

6   Summary

According to the requirements of the prefabricated building PC component visualization platform, the platform is designed by using a Web Browser as the client-side and adopting Pychar compiler, Python programming language, and MySQL database based on the

Django framework and B/S architecture. Combining the relevant information of PC components with the current popular programming language Python and web application framework Django, the data information can better displayed intuitively in front of the public through data visualization and other technologies, thus reflecting the data information and operation status of PC components in real time, and ensuring the normal operation of PC components.

References

1. Liu, R.N., Zhang, J., Wang, Y., et al.: The development background and status quo of prefabricated buildings in China. Hous. Real Estate **32**, 32–47 (2019)
2. Chen, X.Z., Zhang, D.H.: Exploration on Training new talents for construction industrialization of Civil engineering major. J. Nanyang Inst. Technol. **9**(5), 50–52 (2017)
3. Peng, X., Xu, Y.M., Wang, L.F.: Research on the path of prefabricated building helping to achieve the goal of carbon peak and carbon neutrality. In: Proceedings of the 2022 Industrial Architecture Academic Exchange Conference, vol. 2 (2022)
4. Fuyu, W., Zhang, S.Q., Chen, M.K.: Production and Operation Management of PC components in prefabricated buildings based on third-party insurance platform, **11**(2), 1–5 (2019)
5. Liu, B.X.: Discussion on the production and construction technology of new prefabricated building PC components. Constr. Budget (2) (2022)
6. Lu, H.: Analysis of carbon emission evaluation indicators during the production stage of prefabricated building components. Green Environ. Prot. (8), 152–154 (2022)
7. Li, L.L.: Design of an online monitoring system for prefabricated building PC production line based on digital technology. Agric. Mach. Equip. (11), 115–117 (2022)
8. Wang, D.: Research on Track Monitoring System of Prefabricated Building Components Based on Beidou High Precise Positioning. Harbin University of Science and Technology, Harbin (2021)
9. Michail, D., Kinable, J., Naveh, B., Sichi, J.V.: Jgrapht.A java library for graph data structures and algorithms. ACM Trans. Math. Softw. **46**(2), 1–29 (2020)
10. Wu, J.H.: Development and Application of Django-Based Web Automated Testing Platform. Southwest University, Chongqing (2022)
11. Wang, D.D.: Design of Python course online education Platform based on Django framework. Inf. Educ. **12**, 242–244 (2023)

Exploring the Application of Prefabricated Construction Technology Under the Concept of Green Building

Jingyi Zhou and Junhua Zhou[✉]

Chongqing Vocational College of Architecture Science and Technology, Chongqing, China
22088786@qq.com

Abstract. In recent years, the development of the construction industry has been accelerating. As a typical product in the process of industrialization, prefabricated building has become a new mode of construction. However, due to many factors, there are a series of problems in the construction of prefabricated building, such as environmental pollution and resource waste, which have a great impact on its sustainable development. Therefore, in developing prefabricated building buildings, we should actively introduce the concept of green building, observe the working principle of energy conservation and environmental protection, apply advanced technology and constantly explore new technology applications. This paper explores the application of the green building concept from four dimensions: design principles, technical advantages, main technical analysis and practical application, and the prefabricated building of the green building concept.

Keywords: Concept of green building · Prefabricated construction technology · Application

1 Introduction

The construction industry is an important pillar of a country, and with the rapid development of science and technology, the concept of green buildings has been effectively implemented. This new type of prefabricated building technology has been widely applied. Compared with traditional construction methods, this technology has significant advantages in engineering quality, environmental friendliness, standardized production processes, and controllable construction progress. Therefore, it has become an important research topic for the development of the construction industry in the new era [1].

2 Principles of Applying Prefabricated Construction Technology Under the Concept of Green Building

2.1 Effective Utilization of Renewable Energy

During the construction of the new concept, green prefabricated building pay more attention to the use of green and environmentally friendly new energy sources. By collecting and utilizing solar energy, water energy and wind energy, and properly using the existing

P. Xiang and L. Zuo (Eds.): PBSFTT 2023, LNCE 382, pp. 548–555, 2024.
https://doi.org/10.1007/978-981-97-5108-2_58

resources around, not only can further solve the building's energy needs, but also in a certain sense can alleviate environmental pollution, thus effectively reduce the damage to the natural environment.

2.2 Principle of Effective Environmental Protection

Environmental assessment holds a relatively important position in both domestic and international building evaluation systems. This indicator encompasses not only the waste generated by construction projects but also the environmental pollution caused by them. In modern green buildings, waste recycling pools can be used to recycle waste materials. Advanced technologies such as mobile jet engines can even be employed to recycle these waste materials, thereby reducing environmental pollution.

2.3 The Principle of Standardization

In the process of architectural design, we should carry out the concept of "Standardization", especially green building. In traditional building engineering, waterproof and thermal insulation will affect the quality of the whole building, construction technology can not guarantee the quality of the building. The use of prefabricated building technology makes it possible to standardize the production of waterproof and thermal insulation materials, including waterproofing, insulation and main structures, in the processing of components of external walls and roofs of buildings, can make waterproof layer and insulation layer more perfect, effective quality control [2].

3 Effective Advantages of Applying Prefabricated Construction Technology Under the Concept of Green Building

3.1 Improved Energy Efficiency and Reduced Pollution Emission

In prefabricated building technology, the building materials and components used are more environmentally friendly. They are all environmentally friendly materials advocated by the state. Standardized modules are used for production and construction, can further reduce industrial waste, greatly reduce the actual construction in the construction of building materials, electricity, water and other resources consumption, greatly improve the building's resource efficiency. At the same time, because of greatly improving the utilization rate of resources and reducing the waste of resources, the quantity of construction waste and the emission of pollution will be significantly reduced. In addition, can further reduce the construction process of dust and other emissions, after the demolition of part of the assembly-type accessories and building materials can be re-used.

3.2 Enhanced Safety Performance of Building Projects

Through the reasonable design of assembly structure, it can effectively solve the contradiction between high-strength concrete technology and precast prestressing technology, therefore, its anti-seismic effect is also better, the safety performance is also better. At the same time, the materials of the prefabricated components used in the prefabricated building are able to effectively control their ignition point and, in the event of a fire, to effectively isolate the fire by controlling the rapid spread of the fire, it can improve the fire protection ability of the building and greatly improve the safety level of the building.

3.3 Upgraded Development of Green Building Concepts

Most of the building materials used in the construction of prefabricated building projects are green, while a small amount of light construction materials can effectively control the overall quality of the house. In addition, it also has strong corrosion resistance, fire resistance and sound insulation effect, through the appropriate adjustment and combination of prefabricated components, can achieve an effective connection between the components, thus improving the overall performance of the building. Moreover, the low thermal conductivity of the materials used in prefabricated construction results in strong insulation performance. Hence, prefabricated construction technology aligns effectively with the principles of green building [3].

3.4 Reduction of Construction Time

Since the structures used in the prefabricated building are manufactured in the factory, the construction process can be completed simultaneously with the assembly process through a scientific and rational design of the construction plan, thus the construction time is effectively reduced. The work efficiency of engineering installation can be greatly improved by transporting the pre-prepared construction parts to the site and then assembling the pre-fabricated parts according to the installation construction drawings and using appropriate mechanical equipment, the hidden trouble of construction quality and safety risk caused by human factors are reduced, and the construction period of construction project is shortened [4].

4 Analysis of Key Technologies in Prefabricated Construction Under the Concept of Green Architecture

4.1 BIM Technology

The application of BIM technology can optimize the design of the prefabricated building. It can provide a more comprehensive and detailed picture of the construction project in the form of a three-dimensional visual model, this is very important to improve the quality of the building. Manufacturers can calculate the size and number of prefabricated components needed for a building based on a three-dimensional model, which not only increases the abundance of the material itself, but also reduces the waste of resources, meet the needs of green building development. In addition, the correct application of BIM technology can also reduce the investment in the operation of existing technology, which is very important to improve the level of construction. In the past, when a building was under construction, builders often had to consult a lot of paper documents to more accurately assess whether the deliverables met the needs of major projects, some projects also encounter problems with changes to data and drawings. If the data can not be collected scientifically and effectively, it will inevitably have a negative impact on the normal operation of the follow-up work. With the application of BIM technology, it can help the staff to collect all kinds of information of the project, and build the model on this basis, so as to facilitate the construction operation and management [5].

4.2 Sleeve Grouting Technique

Sleeve grouting technology can effectively improve the overall quality of joints, is an indispensable technology in the construction of assembly-type housing. In this technique, no welding is required, but grout is injected into a cylinder with a raised shape to hold the rebar tightly together. Sleeve grouting is a new technology of compression and tension resistance, which can improve the joint of steel bar and the whole performance of structure. When adopting this technology, the grouting method can be used to make the casing and the reinforcing bar tightly combined, thus effectively improving the overall quality of the joint. The solidified cement slurry can meet the requirement of grade 1 welding. In the construction of Assembly type house, the whole quality of joint is an important link that can not be ignored. Meanwhile, the rigid connection is formed at the grouting hole, which improves the quality of composite structure. However, the construction process is complex, and it is difficult to effectively control the actual quality of grouting casing, and the project cost is high. For this technology, the quality of grouting tube must be strictly controlled to achieve better results as shown in Figs. 1 and 2.

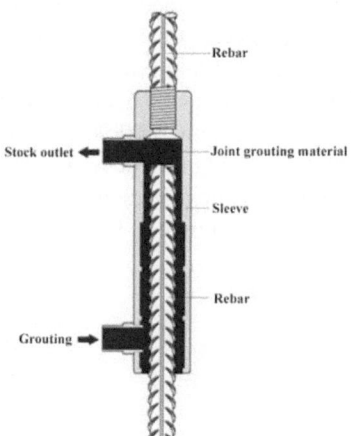

Fig. 1. Semi-sleeve grouting

4.3 Optimization Technology of Reinforced Concrete Structure

In prefabricated construction, each floor of the building is considered as a relatively independent building unit, and the processing specifications of the steel components

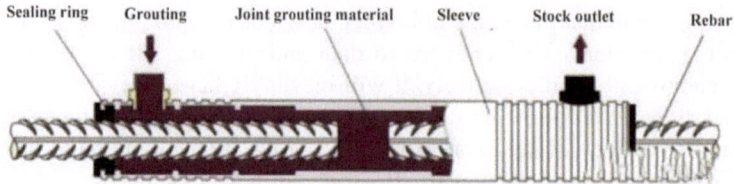

Fig. 2. Full-sleeve grouting

are determined, which can reduce the waste of materials and increase the efficiency of resource use. When the concrete is poured, the construction personnel may, according to the situation on the spot, appropriately choose the method of division pouring, make clear the process of the pouring work, and record the specific quantity of pouring, in the early stage of construction, the amount of pouring materials to reasonable control, so as to reduce the root causes of construction problems in the probability [6].

5   Application of Prefabricated Construction Technology in Real Life Under the Background of Green Building

5.1   Applications in Assembly Customization

In traditional building assembly, there is usually a fixed form of assembly, however, in the concept of green building, assembly construction technology has broken this pattern. According to the special needs of customers, wall panels, balconies, beams, columns, stairs and other related components and connectors, in advance in the construction of the manufacturing base of prefabricated components to prefabricate, at the same time, it is also necessary to select a more reasonable mixture proportion according to the specific construction site and characteristics, so as to better prevent pitting and cracks during pouring. After finally arriving at the construction site, all the prefabricated building components are spliced together by using the method of building blocks splicing. Thus in the premise of standardization, to achieve personalized custom assembly construction technology, efficient and economic as shown in Fig. 3.

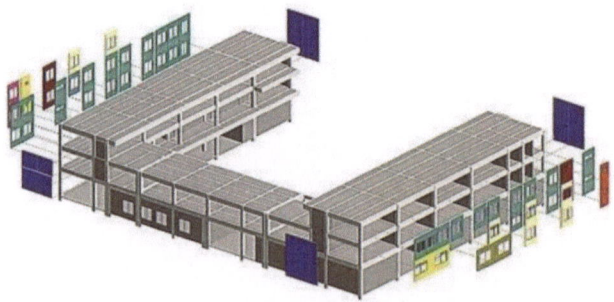

Fig. 3. Prefabricated building structure

5.2   The Application in the Optimization Design of PC Component

The application of PC component is very common in assembly structure. However, there are some problems in the application of this method, such as resource waste and environmental pollution, so it is necessary to optimize the PC component. In practical application, combined with lifting force and installation position of tower crane equipment, the weight of prefabricated steps is reduced, and the balcony of prefabricated steps is changed into a stacking terrace, thus reducing the self-weight of buildings, reduce the use of specific materials. In this way, some common quality problems can be prevented. Among them, the stairs, bay windows, balconies and other relevant parts of the design, to combine the shutter, air conditioning and other specific parts, using prefabricated methods. In practice, CAD technology can be used to build models of various installation nodes of prefabricated PC components to improve the accuracy of design through scientific ways, it promotes the cooperation of installation and civil prevention, realizes the visualization of construction process, realizes the effective control of project quality, and improves the environmental protection performance of work system. In order to effectively reduce energy consumption in transport and minimize the distance between buildings and production sites, Additionally, during the prefabrication stage, the use of permeable temporary brick materials and proper handling of residual concrete materials should be considered to avoid resource waste. Furthermore, the adoption of PC technology for prefabricated components in decoration work effectively enhances construction quality and efficiency while reducing project costs as shown in Fig. 4.

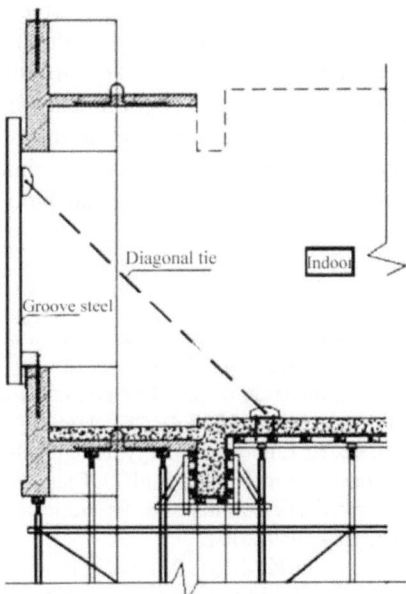

Fig. 4. Optimization of prefabricated building PC component hoisting technology

5.3 Application of Prefabricated Components in Building Doors and Windows

In traditional construction projects, designing and installing doors and windows typically involve leaving openings and adding them later once the doors and windows are ready. This approach often leads to issues such as leaks and aesthetic concerns. However, the application of green and sustainable prefabricated construction methods effectively addresses this problem. In prefabricated wall panels, the corresponding positions for doors and windows are already pre-determined during the manufacturing process. Additionally, during the design and production of features like drip lines, drainage slopes, and joints for doors and windows, they can be integrated with the pre-designed openings and cast in one go. This approach allows for more precise control over the dimensions of the prefabricated components, thereby avoiding water leakage issues in the later stages. In prefabricated construction, once the prefabricated door and window components are completed, they can be directly assembled with the main structure at the construction site. This eliminates the need for additional on-site adjustments and ensures a more efficient and streamlined installation process [7].

6 Conclusion

In the current era, China advocates for green development as a fundamental national policy, and all industries should actively respond to it. It is necessary to not only promote industrial development but also prioritize environmental protection. Particularly in popular industries like construction, it is crucial to set an example and contribute to the country's environmental efforts. In the context of green development, the use of prefabricated construction techniques is a response to the new era and new policies. Prefabricated construction technology not only greatly improves the quality and speed of construction but also achieves resource and environmental conservation, significantly reducing the environmental impact of the construction industry. Therefore, under the concept of green architecture, prefabricated construction technology is a response to the needs of the times and should be strongly advocated and supported.

References

1. Yeh, Y.: Exploration of the application of prefabricated construction technology under the green building concept. Ceramics **2022**(10), 175–177 (2022)
2. Xiang, D.: Discussion on the application of prefabricated construction technology based on the green building concept. Model World **2020**(4), 214–216 (2020)
3. Tavares, V., Lacerda, N., Freire, F.: Embodied energy and greenhouse gas emissions analysis of a prefabricated modular house: the moby case study. J. Clean. Prod. **212**, 1044–1053 (2019)
4. Kasperzyk, K., Kim, M.K., Brilakis, I.: Automated re-prefabrication system for buildings using robotics. Autom. Constr. **83**, 184–195 (2017)
5. Ilson, C., Smith, B., Dunn, P.: ECOHOMES XB: a guide to the eco homes methodology for existing buildings. Sustainable Homes Ltd, Kingston upon Thames **2017**(8), 45–66 ((2017))
6. Xi, P.: Exploration of the application of prefabricated construction technology under the green building concept. Urban Inf. **2020**(23), 241–242 (2020)
7. Hu, Y.: Application of prefabricated construction technology under the green concept. New Mater. Decorat. **3**(22), 1–3 (2021)

Experimental Research on Full-Scale Umbrella Shaped Structure

Jianjun He[1], Guanghui Zheng[1], Jun Deng[2(✉)], and Zebin Hu[2]

[1] Guangdong Provincial Government Loan Repayment Expressway Management Center, Guangzhou 510030, China
[2] School of Civil Engineering, Guangzhou University, Guangzhou 510006, China
dengjun@gzhu.edu.cn

Abstract. Umbrella-shape structure (US) were designed and used in the Jixiang service area of expressway, Heyuan Ctiy. To verify the safety of the US, the test of the full-scale US was done. The tested results indicated that the maximum tensile of cantilever steel was 32.96 MPa, which was smaller than the yield strength of steel; the maximum compressive strength of concrete was occurred in root of cantilever concrete beams with 4.98 MPa, which was also smaller than the concrete design strength; the US was in the elasticity stage under 338 kN.

Keywords: Umbrella-shape structures (US) · Full-scale test · Experimental research

1 Introduction

1.1 Background

Umbrella-shape structure (US), innovated building form, usually uses in train station, markets and airports. Traditionally, the US is made up with the column and the umbrella surface.

The first US called "mike umbrella" was built in 1929 [1]. The "mike umbrella" was fabricated by the materials of concrete. Due to the limitation of the structural theory, the "mike umbrella" only used in pavilion [2]. With the development of the structural theory and materials, the US was adopted as the long-span structures like terminal and high-railway station [3]. Although the US widely used in the long-span structures, there were few US in the building. Some of them were used in the museum or memorial hall [4–6]. For the almost highway service area in China, the traditional building was adopted. The single building style caused aesthetic fatigue, as well as no conductive. However, the highway service area was the visiting card to the local city, the single building style of the highway service area may constrain the advertisement of the city.

P. Xiang and L. Zuo (Eds.): PBSFTT 2023, LNCE 382, pp. 556–564, 2024.
https://doi.org/10.1007/978-981-97-5108-2_59

1.2   Project Overview

Jixiang highway service area, locates in Longxun Highway, Heyuan City, is adopted the US as the design scheme. Specifically, the architectural design ideas of the Jixiang highway service area are mainly based on the flower called "Bougainvillea" in Heyuan City, Guangzhou Province, which is integrated with hexagonal umbrella-shape structures and is prefabricated using a "Lego-like" preassembly method (Prefabricated building). The overall design reflects the design ideas of environmentally friendly, green, environmental protection, and low-carbon, and the buildings of the service area are intermixed with clusters of the tufted rhododendron bush. While highlighting the theme of green and environmental protection, it also contains strong local cultural color, and has great visual effect (as shown in Fig. 1). Jixiang service area is being built into the most representative, interesting, innovative and avant-garde expressway service area building quality.

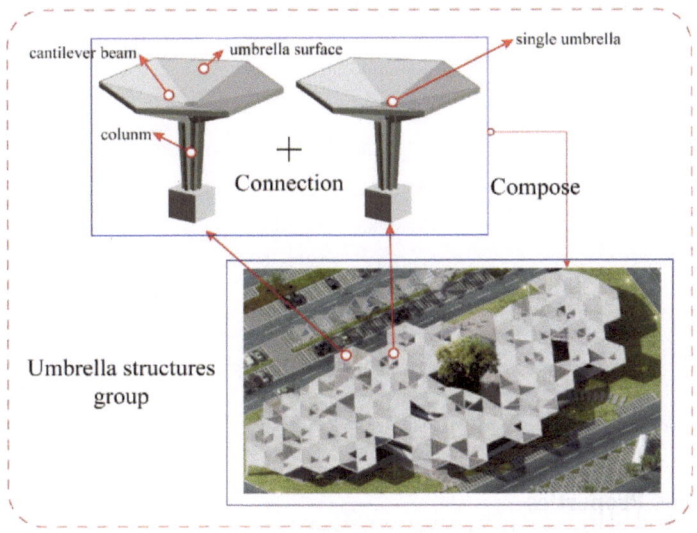

Fig. 1. US in Jixiang service area

US in Jixiang service area is constructed by the method of prefabrication. Different parts of US are designed accordingly, and the parts of US are the concrete-steel combination structures. These parts are fabricated in the manufactory, which could ensure the quality of components. After casting, the components are conserve under controlled condition. Different parts are transported to construction site and assembled.

The construction method of the US is prefabrication, and the similar US was no existing. Thus, to verify the safety of the US, the test of the full-scale US is necessary. The load is applied to the top of the full-scale testing, and the structural response are recorded and analyzed to justify the safety of the US.

2 Experiments Program

2.1 Details of Umbrella-Shape Structure

The dimensions of the US were shown in Fig. 2. The horizontal length of the cantilever beam was 6000 mm, while the heigh of the column was 4740 mm. The steel tube column was combined with different thickness of steel tube. The thickness of top steel tube and of bottom was 14 mm and 5 mm, respectively.

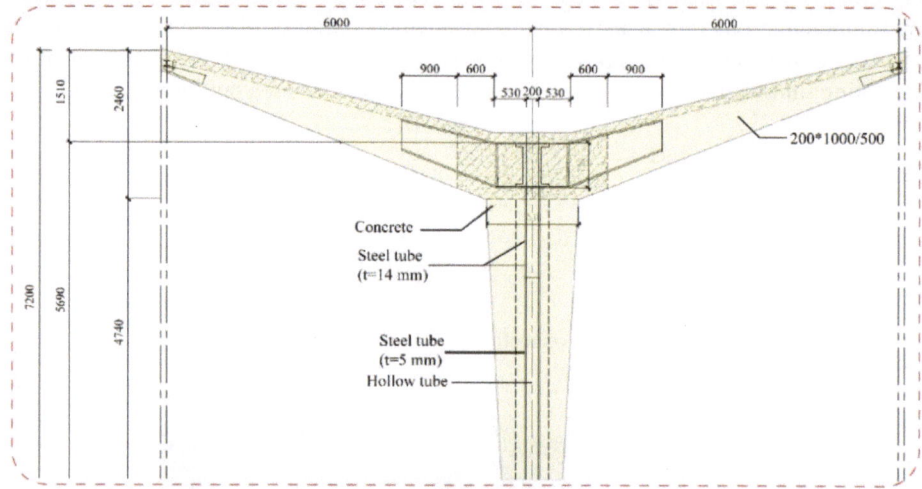

Fig. 2. Details of the US

2.2 Materials Properties

Table 1 listed the materials of US. The designed concrete degree was C35, while the rebars was HRB400. The designed strength of steel was Q355b.

Table 1. Materials of the US

Type	Materials properties
Concrete	C35
Steel	Q355b
Rebars	HRB400

2.3 Testing Setup and Instrumentation

A testing setup as illustrated in Fig. 3. Considering the difficult of applying loads, a total of 13 bundle of rebars was adopted to load, each weight of bundle of rebars was 26 kN. The loading class of the test was listed in Table 2.

Fig. 3. Typical experimental setup of US

Layout of measuring points of US was shown in Fig. 4. The vertical displacement of the end of cantilever beams was measured by the total-station. The strain gauge was arranged in the cantilever steel beam, concrete cantilever beams, and concrete column respectively.

3 Results and Discussion

Table 2 summarized the maximum compressive/tensile stress of steel cantilever beam, maximum compressive stress of concrete columns, maximum compressive/tensile stress of concrete cantilever beams, and vertical displacement of the cantilever beam.

It can be seen that under the maximum loading load (338 kN), the maximum tensile stress of the cantilever steel beam was 32.96 MPa, and the maximum compressive stress was 19.98 MPa; The maximum compressive strain of the concrete column was 4.23 MPa; The measuring points of the concrete beam were all in a compressive state, with a maximum compressive strain of 4.98 MPa. All the strength of concrete and steel of US were small than their design strength respectively.

3.1 Load-Displacement Curve of US

The load-displacement curve of US was shown in Fig. 5. It can be seen that the maximum displacement of the US was 4.9 mm when applied the load of 338 kN. According to

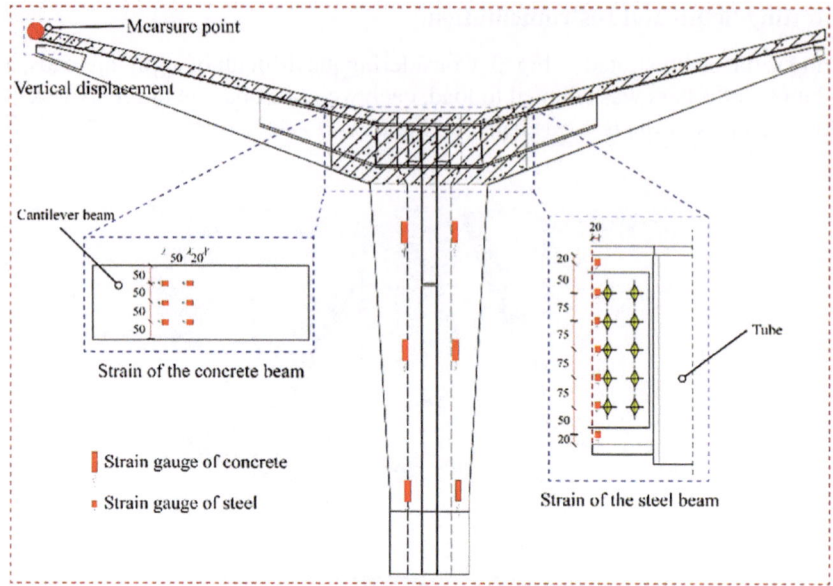

Fig. 4. Layout of measuring points

Table 2. Summary of the tested results

Loading type	Loading class (kN)	Vertical displacement (mm)	Steel (MPa)		Concrete column (MPa)	Concrete beams (MPa)	
			C	T	C	C	T
Gradation loading	52	1	−7.42	1.64	−1.23	0	–
	104	1.2	−11.33	10.3	−2.13	−1.08	–
	156	2.7	−12.36	10.6	−2.79	−2.1	–
	208	3.2	−14.4	18.54	−3.75	−3	–
	234	3.8	−18.13	20.09	−3.90	−3.72	–
	286	4.5	−18.54	32.34	−4.14	−4.47	–
	338	4.9	−19.98	32.96	−4.23	−4.98	–

Note: the stress of steel plate and concrete is obtained by converting the strain into stress; The Modulus of elasticity of concrete is 3.0×10^4 MPa, and the Modulus of elasticity of steel is 2.06×10^5 MPa; Compression is represented by C, while the tensile is represented by T

the Fig. 5, the linear relationship between the load and displacement maintained under loading. It was indicated that the US was in elasticity stage when the load reached 338 kN.

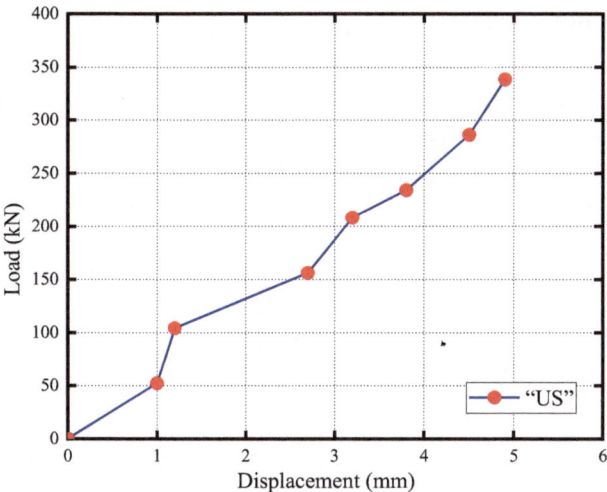

Fig. 5. Load-displacement curve of US

3.2 Strain of Cantilever Steel Beam

The load-strain curve of cantilever steel beam was illustrated in Fig. 6. It can be seen that the load-strain curve of the cantilever steel beam was in a linear stage. When the load reached 104 kN, the compressive strain in the lower part of the cantilever steel beam increased faster than the tensile strain in the upper part. However, when the load was greater than 104 kN, the growth rate of tensile strain was significantly faster than that of compressive strain. Finally, the maximum compressive strain of the cantilever steel beam was 100 με, while the maximum tensile strain was 159 με.

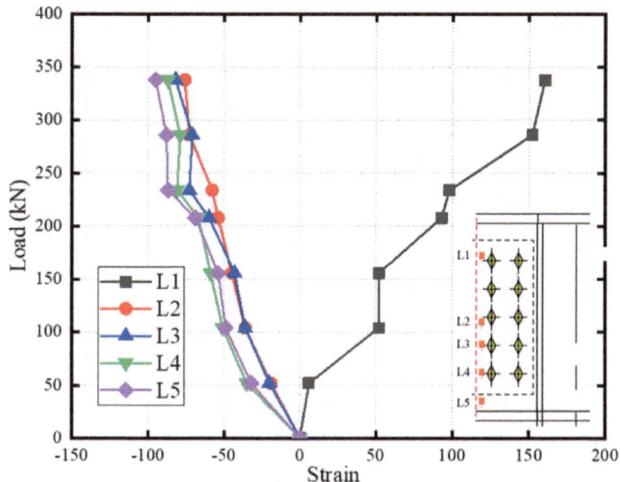

Fig. 6. Load-strain curve of cantilever steel beam

3.3   Strain of Concrete Column

The load-strain curve of concrete column was illustrated in Fig. 7. It can be seen that before the load reached 208 kN, there was a linear relationship between the load and the strain of the concrete column. However, after 208 kN, the strain of the concrete column increased slowly, and the strain at the top of the concrete column at measurement point 4 even shown rebound.

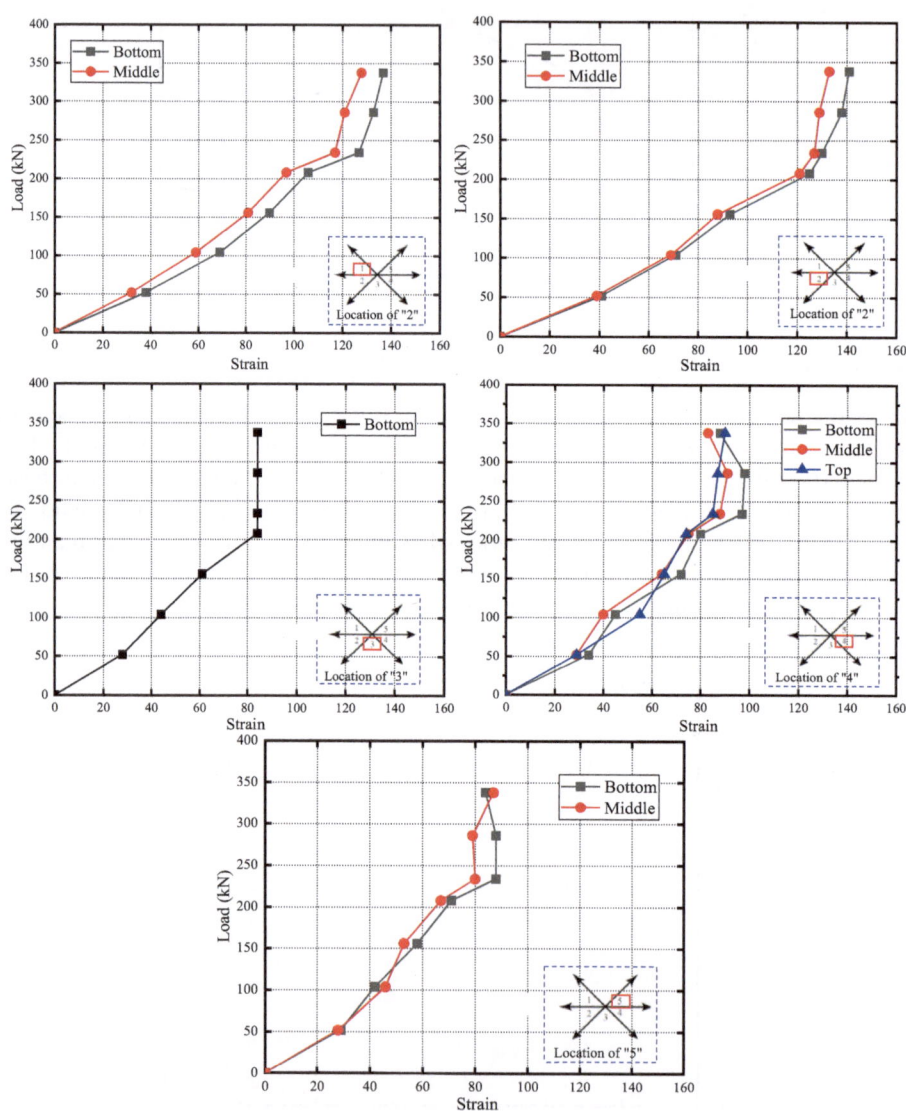

Fig. 7. Load-strain curve of concrete column

3.4 Strain of Concrete Cantilever Beam

The load-strain curve of concrete cantilever beam was illustrated in Fig. 8. It can be seen that under different load levels, the part near the concrete column at the root of the concrete beam was in a compressive state, and the maximum compressive strain occurs at the bottom of the beam, which was 184 με.

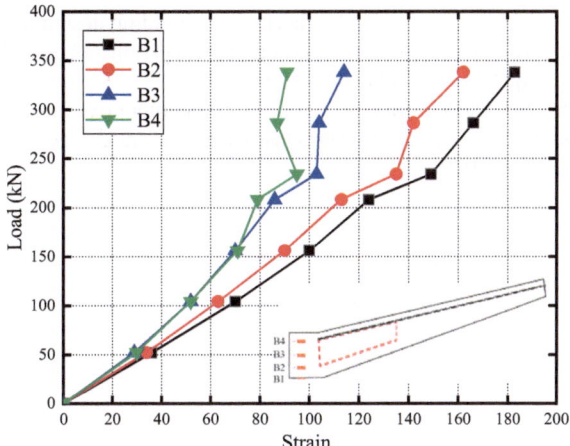

Fig. 8. Load-strain curve of concrete cantilever beam

4 Conclusion

1. The US was in elasticity stage when loading, which Was reflected in load-displacement curve.
2. Under the final load (338 kN), the maximum compressive strain of the cantilever steel beam was 19.98 MPa, and the maximum tensile strain was 32.96 MPa, both of which were smaller than the yield strength of Q355b (355 MPa); The maximum compressive stress of concrete columns was 4.23 MPa, which was less than the strength of C35 concrete; The maximum compressive stress of the concrete at the root of the cantilever beam was 4.98 MPa, which was less than the strength of C35 concrete. The experimental results shown that under the final load, the internal forces of the steel plates, cantilever beams, and concrete columns of the structure were all less than their corresponding resistance, and the structure was in a safe state.

References

1. Slak, T., Kilar, V.: The Design of Wooden and Steel Structures for Earthquake Areas, pp. 46–51. Arhitektura, Raziskave (2005)

2. Liu, C., Wang, F., He, L., et al.: Experimental and numerical investigation on dynamicresponses of the umbrella membrane structure excited by heavy rainfall. J. Vib. Control **27**(5–6), 675–684 (2021)
3. Jaksch, S., Sedlak, V.: A foldable umbrella structure-developments and experiences. Int. J. Space Struct. **26**(1), 1–18 (2011)
4. Espinoza, R.G.P.: Fundamentos Geométricos De Las Superficies De Parábolas Invertidas, vol. 6, p. 75. Caracas:Universidad Central De Venezuela,Facultad De Arquitectura Y Urbanismo (2013)
5. Zhang, S., Liu, N., Li, W., Yan, S.: Umbrella structure building design method via case-based design and statistical analysis of structural morphological parameters. J. Build. Eng. **45**, 103542 (2022). https://doi.org/10.1016/j.jobe.2021.103542
6. Shuyang, Z., Yang, W.: The historical evolution and contemporary value of umbrella structure: cases study based on the practice in Western Europe and Latin America since the 20th century. South Arch. **06**, 22–29 (2020)

Research on the Promotion of Characteristic Prefabricated Buildings Under the Background of Rural Revitalization

Yuan Dong[✉] and Yunlai Zhang

Chongqing Vocational College of Building Technology, Shapingba, Chongqing, China
1273421875@qq.com

Abstract. New rural construction is a key part of the rural revitalization strategy. Prefabricated buildings can greatly improve the living quality of buildings and meet the needs of production, life and environmental protection. This paper puts forward the dilemma of traditional architecture and current prefabricated buildings in rural development, puts forward how to promote and develop prefabricated buildings in rural areas with a problem-oriented approach, and puts forward suggestions in terms of economy, technology, policy and publicity guidance. In order to promote the development of characteristic prefabricated farm house building to provide beneficial ideas.

Keywords: Prefabricated building · Rural revitalization · Rural prefabricated housing

1 Problems Existing in Traditional Architecture

1.1 Building Materials Do not Meet the Requirements of Sustainable Development

Due to the terrain, the building forms are dry column type, courtyard type and earthen house type. Most of the traditional rural houses are brick-concrete and brick-wood structure, and some are still mud wall bungalows [1]. The materials are solid clay bricks and wood, which cause damage to cultivated land resources and produce energy consumption and waste of resources. In addition, due to the low cultural level of villagers, they will choose cheaper building materials, resulting in a decline in the quality of rural housing and low safety performance.

1.2 Lack of Professional Design Drawings

Traditional rural housing has security risks, only more than 10% of villagers ask professional designers to design the structure and appearance of rural housing, most villagers blindly build their own, there are structural risks. According to the statistics of relevant institutions, about 76% of the houses built in rural housing do not use formal drawings, and most of the construction of rural housing is completed by relying on the experience of the construction team, and in the construction process to meet the needs of villagers on the arbitrary demolition of housing, seriously reducing the quality of housing performance.

P. Xiang and L. Zuo (Eds.): PBSFTT 2023, LNCE 382, pp. 565–571, 2024.
https://doi.org/10.1007/978-981-97-5108-2_60

1.3 Lack of Professional Construction Team

Due to the small scale of construction, self-built houses are common, and the lack of professional construction teams leads to more serious illegal construction and unreasonable design. The crowded layout affects the lighting and ventilation between the houses, which seriously reduces the living comfort and greatly increases the energy consumption of the building. Thermal insulation, shading and other energy saving measures are poor. It makes the security, applicability and durability of rural housing lack of guarantee.

1.4 Construction Waste Pollution is Serious

During the construction process, only the construction waste will produce 500–600 tons/10,000 square meters, and the dismantled old buildings will produce 7000–12,000 tons/10,000 square meters of construction waste. The long-term and high-frequency self-built house construction in rural areas will produce a large amount of construction energy consumption and construction waste every year. Most villages have not set up unified landfill sites, and some farmers who lack environmental awareness use village roads, rivers and lakes, and even their own farmland as landfill or stacking sites for construction waste, causing serious environmental pollution.

2 The Promotion of Prefabricated Buildings is an Effective Means to Implement the Strategy of Rural Revitalization

2.1 We Will Revitalize Rural Tourism

Rural revitalization depends on solving rural environmental problems, and the way out is the transformation and upgrading of rural industries. Relying on terrain features, we will vigorously develop environment-friendly rural tourism industry, transform environmental resource advantages into industrial development advantages, implement leisure agriculture and rural tourism quality projects, develop rural sharing economy, creative agriculture, and characteristic cultural industries, and build quality villages and towns suitable for living, working and traveling.

2.2 Improve Living Conditions in Rural Areas

Ecological environment damage is increasingly serious, building energy consumption continues to rise, in the construction and use of houses to achieve ecological livable, reduce energy consumption is imminent. The development of green farming in rural areas is not only to protect resources and reduce energy consumption, but also to meet the needs of ecological, social and economic sustainable development. Prefabricated buildings should highlight the characteristics of people's livelihood and share the achievements of urbanization civilization.

2.3 Protect the Environment and Save Energy

According to incomplete statistics, every 10,000 square meters of construction projects in China produce 500–600 tons of garbage, dust, noise and waste pollution seriously threaten people's health. Prefabricated buildings save resource consumption in form work, insulation materials, industrial water and electricity, and can reduce about 80% of construction waste and 90% of dust and noise pollution, to achieve the construction goal of low energy consumption and low emissions.

3 Prefabricated Building in the Rural Development of the Current Dilemma

3.1 Lack of Perfect Standard

The lack of policy support in rural areas, the pilot work of rural housing construction has been officially launched since February 2019, and the housing construction standards have been preliminaries improved, but for rural prefabricated housing from the design stage to the final completion and acceptance stage, there are neither strict standards nor a regulatory system.

3.2 Lack of Professional and Technical Personnel

Prefabricated building as a new building structure, PC components lifting, installation and node connection pouring, each link requires the participation of professional and technical personnel, the current talent is mainly concentrated in the traditional reinforced concrete cast-in-place structure, in the prefabricated building lack.

3.3 Limitation of Transport Conditions

PC components have high requirements for accuracy, bumps and bumps in the transportation process will affect the next process, and then reduce the quality of rural prefabricated housing [2]. For different types of PC components, it is necessary to choose the appropriate means of transportation, and at the same time, it is necessary to consider the distance between the prefabrication plant and the construction site, and choose the appropriate transportation route.

3.4 High Cost of Construction

The cost of component customization, long-distance transportation, factory management and personnel social security is far higher than the labor cost of traditional construction methods. The splicing and assembling work of PC components requires large mechanical equipment and increases the construction cost.

4 Characteristic Prefabricated Building to Promote the Implementation Strategy

4.1 "Bamboo Instead of Wood" Green Assembly

Promote the development and utilization of bamboo in the prefabricated building system. It can not only alleviate the contradiction between supply and demand of wood, but also promote the scientific development of bamboo industry and rural revitalization, save and protect environmental resources, and realize the green and sustainable development of the construction industry. Bamboo structure has great toughness and strong resistance to instantaneous impact load and periodic fatigue failure [3]. Chen Guo detailed the relationship between the performance changes of each component and connector and the overall force transmission under the action of earthquake and lateral force resistance [4].

4.2 Intelligent Assembly Strategy Using BIM Technology

BIM technology plays an active role in the development, design, construction and management process of construction projects. Based on BIM technology, professional models such as architecture, structure and machinery can be integrated, project progress and process flow can be entered into the system before construction [5], the progress of hoisting operation can be simulated, the safety load can be calculated and the actual construction situation can be analyzed, and the danger sources can be timely warned and accurately positioned. Combined with RFID chip technology, real-time tracking can control equipment and components to prevent security problems [6].

4.3 Reasonable Split Component Strategy

The lifting point position of the component can ensure the integrity of the component [7]. The traditional prefabricated component is divided according to the distance of the shaft network, and the component types are many, the standardization degree is not high, and the mold investment is large, resulting in low utilization rate. Based on the idea of component grouping, the mold is optimized as an important production resource, combined with production, transportation and construction needs, and the mold utilization rate and turnover times are increased as much as possible, the production cost is reduced, the hoisting times are reduced, and the construction cost is reduced (see Fig. 1).

4.4 Characteristic Strategies to Adapt to the Environment

Standardization is the method and process, diversity is the result, components are fixed, but the collocation of components is ever-changing [8]. The protection of local characteristics is not in contradiction with the characteristics of the era of the building. Through the highly abstract modularization of prefabricated buildings, innovation is carried out on the basis of maintaining the original characteristics of the city.

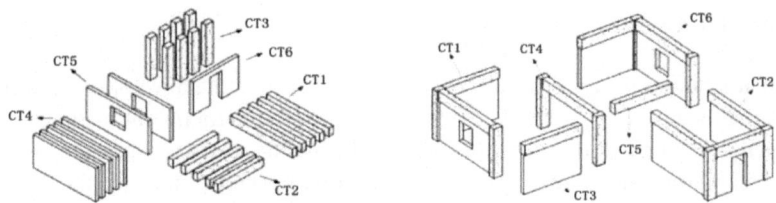

Fig. 1. Separate and compare different components.

5   Proposals to Enhance the Development Potential of Prefabricated Buildings

5.1   Technology

Deepen the training of existing technical personnel, create better working conditions, and invest more innovative talents in the development of prefabricated building technology. The government supports the technological innovation of prefabricated buildings and has a superior technological environment.

5.2   Optimize the Whole Industrial Chain of Prefabricated Buildings

The government should do a good job of policy guidance, optimize the industrial structure, encourage enterprises to develop in the blank field of the whole industrial chain of prefabricated buildings, increase the types and coverage of industries, and encourage the diversified development of enterprises [9].

5.3   Step up Publicity for Prefabricated Buildings

The government should promote prefabricated buildings through multiple channels, deepen the contrast between traditional buildings and prefabricated buildings by enterprises and consumers. Form a social form in which enterprises take the initiative to develop and consumers voluntarily purchase prefabricated building products.

5.4   Advocate Design into the Village

We will set up pilot and demonstration rural housing projects and encourage designers to go to rural areas [10]. Master Chen, a resident and structural designer, returned to his hometown from Wuhan, Hubei Province, under the call of rural revitalization, and led local residents to design and build the first prefabricated residential house in the village with an assembly rate of 80%. The entire construction cycle was only 58d, which truly met the requirements of building energy saving, green and environmental protection.

6   Conclusion

In conclusion, under the background of rural revitalization, the development of prefabricated residential buildings in rural areas has great development potential, give play to the government's leadership, optimize the industrial structure, increase technical investment, promote "bamboo instead of wood", do a good job of policy guidance, and increase the training of rural prefabricated talents. With the active role of BIM technology, reasonable separation of prefabricated components is in line with characteristic development. Strengthening the publicity work of prefabricated buildings in order to promote the high-quality development of rural prefabricated buildings can not only improve rural economic growth and improve the quality of life, but also help promote the construction of a beautiful new countryside, which is in line with the long-term needs of the rural revitalization strategy.

References

1. Mu, L.: Prefabricated construction in the application and development of housing construction in Shanxi Province. J. Shanxi Build. **47**(08), 28 and 30 (2021). https://doi.org/10.13719/j.car olcarroll nki.1009-6825.2021.08.010
2. Wang, X., Zouchao, Wu, G., et al.: Research on construction technology application of prefabricated components in prefabricated residential buildings. Sci. Technol. Innov. Appl. **13**(27), 160–164 2023. https://doi.org/10.19981/j.CN23-1581/G3.2023.27.038
3. Li, H., Xuan, Y., Xu, B., et al.: The application of bamboo in the field of civil engineering. J. Forestry Eng. **5**(6), 1–10 (2020). https://doi.org/10.13360/j.iSSN.2096-1359.202003001
4. Chen, G., Xiao, Y., Shan, B., et al.: Modern bamboo structure housing design and engineering application. Ind. Constr. **9**(4), 66–70+74 (2011). https://doi.org/10.13204/j.gyjz2011.04.019
5. Spcswinson. "double carbon" prefabricated construction technology in development research under the targets. J. Intell. Build. Intell. City (9), 115–117 (2023). https://doi.org/10.13655/j.carolcarrollnkiibci.2023.09.034
6. Wang, Y., Yao, S.: Research on life cycle safety management of prefabricated buildings based on BIM technology. Shanghai Real Estate (9),15–18 (2023). https://doi.org/10.13997/j.carolc arrollnkicn31-1188/f2023.09.006
7. Yang, B., Jiang, S., Dong, M., et al.: Graph database and matrix-based intelligent generation of the assembly sequence of prefabricated building components. Appl. Sci. **13**(17), 9834 (2023)
8. Lupíek, A., Vaculíková, M., Maník, T., et al.: Design strategies for low embodied carbon and low embodied energy buildings: principles and examples. Energy Procedia **83**, 147–156 (2015)
9. Yuan, Z., Sun, C., et al.: Design for manufacture and assembly-oriented parametric design of prefabricated buildings. Autom. Constr. **88**, 13–22 (2018)
10. Li, C.Z., Hong, J., Xue, F., Shen, G.Q., Xu, X., Luo, L.: SWOT analysis and internet of things-enabled platform for prefabrication housing production in Hong Kong. Habitat Int. **57**, 74–87 (2016)

A Prefabricated Wall with Automatic Air-Circulation and Dust-Removal Function Based on TRIZ Theory

Yingjia Wang, Yanan Yi[✉], and Xianhui Man

Chongqing College of Architecture and Technology, Mindestr. 3, Chongqing 400000, China
504064378@qq.com

Abstract. In response to the problems of high energy consumption, high maintenance costs, and large space occupation of traditional dust removal equipment, this work utilizes TRIZ theory to innovate and invent a prefabricated wall that automatically circulates air to achieve dust removal function. The wall includes three parts: a solar thermal module, an air circulation dust removal module, and a building solid waste regeneration material insulation module. It uses solar energy to provide power, drive air circulation exchange, and achieve the application of clean energy to reduce dust in the factory building, The effect of reducing summer room temperature and low-carbon heating in winter.

Keywords: Solar Energy · Prefabricated Walls · TRIZ Theory · Material Field Analysis · Contradiction Matrix

1 Introduction

Dust pollution is the main cause of high incidence of occupational diseases such as pneumoconiosis and silicosis among workers, and is also the primary air pollutant in most cities in China. With the increasing efforts of the country in environmental protection and stricter control of occupational diseases among workers, the cost of using dust removal products in factories is also accounting for a higher proportion of production costs. Whether to control and reduce the cost of using and maintaining dust removal products has become a key factor for factories to choose dust removal products.

The traditional dust removal methods mainly include gravity dust removal, filtration dust removal, wet dust removal, and electrostatic dust removal. By comparing and analyzing the market share, dust removal efficiency, dust removal efficiency, preliminary procurement costs, energy consumption costs, and maintenance costs of the three traditional dust removal methods, namely bag filter, inertial dust removal, and spray tower, which have the highest market share, it can be concluded that low-cost equipment has poor dust removal efficiency, while equipment with good dust removal efficiency has higher procurement and maintenance costs [1], as shown in Table 1.

Emerging dust removal methods include photovoltaic dust removal, solar thermal dust removal, etc. The haze removal tower in Xi'an, China, with a height of 60 m and an

P. Xiang and L. Zuo (Eds.): PBSFTT 2023, LNCE 382, pp. 572–581, 2024.
https://doi.org/10.1007/978-981-97-5108-2_61

Table 1. Comparison table of traditional dust removal equipment.

Type	Market share (%)	Dust removal efficiency (%)	Minimum capture particle size (μm)	Pressure loss (KPa)	Equipment cost (10000 yuan/year)
bag filter	47	95–99	0.5	20	20
inertia filter	32	50–70	20–50	8	8
spray tower	21	70–85	10	6	6

inner diameter of 10 m, utilizes solar energy to provide heat and drive air circulation, with a haze removal area of 10 km^2; at the FedEx in Denver, Colorado, USA, due to the need for a large number of trucks to pass by during work and the high dust emissions, a perforated solar wall panel was installed on the south exterior wall of the factory building, which can provide 76000 m^3/h fresh preheated air, saving approximately $12000 annually [2], as shown in Fig. 1.

Fig. 1. Haze removal tower and solar collector wall.

2 TRIZ Problem Analysis Model

The TRIZ theory uses innovation laws to reveal the general laws and basic principles of the process of creation and invention, mainly to clarify and emphasize the contradictions in the system, with the ultimate goal of completely solving the contradictions in the system and obtaining the final ideal solution [3]. After years of development and dissemination, TRIZ theory has been widely applied in various industries and has long been used as a systematic, knowledge-based, and people-oriented method to solve invention and creation problems [4].

2.1 Current Technical Issues

No matter which traditional dust removal method is used, it is necessary to use electricity to drive the equipment to operate, achieve the effect of air circulation and filtration dust removal, which inevitably requires small and medium-sized enterprises to invest a large amount of electricity bills for dust removal in the factory. In addition, although traditional dust removal equipment has been improved, the problem of occupying a large area remains unresolved.

2.2 Functional Analysis

Firstly, it is necessary to summarize which components a technical system is composed of, including system objects, technical system components, subsystem components, and metasystem components that interact with system components [5]. The functional model obtained by analyzing the interaction relationship between various components of the device is shown in Fig. 2. From the figure, it can be seen that the current system has the following problems:

(1) The operation of the equipment can meet the dust removal needs of the factory and ensure the health of workers. However, due to high electricity consumption during use, it can lead to an increase in electricity costs.
(2) Each part of the equipment requires shell support, and to ensure sufficient air circulation and flow, it needs to occupy a large workshop space, sometimes even affecting the lighting and ventilation needs of the factory.
(3) The accumulation of dust on the bag of a bag filter can lead to a decrease in dust removal efficiency; For factories, the replacement of filter bags is also a significant expense.

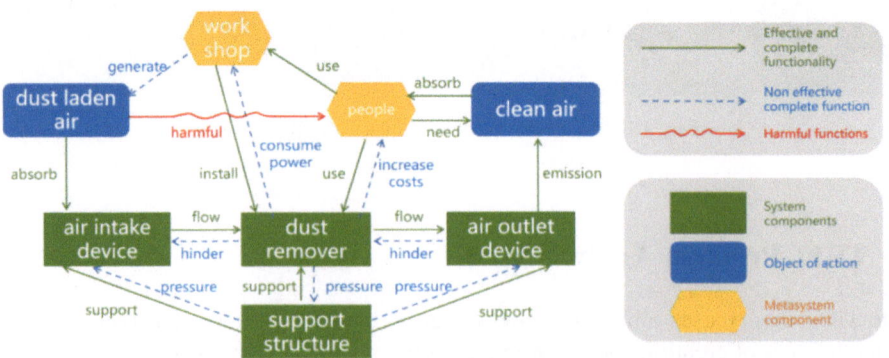

Fig. 2. Functional analysis model diagram of traditional bag filter.

2.3 Causal Analysis

Causal analysis is a method of studying the relationship between the results of the development of things and the causes of their occurrence, and analyzing the factors that affect the causal relationship [6]. Firstly, identify the factors that make the use and maintenance costs of traditional dust removal equipment high, and further evaluate the likelihood of these reasons occurring, using V (very likely), S (some possible), and N (unlikely) as indicators. Evaluate the possibility of solving the reasons marked with V and S, and use three types to indicate them: V (very easy to solve), S (relatively easy to solve), and N (not very easy to solve). Further evaluate the difficulty of implementing corrective measures for reasons marked with VV, VS, SV, and SS, using V (very easy to verify), S (relatively easy to verify), and N (not very easy to verify) as indicators. Draw the above analysis into a causal analysis fishbone diagram [7] (Fig. 3), and from the fishbone diagram, it can be seen that:

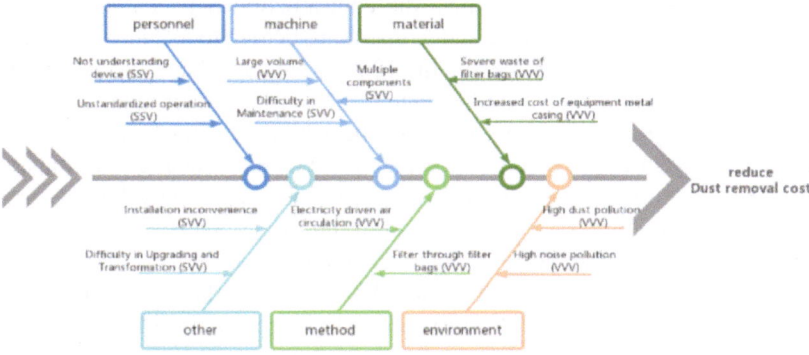

Fig. 3. Causal axis analysis diagram.

3   Apply TRIZ Tool to Solve Problems

In the TRIZ tool, in order to facilitate the definition of contradictions, Archishuler analyzed a large number of patents and successively summarized and extracted 39 general engineering parameters [8]. Using these parameters is sufficient to describe the vast majority of contradictions that appear in the engineering field. Through functional analysis and causal analysis, it has been found that there are three contradictions that make traditional dust removal equipment expensive to use, expensive to maintain, and large in terms of floor space:

(1) The contradiction between power and power.
(2) The contradiction between shape and force.
(3) The contradiction between shape and temperature.

3.1   Physical Contradiction Method

Among the three sets of contradictions, the contradiction between power and power belongs to the physical contradiction [9]. The contradiction between power and power is mainly reflected in the need for dust removal equipment to meet the requirements of dust removal function and dust removal efficiency. It is necessary to use energy to drive air circulation and flow, allowing dusty air to enter the interior of the dust removal equipment and achieve dust removal function. However, the consumption of electrical energy is the main reason for the high cost of use. Simply put, this contradiction is: equipment cannot run without electricity, and electricity is expensive.

In response to this contradiction, the "conditional separation" method among the four separation methods of TRIZ is adopted to separate the electrical energy, change the energy supply, and achieve the purpose of dust removal.

3.2   Technical Contradiction Method

Among the three sets of contradictions, the contradiction between shape and force, and the contradiction between shape and temperature belong to technical contradictions [10]. Among the 39 general parameters, the 12th parameter "shape" is an improvement parameter. We can improve the performance of the product through shape, such as reducing volume to reduce land occupation; The 10th parameter "force" and the 17th parameter "temperature" are deterioration parameters. Due to changes in shape, it may lead to a decrease in the power of air flow required for dust removal and a change in air circulation temperature, greatly affecting dust removal efficiency. Find the parameter No.12 shape that the system needs to improve in the first column of the contradiction matrix, find the deteriorating parameter No.10 force and No.17 temperature in the first row, and find the corresponding invention principle, as shown in Fig. 4.

Improvement parameter	Deterioration parameter	Applicable principle
No.12 shape	No.17 temperature	No.22 Principle of turning harm into profit
		No.14 Principle of Curvification
		No.19 Principle of Periodic Action
		No.32 Color Change Principle
No.12 shape	No.10 force	No.35 physical or chemical parameters Change principle
		No.10 Principle of Pre action
		No.37 Thermal Expansion Principle
		No.40 Principles of Composite Materials

Fig. 4. Using Contradiction Matrix to Derive Invention Principle Diagram.

(1) According to Article 22 of the Invention Principle, the principle of turning harm into profit is to replace the equipment structure that originally occupied a large area of the factory area with a wall. The wall itself is a necessary enclosure structure for the factory building, and replacing the equipment with it can greatly reduce the land occupied due to dust removal needs.
(2) According to Article 32 of the Invention Principle, the principle of color change, the efficiency of coating solar thermal conversion is the source of aerodynamic force, and the higher the conversion efficiency, the greater the power obtained by the air.
(3) According to Article 37 of the Invention Principle, the principle of thermal expansion, we simulate and adjust the size of the wall through software to obtain the optimal wall height to thickness ratio. The air inside the wall can be fully heated by solar thermal energy, maximizing the power of air circulation.

(4) According to Article 40 of the Invention Principle, the principle of composite materials, a new sandwich composite wall is formed by combining solar photovoltaic panels with the wall.

3.3 Material Field Analysis

Although photovoltaic power supply can solve the problem of high electricity bills for traditional dust removal equipment, there are still the following problems:

(1) Susceptible to dust pollution, long-term use of surface ash affects photoelectric conversion efficiency.
(2) Each square meter of photovoltaic panels costs about 900 RMB, and it is not a small cost to use them on a large area outside the factory building.
(3) During long-term use, the voltage board may face surface damage and require investment in replacement.

In response to the most serious issue of photoelectric conversion efficiency in the product, the TRIZ method was used to analyze the material field of the product, and solution 6 was used. Firstly, it was determined that the low photoelectric conversion efficiency was insufficient. Therefore, we searched for solutions in the second and third standard solutions. The final selection number 2.2.1, the 16th standard solution, is to replace the previously difficult to control field with a field that is easier to control, or to superimpose it on the field that is not easy to control. Using "solar thermal conversion" as a "more easily controllable field", solar energy is directly converted into thermal energy. By utilizing the chimney principle of hot air rising and cold air falling, solar energy directly drives air circulation.

4 Scheme Description

By continuously developing and iterating with TRIZ tools, we have innovatively invented an air circulation self dust removal prefabricated wall, as shown in Fig. 5.

The wall is composed of three parts: outer wall, middle cavity, and inner wall.

(1) The exterior wall is a solar thermal module, composed of organic heat absorbing materials and aluminum based thermal conductive materials developed by us with independent knowledge products. It absorbs solar energy, heats the air in the cavity, and promotes air circulation and exchange.
(2) The middle chamber is an air circulation dust removal module that uses high-voltage static electricity to form an electrostatic field to adsorb dust in the air inside the chamber.
(3) The inner wall is a thermal insulation module made of recycled building solid waste materials. It utilizes recycled building waste materials, which are lightweight, have a large pore ratio, and have good insulation effects. It plays a role in isolating temperature transmission and ensuring room temperature.

The working principle is shown in Fig. 5: When sunlight shines on the outer side of the factory building, the sunlight is captured by the external wall photothermal module, which heats the external wall and conducts convection and radiation heat exchange with the dust containing air in the dust removal cavity. At this time, the dust containing air in the dust removal cavity is rapidly heated and rises, causing negative pressure at the bottom of the wall, causing the bottom of the inner wall to suck in the dust containing air. The dust in the dust containing air is captured by the electrostatic dust removal device, and the remaining clean air rises along the wall, Finally, it is discharged from the exhaust outlet on the top of the exterior wall.

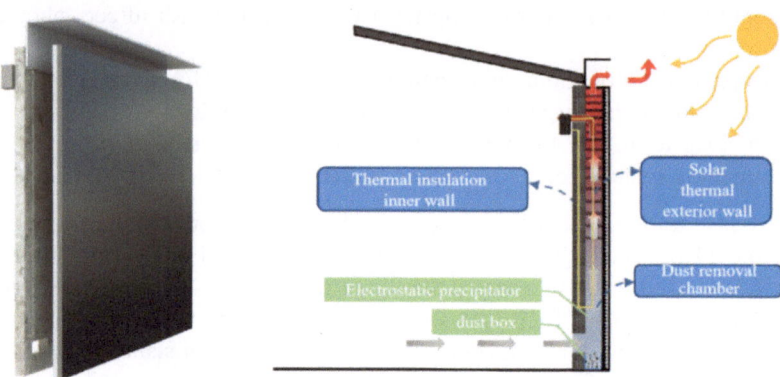

Fig. 5. Product model and schematic diagram.

Install the product with a length of 5 m and a height of 2.6 m on the outer side of the factory building with a length of 50 m, a width of 20 m, and a height of 9 m. Under the condition of solar irradiance of about 600 W/m^2, conduct a test on the dust removal effect of traditional bag filter and this air circulation self dust removal assembly wall. The experiment has shown that in the first 20 min, the dust removal efficiency of the prefabricated dust removal wall is not as good as that of the traditional bag filter, but after 20 min, the dust removal efficiency is basically the same as that of the bag filter (Fig. 6).

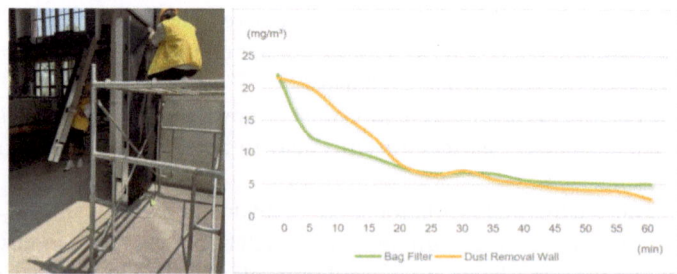

Fig. 6. Dust removal effect test and relationship diagram between indoor dust concentration and dust removal time.

In addition, the air flow rate and flow rate in the dust removal chamber show an upward trend with the increase of solar radiation intensity. The higher the solar radiation intensity, the more energy is absorbed by the surface of the collector wall, resulting in a greater difference between the hot air density in the dust removal chamber and the air density in the external environment, resulting in a greater suction force that increases the air flow rate in the dust removal chamber. By detecting and analyzing the wind speed at the bottom entrance of the inner wall under different illuminances (Fig. 7), it is found that when the solar irradiance is 600 W/m² or above, the air flow rate can meet the dust removal requirements.

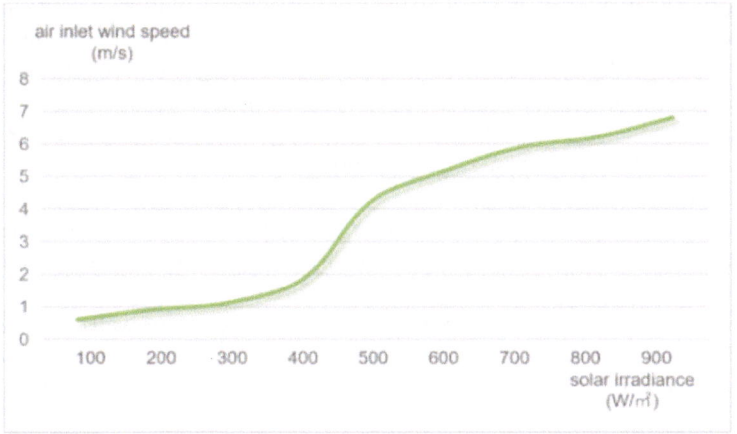

Fig. 7. Relationship between air inlet wind speed and solar illumination.

The solar irradiance is the key to ensuring the dust removal effect of the product. However, when the solar irradiance is insufficient on rainy days, it is still necessary to use photovoltaic dust removal walls for assistance. Firstly, solar energy is converted into electrical energy and stored in batteries. When the solar irradiance is insufficient, air circulation power is provided for the dust removal walls. The configuration ratio of the solar thermal dust removal wall and the photovoltaic dust removal wall is adjusted according to the geographical location of the factory building.

In addition, the intensity of solar radiation has a pattern of high in summer and low in winter, and summer is the peak of electricity consumption. For factories, relying solely on industrial electricity for dust removal in their factories will not only greatly increase the cost of electricity, but also make it impossible to use traditional dust removal equipment due to peak electricity usage. However, for solar powered dust removal walls, there is no such problem, as higher solar radiation intensity can provide strong power for air circulation. In summer, the filtered clean air with higher temperature is discharged from the top of the exterior wall, and the thermal insulation inner wall composed of recycled building waste materials can effectively isolate outdoor high temperatures. In this way, the solar dust removal wall not only has dust removal function, but also effectively reduces room temperature. In winter, the filtered clean air with higher temperature returns from the top of the inner wall to the inside of the factory building, playing a heating role.

5 Conclusion

(1) Establish a functional model of traditional dust removal products using the functional analysis in TRIZ theory; Using causal analysis fishbone diagram to identify the key issues that lead to high energy consumption and maintenance costs of traditional dust removal equipment.

(2) Use technical contradictions, physical contradictions, and contradiction matrices to find the corresponding invention principles to solve the problem, and propose a solution to use solar energy to provide power, replacing the metal shell of traditional dust removal equipment with a wall.

(3) Using physical field analysis and the 16th standard solution out of 76 standard solutions, replacing solar photovoltaic panels with solar thermal coatings can solve the problems of low photovoltaic conversion efficiency and maintenance costs.

(4) The test on the air circulation self dust removal prefabricated wall shows that when the solar irradiance is 600 W/m^2 or above, the product can meet the dust removal needs of small and medium-sized enterprise factories. This provides a foundation for subsequent analysis.

(5) Due to the significant impact of weather on the performance of solar thermal dust removal walls, it is still necessary to retain some photovoltaic dust removal walls and use solar energy to provide electricity to drive air circulation.

(6) In summer, hot air is discharged from the top of the exterior wall to achieve indoor cooling; In winter, hot air returns from the top of the inner wall to the interior, achieving heating.

References

1. Muratova, K.M., Makhnin, A.A., Volodin, N.I., Chistyakov, Y.V.: Cleaning of industrial dust-and-air flows in centrifugalinertial type devices. Khimicheskoe I Neftegazovoe Mashinostroenie **3**, 31–35 (2017)
2. Alqaed, S., Mustafa, J., Sharifpur, M.: Numerical study of the placement and thickness of blocks equipped with phase change materials in a Trombe wall in a room-thermal performance prediction using ANN. Eng. Anal. Boundary Elem. **141**, 91–116 (2022)
3. Souili, A., Cavallucci, D., Rousselot, F.: Identifying and reformulating knowledge items to fit with the inventive design method (IDM) model for a semantically-based patent mining. Procedia Eng. **131**, 1130–1139 (2015)
4. Delgado-Maciel, J., Robles, G., Sanchez-Ramirez, C., García-Alcaraz, J.: The evaluation of conceptual design through dynamic simulation: a proposal based on TRIZ and system dynamics. Comput. Ind. Eng. **149**, 106785 (2020)
5. Bertoncelli, T., Mayer, O., Lynass, M.: Creativity, learning techniques and TRIZ. Procedia CIRP **39**, 191–196 (2016)
6. Du, L., Dinçer, H., Ersin, İ., Yüksel, S.: IT2 fuzzy-based multidimensional evaluation of coal energy for sustainable economic development. Energies **13**(10), art. no. 2453 (2020)
7. Mangla, S.K., Luthra, S., Jakhar, S., Gandhi, S., Muduli, K., Kumar, A.: A step to clean energy - sustainability in energy system management in an emerging economy context. J. Clean. Prod. **242**, art. no. 118462 (2020)

8. Zhong, J., Hu, X., Yuksel, S., Dincer, H., Ubay, G.G.: Analyzing the investments strategies for renewable energies based on multi-criteria decision model. IEEE Access **8**, art. no. 9125895, 118818–118840 (2020)
9. Liu, L., Fang, M., Xu, S., Guo, D.: TRIZ-based design innovative solutions for carbon capture process. Zhongguo Dianji Gongcheng Xuebao/Proc. Chin. Soc. Electr. Eng. **40**(20), 6625–6632 (2020)
10. Abdala, L.N., Fernandes, R.B., Ogliari, A., Löwer: Creative contributions of the methods of inventive principles of TRIZ and BioTRIZ to problem solving. Jo. Mech. Des. **139**(8), art. no. 082001 (2017)

Encourage the Use of Prefabricated Building Technology to Improve the Quality of Life in Rural Areas During the Process of Rural Revitalization

Qi Wang[1(✉)], Yu Li[2], and Fei Wang[1]

[1] Xi'an Jiaotong University, No. 28 Xianning West Road, Xi'an 710049, Shaanxi, People's Republic of China
wang.jing.qi@xjtu.edu.cn
[2] Xi'an University of Architecture and Technology, No. 13 Yanta Road, Xi'an 710055, Shaanxi, People's Republic of China

Abstract. The connotation of rural revitalization not only includes rural economic development and rural landscape reconstruction, but more importantly, the return of rural family integrity and the reshaping of rural pastoral life. The use of prefabricated building technology can quickly improve the quality of rural life at a low cost. Rooms, stairs, kitchens, bathrooms, and various building components can be prefabricated in the factory, while on-site workers are only responsible for overall assembly. Made of fiberglass material, the cost of prefabricated septic tanks and biogas digesters can be greatly reduced, and replacement and recycling are convenient, all of which can be completed in the factory. In addition to creating job opportunities, prefabricated building technology can also attract urban populations to settle in rural areas, creating a happy Chinese style rural pastoral life together.

Keywords: Rural revitalization · Prefabricated building technology · Rural pastoral life

1 Introduction

The development process of the People's Republic of China is a process of rapid industrialization of an agricultural country. Whether relying on the aid of the former Soviet Union to develop the military heavy industry in the early stage or undertaking the western labor-intensive low-end consumer goods industry in the later stage, it is at the cost of extracting a large amount of agricultural surplus value. Therefore, when China has basically completed industrialization, it is necessary to give back and feed back to agriculture and rural areas. The strategy of rural revitalization is the compensatory investment and modernization of agriculture and rural areas, and urban agriculture with short cycle and quick effect will be the entry point and breakthrough to realize the strategy of rural revitalization [2]. The rural revitalization strategy is also to find a new investment channel

P. Xiang and L. Zuo (Eds.): PBSFTT 2023, LNCE 382, pp. 582–588, 2024.
https://doi.org/10.1007/978-981-97-5108-2_62

for the surplus urban financial capital and the surplus urban industrial capital. According to the concept of rural revitalization strategy, the scope of projects that can be operated in rural areas is very broad. In addition to traditional agriculture, planting, aquaculture, and agricultural by-product processing industries, tourism, leisure, culture, specialty cuisine, specialty processing and production, financial services, logistics services, real estate services, etc. [6] can also be developed in combination with local characteristics and specialties, which can attract a large number of local labor employment, and extend the industrial chain in both directions.

The rapid industrialization process, especially the labor-intensive industry, requires massive labor input. Therefore, most of the rural middle-aged peoples are attracted to work in cities, which has a huge impact on rural life, and even devastating in some areas. In the countryside, most families are left with only young children and elderly grandparents, and the elderly grandparents can't take care of agricultural production and the normal growth of children at all, so rural life has lost its normal state, agricultural production can only be maintained at a low level, and children's growth is also lack of parental care and discipline. This is extremely detrimental to children's physical and psychological growth, and life without parents is not real life. Therefore, the connotation of rural revitalization is not only the rural economic development and rural landscape reconstruction, but also the integrity of rural families. The real core is the return of rural family integrity and the remodeling of rural pastoral life [3]. It is necessary to create enough employment opportunities to attract parents to return to reunite with children and live a happier rural pastoral life than before.

The return of rural family integrity and the reshaping of rural pastoral life should be based on a higher quality of life, otherwise it will not be attractive. Rural areas do not lack fresh air, bright sunshine, sweet springs, and beautiful scenery. What rural areas lack is clean and tidy living facilities and a high level of hygiene. Therefore, how to make up for the shortcomings in living facilities and hygiene is the first problem faced by rural revitalization. After cost-effectiveness comparison, adopting mature and affordable prefabricated building technology can significantly improve the living facilities and hygiene level in rural areas in a short period of time, providing clean and comfortable indoor and outdoor living environments for rural families. When the quality of life in rural areas approaches or is on par with the quality of life in urban areas, it will undoubtedly significantly weaken the urban-rural gap, enhance the flow of urban-rural personnel, promote more people to stay in rural areas, and directly assist in rural revitalization.

2 Countermeasures to be Taken in the Revitalization of Rural Areas in China

The long-term dual structure of urban and rural areas has to some extent restricted the development of rural and agricultural areas in China. While the rapid industrialization and urbanization process has drained a large number of rural labor, agriculture has actually been weakened, and obviously there is no need to pay attention to the development of agriculture in urban areas. After China has basically completed the industrialization process and rural poverty alleviation tasks around 2020, rural revitalization has been put on the government's agenda, and improving the quality of life in rural areas is an important task.

2.1   Adopting Low-Cost Prefabricated Building Technology can Quickly Improve the Quality of Rural Life

The vast rural areas of China have made significant contributions and sacrifices to China's industrialization process. Rural revitalization is the feedback of industrialization on rural society. For China, as the world's factory, it is not difficult to improve the quality of life in rural areas with mature and affordable industrial products. In the field of architecture, prefabricated buildings and their technologies are the best choice for quickly improving the quality of rural life.

Traditional rural life in China is garbage free, and all food residues and human and animal excrement can be used as fertilizer to return to the field. However, the process of returning to the field is considered unhygienic, has an unpleasant odor, and can produce unpleasant experiences [4]. It is considered one of the important labels of the low quality of rural life. And prefabricated toilets and prefabricated septic tanks can completely change this perception, making the quality of life of rural households comparable to that of cities [7] (see Fig. 1). Combined with sewage pumps, the residue and liquid from septic tanks can be conveniently transported to the fields and used as organic fertilizers. Poultry and livestock manure can be put into prefabricated biogas digesters to produce biogas. After being liquefied and filled, biogas can be used as household fuel, and the residue can still be used as organic fertilizer (see Fig. 2).

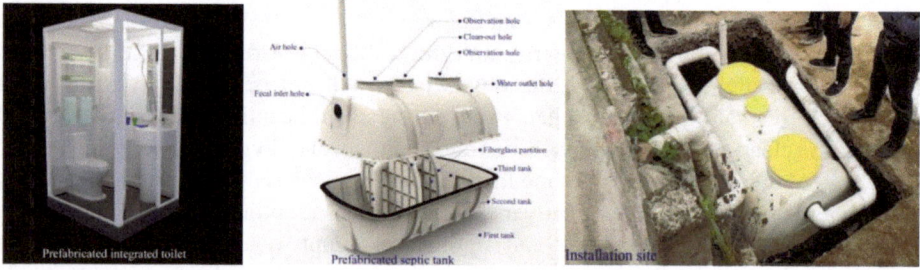

Fig. 1. Prefabricated integrated toilet/Prefabricated septic tank/Installation site.

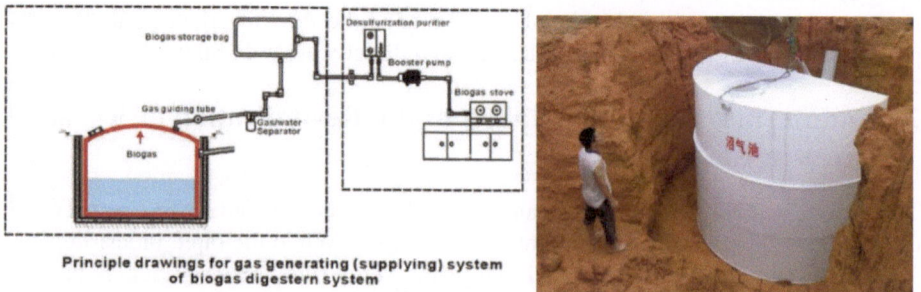

Fig. 2. Installation of prefabricated biogas digesters.

Made of fiberglass material, the cost of prefabricated septic tanks and biogas digesters can be greatly reduced, and replacement and recycling are convenient, all of which can be completed in the factory [8]. The fiberglass material is light and hard, non-conductive, stable in performance, high in mechanical strength, pollution-free, corrosion-resistant, and inexpensive, making it particularly suitable for the manufacturing of concealed prefabricated components.

Due to the relatively small scale of rural residential buildings, it is entirely possible to adopt prefabricated overall construction [5]. Rooms, stairs, kitchens, bathrooms, and various building components can be prefabricated in the factory, while on-site workers are only responsible for overall assembly, which can greatly improve efficiency and reduce costs [1] (see Fig. 3).

Fig. 3. Construction of prefabricated rural residential buildings.

2.2 Prioritize Planning and Design Work Based on Local Conditions

Prefabricated building technology can provide support for rural revitalization from a technical perspective, while planning and design are needed to address project operations. Rural revitalization is aimed at achieving a better rural life, rather than continuing to destroy the countryside. Therefore, it cannot be constructed in a disorderly manner and must be comprehensively considered. Planning and design must take precedence [11]. For example, in Wangxian Valley, Shangrao, Jiangxi Province, the tourism and tourism industry chain of "Ecological Fairy Tour" has been planned and designed, driving the extension of industries such as humanities, folk customs, agriculture, food, vacation, and research (see Fig. 4). It provides more than 2000 direct employment positions and more than 30000 surrounding employment, promoting the return of villagers in hollow villages such as Wangxian, Shangzhen, Daji, and Gelu, promoting the development of all industries and achieving genuine rural revitalization. (Data cited from Jiangxi Daily).

Due to the different economic levels, development conditions, and terrain and climate across the country, it is necessary to fully leverage the subjective initiative of each region, formulate development plans tailored to local conditions, and find the best development path suitable for local rural revitalization.

2.3 Pay Attention to Landscape and Ecology

Rural revitalization projects should be conducive to the protection of the ecological environment and the sustainable development of the landscape. While addressing the environmental pollution and landscape damage caused by industrialization, it is also necessary to handle the interactive relationship between the environment and people. Rural

Fig. 4. Wangxian Valley, Shangrao, Jiangxi Province.

revitalization should be a dual improvement of the natural environment and living environment, and a psychological experience with good visual and cultural aesthetics should be formed. This will facilitate the formation of unique rural landscapes and tourism and leisure industries, promote the formation and development of local characteristic cultures, extend the ecological industrial chain, and promote harmonious coexistence between humans and nature.

2.4   Pay Attention to Information and Logistics

Rural revitalization can utilize online platforms and video live streaming to communicate in real-time with end consumers, accurately grasp the supply and demand relationship, and meet the diverse needs of consumers. In order to better establish direct contact with consumers, various network application platforms should be fully utilized to form information groups, dynamically publish and receive information in real time, and establish fast logistics channels to complete transactions in a timely manner, maintain stable operations and secure cash flow [9]. Information and logistics are necessary support and reliable guarantee for the sustainable and healthy development of rural revitalization projects.

3   Conclusion

Rural revitalization is a complex and systematic task, with the most important technical support being prefabricated building technology. Due to the generally small volume and scale of rural buildings and landscapes, which do not meet the high requirements of cities, prefabricated building technology applied in rural areas not only has the characteristics of fast construction, low cost, durability, energy conservation and environmental protection, but also has considerable flexibility, which can adapt well to various rural planning and design needs. For special needs in special regions and climatic conditions, prefabricated building technology can also be well met, which can be simulated and verified in factory workshops. At the planning and design level, the use of prefabricated building technology should be actively encouraged. Not only can building components and functional rooms be prefabricated and assembled, but even the entire house can be prefabricated and assembled. The workers on the construction site are only responsible for firmly connecting to the foundation. This not only saves construction costs and time costs, but also greatly enhances the replaceability of buildings and the flexibility of planning and design, showcasing different architectural styles and landscape features in different seasons and periods, creating a variety of visual and cultural experiences. This will greatly promote the formation and extension of the rural industrial chain, increase rural employment opportunities, meet the needs of family members to live together [10], build a complete and comprehensive beautiful rural life, and achieve true rural revitalization.

Acknowledgements. Thanks for the support of the General Program of the National Natural Science Foundation of China (Project approval number: 52078403) and the Natural Science Foundation of Shaanxi Province (2021JM-013) and the Social Science Planning Foundation of Xi'an (23LW99) and China Scholarship Council (No. 202006285048).

References

1. Wang, Y.L., Li, Y.L., Tuo, M.B.: Research on the factors influencing the cost of assembled steel structure building components from the perspective of smart production. J. Build. Econ. **2**, 89–97 (2023)
2. Gulyas, B.Z., Edmondson, J.L.: Increasing city resilience through urban agriculture: challenges and solutions in the Global North. Sustainability **13**, 1465 (2021)
3. Liu, Y.L.: Research on rural governance issues from the perspective of rural revitalization. J. New West **34**, 59–62 (2018)
4. Guan, C.C.: Research on the construction of agricultural non point source pollution control system from the perspective of rural revitalization strategy. J. China Collect. Econ. **1**, 6–7 (2019)
5. Wang, S.T.: Innovative application of intelligent construction technology in engineering construction management. J. Build. Econ. **42**(4), 49–52 (2021)
6. Le, H.T., Vo, T.H.L., Chau, N.H.: The role of urban agriculture for a resilient city. J. Vietnam. Environ. **12**, 148–154 (2020)
7. Xu, Q.J., Zhao, Z.W., Yuan, W.L.: Application of prefabricated and assembled septic tank in underground pipe network maintenance and transformation engineering. J. Eng. Constr. Des. **3**, 121–122 (2019)
8. Miao, Q.: Research on types, system collaboration, and interface connection of prefabricated toilets. J. Archit. Technol. **9**, 1231–1235 (2022)
9. Górna, A.: Urban agriculture: an opportunity for sustainable developmentIn: Czerny, M., Serna Mendoza, C.A. (eds.) Globalización y Desarollo Sostenible, pp. 129–142. Wydawnictwa Uniwersytetu Warszawskiego: Warszawa, Poland (2018)
10. Liu, X.J.: Reflections on the path of effective rural governance in the context of rural revitalization. J. Econ. Res. Guide **35**, 25–27 (2018)
11. Chen, J.: Research on the Ways to Revitalize Rural Areas through Art. China Textile Press (2019)

Research on the Path of Green Finance to Help Green Building Development Under the Background of "Double Carbon"

An Wang(✉), Chunli Wang(✉), and Zhijiao Chu

Chongqing College of Architecture and Technology, Chongqing, China
382834984@qq.com, 119072680@qq.com

Abstract. Under the background of China's "dual carbon" target, green development has become an important development direction of the construction industry. For a long time, the development of China's green building industry mainly relies on the support of administrative means and financial funds, and the role of green finance and other market mechanisms has not been fully brought into play. This paper analyzes the current situation and problems of green finance supporting new green buildings.

Keywords: green building · Green finance · Path of development

1 Introduction

The industrial revolution not only contributed to the great progress of social and economic development, but also caused great damage to the ecological environment. Since the 1990s, in order to control and reduce carbon dioxide emissions, the concept of "carbon neutrality" came into being. So far, more than 370 countries have made carbon neutrality commitments, and countries around the world have made positive progress in addressing climate change and promoting carbon emission reduction. This commitment not only reflects China's determination and confidence to firmly take the path of green development as a responsible major country, but also is an inherent requirement of China's sustainable development. China hopes that by accelerating the quantitative target of green economic development, it can realize the transformation of old drivers of growth from new drivers, take this as an important starting point to promote economic transformation and upgrading, and point out the direction for China's green and low-carbon development and ecological civilization construction.

As one of the three major energy consumption fields in China, the green development of the construction industry has become an important support to achieve the goal of "dual carbon". According to the Research Report on Energy Consumption and Carbon Emissions of Buildings in China (2022), the total carbon emissions from the whole process of buildings in China in 2020 were 5.08 billion $tCO2$, accounting for 50.9% of the national carbon emissions, accounting for more than half of the total. Among them, the carbon emission in the production stage of building materials is 2.82 billion $tCO2$, accounting

P. Xiang and L. Zuo (Eds.): PBSFTT 2023, LNCE 382, pp. 589–599, 2024.
https://doi.org/10.1007/978-981-97-5108-2_63

for 28.2% of the total carbon emission in China; The carbon emission in the construction phase was 100 million tCO2, accounting for 1.0% of the total carbon emission in the country; The carbon emission of buildings in the operation stage is 2.16 billion tCO2, accounting for 21.7% of the total carbon emission in China. According to the constraint target of carbon emission peak and carbon intensity decline set by our country, the problem of carbon emission reduction in the construction field has reached the point of urgency. Promoting the green development of buildings has become an inevitable demand of China's economic and social development, and the state has issued a series of policies and measures for this purpose. Development cannot be separated from finance, and green development cannot be separated from green finance. Ma (2018) believed that green building standards should be transformed into green financial standards and green building insurance should be used to realize the support of green finance for green buildings [1]. Ma (2018) believed that green bonds would play a positive role in promoting green buildings [2]. Therefore, carrying out research on green finance to promote the green development path of the construction industry is conducive to the realization of the "dual carbon" development goal.

2   Connotation of Green Finance

2.1   Definition of Green Finance

As the name suggests, green finance is a kind of "green" finance, which is the financial behavior that supports the development of green economy and supports environmental and climate-friendly economic activities. The term Green Finance first appeared in the 1990s, when some scholars proposed that green finance was the product of combining the financial industry with the green industry and incorporating environmental benefits into financial innovation [3]. In recent years, green finance has been clearly defined by some official institutions and international groups. In the G20 Green Finance Report, green finance is defined as "investment and financing activities that generate environmental benefits to support sustainable development" [4], while in China's Guiding Opinions on Building a Green Finance System, green finance is defined as: Economic activities that support environmental improvement, climate change response and resource conservation and utilization mainly include financial services provided for project investment and financing, project operation and risk management in the fields of environmental protection, energy conservation and green building (Fig. 1) [5].

2.2   Importance of Green Finance

Since the 1970s, a series of major environmental pollution has prompted people to reflect, and the value concept of green, low-carbon and environmental protection has entered people's vision. As the problem of climate change has been paid more and more attention, we have come to realize that it is not only an environmental problem, but also a development problem. With the deepening of the concept of sustainable development and building a community with a shared future for mankind, the development of green finance has emerged. Jeucken (2001) Green finance is an objective requirement for the

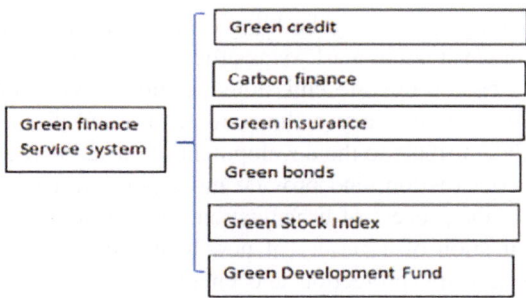

Fig. 1. Green financial service system

financial industry to achieve sustainable development, which makes banks and other financial institutions actively promote green finance business, increase the proportion of green finance in investment and financing activities, help financial institutions improve their own credibility, and effectively promote risk management [6]. When the discussion on financial support to reduce carbon dioxide and other greenhouse gas emissions and deal with climate change gradually appeared in people's life, a series of concepts such as climate finance and carbon finance were also derived. People began to use the corresponding financial means to support the activities of protecting the environment. Abroad, with the European Union as a pioneer, the United States as an innovation leader, and Japan as a participant, with the help of the top-level design of green finance and the promulgating of relevant laws and policies, after decades of development, it has formed a market-led and government-guided development model, and its green finance system is becoming increasingly mature. China has started to build a green finance policy framework and explore the development mode of green finance on a pilot basis.

3   Green Buildings and Their Development

Green building refers to the building that saves resources to the greatest extent during the whole life period, including energy saving, land saving, water saving and material saving, so as to protect the environment and reduce pollution, provide people with green, healthy, comfortable and convenient activity space, and coexist in harmony with nature. Compared with ordinary buildings, green buildings will generate a certain amount of additional investment, but foreign practice shows that green buildings will effectively solve the problem of environmental deterioration and resource bottleneck by its good environmental protection, energy saving and emission reduction, and transform the low-carbon, green and sustainable development of buildings into feasible. The United States Green Building Council (USGBC) estimates that green buildings consume 32% less electricity and 36% less total energy than ordinary buildings. From the results of saving a lot of energy consumption and reducing greenhouse gas emissions, choosing green buildings is a wise choice for both the national economy and families, and the promotion of green buildings in developed countries began as early as the 1980s. But in our country the start time of green building is relatively late, our country in June 2006 before the official implementation of the "green building evaluation" standard, the first batch of "green building design evaluation mark" was officially announced in August 2008.

In 2021, China's carbon dioxide emissions per unit of GDP decreased by 50.3% from 2005. Therefore, the advocacy of low-carbon life and the development of low-carbon economy will contribute to the development of our urbanization to the direction of low-carbon city construction, and provide more possibilities for the development of green buildings. In the process of urbanization, the development of green buildings will realize the application of a variety of new technologies such as comprehensive integration of environmental protection and energy saving in the construction field, promote the transformation of construction production mode, promote the new energy-saving and environmental protection building materials, new energy and other related industries, and complete the optimization and upgrading of industrial structure. It is of great significance to promote sustainable economic and social development.

Improving the investment and financing mechanism of green building and correctly evaluating its economic benefits are the key to the development of green building at present. At present, although some green buildings can obtain certain subsidies and other channels of financing, these funds are often not directly related to the energy conservation and emission reduction effect of green buildings. It wants to promote the innovation of building energy-saving technology and promote green and low-carbon buildings, so as to make living more comfortable and achieve energy conservation and emission reduction. It is necessary to use various market means including carbon finance to quantify the emission reduction effect of green buildings through the pricing of carbon emissions, and incorporate the emission reduction into the carbon market for trading, so as to realize the organic correlation between economic benefits and the emission reduction achieved by green buildings, and consider the energy conservation and emission reduction benefits when evaluating the economic benefits of green buildings. Providing more financial services to green building enterprises makes the investment and financing mode of green building operable and attracts social capital and financial capital to enter the field of green building.

4 Current Situation and Existing Problems of Green Finance in Helping Green Buildings

4.1 Development Status of Green Building Industry Aided by Green Finance

Under the background of "dual carbon" development, all parts of the country are actively promoting the green transformation and development of the building field, and the development of green building and green finance is identified as an important direction of development. At present, the collaborative development of green finance and green building in China is still dominated by local exploration, and some replicable experiences have been obtained in the process of exploration and practice. The People's Bank of China and other departments have also taken green buildings as an important support area (see Table 1), but the support of green finance for green buildings is still sufficient in terms of scale.

As the first pilot city in China, Huzhou, Zhejiang Province, took the lead in issuing local standards such as Opinions on Accelerating the Development of Green Building

Quality Improvement and Guidelines for Green Building Evaluation of Huzhou City. As the first municipal green finance promotion regulation in China, it has effectively linked the development of green buildings with green financial products and services, forming a system and framework. It has developed 114 green finance related products such as "green loan" and "green purchase and construction loan", built the first green financial service platform "Green Loan", and built the first "carbon neutral" sub-branch. It has promoted the green development of the whole industry chain of "green construction, green building and green building materials" in the construction industry [7], formed a "five-step" system of policy guidance, standard first, product innovation, science and

Table 1. Green building support projects in each green finance standard

Introduction time	Green finance standard	Issuing department	Green building related projects
2019	*Special Statistical System for Green Loans*	People's Bank of China	Design and construction of green buildings
2020	*Green Financing Statistical System*	China Banking and Insurance Regulatory Commission	1. Design, construction, operation and maintenance projects of green buildings 2. Purchase green buildings that comply with green building regulations
2021	*List of projects supported by Green Bonds (2021 Edition)*	People's Bank of China, National Development and Reform Commission, China Securities Regulatory Commission	Design and construction according to the relevant national green building codes and standards, the pre-evaluation of building construction drawings reaches the green building star standard during the validity period, and the construction and purchase of various civil and industrial buildings that meet the relevant national green building operation evaluation star standard during the validity period are constructed according to the green building star standard

technology support and demonstration drive, which has achieved remarkable results and explored an effective practical path for realizing carbon peak and carbon neutrality [8].

4.2   There are Problems in Green Finance Helping Green Building Industry

Green buildings need to implement the concept of energy conservation and emission reduction into the whole process of building design, construction mode, operation and management, recycling and disposal, and achieve harmonious coexistence with nature through improving the standard system and innovative technology. Compared with traditional construction enterprises, green construction enterprises need to incorporate the social responsibility for environmental protection into their business decisions and cost input. At present, green finance faces a variety of problems in the process of helping the development of green buildings, including at least: Lack of clear and detailed implementation policies, lack of management methods and mechanisms, financing difficulties and maturity mismatch, consumer demand for green buildings has not been activated, supporting financial products and services need to be improved and innovated, which also make green buildings cannot be recognized at both ends of supply and demand.

Lack of Clear and Detailed Implementation Policies

The central bank has only made preliminary guidance and norms on the definition of green industry projects, the investment of raised funds, the management of funds during their existence, information disclosure and the evaluation or certification of independent institutions, and other relevant policies still need to be further established and improved. At the same time, since 2019, the central government has made a series of statements on the real estate industry, indicating to the public its resolute attitude on real estate regulation. In view of the large amount of financial resources occupied by the real estate industry, relevant departments have clearly proposed to control real estate credit and non-bank financing. This makes the green building in the national macro adjustment is also affected. Although it is a green industry encouraged by the state, the "one-size-fits-all" strict regulation policy does not distinguish green building from general real estate development. In turn, many commercial banks and non-bank financial institutions are afraid, unable or unwilling to support green building developments when it comes to obtaining financing. It worries that its support for green buildings will be "identified" as an opportunity to support property development and hit the red line of national regulation policy. The development and operation of green building projects are restricted by real estate regulation policies, which makes construction enterprises increasingly take a "wait-and-see" attitude towards the development of green building.

Lack of Management Methods and Mechanisms

At present, the total amount of green building projects in China is relatively small, and the proportion of high-quality identification is relatively low. According to the provisions of the "new National Standard" for green building in 2019, after the construction drawing design of the construction project is completed, green financing can be obtained from financial institutions and the market according to the pre-evaluation (pre-evaluation does not grant identification). However, the management method of green building operation identification after the completion of the project and the evaluation mechanism after the completion of the project are currently missing. As for the green building industry in the operation stage, the regulatory agencies have not been able to follow up, evaluate and supervise, and the information disclosure mechanism of green buildings has not been established. Once the "green" benefits of green buildings in the operation stage are not guaranteed, the motivation of financial institutions to support green buildings and the interest of owners/consumers to purchase and rent green buildings will be greatly reduced, and the market will lose confidence in the development of green buildings.

Facing Financing Difficulties and Maturity Mismatch
The purpose of green finance to promote the development of green buildings is to solve the financing problem. Due to the lack of effective evaluation of green benefits of green buildings and timely information disclosure mechanism, it is difficult for financial institutions and guarantee institutions to directly measure the value of green technology of small and micro enterprises in the industrial chain of green building industry. The credit risk assessment for these enterprises is also difficult due to the imperfect evaluation system, resulting in the inability of small and micro enterprises in the color construction industry chain to obtain financing or guarantee from financial institutions in a timely manner. Financing difficulties also hinder the long-term development of the entire green building industry. In addition, the incremental cost caused by the use of green technology, equipment and materials, the cost of project design, simulation and demonstration, and the additional cost of applying for green building identification make the incremental cost account for 2.7%–9.3% of the overall cost of green building, and the payback period of investment is long [9]. When developers try to solve the maturity mismatch between incremental costs (incurred in the short term) and incremental benefits (incurred in the long term) by raising housing prices or rents, the problem is solved. However, due to the low willingness of consumers and the lack of incentives from financial institutions and the government (such as price limit relaxation, floor area ratio relaxation, etc.), it is difficult to achieve. Finally, the amount of credit invested in green building accounted for only 1.59% of the total green credit in China; green bonds started late and 30% were not labeled; and green funds lacked mature cases, making it difficult to match the huge financing needs of green real estate enterprises.

The Demand for Green Buildings on the Consumer End is not Activated
Green building itself started late in our country. As far as most consumers are concerned, they lack awareness of the concept of green building and its benefits (energy saving benefits, environmental improvement and health impact, etc.). In addition, the majority of consumers are still sensitive to housing prices. Even if there is publicity about the benefits of green buildings by developers, it is difficult to directly understand the economic benefits of energy saving and water saving of green buildings. At the same time, due to the lack of corresponding monitoring and information disclosure mechanism in the later green building operation and maintenance stage, consumers are not willing to bear relatively high prices for green buildings. Therefore, there is no spontaneous demand for green buildings on the consumer side, which also brings uncertainty to real estate developers' investment in green building projects. However, in the actual use process, consumers cannot intuitively understand the energy saving benefits of green buildings.

Due to the above reasons, the consumer market of green buildings in China has neither obvious price advantage nor effective activation. This has become one of the important factors hindering real estate developers from investing in green buildings, and has also directly and indirectly led to the lack of demand for related financial products.

Supporting Financial Products and Services Need to Be Improved
Green financial products have not yet been able to well meet the needs of the development of China's green building market. The lack of data on indicators of green building-related financial products makes it impossible for some commercial banks to know the default rate, default loss rate and "carbon emission reduction" index of green bonds, which hinders the formation of pricing mechanism and product innovation of green building-related financial products. In addition, the existing green building-related financial products are not highly differentiated in terms of credit conditions, cycles and interest rate differences.

5   Path Suggestions

Based on the above analysis of the problems existing in green finance in supporting the development of green buildings, this part puts forward the following preliminary ideas on promoting the development of green building market through green finance from the perspectives of macro-control, industry supervision, consumer market and financial product innovation (Fig. 2).

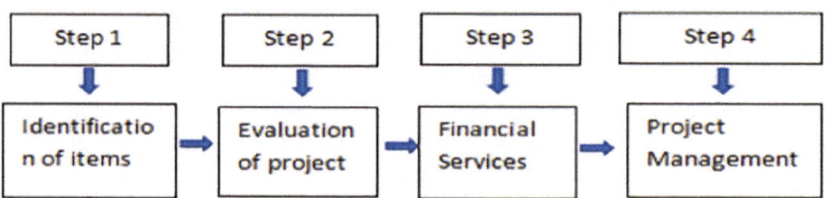

Fig. 2. Green finance boosts green building operation processes

5.1   Further Improve Policies Related to Green Buildings

At present, the government's regulation of the real estate industry is a "one-size-fits-all" restrictive measure in financing. This paper suggests that to distinguish green from non-green construction enterprises, key restrictions should be imposed on "non-green" real estate enterprises and projects, relatively loose financing conditions should be given to those qualified green developers and green projects, and "structural adjustment" under the requirement of "total volume control" should be realized. So as to stimulate the endogenous power of industrial transformation of real estate enterprises. We will increase policy support for green buildings from both the supply side and the demand side. Housing developers can be encouraged to participate in green building development from the supply side through floor area ratio incentive, financial subsidy, tax reduction and other ways. On the demand side, consumers' consumption costs of green buildings should be reduced through deed tax incentives and other means to guide them to purchase and lease green buildings.

5.2 Improve the Standard System

Further improve the green building evaluation system, improve the evaluation and supervision of the building operation stage and the corresponding trust-breaking punishment mechanism, to ensure that the building with green building logo "always green". From the design stage, it is necessary to make clear project evaluation and strengthen the relevance with green financial services. Establish green building project management system and green financial service management mechanism to ensure the promotion of projects, the standardization of financial services and capital risk supervision, and carry out the compilation of green financial service standards. Establish a complete green financial service process of application acceptance, preliminary screening, due diligence, project evaluation, loan review, loan approval, loan issuance and post-loan management [10].

5.3 Activate the Demand for Green Buildings

By building a platform for exchange, discussion and display of the integration of green finance and green building, strengthen the publicity of the coordinated development of green finance and green building with the help of various media prices, and improve the public's awareness of green finance and green building; Through the development and holding of relevant lectures and training, the publicity and promotion of green finance and green buildings to eliminate consumers' doubts, while strengthening the establishment of supporting measures such as information disclosure of green buildings and punishment for trust-breaking, so as to realize the guarantee of "green benefits" that consumers deserve. In order to make consumers feel the "green benefits" brought by the choice of green buildings, financial institutions can provide them with calculation tools such as the "electricity and water Saving Calculator" APP to take consumers to calculate the benefit account and help them intuitively understand the benefits brought by the choice of green buildings due to energy saving in the future.

5.4 Innovate Green Financial Products and Services

Combined with the characteristics of the green construction industry to innovate and develop a number of green financial service products, effectively expand the financing channels of green building. Encourage the development of "green development loan", "mortgage loan", "securitization products of building photovoltaic assets", "green building insurance" products, as well as trading mechanisms related to carbon emission reduction of green buildings. With the improvement of the carbon financial system, carbon emission quota and certified emission reduction are no longer just tools for the performance of the carbon market. In terms of credit rating index design, differentiation should be implemented for construction enterprises that develop more green buildings, and their financing costs should be moderately reduced.

6   Conclusion

As an industry with high carbon emissions, if the construction industry can realize the transformation to green building industry, it will reduce the carbon emissions by 90% compared with the traditional construction industry, and green building has become an inevitable choice. Based on the current situation and existing problems of the development of green buildings, this paper carries out research work, and puts forward development suggestions from the aspects of improving policies, improving the standard system, and innovating financial products, so as to further enrich and improve the development path of green buildings promoted by green finance, and help the green and low-carbon development in the construction field.

Acknowledgement. Fund project: Support by the Science and Technology Research Program of Chongqing Municipal Education Commission (Grant No. KJQN202305201), Study on the optimization of carbon financial tools to promote the development of green buildings in Chongqing from the perspective of carbon peaking and carbon neutrality goals.

References

1. Ma, J.: Building China's Green Financial System. China Finance Press (2015)
2. Ma, X.: Green bonds help green building development. Environ. Econ. **13**, 52–53
3. Chen, B., Tao, J.: Review and prospect of green finance research. Oper. Manag. (2), 157 (2021)
4. G20 Green Finance Study Group, "G20 Green Finance Synthesis Report". http://www.pbc. gov.cn/goutongjiaoliu/113456/113469/3142307/2016091419074561646.pdf
5. People's Bank of China, Ministry of Finance, National Development and Reform Commission, Ministry of Environmental Protection, China Banking Regulatory Commission, China Securities Regulatory Commission, China Insurance Regulatory Commission. Guiding Opinions on Building a Green Financial System (2016)
6. Jeucken, J.: Sustainable Finance and Banking. The Earths CanPublication, USA (2006)
7. Gong, X.T., Liu, D.G., Cen, Q.Y., et al.: Practice of green building and green finance coordinated development: a case study of Huzhou pilot city in Zhejiang Province. Constr. Sci. Technol. (20), 48–52 (2020)
8. Gong, X.T., Liu, D.G., Wu, J.D.: Huzhou: "Five-step method" to boost the coordinated development of green building and green finance. Urban Rural Constr. (14), 80–81 (2021)
9. Jin, Z., Tian, Y., Wu, P., He, B.: Research on credit enhancement mechanism of green building investment and financing insurance. Constr. Sci. Technol. (09), 38–39 (2017)
10. Housing and Construction Department of Hebei Province, Shijiazhuang Central Sub-branch of People's Bank of China, Hebei Regulatory Bureau of China Banking and Insurance Regulatory Commission, Hebei Local Financial Supervision Administration. Notice on Orderly Implementation of Green Finance to Support Green Building Development (2022)

Author Index

P. Xiang and L. Zuo (Eds.): PBSFTT 2023, LNCE 382, pp. 601–603, 2024.
https://doi.org/10.1007/978-981-97-5108-2